Thieme Foundations of Organic Chemistry Series

Series Editors: D. Enders, R. Noyori, B. M. Trost

Spectroscopic Methods in Organic Chemistry

Related Thieme Titles of Interest

Organic Chemistry Monograph Series

A. Hirsch
The Chemistry of the Fullerenes

T. Eicher, S. Hauptmann
The Chemistry of Heterocycles

H. Scherz and G. Bonn
The Analytical Chemistry of Carbohydrates
(in preparation)

Foundations of Organic Chemistry Series

P. J. Kocienski
Protecting Groups

H. B. Kagan
Asymmetric Synthesis (in preparation)

Spectroscopic Methods
in Organic Chemistry

Manfred Hesse
University of Zürich

Herbert Meier
University of Mainz

Bernd Zeeh
BASF Limburgerhof

Translated by Anthony Linden and Martin Murray

221 Figures, 100 Tables

1997

Georg Thieme Verlag Stuttgart · New York

Prof. Dr. Manfred Hesse
Institute of Organic Chemistry
University of Zürich
Winterthurerstraße 190
CH-8057 Zürich
Switzerland

Prof. Dr. Herbert Meier
Institute of Organic Chemistry
Johannes-Gutenberg-University
Johann-Joachim-Becher-Weg 18-20
D-55099 Mainz
Germany

Dr. Bernd Zeeh
BASF Aktiengesellschaft
Landwirtschaftl. Versuchsstation
D-67117 Limburgerhof
Germany

Translated by
Dr. Anthony Linden
Institute of Organic Chemistry
University of Zürich
CH-8057 Zürich
Switzerland

Dr. Martin Murray
School of Chemistry
University of Bristol
Cantock's Close
Bristol BS8 1TS
UK

This book is an authorized translation of the German edition published and copyrighted 1979, 1984, 1987, 1991, 1995 by Georg Thieme Verlag, Stuttgart, Germany. Title of the German edition: Spektroskopische Methoden in der organischen Chemie

Library of Congress Cataloging-in-Publication Data

Hesse, Manfred, 1935
([Spektroskopische Methoden in der organischen Chemie. English]
Spectroscopic Methods in Organic Chemistry / Manfred Hesse, Herbert Meier, Bernd Zeeh; translated by Anthony Linden und Martin Murray.
    p. cm. -- (Thieme foundations of organic chemistry series)
Includes bibliographical references and index.
    1. Spectrum analysis. 2. Organic compounds -- Analysis.
    I. Meier, H.   II. Zeeh, Bernd   III. Title   IV. Series.
QD 272.56H4713 1996
547.3'0858 -- dc20   96-35773 CIP

Die Deutsche Bibliothek – Cataloging-in-Publication Data

Hesse, Manfred:
Spectroscopic methods in organic chemistry/Manfred Hesse; Herbert Meier; Bernd Zeeh. Transl. by Anthony Linden and Martin Murray. - Stuttgart: New York: Thieme; New York: Thieme New York, 1997
    (Thieme foundations of organic chemistry series)
    Einheitssacht.: Spektroskopische Methoden in der organischen Chemie–
    <engl.>
NE: Meier, Herbert:; Zeeh, Bernd:

© 1997 Georg Thieme Verlag,
Rüdigerstraße 14, D-70469 Stuttgart
Printed in Germany by Offizin Andersen Nexö, Leipzig

Georg Thieme Verlag, Stuttgart
ISBN 3 13 106041 7 Flexicover
ISBN 3 13 106061 1 Hardcover

Thieme, New York
ISBN 0 86577 667 9 Flexicover
ISBN 0 86577 668 7 Hardcover   1 2 3 4 5 6

Foreword

Spectroscopic methods were introduced into organic chemistry laboratories around 1960. The prerequisite for this was the commercial availability of relatively easy to use and reliable instrumentation. Measurements in the ultra-violet and visible regions of the spectrum were possible even earlier.

After initial delays, which were caused principally by the need to obtain funding for the purchase of the expensive instruments, the use of spectroscopic methods spread rapidly throughout the world. As a result, it became possible to carry out methodically the structure elucidation of previously unknown compounds. Until then, chemical degradation and the wearisome characterisation of the cleaved pieces by melting and boiling points had been the principle means of structural analysis. These procedures were superseded by spectroscopy, which allowed us to recognise quickly functional groups or even complete structural units. This change was also supported by the introduction at about the same time of the chromatographic methods: thin-layer, gas and, later, high performance liquid chromatography.

Even in 1962, as was characteristic at this revolutionary time, the 72 year old Nobel Prize winner, Paul Karrer, in a working discussion, requested one of the authors of this book (M.H.) to synthesise the corresponding oxime or semicarbazone derivative of a rare alkaloid in order to prove the presence of either an aldehyde or a hydroxy group. "A difference of two mass units is insufficiently specific", he said. Within a few years the chemistry of the recognition of functional groups by the synthesis of derivatives was at an end. The classical degradation reactions, whose transformation products allowed such logical conclusions to be drawn during the structural elucidation, were no longer needed. This epoch, which had lasted for about 130 years, encompassed the origin and development of a substantial part of organic chemistry.

The further development of the spectroscopic methods, particularly with regard to instrumentation, has enabled us today to advance into areas that were completely inconceivable in the 1960's: e.g. the analysis of aqueous solutions in a mass spectrometer, or the structure elucidation, including the spatial arrangement, of the smallest quantities of large molecules by means of NMR spectroscopy. In order to aquire a comprehensive knowledge of these methods and particularly to learn the way one should think about this subject, the student is directed to introductory literature, which can also be used as reference material in the research laboratory.

This book is a translation of the 5th German edition. Each new edition is thoroughly updated, so that this volume describes the current state of the art in spectroscopic methods. We are most grateful to Drs. Anthony Linden and Martin Murray for carrying out the translation in such an exemplary manner. Our thanks also go to our co-workers, who have contributed to the success of this book, and to Georg Thieme Publishers for a successful collaboration.

Zürich, Mainz and Ludwigshafen,
August 1996

Manfred Hesse
Herbert Meier
Bernd Zeeh

Contents

Chapter 5
Combined Examples

1 UV/Vis Spectroscopy

1. Theoretical Introduction

1.1 Electronic Transitions

Electromagnetic radiation is characterised by the **wavelength** λ or the **frequency** ν. These values are connected with each other by the equation

$$\nu \cdot \lambda = c$$

c is the **velocity of light** (in vacuo $\approx 2.99 \cdot 10^{10}$ cm $\cdot$ s^{-1}). A quantum of light with frequency ν has the *Energy*

$$E = h\nu.$$

Planck's constant h has the value $\approx 6.63 \cdot 10^{-34}$ Js. The interaction of electromagnetic waves and molecules leads in the case of absorption of ultraviolet and visible light to the excitation of electrons, generally valence electrons. Fig. 1.1 illustrates the relevant regions of the electromagnetic spectrum. The region of light visible to the human eye (Vis) is followed below $\lambda =$ 400 nm by the UV region. Based on the different biological activities a subdivision is made into UV-A (400–320 nm), UV-B (320–280 nm) and UV-C (280–10 nm).

The wavelength was earlier frequently given in Ångström, nowadays nanometres (nm) are usually used (1 nm $= 10^{-7}$ cm). Instead of giving frequencies in s^{-1}, it is customary to quote the **wavenumber** $\tilde{\nu}$ in cm^{-1}.

$$\tilde{\nu} = \frac{1}{\lambda} = \frac{\nu}{c}$$

If the energy is based on a quantum or an individual atomic or molecular process, the customary unit is the electron volt (eV). For a mole, i.e. $6.02 \cdot 10^{23}$ quanta of light, the energy is given in kJ. Energy and wavenumber are directly proportional to each other. For conversions the following relationships are recommended:

$$1\,\text{eV} \equiv 23\,\text{kcal} \cdot \text{mol}^{-1} = 96{,}5\,\text{kJ} \cdot \text{mol}^{-1} \equiv 8066\,\text{cm}^{-1}$$
$$1000\,\text{cm}^{-1} \equiv 12\,\text{kJ} \cdot \text{mol}^{-1}$$
$$1\,\text{kJ} \cdot \text{mol}^{-1} \equiv 84\,\text{cm}^{-1}$$

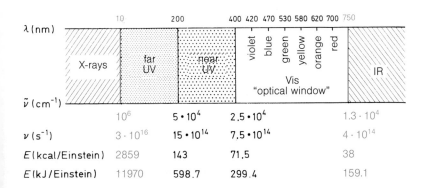

Fig 1.1 UV/Vis region of the electromagnetic spectrum
(1 Einstein = 1 quantum of light)

If light with the appropriate frequency v meets a molecule in the **ground state** ψ_0, it can be **absorbed** and raise the molecule to an **electronic excited state** ψ_1. By **spontaneous emission** or by **stimulated emission**, caused by the light rays, the system can return to the ground state. The word "can" in these senses expresses the **transition probability** of the two radiative processes, absorption and emission (Fig 1.2).

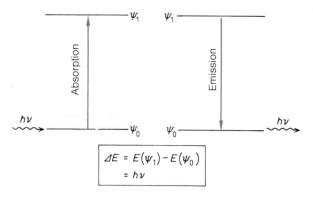

$$\Delta E = E(\psi_1) - E(\psi_0)$$
$$= h v$$

Fig. 1.2　Electronic transitions and radiative processes

The connection with the orbitals involved in the electronic transition is shown in Fig. 1.3. The difference in energy between the lowest unoccupied orbital (LUMO) and the highest occupied orbital (HOMO) is considerably greater than the activation energy A for the transition from the singlet ground state S_0 to the singlet excited state S_1. The difference arises from the different electronic interactions (Coulomb term J, exchange term 2K). The singlet-triplet splitting in this approximation is 2K. Since K > 0 the lowest triplet state T_1 is always below S_1. As a result of the configurational interaction the HOMO–LUMO transition is not necessarily the lowest transition $S_0 \to S_1$.

A measure for the transition probability is the **oscillator strength** f_{01}, a dimensionless quantity which classically repre-

sents the fraction of negative charge (electrons) which brings about the transition (by oscillation). The quantum mechanical equivalent of f is the vector of the **transition moment** M_{01}, which represents the change of the **dipole moment** during the transition. The **dipole strength** $D_{01} = |M_{01}|^2$ is directly proportional to f_{01}. If $D_{01} = M_{01} = f_{01} = 0$ then even if the **resonance condition** $\Delta E = h v$ is fulfilled, no transition is possible. When the f-value is small the term **forbidden transition** is used, if the f-value is close to 1 the term **allowed transition** is used.

For diatomic or linear polyatomic molecules, as with atoms, **selection rules** for the allowed transitions between two different electronic states can be established based on the rule of the conservation of angular momentum. For other molecules, which constitute the overwhelming majority, these rules result in **transition exclusions**.

The **spin exclusion rule** states that the **total spin** S and the **multiplicity** M = 2S + 1 may not change during a transition, i.e. singlet states, for example, may undergo transitions to singlet states by absorption or emission, but not to triplet states. M_{01} can also be zero because of the symmetry of the orbitals (which are described by the wave functions φ_0 and φ_1 and represent the electronic part of the total wave functions ψ_0 and ψ_1). This is described as a **symmetry exclusion**. An easily comprehensible special case occurs for molecules containing a centre of symmetry, whose wave functions are either symmetric (g, gerade) or antisymmetric (u, ungerade). The symmetry exclusion states in such cases that electronic transitions between orbitals of the same **parity** are forbidden (parity exclusion, Laporte's rule).

allowed:　g $\longrightarrow$ u　　forbidden:　g $\longrightarrow\!\!\!\!/\,$ g
　　　　　u $\longrightarrow$ g　　　　　　　u $\longrightarrow\!\!\!\!/\,$ u

Movement of nuclei can reduce the symmetry, so that symmetry-forbidden transitions can in fact be seen. (An example of a vibrationally allowed transition is the long-wavelength absorption band of benzene; cf. p. 14.)

A further possible cause for the disappearance of the electronic transition moment is the so-called **overlap exclusion**. It takes effect when the two orbitals which are taking part in the electronic transition overlap poorly or not at all. That is quite clearly the case in a **charge-transfer transition** where the electronic transition takes place from the donor- to the acceptor-molecule. There are also numerous intramolecular examples of the overlap exclusion. (Compare the $n \to \pi^*$ transition of carbonyl compounds, p. 17.)

If the possibilities of transitions between two orbitals of a molecule are worked out, it becomes apparent that exclusions become the rule and allowed transitions are the exceptions. However, forbidden transitions frequently occur, albeit with

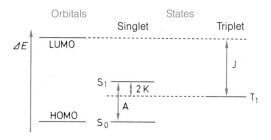

Fig. 1.3　Energy scheme for the electronic transition between HOMO and LUMO

low transition probability, i.e. a low f value $(10^{-1} \geq f \geq 10^{-6})$. The spin exclusion is the most effective. Even spin-forbidden transitions can however be observed in cases of effective spin-orbit coupling (e.g. by heavy atoms) or in the presence of paramagnetic species.

If the molecule under investigation is considered in a Cartesian co-ordinate system whose axes are established with reference to the molecular axes, the vector M_{01} can be separated into its spatial components M_x, M_y, and M_z. For $M_{01} \neq 0$ at least one of the three components must be non-zero. When $M_x = M_y = 0$ and $M_z \neq 0$ the absorbed or emitted radiation is **polarised** in the z direction. This optical anisotropy of the molecule cannot normally be observed, since the molecules are present in a random orientation. Polarisation measurements are carried out on single crystals or stretched plastic films.

The statements in Sect. 1.1 apply to single photon transitions. With the use of lasers **two photon spectroscopy** has been developed. High photon densities allow the simultaneous absorption of two photons. This leads to altered selection rules; thus transitions between states of the same **parity** are allowed $(g \rightarrow g, u \rightarrow u)$ and transitions between states of opposite parity are forbidden. The **degree of polarisation** in solution can also be determined. Two photon spectroscopy thus provides useful extra information in studies of electronically excited molecules.

At the end of this section the photophysical processes of electron transitions are summarised in a modified Jablonski term scheme. From the ground state, which in general is a singlet state S_0, absorption leads to higher states S_1, S_2, etc. The return to S_0 from S_1, and more rarely from higher singlet states S_n, can occur by the emission of radiation, known as **fluorescence**, or by non-radiative deactivation **(internal conversion)**. Non-radiative spin-inversion processes **(intersystem crossing)** lead to

triplet states T, which can return to S_0, disregarding the spin exclusion, either by emission of radiation, known as **phosphorescence**, or by renewed intersystem crossing (Fig 1.4).

"True" two photon absorptions must be differentiated from processes in which two photons are absorbed one after the other. At high light intensities populations of excited states can be attained which allow further excitation; for example the process $S_0 \rightarrow S_1 \rightsquigarrow T_1$ can be followed by a triplet-triplet absorption $T_1 \rightarrow T_2$.

In contrast to atoms the various electronic states of molecules have rather broad energies because of the added effect of vibrational and rotational levels. Each term in Fig. 1.4 is therefore split into many energy terms, as shown schematically in Fig. 1.5. A specific energy level $E_{tot.}$ therefore corresponds to a particular vibrational and rotational state of the molecule.

To a first approximation the three energy components can be separated

$$E_{tot.} = E_{electr.} + E_{vibr.} + E_{rot.}$$

For an electronic transition it follows that

$$\Delta E_{tot.} = \Delta E_{electr.} + \Delta E_{vibr.} + \Delta E_{rot.}$$

The electronic part is always much larger than the vibrational part, which in turn is much larger than the rotational part. The **relaxation** R (see Fig. 1.5) is an additional non-radiative deactivation within each electronic state. In addition to the monomolecular processes described here it should also be noted that **bimolecular physical processes** (energy transfer: **sensitisation, quenching**) and **primary photochemical processes** can occur.

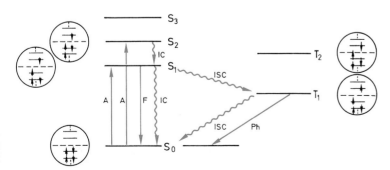

Fig. 1.4 Jablonski term diagram with a visual representation of the electronic configurations

Radiative processes: →
A   Absorption
F   Fluorescence
Ph   Phosphorescence

Non-radiative processes: ⤳
IC   internal conversion
ISC intersystem crossing

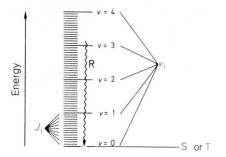

Fig. 1.5 Schematic representation of the superimposition of electronic, vibrational, and rotational states; n_i vibrational quantum numbers, J_i rotational quantum numbers

1.2 Light Absorption and the Spectrum

If a beam of light of intensity I_0 falls on a layer of homogeneous, isotropic material of thickness d, then apart from losses through reflection or diffraction it can be weakened by absorption. The intensity I of the emerging beam (transmission) is then given by:

$$I = I_0 - I_{abs.}.$$

The differential equation for the reduction of the intensity dI by an increment dx of the width of the absorbing layer

$$dI = -\alpha \cdot I \, dx$$

and evaluation of the integral

$$\int_{I_0}^{I} \frac{dI}{I} = -\int_0^d \alpha \, dx$$

yields the function

$$I = I_0 \cdot e^{-\alpha d}.$$

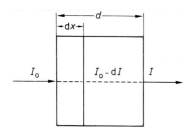

α is a characteristic absorption coefficient for the medium. If consideration is restricted to dilute solutions, where only the solute, of concentration c, absorbs, then α can be replaced by $2.303 \cdot \varepsilon \cdot c$ to give

$$\ln \frac{I_0}{I} = 2.303 \cdot \varepsilon \cdot c \cdot d \quad \text{or} \quad A = \log \frac{I_0}{I} = \varepsilon \cdot c \cdot d.$$

The **absorbance** A has no dimensions. The thickness d of the layer is measured in cm, the concentration c in $mol \cdot l^{-1}$. The molar **absorption coefficient** ε has the dimensions $1000 \; cm^2 \cdot mol^{-1} = cm^2 \cdot mmol^{-1}$, but is often quoted without dimensions. This law, which was developed by Bouguer (1728), Lambert (1760), and Beer (1852), is valid for monochromatic light and dilute solutions ($c \leq 10^{-2} \, mol \cdot l^{-1}$). The absorbance is, with a few exceptions, an additive property. For n absorbing species therefore

$$A_{tot.} = \log \frac{I_0}{I} = d \sum_{i=1}^{n} \varepsilon_i c_i$$

Especial care is necessary when entering values for the concentrations for compounds which undergo a chemical change when dissolved, e.g. dissociation, dimerisation, etc.

If the absorbance is determined for all λ or $\tilde{v}$ and from that the substance-specific value ε, the absorption plot $\varepsilon(\tilde{v})$ or $\varepsilon(\lambda)$ can be obtained and thus the UV or UV/Vis spectrum. As a consequence of the width (in energy terms) of the electronic states it is a **band spectrum**. The individual bands are characterised by their properties of **position, intensity, shape**, and **fine structure**.

A classification of the electronic transitions (bands) can be made from a knowledge of the molecular orbitals (MO's) involved. From occupied **bonding** σ- or π-orbitals or from **non-bonding** n-orbitals (lone pairs of electrons) an electron can be raised to an empty **anti-bonding** π^*- or σ^*-orbital. Correspondingly the electron transitions (bands) are indicated as $\sigma \rightarrow \sigma^*$, $\pi \rightarrow \pi^*$, $n \rightarrow \pi^*$, $n \rightarrow \sigma^*$ etc. (Fig. 1.6).

Apart from this nomenclature based on a simplified MO-description there are several other conventions for the specification of electronic states and the possible transitions between

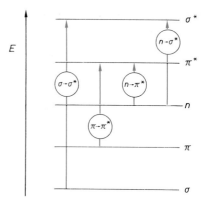

Fig. 1.6 Molecular orbitals and electron transitions

Tab. 1.1 Nomenclature for electron transitions

System	Term symbol	State	Examples of electron transitions
enumerative	S_0	singlet ground state	$S_0 \rightarrow S_1$
	$S_1, S_2, S_3, \ldots$	higher singlet states	$S_0 \rightarrow S_2$ $S_0 \rightarrow S_3$
	$T_1, T_2, T_3, \ldots$	triplet states	$T_1 \rightarrow T_2$
according to Mulliken	N	ground state	$V \leftarrow N$
	Q, V, R	excited states	$Q \leftarrow N$
according to Platt	A	ground state	$B \leftarrow A$
	B, C, L	excited states	$C \leftarrow A$
			$L \leftarrow A$
according to Kasha	σ, π, n	orbital of origin	$\sigma \rightarrow \sigma^*$
	σ^*, σ^*	orbitals of the excited electrons	$\pi \rightarrow \pi^*$
			$n \rightarrow \pi^*$
			$n \rightarrow \sigma^*$
group theory	symbols of the symmetry classes*		$^1A_2 \leftarrow {}^1A_1$
	A: sym. $\Big\}$ B: antisym.	related to rotation about the rotational axis (axes) C_n of maximum order	$^1B_{1u} \leftarrow {}^1A_{1g}$ $^1B_{2u} \leftarrow {}^1A_{1g}$ $^1E_{1u} \leftarrow {}^1A_{1g}$
	E: doubly degenerate state		
	T: triply degenerate state		
	Indices:		
	g: sym. $\Big\}$ u: antisym.	with respect to inversion	
	1: sym. $\Big\}$ 2: antisym.	with respect to. C_2 axes which are perpendicular to C_n	
	': sym. $\Big\}$ ": antisym.	with respect to plane of symmetry σ_h (perpendicular to C_n)	

* see for example Jaffé, H. H., Orchin, M. (1973), Symmetry in Chemistry, Alfred Hüthig Verlag, Heidelberg

them, of which especially the last one in Tab. 1.1 is to be recommended.

From Fig. 1.6 it follows that the positions of the absorption bands depend on the nature of the electron transitions. For isolated chromophores Fig. 1.7 gives a guide. The position of the absorptions are however strongly influenced by steric, inductive, and resonance effects – the latter being particularly strongly affected by inclusion of the chromophore in large conjugated systems (Fig. 1.7).

For certain chromophores the solvent also has a characteristic influence (see Fig. 1.25).

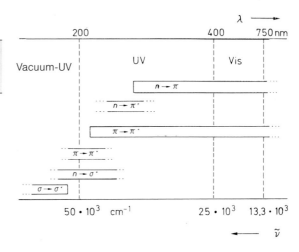

Fig. 1.7 Absorption regions of various electron transitions

A shift to longer wavelengths (red shift) of a transition is called a **bathochromic effect**, a shift to shorter wavelengths (blue shift) a **hypsochromic effect**.

The term **hyperchromic effect** is used to describe an increase in intensity. **Hypochromic** means the opposite, a decrease in intensity.

As described above, the transition moment $|M|$ or the oscillator strength f is a measure of the intensity of a transition. An alternative measurement for the intensity is the area S

$$S = \int_{(-\infty)}^{(+\infty)} \varepsilon \, \mathrm{d}\tilde{\nu}$$

The relationship for a denominator of $n \cong 1$ is given by:

$$f \approx \frac{m \cdot c^2}{N_A \pi e^2} 10^3 \, (ln \, 10) \, S$$

$$f \approx 4.32 \cdot 10^{-9} \, S$$

m mass of an electron
e charge of an electron
N_A Avogadro's constant
c speed of light

S can often be determined by graphical integration or estimated very roughly from approximations such as

$$S = \varepsilon_{max} \cdot b.$$

where b is the width of the band at half height (Fig. 1.8).

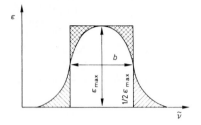

Fig. 1.8 True and approximated area of an absorption band

The higher the transition probability is, the shorter the radiation lifetime τ_0 of an excited state. τ_0 can be calculated from f and thus from S

$$\tau_0 = \frac{c^3 m}{8\pi^2 v^2 e^2} \cdot \frac{1}{f}.$$

As an approximation τ_0 is given in seconds by

$$\tau_0 \approx \frac{1}{10^4 \cdot \varepsilon_{max}}.$$

Usually the intensity of a band is judged simply by ε_{max}. The following assignments have become customary:

$\varepsilon \leq 10$	Transition: forbidden
$10 < \varepsilon < 1000$	weakly allowed
$1000 < \varepsilon < 100\,000$	allowed
$\varepsilon \geq 100\,000$	strongly allowed

Further important properties of absorption bands are shape and fine structure. Even if the different kinetic energies of individual molecules are ignored, an electronic state does not have a uniform energy. Instead the overlying molecular vibrations and rotations must be taken into account, as described above. From **Boltzmann statistics** it will be appreciated that in the ground state S_0 the lowest vibrational level ($v = 0$) is almost exclusively populated. For two states with an energy difference ΔE the ratio of populations (N) is given by

$$\frac{N_i}{N_j} = e^{-\Delta E/kT}$$

Since the Boltzmann constant $k = 1.38 \cdot 10^{-23}\,\mathrm{JK^{-1}}$ it can be seen that in the wavenumber scale $kT \approx 200\,\mathrm{cm^{-1}}$ at room temperature. For a typical vibration in the IR-region with $\tilde{v} = 1000\,\mathrm{cm^{-1}}$

$$\frac{N_i}{N_j} = e^{-1000/200} = e^{-5} = 0.0067$$

The energetically higher vibrational level therefore has a population of less than 1%. Higher rotational levels on the other hand are appreciably populated. Thus for rotations about single bonds with $\tilde{v} = 50\,\mathrm{cm^{-1}}$ the Boltzmann distribution at room temperature gives

$$\frac{N_i}{N_j} = e^{-50/200} = 0.78 = 44 : 56$$

The transition to S_1 leads to vibrational states with $v' = 0, 1, 2, 3 \ldots$ Because of very rapid relaxation to $v' = 0$ the fluorescence starts entirely from $v' = 0$ and leads to S_0 with $v = 0, 1, 2, 3 \ldots$

Fig. 1.9 shows the situation schematically.

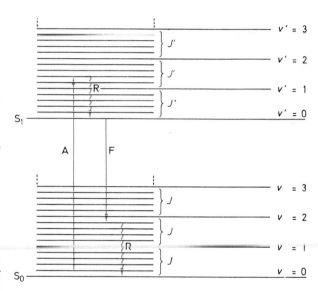

Fig. 1.9 Absorption and fluorescence as transitions between electronic, vibrational, and rotational levels

Spectra measured in solution do not show rotation lines – the electronic bands are composed of **vibrational bands**. The degree of structure observed in the absorptions depends on the substance. Vibrational fine structure is most likely to be seen in rigid molecules. In polyatomic molecules the vibrational levels lie very close together. Restricted rotation in solution and line broadening due to local inhomogeneities in the solvation result in unstructured bands. The measurement conditions can also play an important role. Fig. 1.10 shows the reduction in structure with increasing interaction with the solvent and under the influence of temperature.

In line with the **Franck-Condon principle** the absorption probability is largest for a vertical transition from the energy hypersurface of the ground state into that of the electronically excited state, i.e. all molecular parameters (bond lengths and

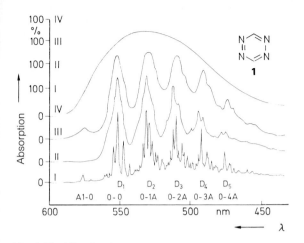

Fig. 1.10   Vibrational structure of the $n \rightarrow \pi^*$ absorption of 1,2,4,5-tetrazine **1** (from Mason, S.F. (1959), J. Chem. Soc., 1263)
I   Vapour spectrum at room temperature (with vibrational modes)
II   Spectrum at 77 K in an isopentane/methylcyclohexane matrix
III   Spectrum in cyclohexane at room temperature
IV   Spectrum in water at room temperature
The λ-Scale is referenced to I; II is shifted by 150 cm^{-1}; III by 250 cm^{-1} to higher wavenumbers; IV by 750 cm^{-1} to lower wavenumbers

angles, conformation, solvation cage, etc.) remain unchanged during the transition.

From the expression for the vibrational component of the transition moment M it can be seen that the transitions from the lowest vibrational level of the ground state ($v = 0$) to the various vibrational levels of the excited state ($v' = 0, 1, 2, \ldots$) do not have equal probability. Two extreme cases can be proposed for the overlay of the vibrational bands; these can be recognised from the shape of the absorption. In Fig. 1.11 this is demonstrated for a diatomic molecule, using the simplification of the multi-dimensional energy hypersurface to the so-called Morse function $E_{pot} = f(r)$. The shape of the band is determined by whether the Morse function of the excited state is only vertically shifted or is additionally shifted to other r values (Fig. 1.11 a or 1.11 b, respectively).

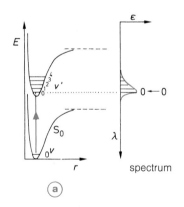

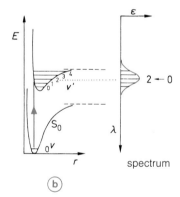

Fig. 1.11   Composition of an absorption band from different vibrational bands in a diatomic molecule; r interatomic distance; E energy

a unsymmetric band with intense $0 \leftarrow 0$ transition

b symmetric band with intense $2 \leftarrow 0$ transition

2.   Sample Preparation and Measurement of Spectra

For analytical purposes UV/Vis spectra are normally measured in solution. **Optically pure solvents**, available commercially, are used and allowed transitions measured at **concentrations** of about 10^{-4} mol·l^{-1}. For the weak bands of forbidden transitions the concentration must be increased appropriately.

(As a guide the absorbance E should be ≈ 1. For a layer thickness – length of the wave path through a quartz cell – of 1 cm it follows from the Beer-Lambert law that $c \cdot \varepsilon \approx 1$. If $\varepsilon_{max} = 10^n$ the measurement should therefore be made at a concentration of 10^{-n} mol·l^{-1}.)

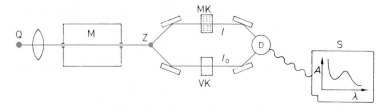

Fig. 1.12 Schematic diagram of a double-beam spectrometer
Q   radiation source (UV: Hydrogen or deuterium lamp, Vis: Tungsten-Halogen lamp)
M   (Double) monochromator using prisms and/or grating for spectral dispersion
Z   Beam splitter (rotating mirror)
Mk Measurement cell with solution
Vk Control cell with pure solvent
D   Detector (photoelectron multiplier)
S   Computer/Display/Printer, which records the transmission or absorption

Solvents with their own absorptions in the measurement region are unsuitable. The best transparency down to the vacuum-UV region is shown by perfluorinated alkanes like perfluorooctane. Sufficiently transparent down to 195 nm (for $d=1$ cm) are the saturated hydrocarbons **pentane**, **hexane**, **heptane**, or **cyclohexane** and the polar solvents **water** and **acetonitrile**. **Methanol**, **ethanol**, and **diethyl ether** are useable down to ca. 210 nm. In order of increasing lower measurement limit then follow **dichloromethane** (220 nm), **chloroform** (240 nm), and **carbon tetrachloride** (250 nm). **Benzene, toluene**, and **tetrahydrofuran** are generally only useable above 280 nm. An increase in the interaction between the compound being measured and the solvent leads to the loss of fine structure. It is therefore recommended to use non-polar solvents wherever possible. The effect of solvent polarity on the position of absorption bands is discussed in Sect. 3.4 (see p.17) using the case of ketones as an example. In the customary double beam spectrometers the cell with the solution to be measured is placed in one beam and a cell with the pure solvent in the other beam. The intensities are then compared over the whole spectral region. Fig. 1.12 shows schematically the construction of a double beam spectrometer.

Most instruments show the absorption A as a function of the wavelength λ. In contrast to A the extinction coefficient ε depends on the substance. It is therefore better to record a plot of ε against λ or even better against the wave number $\tilde{\nu}$. $\tilde{\nu}$, unlike λ, is proportional to the energy. In the long wavelength region, spectra which have a linear λ-scale are expanded, in the short wavelength region compressed. If strong and weak bands occur in the same spectrum, it is better to have log ε on the ordinate. Fig. 1.13 shows a comparison of the four frequently used ways in which UV/Vis spectra are commonly displayed.

A special form of measurement is the recording of the fluorescence as a function of the wavelength of the excitation. The **excitation spectra**, thus obtained, are not always identical with the absorption spectra. Even with a very pure compound the participation of different rotational isomers can result in different spectra. Two photon spectroscopy is often carried out by this technique.

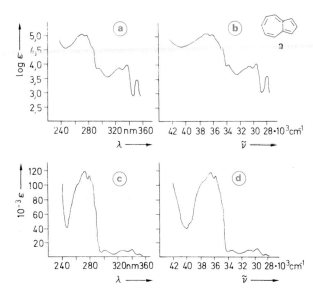

Fig. 1.13   UV spectra of azulene (**2**) in cyclohexane

a log $\varepsilon = f(\lambda)$ **b** log $\varepsilon = f(\tilde{\nu})$ **c** $\varepsilon = f(\lambda)$ **d** $\varepsilon = f(\tilde{\nu})$

The blue colour of this hydrocarbon arises from an absorption in the visible region of the spectrum, not shown in the above spectra

3. Chromophores

3.1 Individual Chromophoric Groups and their Interactions

As shown in Sect. 1.2, the position of an absorption band depends on the nature of the electronic transition involved. Tab. 1.2 gives a list of the excitation energies of σ-, π-, and n-electrons in various isolated **chromophoric groups**.

If a molecule has several π- or n-orbitals, which do not interact with each other, the spectrum will usually be the sum of absorptions assigned to the individual isolated chromophores. Steric effects, ring strain, etc., can lead to exceptions. Non-conjugated chromophores can also interact if they are near to each other, causing a shift or splitting of the bands (Davidov splitting). Where there are two identical chromophores there will often be two bands instead of the expected one, one of higher energy, one of lower energy than the energy of the isolated chromophore. This is shown by the examples of 1,4-pentadiene (**3**) and norbornadiene (**4**).

3 (λ_{max} = 178 nm, ε_{max} = 17 000)

4 (λ_{max} = 205 nm, ε_{max} = 2 100)

The homoconjugation in (**3**) is hardly noticeable. The absorption band starts at 200 nm and extends (as for a monoolefin) into the vacuum-UV with a maximum at λ = 178 nm (ε_{max} = 17 000). The absorption of norbornadiene (**4**) on the other hand starts at 270 nm, has a shoulder at 230 nm, and a structured absorption between 226 and 199 nm with a maximum at 205 nm (ε_{max} = 2100). The two non-conjugated double bonds in (**4**) therefore show a strong interaction, in contrast to (**3**).

Conjugated chromophores are of special importance for UV/Vis spectroscopy. Classical examples are the polymethine dyes. As the conjugated system becomes larger so the lowest energy $\pi \to \pi^*$ transition moves to longer wavelengths and becomes more intense; however a convergence limit is reached for series of oligomers. A bathochromic and hyperchromic effect is in general also observed when atoms or groups with n-orbitals ($-\overline{O}H$, $-\overline{O}R$, $-\overline{N}H_2$, $-\overline{N}HR$, $-\overline{N}R_2$, $-\overline{S}H$, $-\overline{S}R$, $-H\overline{al}|$ etc.) are directly bound to a chromophoric group. In this context the term **auxochromic group** is used.

[a] The λ_{max} and, to a lesser extent, the ε_{max} values depend on the solvent used (see Sec. 3.3, p. 17)

Tab. 1.2 Absorptions of isolated chromophoric groups (lowest energy transitions)

Chromophore	Transition	Example	λ_{max}^{a} (nm)	ε_{max}^{a}
C–H	$\sigma \to \sigma^*$	CH_4	122	strong
C–C	$\sigma \to \sigma^*$	H_3C—CH_3	135	strong
$-\overline{O}-$	$n \to \sigma^*$	H_2O	167	1 500
	$n \to \sigma^*$	H_3C—OH	183	200
	$n \to \sigma^*$	C_2H_5—O—C_2H_5	189	2 000
$-\overline{S}-$	$n \to \sigma^*$	H_3C—SH	235	180
	$n \to \sigma^*$	H_3C—S—CH_3	228	620
	$n \to \sigma^*$	C_2H_5—S—S—C_2H_5	250	380
$-\overline{N}-$	$n \to \sigma^*$	NH_3	194	5 700
	$n \to \sigma^*$	C_2H_5—NH_2	210	800
	$n \to \sigma^*$	C_2H_5—NH—C_2H_5	193	3 000
	$n \to \sigma^*$	$(C_2H_5)_3N$	213	6 000
— Hal	$n \to \sigma^*$	H_3C—Cl	173	200
	$n \to \sigma^*$	H_3C—Br	204	260
	$n \to \sigma^*$	H_3C—I	258	380
	$n \to \sigma^*$	CHI_3	349	2 170
$\diagdown C{=}C \diagup$	$\pi \to \pi^*$	$H_2C{=}CH_2$	165	16 000
	$\pi \to \pi^*$	C_2H_5—$CH{=}CH$—C_2H_5	185	7 940
$-C{\equiv}C-$	$\pi \to \pi^*$	$HC{\equiv}CH$	173	6 000
	$\pi \to \pi^*$	H—$C{\equiv}C$—C_2H_5	172	2 500
$\diagdown C{=}\overline{O}$	$n \to \pi^*$	H_3C—$CH{=}O$	293	12
	$\pi \to \pi^*$	H_3C—$\overset{\overset{O}{\|}}{C}$—$CH_3$	187	950
	$n \to \pi^*$	H_3C—$\overset{\overset{O}{\|}}{C}$—$CH_3$	273	14
$\diagdown C{=}\overline{S}$	$n \to \pi^*$	H_3C—$\overset{\overset{S}{\|}}{C}$—$CH_3$	460	weak
$\diagdown C{=}\overline{N}-$	$\pi \to \pi^*$	H_3C—$CH{=}N$—OH	190	8 000
	$n \to \pi^*$	H_3C—$CH{=}N$—OH	279	15
$-\overline{N}{=}\overline{N}-$	$n \to \pi^*$	$\underset{N\,=\,N}{H_3C\diagdown \quad \diagup CH_3}$	353	240
		$\underset{\qquad \diagdown CH_3}{\overset{H_3C\diagdown}{N\,=\,N}}$	343	25
$-\overline{N}{=}\overline{O}$	$n \to \pi^*$	$(H_3C)_3C$—NO	300	100
		$(H_3C)_3C$—NO	665	20
$-NO_2$	$\pi \to \pi^*$	H_3C—NO_2	210	10 000
	$n \to \pi^*$		278	10

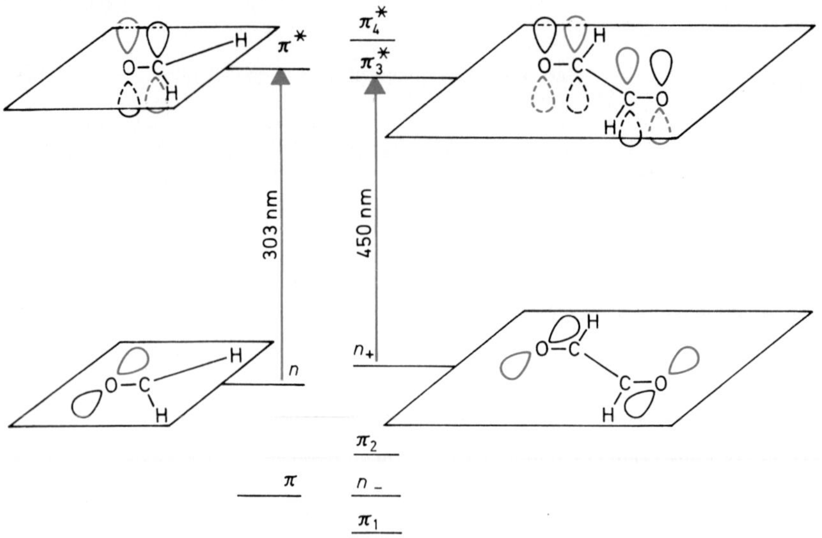

303 nm

450 nm

π_4^*

π_3^*

π^*

n

n_+

π_2

π

n_-

π_1

Fig. 1.14 Long wavelength electronic transitions $S_0 \to S_1$ in formaldehyde and glyoxal

Interactions between several chromophores or chromophores and auxochromes will be extensively discussed in the following chapters. As explicit examples formaldehyde (**5**) and glyoxal (**6**) will be treated here.

The forbidden $n \to \pi^*$ transition of formaldehyde gives a band with extensive fine structure in the gas phase with a maximum at 303 nm.

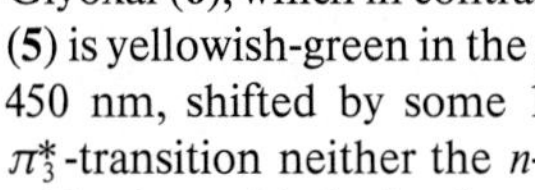

$$\begin{array}{c}\text{H}\\ \quad \diagdown\\ \quad \quad \text{C=O}\\ \quad \diagup\\ \text{H}\end{array}$$

λ_{max} = 303 nm

ε_{max} = 18

5

$$\begin{array}{c}\text{H}\\ \diagdown\\ \text{O=C}\\ \quad \diagdown\\ \quad \quad \text{C=O}\\ \quad \diagup\\ \text{H}\end{array}$$

λ_{max} = 450 nm

ε_{max} = 5

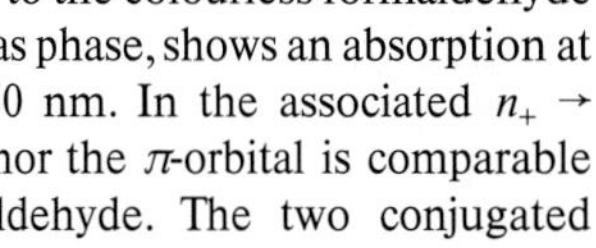

6

Glyoxal (**6**), which in contrast to the colourless formaldehyde (**5**) is yellowish-green in the gas phase, shows an absorption at 450 nm, shifted by some 150 nm. In the associated $n_+ \to \pi_3^*$-transition neither the n- nor the π-orbital is comparable with the orbitals in formaldehyde. The two conjugated π-bonds in glyoxal are described by the bonding orbitals π_1 and π_2 and the two antibonding orbitals π_3^* and π_4^*; the latter are empty in the ground state. The two lone pairs of electrons (with p-character) also interact and split into n_+ and n_-, with the symmetric combination n_+ having the higher energy (Fig. 1.14).

3.2 Olefins, Polyenes

The $\pi \to \pi^*$ transition of ethylene lies in the vacuum-UV with an intense band at $\lambda_{max} = 165$ nm ($\varepsilon_{max} = 16\,000$). If a hydrogen atom is substituted by an auxochromic group, a bathochromic shift is observed. This results from the interaction of a lone pair of the auxochromic group with the π-bond. From consid-

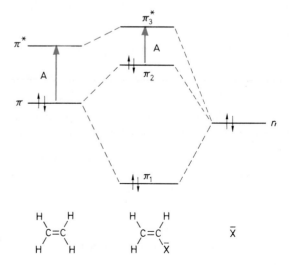

Fig. 1.15 Schematic energy diagram explaining the bathochromic shift of the $\pi \to \pi^*$ transition of ethylenes with auxochromic groups X

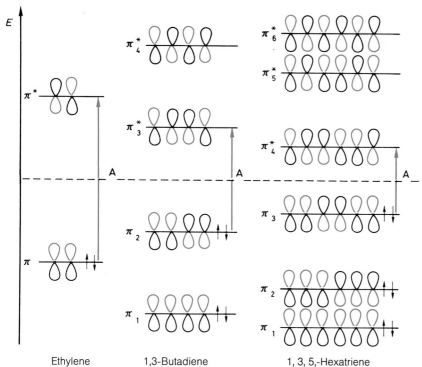

Ethylene
($\lambda_{max} = 165$ nm)

1,3-Butadiene
($\lambda_{max} = 217$ nm)

1, 3, 5,-Hexatriene
($\lambda_{max} = 258$ nm)

Fig. 1.16 HOMO–LUMO transitions in ethylene, 1,3-butadiene, and 1,3,5-hexatriene. The absorption coefficients ε_{max} increase in parallel with λ_{max}

eration of the resonance and inductive effects three new orbitals π_1 to π_3^* are predicted, as shown in Fig. 1.15. The shift to longer wavelength of the absorption results from the reduction of the energy difference ΔE between the HOMO and the LUMO.

The introduction of alkyl groups also leads to a shift of the $\pi \to \pi^*$ absorption. This effect is frequently explained on the basis of hyperconjugation.

When two or more olefinic double bonds are conjugated the centre of gravity of the π-orbitals is indeed reduced by the mesomeric effect, but the energy difference between the HOMO and LUMO gets less with increasing chain length, as shown in Fig. 1.16 and Tab. 1.3.

Accurate calculations demonstrate, in agreement with the observations, that λ_{max} approaches an eventual limiting value ($n \to \infty$).

Remarkably, the lowest excited state S_1 of linear *all-trans*-polyenes is not the optically allowed ^1B_u state, reached by the HOMO–LUMO transition, but a forbidden ^1A_g state. A doubly excited configuration makes a considerable contribution to this state. This was first predicted by quantum mechanical calculations taking into account the interactions between configurations, and has been experimentally confirmed by two photon spectroscopy.

Tab 1.3 Longest wavelength absorptions of conjugated *all-trans*-polyenes

$$R—(CH=CH)_n—R$$

n	$R = CH_3$		$R = C_6H_5$	
	λ_{max}[a]	ε_{max}	λ_{max}[b]	ε_{max}
1	174	24 000	306	24 000
2	227	24 000	334	48 000
3	275	30 200	358	75 000
4	310	76 500	384	86 000
5	342	122 000	403	94 000
6	380	146 500	420	113 000

[a] measured in petroleum ether or diethylether
[b] measured in benzene

There are cases (porphyrins, polymethine dyes, etc.) where the long wavelength absorption reaches as far as the **near infrared** (NIR). If for example a poly(phenylvinylene) system (PPV) is doped with an oxidising agent, then an electronic transition can take place which leads to polymeric radical ions and doubly charged ions (polarons, bipolarons). The insulator **(7)** is thus converted into a semiconductor **(8)**.

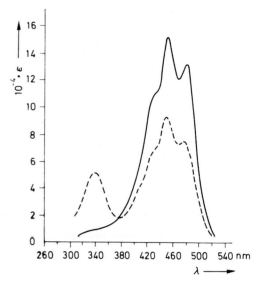

In the solution state absorption spectra the doping results in the appearance of new bands. The lowest energy transition can be shifted well beyond the visible wavelength region ($\lambda_{max} \approx 2000$ nm); it can only be observed with spectrometers specially equipped to observe the NIR region.

The configuration of olefins also affects the position and intensity of absorptions. (*Z*)-Stilbene absorbs at slightly shorter wavelength and less intensively than the (*E*)-isomer (Fig. 1.17).

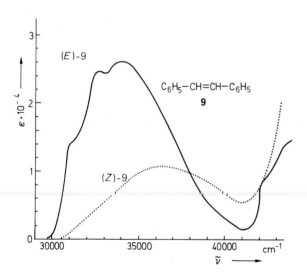

β-Carotenes

10　all (*E*)　——

11　15(*Z*)　- - -

Fig. 1.18　Absorption spectra of β-carotenes of different configurations (**10, 11**)

Fig. 1.17　UV spectrum of (*Z*)- and (*E*)-stilbene **9** at 295 K in methylpentane (from Dyck, R.H. and McClure, D.S. (1962), J.Chem.Phys. **36**, 2336)

The (*Z*)- or (*E*)-configuration has particular influence on the higher energy transitions of polyolefins. The first overtone in β-carotene lies at 340 nm. In the *all*-(*E*)-configuration it is symmetry forbidden (cf. parity rule). On inclusion of a (*Z*)-double bond the symmetry is changed. The transition is allowed and leads to the so-called (*Z*)-peak of the carotenes (Fig. 1.18).

Empirical rules for the absorption maxima of the longest wavelength $\pi \to \pi^*$ transitions of dienes and trienes were established by Woodward in 1942 and later and independently by Fieser and Scott. These start from specific base values for open-chain, homo- or heteroannular dienes with s-*cis* or s-*trans* configurations, increments being added for the various substituents (Tab. 1.4).

The last two examples in Tab. 1.5 show that these rules break down when strong steric effects are present. The influence of steric factors on the 1,3-diene chromophore is well demonstrated by the series of (*Z,Z*)-1,3-cycloalkadienes (Tab. 1.6).

The incremental rules also break down when special electronic effects are present, as in the case of annulenes (Tab. 1.7). Here there may be **aromatic** $(4n + 2)$-π-electron systems, **antiaromatic** $4n$-π-systems and non-planar molecules with so-called **non-aromatic** (olefinic) character. The similarity of the

Tab. 1.4 Incremental system for the calculation of the long wavelength absorption maximum of dienes and trienes

preferred s-trans (e.g. acyclic) 217 nm	s-cis (homoannular) 253 nm	s-trans (heteroannular) 214 nm

Increments

for each further conjugated double bond	+30 nm
for each exocyclic position of a double bond	+ 5 nm
for each alkyl or aryl group	+ 5 nm
for each auxochromic O-alkyl	+ 6 nm
group: O-acyl	± 0
S-alkyl	+30 nm
N(alkyl)$_2$	+60 nm
Cl	+ 5 nm
Br	+ 5 nm

Tab. 1.5 Examples of the calculation of λ_{max} values of conjugated dienes and trienes

Compound	λ_{max} (nm) observed	(nm) calculated
H$_3$C–CH=CH–CH=CH–CH$_3$	227	$217 + 2 \cdot 5 = 227$
=CH$_2$ (cyclohexene)	231	$214 + 2 \cdot 5 + 5 = 229$
(steroid)	282	$253 + 4 \cdot 5 + 2 \cdot 5 = 283$
(steroid)	234	$214 + 3 \cdot 5 + 5 = 234$
H$_3$C–C(O)–O– (acetate)	306	$253 + 30 + 3 \cdot 5 + 5 = 303$
(dimethylenecyclohexane)	220	$253 + 2 \cdot 5 + 2 \cdot 5 = 273$
(bicyclic)	246	$214 + 2 \cdot 5 + 5 = 229$

Tab. 1.6 Long wavelength UV absorptions of homoannular 1,3-dienes

Compound	λ_{max} (nm)	ε_{max}
Cyclopentadiene	238	3400
1,3-Cyclohexadiene	256	8000
1,3-Cycloheptadiene	248	7500
1,3-Cyclooctadiene	228	5600

Tab. 1.7 Absorptions of annulenes

Compound	λ_{max}	lg ε	Solvent	Colour of the solution	Character
Cyclobutadiene	305	2.0			anti-aromatic
Benzene	262 208 189	2.41 3.90 4.74	Hexane	colourless	aromatic
Cyclooctatetraene	285	2.3	Chloroform	yellow	non-aromatic
[10] Annulene	265 257	4.30 4.46	Methanol	yellow	non-aromatic
[14] Annulene	374 314	3.76 4.84	Isooctane	red-brown	aromatic
[16] Annulene	440 282	2.82 4.91	Cyclohexane	red	anti-aromatic
[18] Annulene	764 456 379	2.10 4.45 5.5	Benzene	yellow-green	aromatic
[24] Annulene	530 375 360	3.23 5.29 5.26	Benzene	violet	(anti-aromatic)

UV/Vis spectra of the aromatic [18]annulene and [6]annulene ($\equiv$ benzene) serves as an introduction to the next chapter.

3.3 Benzene and Benzenoid Aromatics

In contrast to 1,3,5-hexatrienes (see p. 11) the π_2/π_3 and π_4^*/π_5^* orbitals of **benzene** form pairs of degenerate (i.e. equal energy) orbitals. As can be theoretically demonstrated, the four conceivable $\pi_{2/3} \rightarrow \pi_{4/5}^*$ transitions lead from the $^1A_{1g}$ ground state to the excited singlet states $^1B_{2u}$, $^1B_{1u}$, and $^1E_{1u}$. (The latter state is, as the symbol E implies, a degenerate state.) Because of the electron correlation the three excited states and therefore the three transitions are of different energy (Fig. 1.19a and b).

In the UV spectrum of benzene (Fig. 1.20) the highly structured α-band and the p-band correspond to symmetry forbidden transitions. The p-band, which appears as a shoulder, "borrows" intensity from the neighbouring allowed transition (β-band). Because of the symmetry prohibition there is no $0 \leftarrow 0$-transition in the α-band. The ν'_A vibration distorts the hexagonal symmetry and leads to the longest wavelength vibrational band. Further vibrational bands follow, separated by the frequency of the symmetric breathing vibration ν'_B (Fig. 1.20).

The introduction of a substituent reduces the symmetry of benzene, enlarges the chromophoric system, and changes the orbital energies and thus the absorptions, so that the p-band can overtake the α-band. The α-band, sometimes also called the B-band, gains intensity and often loses its fine structure; because of the reduction in symmetry its $0 \leftarrow 0$ transition becomes visible.

An overview of **monosubstituted benzenes** is given in Tab. 1.8.

Tab. 1.8 UV Absorptions of monosubstituted benzenes C_6H_5-R

Substituents R	Long wavelength, stronger transition λ_{max} (nm) ε_{max}		Long wavelength, (forbidden) transition λ_{max} (nm) ε_{max}		Solvent
H	204	7400	254	204	Water
	198	8000	255	230	Cyclohexane
CH_3	207	9300	260	300	Ethanol
C_2H_5	200	31600	259	158	Ethanol
$CH(CH_3)_2$			251	250	Hexane
F			259	1290	Ethanol
Cl	210	7400	264	190	Water
Br	210	7900	261	192	Water
I	207	7000	257	700	Water
OH	211	6200	270	1450	Water
O^-	235	9400	287	2600	Water
OCH_3	217	6400	269	1480	Water
OC_6H_5	255	11000	272	2000	Water
			278	1800	
NH_2	230	8600	280	1430	Water
NH_3^+	203	7500	254	160	Water
$N(CH_3)_2$	251	12900	293	1590	Ethanol
NO_2			269	7800	Water
$CH{=}CH_2$	244	12000	282	450	Ethanol
$C{\equiv}CH$	236	12500	278	650	Hexane
$C{\equiv}N$	224	13000	271	1000	Water
$CH{=}O$	242	14000	280	1400	Hexane
			330	60	
$CO{-}CH_3$	243	13000	278	1100	Ethanol
			319	50	
COOH	230	11600	273	970	Water
COO^-	224	8700	268	560	Water
SO_3H	213	7800	263	290	Ethanol

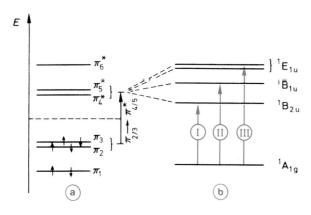

Fig. 1.19 **a** Energy scheme of the π-orbitals of benzene **b** Electronic excitations in benzene

		λ_{max}:	ε_{max}:
I	$^1B_{2u} \leftarrow {}^1A_{1g}$ according to Platt $^1L_b \leftarrow {}^1A$ according to Clar: α-band	256 nm	204
II	$^1B_{1u} \leftarrow {}^1A_{1g}$ according to Platt $^1L_a \leftarrow {}^1A$ according to Clar: p-band	203 nm	7400
III	$^1E_{1u} \leftarrow {}^1A_{1g}$ according to Platt $^1B \leftarrow {}^1A$ according to Clar: β-band	184 nm	60000

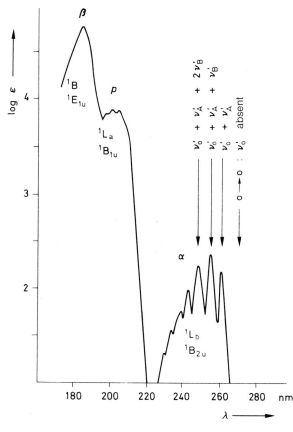

$\nu'_A = 520\ cm^{-1}$ $\nu'_B = 923\ cm^{-1}$

Fig. 1.20 Absorption spectrum of benzene

The change in the spectra caused by the introduction of two or more substituents into the benzene ring, compared to the monosubstituted derivatives, is particularly marked in those cases where both an electron-withdrawing group and an electron-donating group are present (Tab. 1.9).

In these cases the increase in the size of the chromophore is linked to the possibility of an **intramolecular charge transfer**:

Tab. 1.9 Long wavelength absorptions λ_{max} (nm) of some *para*-disubstituted benzenes $X^{1'}–C_6H_4–X^2$ (in water)

X^1	$X^2 = H$		OH		NH_2		NO_2	
	λ_{max}	$\log\varepsilon$	λ_{max}	$\log\varepsilon$	λ_{max}	$\log\varepsilon$	λ_{max}	$\log\varepsilon$
H	254	2.31						
OH	270	3.16	293	3.43				
NH_2	280	3.16	294	3.30	315	3.30		
NO_2	269	3.89	310	4.00	375	4.20	267	4.16

The effect should be even stronger in the *p*-nitrophenolate anion than in *p*-nitrophenol itself (**12**; Fig. 1.21a). This is confirmed by Fig. 1.21b, which however also shows that a similar effect occurs where there is *m*-substitution, and hence independently of the participation of quinonoid resonance structures.

The solvent can have a particularly strong effect in such cases, and even change the energetic order of the states. A good example of this is dimethylaminobenzonitrile **13**.

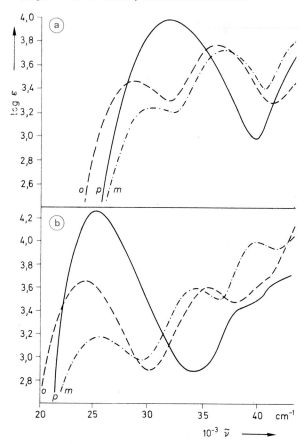

Fig. 1.21 UV/Vis spectra of *o*-, *m*-, and *p*-nitrophenol: **a** in 10^{-2} molar hydrochloric acid, **b** in $5\cdot10^{-3}$ molar sodium hydroxide (from Kortüm, G. (1941) Ber. Dtsch.Chem.Ges. **74**, 409)

12

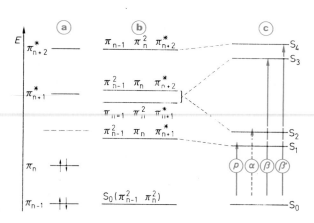

13

The intramolecular charge transfer can be stabilised by a twist about the CN bond. The **TICT state** (twisted intramolecular charge transfer) so formed has a large dipole moment ($\mu =$ 12 D) and is so energetically favoured by polar solvents that it becomes the lowest electronically excited singlet state. Thus the fluorescence occurs from different singlet states in polar and non-polar solvents. The dual fluorescence which is coupled with an intramolecular charge transfer (**ICT**) can also be explained by a solvent induced pseudo-Jahn-Teller effect. The assumption of bond twisting is not necessary; the two singlet states must however have very similar energies.

In the spectra of **condensed benzenoid aromatics** there are many common features. The two highest occupied orbitals π_{n-1} and π_n and the two lowest π^*_{n+1} and π^*_{n+2} are no longer degenerate as they are in benzene. Four electronic transitions are possible between them (Fig 1.22).

Fig. 1.22 Electronic transitions in polyacenes

a Orbital scheme
b State diagram (from HMO theory)
c Electronic transitions taking into account the configurational interactions

With increasing anellation the α-, p-, and β-bands shift to longer wavelengths. For polyacenes the p-band overtakes the weak α-band and is superimposed on it (Fig. 1.23). The intensity of the p-band remains more or less constant. (The increase in the ring size has no effect, because this electronic transition is polarised parallel to the short axis.) The bathochromic shift in the acene series leads to the members from tetracene onwards being coloured:

Benzene, naphthalene, anthracene	colourless
Tetracene	orange-yellow
Pentacene	blue-violet
Hexacene	dark green

If the anellation is non linear, characteristic changes occur in the spectra (cf. for example anthracene and phenanthrene,

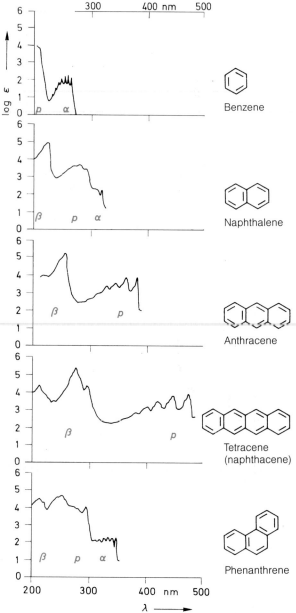

Fig. 1.23 UV/VIS spectra of condensed aromatic hydrocarbons (in heptane)

Fig. 1.23). As well as the linear anellated tetracene the condensation of four benzene rings can lead to four angular systems, benz[a]anthracene, benzo[c]phenanthrene, chrysene, and triphenylene, and the *peri*-condensed system pyrene. Of these only tetracene absorbs in the visible region; the others are colourless, but show coloured fluorescence. The strong band structure of nearly all benzenoid aromatic hydrocarbons is of particular analytical value for the identification of the individual members of the series.

3.4 Carbonyl Compounds

The carbonyl functional group contains σ-, π-, and n-electrons with s-character as well as n-electrons with p-character. This simple picture is based on the assumption of a non-hybridised oxygen atom. But even a detailed consideration of the delocalised group orbitals shows that the HOMO has largely the character of a p-orbital on oxygen. The excitation of an electron can occur into the anti-bonding π^* or σ^* orbital. For saturated aldehydes and ketones the allowed $n \rightarrow \sigma^*$ and $\pi \rightarrow \pi^*$ transitions are in the vacuum UV region. The forbidden $n(p) \rightarrow \pi^*$ transition lies in the region of 275 to 300 nm. The intensity of the $n \rightarrow \pi^*$ band is normally about $\varepsilon = 15 - 30$. (In β, γ-unsaturated ketones it can however be increased by a factor of 10 to 100.)

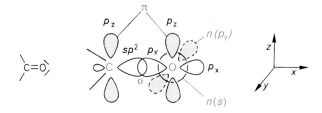

Auxochromes, such as OH, OR, NH_2, NHR, NR_2, Hal, etc., directly bonded to the carbonyl group, act as π-donors to increase the energy of the π^* orbital and as σ-acceptors to lower the n level. The result is a short wavelength shift of the $n \rightarrow \pi^*$ transitions in **carboxylic acids** and their **derivatives** (Tab. 1.10).

Tab. 1.10 $n \rightarrow \pi^*$ Transitions in saturated carbonyl compounds

Compounds	λ_{max} (nm)	ε_{max}	Solvents
Acetaldehyde	293	12	Hexane
Acetone	279	15	Hexane
Acetyl chloride	235	53	Hexane
Acetic anhydride	225	50	Isooctane
Acetamide	205	160	Methanol
Acetic acid ethyl ester	207	70	Petrol ether
Acetic acid	204	41	Ethanol

Conjugation of the carbonyl group with a (C=C) bond leads to a marked shift of the π-level; to a first approximation the n-orbital is unaffected (Fig. 1.24).

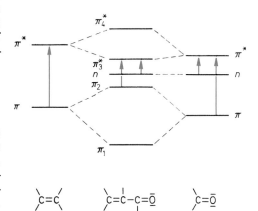

Fig. 1.24   Energy diagram of the electronic transitions in conjugated enones compared to alkenes and saturated carbonyl compounds

With increasing length of the conjugated chain in **enones** the longest wavelength $\pi \rightarrow \pi^*$ transition moves further into the visible region, reaches the position of the $n \rightarrow \pi^*$ band and hides it because of its considerably greater intensity, which also increases strongly with increasing chain length (Tab. 1.11).

Tab. 1.11   Absorption maxima of the long wavelength $\pi \rightarrow \pi^*$ transition in the vinylogous series $C_6H_5-(CH=CH)_n-CO-R$ (in methanol)

n	R = H		R = C_6H_5	
	λ_{max} (nm)	ε_{max}	λ_{max} (nm)	ε_{max}
0	244	12 000	254	20 000
1	285	25 000	305	25 000
2	323	43 000	342	39 000
3	355	54 000	373	46 000
4	382	51 000	400	60 000

The λ_{max} values of the $\pi \rightarrow \pi^*$ transitions of α,β-unsaturated carbonyl compounds can be estimated from the extended Woodward rules (Tab. 1.12).

The agreement between experimental absorption maxima and the values calculated from the increment system is apparent in Tab. 1.13.

Tab. 1.12 Increment system for calculating absorption maxima for α,β-unsaturated carbonyl compounds

$$\delta - C = C - C = C - C = O$$

| | | | | | |
| δ | γ | β | α | X | (in methanol or ethanol) |

Base values

X = H	207 nm
X = alkyl (or 6-membered ring)	215 nm
X = OH, OAlkyl	193 nm

Increments

for each further conjugated (C=C) bond	+ 30 nm
for each exocyclic position of a (C=C) bond	+ 5 nm
for each homoannular diene component	+ 39 nm
for each substituent in	

| | | position | | |
	α	β	γ	δ and higher
Alkyl (or ringrest)	10	12	18	18
Cl	15	12		
BR	25	30		
OH	35	30		50
O-alkyl	35	30	17	31
O-acyl	6	6	6	6
N(alkyl)$_2$		95		

The base values are for measurements in alcohols.
For other solvents the following solvent corrections must be applied

Water	+ 8 nm
Chloroform	− 1 nm
Dioxane	− 5 nm
Ether	− 7 nm
Hexane	− 11 nm
Cyclohexane	− 11 nm

Tab. 1.13 Observed and calculated $\pi \rightarrow \pi^*$ absorptions of some enones (in ethanol)

| Compound | Observed | | Calculated |
	λ_{max} (nm)	ε_{max}	λ_{max} (nm)
$H_3C-CH=CH-C-CH_3$ 3-Penten-2-one	224	9750	215 + 12 = 227
1-Cyclohexene-1-carboxaldehyde	231	13 180	207 + 10 + 12 = 229
1-Cyclohexene-1-carboxylic acid	217	10 230	193 + 10 + 12 = 215
Steroid type	241	—	215 + 10 + 12 + 5 = 242
Steroid type	388	—	215 + 2 · 30 + 5 + + 39 + 12 + 3 · 18 = 385
4, 6, 6-Trimethyl-bicyclo [3. 1.1]-hept-3-en-2-one	253[a]	6460	215 + 2 · 12 = 239[a]

[a] Deviation of the experimental value caused by ring strain

As already mentioned in Sec. 3.1 (p. 9), certain absorptions are very solvent dependent. Such effects have been particularly carefully investigated for ketones. Fig. 1.25 shows the example of benzophenone (**14**).

The electronic states of benzophenone are lowered by solvation, hydrogen bonding in polar, protic solvents being especially effective. The strongest effect is observed with the state of highest polarity, the π,π^* singlet state. Since the doubly occupied *n*-orbital on the oxygen atom is mostly responsible for the hydrogen bonding, the n,π^* singlet state of ketones has much poorer solvation properties (Fig. 1.26).

Similar solvent effects appear in certain heterocycles, azo-compounds, nitroso-compounds, thioketones, etc. However the use of solvent dependence for characterising the $n \rightarrow \pi^*$ and $\pi \rightarrow \pi^*$ transitions should be confined to aldehydes and ketones.

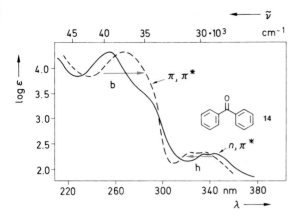

Fig. 1.25 Absorption spectra of benzophenone (**14**)
————— in cyclohexane
— — — — in ethanol
b bathochromic solvent effect (on increasing the solvent polarity)
h hypsochromic solvent effect (on increasing the solvent polarity)

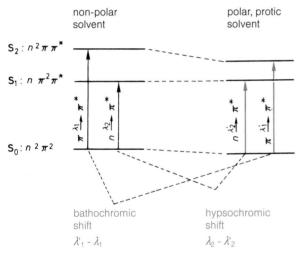

Fig. 1.26 The bathochromic and hypsochromic shifts of the $\pi \to \pi^*$ and $n \to \pi^*$ transitions of ketones on increasing the solvent polarity

The **quinones** represent special "enone"-chromophores. As shown by a comparison of 1,4- (**15**) and 1,2-benzoquinone (**16**), o-quinones absorb at longer wavelengths than the corresponding p-isomers:

15

	1,4-Benzoquinone (yellow)	
λ_{max} (nm)	ε_{max} (in benzene)	
242	24200	(allowed $\pi \to \pi^*$)
281	400	($\pi \to \pi^*$)
434	20	(forbidden $n \to \pi^*$)

16

	1,2-Benzoquinone (red)	
λ_{max} (nm)	ε_{max} (in benzene)	
390	3020	
610	20	(forbidden $n \to \pi^*$)

The reason for this is that the lowest π^*-orbital of the linearly conjugated o-quinone lies lower than that of the cross-conjugated p-quinone. Because of the interaction of the $n(p)$-orbitals of the two oxygen atoms two $n \to \pi^*$ transitions are expected. In general they lie very close together.

The p- and o-quinonoid groups play a decisive role in many **organic dyestuffs**. The acid-base indicator phenolphthalein will be discussed as a typical example. The lactone form (**17**) only contains isolated benzene rings and is therefore colourless. At pH = 8.4 the two phenolic protons are lost, forming the dianion (**18**), which opens the lactone ring to form the red dye (**19**) (λ_{max} = 552 nm; ε_{max} = 31000). With excess alkali the carbinol (**20**) is formed, a trianion in which the chromophoric **meriquinoid** group has disappeared again.

17 (colourless)   +2 OH⁻ / −2 H₂O   **18**

19 (deep red)

+OH⁻

20 (colourless)

4. Applications of UV/Vis Spectroscopy

In combination with other spectroscopic methods UV/Vis spectroscopy can be a valuable method of qualitative analysis and structure determination. In recent years new applications for UV/Vis spectroscopy have been opened up in the study of electronically excited states (photochemistry).

Quantitative analysis (colorimetry, photometry), photometric titration, determination of equilibrium and dissociation constants are other important applications. UV/Vis spectroscopy is also a valuable method in the increasingly important area of trace analysis. An example of photometry is the determination of alcohol in blood. It involves the enzymatic dehydrogenation of ethanol to acetaldehyde. The hydrogen is taken up by NAD (nicotinamide adenine dinucleotide). This conversion can be very accurately assessed by absorption measurements. For chromatographic methods like HPLC (high performance liquid chromatography) measurement of the UV absorption is the commonest detection method. In addition to fixed wavelength photometers photodiode arrays can also be used, thus allowing the measurement of a complete UV spectrum at each time point in the chromatogram.

UV/Vis spectroscopy has a special role as an analytical tool in kinetic measurements. Whereas the measurement of spectra of slowly reacting systems is no problem, and can even be performed in the reaction flask using a light conductor system, special methods are required for fast reactions. The whole spectrum has to be measured as quickly as possible, and stored digitally. Optical multi-channel analysers (OMA) are used. The measurement beam is shone onto a grating monochromator and then onto a two-dimensional array of photodiodes (diode array). The location of each diode corresponds to a particular wavenumber. The information from the individual channels gives the total spectrum.

Even faster spectroscopy is possible with flash laser apparatus. The excitation flash is followed in rapid succession by measurement flashes to establish the chemical intermediates. In this way lifetimes in the ns- and ps-ranges can be measured. Recent advances have reduced the measurable range to the fs-range (1 fs = 10^{-15} s).

In the following sections two simple applications of UV/Vis spectroscopy will be described. UV/Vis spectroscopy can be used for the **determination of the pK value** of a medium strong acid like 2,4-dinitrophenol (**21**) (Fig. 1.27).

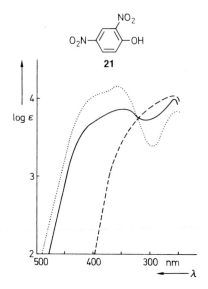

Fig. 1.27   pK value determination of 2,4-dinitrophenol **21**; (from Flexer, L.A., Hammett, L.P., Dingwall, A. (1935), J. Am. Chem. Soc. **57**, 2103)

. . . . . . Solution of **21** in 0.1 molar sodium hydroxide

_ _ _ _ . Solution of **21** in 0.1 molar hydrochloric acid

———— Solution of **21** in an acetate buffer of pH = 4.02

The **dissociation equilibrium** is described by

$$\text{Acid} + H_2O \rightleftharpoons H_3O^+ + \text{Anion}$$
$$A_{\text{tot.}} = d(\varepsilon_s \cdot c_S + \varepsilon_a \cdot c_A) = d \cdot \varepsilon \cdot c$$

The total absorbance with a formal ε and the weighed concentration $c = c_S + c_{A^-}$ is caused by the absorbing species S and A with extinction coefficients ε_s and ε_a. Rearrangement gives

$$\frac{c_S}{c_{A^-}} = \frac{\varepsilon - \varepsilon_a}{\varepsilon_s - \varepsilon} \qquad \varepsilon_s \neq \varepsilon.$$

ε_s and ε_a are obtained from measurements on dilute solutions in strongly acid or alkaline medium, where the concentrations of A$^-$ and S are negligible. The determination of ε is best carried out using buffer solutions with intermediate pH values.

Fig. 1.27 shows three such curves. They all cross in an **isobestic point**. At its wavelength λ_i the two interconvertible species absorbing there, S and A have the same ε value. From the known pH value of the buffer solution:

$$pK = -\log \frac{c_{H_3O^+} \cdot c_{A^-}}{c_S}$$

$$pK = pH + \log \frac{c_S}{c_{A^-}}$$

$$pK = pH + \log \frac{\varepsilon - \varepsilon_a}{\varepsilon_s - \varepsilon}$$

For the determination of the pK value the values determined at different wavelengths should be averaged. For 2,4-dinitrophenol (**21**) this gives p$K = 4.10 \pm 0.04$.

The second example shows a **reaction spectrum** for the photofragmentation of the heterocyclic spirane **22**. In acetonitrile, cyclopentanone (**23**) is formed quantitatively with a quantum yield of 57%.

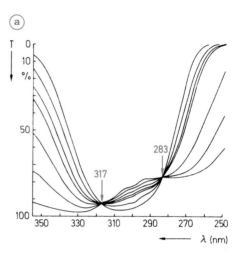

22   **23**

The irradiation is monochromatic at $\lambda = 365$ nm, which is close to the long wavelength absorption of **22** ($\lambda_{max} = 367$ nm, lg $\varepsilon = 2.50$). The reaction spectrum in Fig. 1.28 shows that this band is reduced in the course of the irradiation, i.e. the **transmission** there increases. At $\lambda_1 = 317$ nm this situation is reversed. In the range from $\lambda_1 = 317 > \lambda > 283 = \lambda_2$ the transmission becomes less during the irradiation, since in this region the $n \to \pi^*$ transition of the product **23** builds up. At λ_2 there is a second reversal. λ_1 and λ_2 are the **isobestic points** of this irreversible reaction. The appearance of isobestic points shows the uniformity of the reaction. In particular they rule out the possibility that the fragmentation takes place *via* an intermediate which increases in concentration and absorbs light. The uniformity of the reaction is often even more clearly shown by **extinction difference diagrams.** In such cases $E(\lambda_m)$ must be a linear function of $E(\lambda_n)$; λ_m and λ_n can be any wavelengths in the absorption range. Instead of such **E-diagrams** one can also construct an **ED-diagram.**

Here the **extinction differences** $E(\lambda_m, t) - E(\lambda_m, t=0)$ are plotted against the differences $E(\lambda_n, t) - E(\lambda_n, t=0)$. In Fig. 1.28b ED ($\lambda = 340, 300,$ and 260) are plotted as linear functions against

ED ($\lambda = 360$). The reaction time is the parameter t. If there are two or more independent (thermal or photochemical) reactions the E- or ED-diagram is not linear.

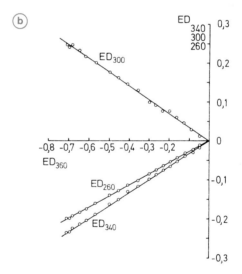

Fig. 1.28 a reaction spectrum measured in % transmission for the photolysis **22** → **23** with monochromatic irradiation ($\lambda = 365$ nm) in acetonitrile; **b** corresponding extinction difference diagram (from Daniil, D., Gauglitz, G., Meier, H. (1977), Photochem. Photobiol. **26**, 225)

5. Derivative Spectroscopy

The recording of the first, second, or nth derivative of an absorption curve is a relatively recent analytical aid, which has gained importance with the development of electronic differentiation. The mathematics of continuous curves establishes the following relationships:

Absorption $A(\lambda)$	Maximum/minimum $\updownarrow$	Inflection $\updownarrow$
1st derivative $\dfrac{dA(\lambda)}{d\lambda}$	Zero crossing	Maximum/ minimum $\uparrow$
2nd derivative $\dfrac{d^2 A(\lambda)}{d\lambda^2}$		Zero crossing

Fig 1.29 shows the long wavelength part ($n \to \pi^*$ transition) of the UV spectrum of testosterone (**24**). The superimposition of the vibrational bands leads to a barely recognisable structure of the band. Above it is shown the first derivative. The broken line joins the absorption maximum with the zero crossing of the first derivative. The dotted lines join the inflection points in the left hand part of the absorption curve with the extremes (maxima and minima) of the first derivative. The structure of the curve of the first derivative is much more obvious. The effect is even stronger in the curve of the second derivative, where zero crossings occur at the positions where extremes occur in the first derivative. Small changes in a spectrum, for example a shoulder, can be emphasised by derivative spectroscopy. The technique is also suited to the solution of difficult quantitative problems, e.g. in trace analysis, and in the following of the progress of reactions.

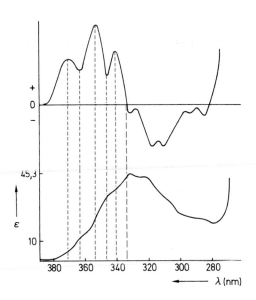

Fig. 1.29 Long wavelength absorption of testosterone (**24**) in diethyleneglycol dimethyl ether and first derivative of the absorption curve (from Olson, E. C., Alway, C. D. (1960), Anal.Chem. **32**, 370)

6. Chirooptical Methods

Chirooptical methods are optical measurements which depend on the **chirality** of the material under investigation. A substance is optically active if it rotates the plane of **linearly polarised light.** As can be seen from Fig. 1.30, this corresponds to the rotation of the vibrational direction of the electrical vector E of the light wave.

The optical rotation results either from a chiral crystal structure, as in quartz or cinnabar, or from the chirality of molecules (or ions). A molecule (object) is chiral if it is not identical with its mirror image. This property requires that the molecule is asymmetric or only possesses symmetry elements which are **symmetry axes** C_n. **Planes of symmetry** σ or **rotation-reflection**

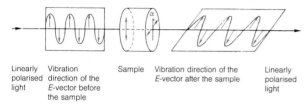

Linearly polarised light | Vibration direction of the E-vector before the sample | Sample | Vibration direction of the E-vector after the sample | Linearly polarised light

Fig. 1.30 Schematic representation of optical rotation

axes S_n, including the **centre of symmetry** $S_2 \equiv i$, must therefore not be present.

The **rotation angle** α measured with a **polarimeter** for a chiral compound in solution is given by the relationship

$$\alpha = [\alpha]_\lambda^T \cdot l \cdot c$$

α in degrees
l layer thickness in dm
c concentration in $g \cdot ml^{-1}$

For the comparison of various optically active compounds the rotation based on the molecular mass M is often preferred

$$[\phi]_\lambda^T = \frac{100\,\alpha}{l \cdot c} = \frac{[\alpha]_\lambda^T \cdot M}{100}$$

Φ, α in degrees
l layer thickness in cm
c concentration in mol/l

The **specific rotation** $[\alpha]_\lambda^T$ depends not only on the compound being measured, but also on the wavelength λ of the monochromatic radiation being employed and on the temperature T. α and $[\alpha]_\lambda^T$ have positive signs when the compound is **dextrorotatory**, i.e. when, viewed against the light beam, E is rotated clockwise. The mirror-image isomer (enantiomer) is then **laevorotatory** (anticlockwise) and has a **negative specific rotation** of the same magnitude.

The measured rotations can therefore be used for the determination of the enantiomeric purity. If for example the dextrorotatory form dominates in a mixture of enantiomers, then by definition

the **enantiomeric excess:**

$$ee = E(+) - E(-)$$

the **enantiomeric purity is the quotient**

$$\frac{E(+) - E(-)}{E(+) + E(-)}$$

and the **optical purity**

$$P = \frac{[\alpha]}{[\alpha]_{max}}$$

with $[\alpha]_{max}$ as the rotation of pure E (+). The ratio E (+)/E(–) is $1 + P/1 - P$, if the two enantiomers behave additively in the polarimetric measurement, otherwise optical purity and enantiomeric purity will not agree. Deviations from additivity are observed for example when there is association through hydrogen bonds.

Table 1.14 gives a collection of some specific rotations, measured in solution at 20 °C with the sodium-D-line (589.3 nm).

The underlying chirality of **optically active compounds** is classified according to **chiral elements** (centres, axes, planes) – see a textbook of stereochemistry. The commonest chiral element is the asymmetric carbon atom with four different ligands.

$$R^1 - \overset{\overset{\textstyle R^2}{|}}{\underset{\underset{\textstyle R^4}{|}}{C}} - R^3 \qquad C_6H_5 - \overset{\overset{\textstyle H}{|}}{\underset{\underset{\textstyle D}{|}}{C}} - CH_3$$

25

The H/D isotope effect is in principle sufficient for a measurable optical activity; thus 1-deuterio-1-phenylethane (**25**) has a specific rotation of 0.5°.

Tab. 1.14 Specific rotations of some optically active compounds

Compound	Solvent	$[\alpha]_D^{20}$	
R-Lactic acid (D-Lactic acid)	Water	−	2.3
S-Alanine (L-Alanine)	Water	+	2.7
S-Leucine (L-Leucine)	6-molar hydrochloric acid	+	15.1
	Water	−	10.8
	3-molar sodium hydroxide	+	7.6
α-D-Glucose	Water	+	112.2
β-D-Glucose	Water	+	17.5
D-Glucose in solution equilibrium (mutarotation)	Water	+	52.7
Sucrose	Water	+	66.4
(1R, 4R)-Camphor (D-Camphor)	Ethanol	+	44.3
Cholesterol	Ether	−	31.5
Vitamin D_2	Ethanol	+	102.5
	Acetone	+	82.6
	Chloroform	+	52.0
(P)-Hexahelicene	Chloroform	+	3707

Nevertheless there are chiral compounds with $[\alpha] = 0$. An example is the enantiomerically pure 1-lauryl-2,3-dipalmityl glyceride (**26**). Although **26**, in contrast to the achiral 2-lauryl-1,3-dipalmityl glyceride has no plane of symmetry, the difference between the 1-lauryl and the 3-palmityl group in relation to the chiral centre at C-2 is too small to lead to an observable rotation.

$$\alpha = \frac{180\,(n_l - n_r)\,l}{\lambda_0}$$

α rotation in degrees
l layer thickness and
λ_0 vacuum wavelength in the same units of length
n_l, n_r refractive indices

The **normal optical rotation dispersion** (ORD) $\alpha(\lambda)$ or $\Phi(\lambda)$ is shown in Fig. 1.32 for several steroids. Characteristic is the monotonic form of the curves.

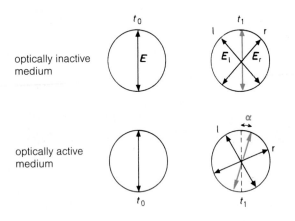

26

For an understanding of optical rotation it is necessary to consider linearly polarised light as consisting of a **right-handed** and a **left-handed polarised wave** of the same amplitude and phase (Fig. 1.31). In an optically active medium the two waves with opposite rotations have different velocities c and in the absorption region have different extinction coefficients ε. The case where $c_l \neq c_r$ leads to a phase difference between the two waves and thus to a rotation of the **E**-vector of the linearly polarised light which is reformed by combination of the two circularly polarised beams (Fig. 1.31).

The rotation angle depends on the wavelength used.

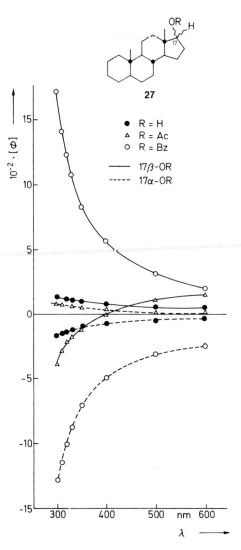

Fig. 1.32 Normal ORD curve of 5α androstanes **27** substituted at the C-17 position (from Jones, P.M., Klyne, W., [1960], J. Chem. Soc., 871)

Fig. 1.31 Decomposition of the linearly polarised light beam into right- and left-handed circularly polarised beams. In an optically inactive medium $c_l = c_r$ and at any time point t the original vibration direction of the **E**-vector is retained. In an optically active medium where $=c_r > c_l$ the vector **E**$_r$ of the right-handed wave has rotated at time t_1 by a larger angle than **E**$_l$. The resultant vibration direction shows a positive rotation α

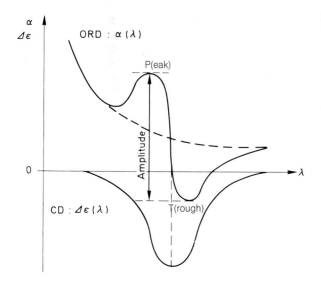

Fig. 1.33 Cotton effect – relationship between ORD and CD curves; in this example the CD is negative and the normal ORD (dotted line) positive

In the region of absorption bands the normal ORD curve is overlaid with an S-shaped component to produce the so-called **anomalous ORD-curve** (Fig. 1.33).

$|E_l| \neq |E_r|$ means that the E-vectors of the two opposed circularly polarised light beams have different lengths after passing through the medium due to differing absorption. Their recombination no longer produces linearly polarised light, instead the vector diagram is elliptical. If the end of the E-vector runs through the ellipse in a clockwise direction, the **circular dichroism (CD)**, $\Delta\varepsilon(\lambda)$, is said to be **positive,** if in an anticlockwise direction it is **negative.** Anomalous optical rotation dispersion and circular dichroism together make up the **Cotton effect.**

From the combination of positive or negative normal ORD curves with positive or negative Cotton effect there are four types of anomalous ORD curves.

$\lambda_{peak} < \lambda_{trough}$ (Fig. 1.33) implies a negative Cotton effect, $\lambda_p > \lambda_T$ a positive effect.

The extreme (maximum or minimum) of the CD curve lies at the same λ value as the crossing point of the anomalous and interpolated normal ORD curves (approximately the inflection point, Fig. 1.33).

In simple cases this λ_{max} value corresponds roughly to the maximum of the normal UV/Vis absorption (Fig. 1.34).

Instead of $\Delta\varepsilon(\lambda)$ the **molar ellipticity** $[\Theta]_M$ is often recorded as a function of the wavelength:

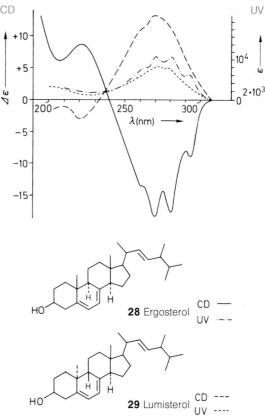

Fig. 1.34 CD curves and UV absorptions of ergosterol **28** and lumisterol **29**

$$[\Theta]_M = \frac{\Theta \cdot M}{100 \cdot c \cdot l} = 3.3 \cdot 10^3 \cdot \Delta\varepsilon$$

$$[\Theta]_M = 10^{-2} \cdot [\Theta] \cdot M$$

$$[\Theta] = \frac{\Theta}{c \cdot l} \qquad\qquad \text{specific ellipticity}$$

Ellipticity Θ in degrees
Concentration c in g/ml
Layer thickness l in dm
Molecular mass M

For the analytical interpretation of the Cotton effect there are a number of rules, theoretical, semi-empirical, and purely empirical. One such is the octant rule for saturated ketones, which have an $n \rightarrow \pi^*$ transition at about 280 nm. The three node planes of the n and π^* orbitals divide the space into eight octants, which can be represented as the octants of a Cartesian x, y, z co-ordinate system. The four octants with positive y values are shown in Fig. 1.35a.

The xy plane corresponds to the σ-bonding plane of the carbonyl function and the carbonyl C atom is assumed to lie on the

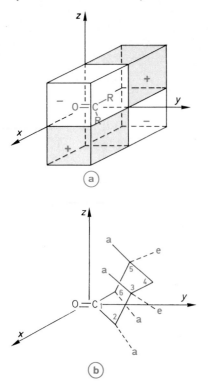

Fig. 1.35 Illustration of the octant rule for a saturated ketone (cyclohexanone derivative)

positive side of the y axis. In the two octants shown in blue the Cotton effect has a positive sign (looking from O to C, upper left and lower right), in the other two octants a negative sign. If a cyclohexanone structure is introduced into this co-ordinate system, as in Fig. 1.35b, then the substituents at C-4 lie in the yz plane and the *equatorial* substituents at C-2 and C-6 approximately in the xy plane; these make no contribution to the Cotton effect. Positive Cotton effects are produced by the *axial* C-2 substituents and by the *axial* and *equatorial* substituents at C-5, negative Cotton effects by the *axial* C-6 and *axial* and *equatorial* substituents at C-3. It must be remembered that only chiral cyclohexanone derivatives need be considered. For other types of compounds similar rules can be established. The reader should refer to the literature quoted in the bibliography. In general it can be established that **polarimetry** is useful for the determination of concentration or purity or, as in carbohydrate chemistry, for the study of rearrangement processes (mutarotation, inversion), whereas **ORD and CD spectra** provide valuable information about structure by characterising absolute configurations, particularly in natural product chemistry.

At the end of this section it should be pointed out that substances in a magnetic field are always optically active **(Faraday effect).** The vibrational plane of linearly polarised light, parallel to the magnetic lines of force, will be rotated by substances which are normally optically inactive. Thus **MORD and MCD** measurements can be a useful supplement to **ORD** and **CD** measurements. Literature references to further information on these techniques are given.

Literature

UV/Vis Spectroscopy

Books

Barrow, G. R. (1962), Introduction to Molecular Spectroscopy, McGraw-Hill, New York.

Dyer, J. R. (1965), Applications of Absorption Spectroscopy of Organic Compounds, Prentice-Hall, Englewood Cliffs.

Ewing, G. W. (1975), Instrumental Methods of Chemical Analysis, McGraw-Hill Book Comp., New York.

Fabian, J., Hartmann, H. (1980), Light Absorption of Organic Colorants, Springer Verlag, Berlin.

Foerst, W. (1966), Optische Anregung organischer Systeme, Verlag Chemie, Weinheim.

Gauglitz, G. (1983), Praxis der UV/Vis-Spektroskopie, Attempto Verlag, Tübingen.

Gillam, A. E., Stern, E. S. (1957), Electronic Absorption Spectroscopy, Arnold, London.

Griffiths, J. (1976), Colour and Constitution of Organic Molecules, Academic Press, New York, London.

Hampel, B. (1962), Absorptionsspektroskopie im ultravioletten und sichtbaren Spektralbereich, Vieweg Verlag, Braunschweig.

Jaffé, H. H., Orchin, M. (1962), Theory and Applications of Ultraviolet Spectroscopy, Wiley, New York.

Klessinger, M., Michl, J. (1989), Lichtabsorption und Photochemie organischer Moleküle, Verlag Chemie, Weinheim.

Maass, D. H. (1973), An Introduction to Ultraviolet Spectroscopy with Problems, in An Introduction to Spectroscopic Methods for the Identification of Organic Compounds (Scheinmann, F., Herausgeb.), Bd. 2, Pergamon Press, New York.

Murell, J. N. (1967), Elektronenspektren organischer Moleküle, Bibliographisches Institut 250/250a*, Mannheim.

Olsen, E. D. (1975), Modern Optical Methods of Analysis, McGraw-Hill Book Comp., New York.

Parikh, V. M. (1974), Absorption Spectroscopy of Organic Molecules, Addison-Wesley, Reading.

Parker, C. A. (1968), Photoluminescence of Solutions, Elsevier, Amsterdam.

Perkampus, H.-H. (1986), UV-Vis-Spektroskopie und ihre Anwendungen, Springer Verlag, Berlin.

Rao, C. N. R. (1961), Ultraviolet and Visible Spectroscopy, Butterworth, London.

Sandorfy, C. (1964), Electronic Spectra and Quantum Chemistry, Prentice-Hall, Englewood Cliffs.

Schmidt, W. (1994), Optische Spektroskopie, Verlag Chemie, Weinheim.

Schulman, S. G. (1993), Molecular Luminescence Spectroscopy, Wiley, New York.

Scott, A. I. (1964), Interpretation of the Ultraviolet Spectra of Natural Products, Pergamon Press, New York.

Snatzke, G. (1973), Elektronen-Spektroskopie, in Methodicum Chimicum (Korte, F.), Bd. 1/1, Georg Thieme Verlag, Stuttgart.

Staab, H. A. (1970), Einführung in die theoretische organische Chemie, Verlag Chemie, Weinheim.

Thompson, C. C. (1974), Ultraviolet-Visible Absorption Spectroscopy, Willard Grant Press, Boston.

Series

UV Spectrometry Group, Techniques in Visible and Ultraviolet Spectrometry, Chapman & Hall, London.

Data Collections, Spectral Catalogues

Hershenson, H. M., Ultraviolet and Visible Absorption Spectra, Academic Press, New York.

A. P. I. Research Project 44: Ultraviolet Spectral Data, Carnegie Institute and U. S. Bureau of Standards.

Phillips, J. P., Feuer, H., Thyagarajan, B. S. (et al), Organic Electronic Spectral Data, Wiley, New York.

Pestemer, M., Correlation Tables for the Structural Determination of Organic Compounds by Ultraviolet Light Absorptiometry, Verlag Chemie, Weinheim.

Lang, L., Absorption Spectra in the Ultraviolet and Visible Region, Academic Press, New York.

UV-Atlas organischer Verbindungen, Verlag Chemie, Weinheim.

Sadtler Standard Spectra (Ultraviolet), Heyden, London.

Chirooptical Methods

Review Articles

Foss, J. G. (1963), J. Chem. Educ. *40*, 592.

Velluz, L., Legrand, M. (1965), Angew. Chem. *77*, 842.

Eyring, H., Liu, H.-Ch., Caldwell, D. (1968), Chem. Rev. *68*, 525.

Snatzke, G. (1968), Angew. Chem. *80*, 15.

Snatzke, G. (1973), in Methodicum Chimicum (Korte, F., Herausgeb.), Bd. 1/1, Georg Thieme Verlag, Stuttgart, p. 426.

Schellman, J. A. (1975), Chem. Rev. *75*, 323.

Monographs

Djerassi, C. (1964), Optical Rotary Dispersion, McGraw-Hill Book Comp., New York.

Velluz, L., Legrand, M., Grosjean, M. (1965), Optical Circular Dichroism, Verlag Chemie, Weinheim.

Snatzke, G. (1967), Optical Rotary Dispersion and Circular Dichroism in Organic Chemistry, Heyden, Canada.

Crabbé, P. (1971), An Introduction to the Chiroptical Methods in Chemistry, Syntex, Mexico City.

Crabbé, P. (1972), ORD and CD in Chemistry and Biochemistry – An Introduction, Academic Press, New York, London.

Olsen, E. D. (1975), Modern Optical Methods of Analysis, McGraw-Hill Book Comp., New York.

Thulstrup, E. W. (1980), Aspects of Linear and Magnetic Circular Dichroism of Planar Organic Molecules, Lecture Notes in Chemistry 14, Springer Verlag, Berlin.

Mason, S. F. (1982), Molecular Optical Activity and the Chiral Discriminations, University Press, Cambridge.

Harada, N., Nakanishi, K. (1983), Circular Dichroic Spectroscopy, University Science Books, New York.

Michl, J., Thulstrup, E. W. (1986), Spectroscopy with Polarized Light, VCH, Weinheim.

2 Infrared and Raman Spectroscopy

1. Introduction

Molecular **vibrations** and **rotations** can be excited by absorption of radiation in the infrared region of the electromagnetic spectrum. This lies to longer wavelengths than the visible region. Infrared radiation is also referred to as radiant heat, since it is detected by the skin as heat.

There are two possible ways to measure such molecular vibrations and rotations:

- directly as absorptions in the infrared spectrum

- indirectly as scattered radiation in the Raman spectrum (see section 15, p. 66)

The position of an absorption band in the IR spectrum can be expressed in units of the wavelength λ (in μ or μm) of the absorbed radiation. The bands of particular use for the structural analysis of organic molecules lie in the region from

$$\lambda = 2,5\ \mu m - 15\ \mu m\ (10^{-3}\ mm = 1\ \mu m = 10^4\ \text{Å}).$$

Nowadays, however, it is usual to quote the reciprocal wavelength, the so-called wavenumber $\tilde{v}$ (cm^{-1}). The numerical value of $\tilde{v}$ (measured in cm^{-1}) indicates how many waves of the infrared radiation there are in one centimetre.

$$\text{Wavenumber:}\quad \tilde{v} = \frac{1}{\lambda}$$

To convert wave**numbers** into wave**lengths** in the usual units the relationship is

$$\text{wavenumber}\quad \tilde{v}\,(\text{cm}^{-1}) = \frac{10^4}{\text{wavelength}\ \lambda\ (\mu m)}$$

Wavenumbers $\tilde{v}$ have the advantage that they are proportional to the frequency v of the absorbed radiation, and therefore also to the energy ΔE. The following relationships apply:

$$\lambda \cdot v = c$$

$$v = \frac{c}{\lambda} = c \cdot \tilde{v}$$

$$\Delta E = h \cdot v = \frac{h \cdot c}{\lambda} = h \cdot c \cdot \tilde{v}$$

$$\Delta E \sim \tilde{v}$$

c speed of light (3×10^{10} cm s^{-1})
h Planck's constant
v frequency (Hz or s^{-1})
λ wavelength (cm)
$\tilde{v}$ wavenumber (cm^{-1})

The normal region of an IR spectrum lies between 4000 and 400 cm^{-1}.

Many functional groups in organic molecules show characteristic vibrations, which correspond to absorption bands in defined regions of the IR spectrum. These molecular vibrations are essentially localised within the functional groups and do not extend over the rest of the molecule. Thus such functional groups can be identified by their absorption bands. This fact, together with the simple measurement technique, makes IR spectroscopy the easiest, quickest, and often most reliable method of assigning a substance to a particular class of compounds. Usually it is possible to decide immediately if an alcohol, amine, or ketone, or an aliphatic or aromatic compound is present. From closer inspection of the position and intensity of the bands it is possible to draw even more detailed conclusions, for example as to the type of substitution of aromatics, or the presence of carboxylic acid, ester, or amide functions. Furthermore there are now large collections of reference spectra available in catalogues or data bases for comparison. Thus it is often possible to unambiguously identify an unknown compound from its IR spectrum alone. The number of IR spectra catalogued or published in the literature currently amounts to ca. 100,000. This immense range of reference material is increasingly becoming available to computerised search methods.

2. Basic Principles and Selection Rules

To comprehend the processes which lead to an IR spectrum it is useful to consider a simple model derived from classical mechanics. If atoms are considered as point masses, the vibrations in a diatomic molecule (e.g. HCl) can be described as in Fig. 2.1. The molecule consists of the masses m_1 and m_2, which are joined by an elastic spring (**a**). If the equilibrium distance r_0 between the masses is increased by the amount $x_1 + x_2$ (**b**) a restoring force K ensues. On release the system vibrates about the equilibrium distance.

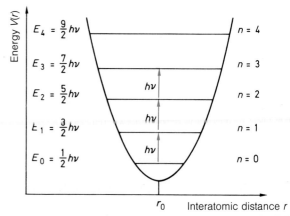

Fig. 2.2 Potential energy curve of a harmonic oscillator with discrete vibrational levels E_i

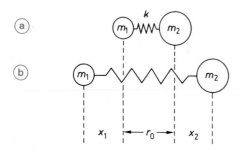

Fig. 2.1 Mechanical model of a vibrating diatomic molecule. (Extension $\Delta r = x_1 + x_2$)

According to Hooke's Law the restoring force is to a first approximation proportional to the extension Δr

$$K = -k \cdot \Delta r.$$

The negative sign arises since the force is opposed to the extension. In the mechanical model the proportionality constant k is the elasticity constant of the spring. In the molecule k (the force constant) is a measure of the bond strength between the atoms.

The energy of the vibration can be derived from the model of a harmonic oscillator (Fig. 2.2). Its potential energy is a function of the internuclear distance r*

$$V(r) = \tfrac{1}{2}k \cdot x^2 = 2\pi^2 \mu v_{\text{osc}}^2 \cdot x^2$$

V potential energy
k force constant
x extension

$$\mu = \frac{m_1 \cdot m_2}{m_1 + m_2} = \text{reduced mass}$$

v_{osc} vibration frequency of the oscillator

---

* see standard physics textbooks

From the above equation the vibration frequency of a diatomic molecule can be calculated from the mechanical model as

$$v_{osc} = \frac{1}{2\pi}\sqrt{\frac{k}{\mu}}.$$

The vibrational frequency v is thus higher, the higher the force constant k is, i.e. the stronger the bond is. And furthermore, the smaller the masses of the vibrating atoms, the higher the frequency.

This relationship is useful for spectral interpretation, since it allows qualitative predictions of absorption frequencies in the IR spectrum. For example from the bond strengths of carbon-carbon bonds it follows that (see also p. 39).

$$k_{c\equiv c} > k_{c=c} > k_{c-c}$$

The application of the mechanical model to diatomic molecules fails to explain certain phenomena, for example the dissociation of the molecule on irradiation with sufficient energy. A better description is the non-harmonic oscillator model (Fig. 2.3). Its potential energy curve is asymmetric, and the vibrational levels are no longer equally separated.

Finally quantum theory must be considered, since at the molecular level energy and the absorption of radiation are quantised. This leads to further rules for the non-harmonic oscillator.

There are only discrete energy levels and therefore vibrational levels. The vibrational state with the quantum number $v=0$ is called the ground state, and has non-zero energy (so-called zero point energy). The amount of energy absorbed for a vibrational transition ΔE_{VIB} is the difference between two adjacent energy eigenvalues E_{n+1} and E_n.

From the Schrödinger equation it follows that:

$$E_{VIB} = hv_{osc}\left(n + \frac{1}{2}\right) = \frac{h}{2\pi}\sqrt{\frac{k}{\mu}}\left(n + \frac{1}{2}\right)$$

$$n = 0, 1, 2, \ldots$$

$$\Delta E_{VIB} = E_{n+1} - E_n = hv_{osc}$$

n Vibrational quantum number
h Planck's constant
E_{VIB} Vibrational energy

The excitation of a vibration corresponds to the absorption of a quantum of light by the molecule, thus raising it from a vibrational level with quantum number n to a higher level, typically with $n+1$. The energy difference between the two levels must be equal to the energy of the light quantum (resonance condition). The separation between adjacent levels becomes smaller as n becomes larger, until finally the dissociation limit is reached (Fig. 2.3).

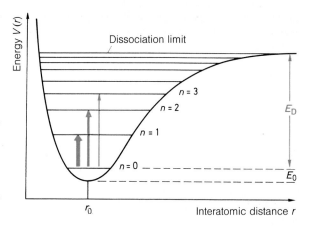

Fig. 2.3 Potential energy plot of an anharmonic oscillator (E_0 zeropoint energy; E_D dissociation energy; the different widths of the arrows indicates the different transition probabilities)

The transition from $n=0$ to $n=1$ is the base transition, from $n=0$ to $n=2$ is the first overtone, which has approximately double the frequency. The probability of these transitions and therefore the intensity of the absorption bands reduces very sharply as the size of the change in n increases.

Apart from the quantum conditions the appearance and intensity of absorption bands depend on the dipole moment of the molecule. Infrared light is only absorbed when the dipole moment interacts with the electrical vector of the light. A simple rule makes it easy to determine whether this interaction can occur: the dipole moment of the molecule at one extremity of the vibration must be different from that at the other extremity. In contrast the Raman effect depends on an interaction between the irradiating light and the polarisability of the molecule. This leads to different selection rules (see p. 67).

The most important consequence of these selection rules is that in a molecule with a centre of symmetry all vibrations which are symmetric with respect to the centre of symmetry are IR inactive (i.e. forbidden), since they produce no change in the dipole moment. These vibrations are however Raman active, since they produce a change in the polarisability. Conversely, those vibrations which are not symmetric about the centre of symmetry are Raman inactive and active in the IR. Raman and IR spectra are thus complementary. The IR spectrum, which is easier to obtain, provides the chemist with more information, however, since the majority of functional groups do not have a centre of symmetry. Therefore IR spectroscopy has a much greater importance for structural determination than Raman spectroscopy.

The symmetry properties of a molecule in a crystal lattice can be different from those in the vapour or in solution. Accordingly the solid state spectrum is then different from the gas-phase or solution spectrum.

3.   IR Spectrometers

Two basically different types of IR spectrometer are currently in use, the traditional grating or prism (*scanning*) instruments having been largely replaced by the more powerful Fourier Transform (FT)-IR spectrometers.

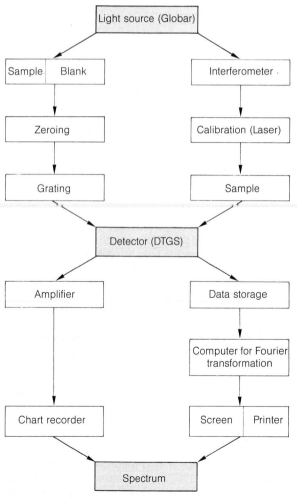

Fig. 2.4   Schematic design of grating (left) and Fourier transform (right) IR spectrometers

Both types work on the same basic principle. An IR source emits radiation, which is reduced in strength as it passes through the sample. This reduction is frequency dependent, corresponding to the excited molecular vibrations. The residual radiation is measured with a detector and electronically converted into a spectrum (Fig. 2.4).

The decisive requirement for the radiation source is that it must constantly emit light covering the total frequency range of interest. Commonly used light sources which fulfil this requirement are a white hot Nernst filament (zirconium dioxide with rare earth additives) or the so-called Globar made from silicon carbide (operating temperature: 1500 K).

The purpose of the detector is to collect the incoming radiation and convert the optical signals into electrical signals. The most common type is the DTGS detector (**d**euteriated **trigly**cine **s**ulfate).

Whereas the radiation source and the detector are identical for both types of spectrometer, the measurement of the frequency dependence of the absorption and the signal processing are fundamentally different.

3.1 Classical (Scanning) IR Spectrometers

These instruments usually operate on the double beam principle: a beam splitter (*chopper*) divides the continuous radiation from the source into two beams of equal intensity. One of the beams is passed through the sample, the other serves as a control beam and usually passes through air, in the case of solution measurements through a cell with pure solvent. After optical comparison in the photometer the light beams are recombined. The monochromator (a prism or diffraction grating) spectrally analyses the resultant radiation. Thus the spectrum can be recorded as a function of wavelength (*scanning*) – at any point in time a single frequency is recorded. After amplification the signals are plotted on a chart recorder as a spectrum (abscissa: wavelength increasing from right to left, ordinate: % transmittance). The recording of a spectrum typically takes about 10 minutes.

3.2 Fourier-Transform IR Spectrometers

The Fourier-transform technique is a development of IR spectroscopy using the possibilities of modern computer technology for the storage and processing of large amounts of data. It has established itself as the standard method and largely superseded conventional spectrometers in the market.

The basic principle is the simultaneous collection of data at all frequencies in the IR spectrum, thus eliminating the time required for scanning through the different frequencies. This is achieved by using an interferometer to convert the intense, multifrequency IR radiation, which is constant with time, into an interferogram, which is not a function of the frequency, but of time (i.e. conversion from a **frequency** domain into a **time** domain). After this "prepared" radiation is passed through the sample the interferogram is converted back to the spectrum (i.e. into the frequency domain) by a mathematical operation, the Fourier transform (see Fig. 2.5).

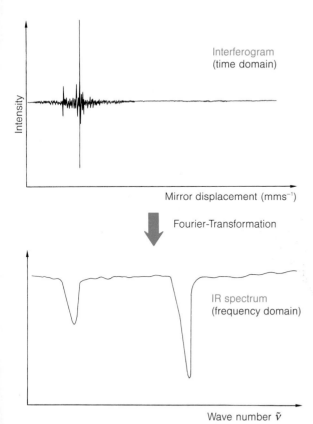

Fig. 2.5   From interferogram to IR spectrum *via* Fourier transformation

This technique requires a totally different design of the spectrometer, as shown in Fig. 2.6. The heart of the instrument is a Michelson interferometer; the IR radiation is led into it through a semi-transparent plate (KBr or CsI coated with germanium) which acts as a beam splitter. One half of the light falls onto a fixed mirror, the other onto a movable mirror, the distance of which from the interferometer can be varied. Both mirrors reflect the light onto the plate, where interference (constructive or destructive) takes place. The signal produced is comparable to the information modulated onto a carrier frequency in a radio transmitter. Since the IR radiation is polychromatic, the interferogram obtained is a superimposition or summation of the interferograms of all the individual frequencies. The modulated radiation is now passed through the sample, where it is selectively absorbed, depending on the vibrations excited in the sample. The detector records the emerging radiation as an interferogram, converts the optical signals into electrical signals and passes them on to the data storage system. A computer extracts the frequency information contained in the interferogram by Fourier transformation and produces the usual, interpretable band spectrum.

Compared to the conventional techniques FTIR spectroscopy offers three advantages:

1) a considerable saving of time: since light of all frequencies is simultaneously recorded in the detector, the measurement time is reduced to a few seconds compared to ca. 10 minutes (multiplex or Fellgett advantage).

2) a better signal to noise ratio: in contrast to the scanning technique, where only one wavelength is recorded at a time (and the rest of the intensity is lost) the whole power of the light source is available at all times (Jacquinot advantage).

3) high wavenumber precision: monochromatic light from a laser source, the frequency of which is very precisely known, can be added to the signal to provide an internal reference (Connes advantage).

These advantages are obtained at the cost of the computationally intensive Fourier transformation. Since this however can be easily achieved with an ordinary personal computer (e.g. 486 PC) in a few seconds, it is no longer a limiting factor.

The FT technique also renders unnecessary the splitting of the light into measurement and comparison beams, a common source of technical problems; FTIR spectrometers are single beam instruments. Comparison and measurement samples are mounted in holders on a sliding mechanism, which measures the samples in the single beam one after another (if air is used as the blank the relevant holder is simply left empty). The spectra are acquired and stored separately and the comparison or background spectrum subtracted from the sample spectrum in the computer.

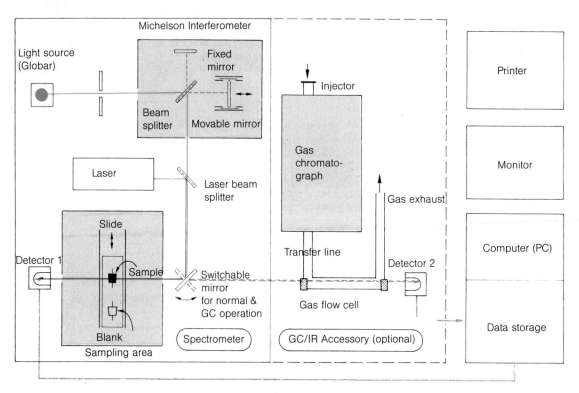

Fig. 2.6 Design of an FTIR spectrometer (with GC/IR accessory)

Because of the speed of FTIR measurements a very useful application for the organic chemist has become available: the coupling of gas chromatography and IR spectroscopy. Gas phase spectra of the fractions eluted from the GC column are measured directly (see also Sec. 4.1). Most FTIR instruments can be fitted with a special detector for these purposes (see also Fig. 2.6). GC/FTIR spectroscopy has become in many cases a genuine alternative or supplement to GC/MS.

4. Sample Preparation

IR spectra can be measured on substances in all three phases of matter (gaseous, liquid, solid) as well as in solution. The choice of the appropriate method depends on the nature and physical properties of the sample, such as melting point and solubility.

4.1 Measurements in the Gas Phase

Gases are measured in gas cells which are fitted with stopcocks and have IR transparent NaCl plates at the ends.

Because of the low density of gases the optical path length through the cell is chosen to be as long as possible (normally 10 cm). Since most organic compounds have relatively low vapour pressures, this method is rarely used.

Cells for coupled GC/IR measurements are designed in basically the same way. The sample is introduced into the cell, which is fitted with entry and exit ports, by a stream of carrier gas (hydrogen or helium). Because of the short dwell time of the sample in the cell and the small amount of substance, GC/IR coupled measurements are only possible with the FT technique (see also Fig. 2.6).

Gas phase spectra show two special features:

1) for certain small molecules (e.g. HCl) rotational fine structure is observable, which arises from excitation of the rotational states of the molecule simultaneously with the vibrational excitation; in larger molecules and in condensed phases these transitions are not resolved; and

2) some bands which are strong in the condensed phase, and get their intensity from intermolecular interactions (e.g. hydrogen bonded OH and NH) are weak, since the molecules in the gas phase are isolated and these interactions do not occur.

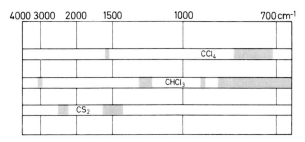

Fig. 2.7   The marked wavenumber regions are where the solvents themselves absorb (cf. Fig. 2.15, spectrum of CHCl$_3$)

4.2 Measurements on Liquids

A drop of the liquid is pressed between two flat sodium chloride plates (transparent in the range 4000 to 667 cm^{-1}). This is the simplest of all methods.

If the liquid absorbs weakly, spacers can be positioned between the plates to increase the thickness of the liquid layer.

Samples with a water content of more than 2% are unsuitable, since the surface of the plates will be damaged; cloudiness of the liquid is also undesirable, since it causes reflections and diffraction of the IR beam which lead to a strong background absorption.

4.3 Measurements in Solution

The compound is dissolved in carbon tetrachloride or for better solubility in alcohol free chloroform (about 1 to 5% solution). This solution is placed in a special sodium chloride cell with a path length of 0.1 to 1 mm. A second cell of the same dimensions, which only contains the solvent, is placed in the path of the other beam of the spectrometer, in order to compensate for the absorption of the solvent. It is generally advisable to record such spectra in non-polar solvents, since intermolecular interactions – which are particularly strong in the crystalline phase – are at a minimum. However many compounds are insoluble in non-polar solvents, and all solvents absorb in the infrared; if the solvent absorbs more than 65% of the incident light a spectrum cannot be measured. In such cases the amount of light transmitted is insufficient for the detector to function properly. Fortunately carbon tetrachloride and chloroform only absorb strongly in **those** ranges (see Fig. 2.7 and 2.15) which are of lesser interest for analytical purposes. Other solvents can of course be used. The region of interest should first be checked carefully with due consideration of the path length of the cell. In exceptional cases even aqueous solutions can be useful, but special cells made from calcium fluoride must be used.

4.4 Measurements in the Solid State

a) As **suspension in oil**. About 1 mg of the solid substance is ground to a fine paste with a drop of paraffin oil (e.g. Nujol) in a small agate mortar. The paste obtained is then pressed between two sodium chloride plates so that a film is formed free from bubbles. If (C–H) vibrations are to be measured, the paraffin oil is replaced by hexachloro- or hexafluorobutadiene.

This method is simple and has the advantage that no interference is likely from the very non-polar paraffin oil, unlike the strongly polar potassium bromide. Especially air and moisture sensitive materials can be easily measured in this way.

b) As **KBr disc**. The solid substance is intimately mixed with the 10 to 100 fold amount of potassium bromide in a small agate mortar and subsequently compressed under vacuum in a hydraulic press. The substance then sinters **under cold flow** conditions to a transparent tablet which appears like a single crystal. If the powder is too fine or too coarse the sintering is incomplete leading to dispersion losses, detectable as a baseline rising to the right of the spectrum.

This technique is the most commonly used for solids. It has the advantage that potassium bromide has no IR bands of its own, and also better spectra can be obtained than with method a). Potassium bromide is however hygroscopic, so that traces of water can rarely be totally excluded during the grinding and pressing. Therefore a weak OH band is usually observed at 3450 cm^{-1}.

Because of intermolecular interactions the positions of the bands in solid state spectra are often different from those observed for the same substance in solution. This is especially true for functional groups which form hydrogen bonds. Also the number of resolved bands in solid state spectra is often higher. If for example a synthetic product is to be compared with a natural product, the measurement is best carried out in the solid state, assuming that both samples exist in the same crystal modification. However a synthetic racemate should be compared with a natural product **in solution**.

5.   The IR Spectrum

In Fig. 2.8 the simple spectrum of Nujol, a paraffin, is given as an example. Nujol is used for the preparation of samples which are to be measured as a suspension (see above); its absorption bands are therefore overlaid on the spectrum of the sample. Fig. 2.8 shows how IR spectra are displayed. The transmittance as a percentage % is the ordinate (y axis). That represents the proportion, by comparison with the reference beam, of the radiation which is transmitted by the sample.

More rarely the percentage absorption (% A) is given instead, where

$$\% T = 100 - \% A$$

The abscissa (x axis) is calibrated both in μm (wavelength λ) and in cm^{-1} (wavenumber $\tilde{v}$). The scale is linear in wavenumbers. This has the advantage that the peaks are symmetrical. Energy differences can be easily recognised.

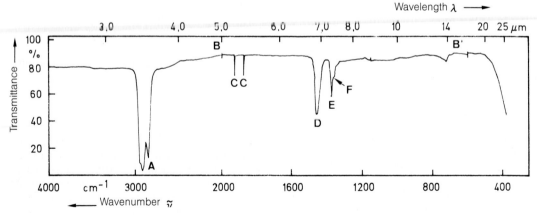

Fig. 2.8   IR spectrum of a paraffin (Nujol measured as a liquid film)

A   Absorption band, i.e. at this wavelength the molecule absorbs the maximum amount of radiation energy. In this case the vibrations responsible are the (C–H) stretching vibrations of the CH$_3$ and CH$_2$ groups.

B,B'   Switching points; at certain frequencies (here 2000 and 600 cm^{-1}) larger instruments change gratings, filters, or the plotting scale; the paper stops moving while the switching takes place. The switching points can act as a check as to whether the paper is inserted properly. They do not occur in FTIR instruments.

C   so called spikes; these are pen deflections caused by uncontrolled voltage irregularities, recognisable by their narrow width. Do not appear with FTIR instruments.

D   (C–H) bending vibrations of CH$_3$ and CH$_2$ groups. For CH$_3$ groups this is the asymmetric (C–H) bending vibration (abbreviated: $\delta_{as}(CH_3)$), for CH$_2$ groups the symmetric (abbrev. δ_s (CH$_2$)). These terms are explained below.

E   symmetric (C–H) bending vibration of CH$_3$ groups ($\delta_s(CH_3)$).

F   Shoulder; arises from the overlap of two or more bands.

In the short wavelength region (left) the scale of the abscissa (from 2000 cm⁻¹) is generally smaller. This form of spectral display has become more and more usual with modern spectrometers. A linear wavelength (λ, µm) scale is most commonly found with older prism instruments. Such spectra may look clearer, but the bands are asymmetrical, energy differences cannot be immediately read off, and the resolution in the short wavelength region is poorer.

The spectrum of this hydrocarbon also shows that the (C–C)-chain does not give noticeable absorptions. They only appear with thicker layers of sample, between 1350 and 750 cm⁻¹. For cyclic hydrocarbons stronger bands often appear, arising from vibrations of the ring.

A complex molecule possesses many possible vibrations. Their number can be determined from a simple rule: a molecule with N atoms has, because of three independent spatial co-ordinates of each atom, a total of $3 \cdot N$ degrees of freedom. Three of these are taken up by the translational movement of the molecule along the three axes, and a further three by rotations about the same three axes. Linear molecules lose one degree of freedom, since the moment of inertia about the linear axis is 0. The actual number n of degrees of freedom thus reduces to:

Degrees of freedom of linear molecules $\quad n = 3N - 5$
Degrees of freedom of non-linear molecules $n = 3N - 6$
(N = number of atoms)

The vibrations calculated from these formulae are called **normal** or **basis** vibrations.

Depending on the **type of vibration** they can be divided into:

— **Stretching** vibrations: these alter the bond lengths and

— **Bending** vibrations (planar or non-planar): these alter the bond angles, while the bond lengths remain essentially unchanged.

A further division is possible on **symmetry grounds**, giving

— symmetric vibrations (Index s): retain the full molecular symmetry;

— antisymmetric vibrations (Index as): disturb one or more symmetry elements;

— degenerate vibrations (Index e): distinct vibrations, which because they have the same energy content absorb at the same frequency and therefore only lead to **one** absorption band.

The most useful vibrations for spectral interpretation are those which to a first approximation only involve one bond in the molecule, i.e. localised vibrations. The methylene group, for example, consisting of three atoms, has the following localised vibrations:

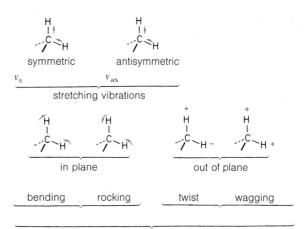

+ = vibration to the front of the plane of the page
− = vibration to the rear of the plane of the page

For the specification of localised vibrations the following symbols are used:

ν = stretching vibration (also called bonding vibration)
δ = deformation vibration (also called bending vibration)
γ = **out of plane** deformation vibration
τ = torsional vibration (alteration of the torsion angle)
      etc.; for example
$\nu_s(CH_2)$ and $\nu_{as}(CH_2)$
      = symmetric and antisymmetric (C–H)-stretching vibrations of a CH_2 group
$\delta_s(CH_3)$ and $\delta_{as}(CH_3)$
      = symmetric and antisymmetric (C–H)-deformation vibrations of a CH_3 group
Many localised vibrations help to identify functional groups.

The skeletal vibrations of a molecule result in absorption bands at relatively low energy (below 1500 cm⁻¹), the positions of which are characteristic for the molecule as a whole. These bands make the assignment of localised vibrations below 1500 cm⁻¹ difficult, since much overlap of bands occurs. Often bands are observed below 1500 cm⁻¹ which cannot be assigned to normal vibrations, but which arise from **overtone** or **combination vibrations**. Overtones occur at double, treble, etc., of the frequency value of the corresponding normal vibration. Combination vibrations occur at frequencies which correspond to the combination of two or more normal vibrations. The bands from overtones or combination vibrations are generally much weaker than those from normal vibrations. Occasionally these bands are of diagnostic value, but in general they are of little use. A special case is the so-called **Fermi** resonance: if an overtone or combination vibration coincidentally has the same frequency as a normal vibration, the two frequencies move apart. Two bands of similar intensity are observed. The bands can then no longer be assigned to individual vibrations.

An IR spectrum therefore exists of two main regions: above 1500 cm^{-1} lie absorption bands which can be assigned to the individual functional groups, whereas the region below 1500 cm^{-1} contains many bands and characterises the molecule as a whole. This region is therefore known as the **fingerprint region**. The use of this fingerprint region to identify a substance by comparison with an authentic sample is in most cases more reliable than using mixed melting points or thin-layer chromatography. The bands within the fingerprint region which arise from functional groups can be used for identification, but such assignments should be considered as only an aid to identification and not as conclusive proof.

The regions in which specific functional groups absorb can be shown by the example of acetone (Fig. 2.9): the stretching vibrations of single bonds to hydrogen (such as C–H, O–H, N–H) absorb at the highest frequencies, as a consequence of the small mass of the hydrogen atom (extreme left-hand region in Fig. 2.9). As the atomic masses increase the absorption bands are shifted to lower wavenumbers, as the following series demonstrates.

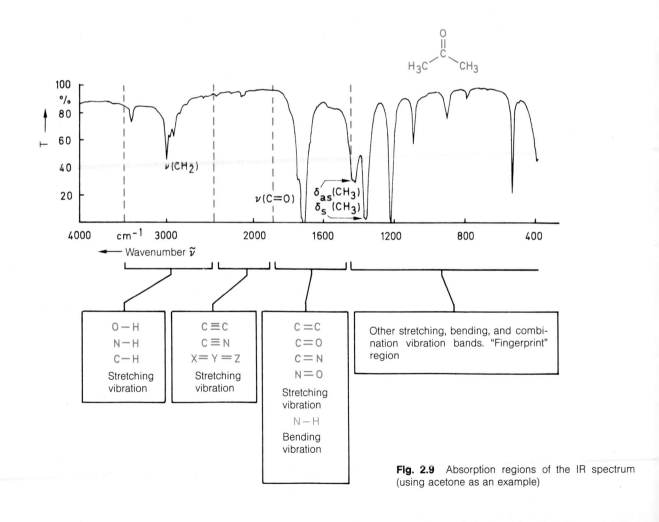

Fig. 2.9 Absorption regions of the IR spectrum (using acetone as an example)

Bond	$\tilde{v}$ (C–X)(cm^{-1})	Atomic mass of X
C—H	≈ 3000	1
C—D	≈ 2100	2
C—C	≈ 1000	12
C—Cl	≈ 700	35

(cf. Sec. 2; wavenumber $\tilde{v}$ and frequency are proportional to one another)

Otherwise the frequencies of stretching vibrations follow the rule: the stronger the bond between two atoms, the higher the vibrational frequency. Triple bonds therefore absorb at higher wavenumbers than double and single bonds:

$$\tilde{v}(C\equiv C) \quad \approx 2200 \text{ cm}^{-1}$$
$$\tilde{v}(C=C) \quad \approx 1640 \text{ cm}^{-1}$$
$$\tilde{v}(C-C) \quad \approx 1000 \text{ cm}^{-1}$$

Bending vibrations only alter bond angles, not bond lengths. These vibrations occur at lower wavenumbers, generally in the fingerprint region below 1500 cm^{-1}. An exception is the (NH)-bending vibration, which appears in the region around 1600 cm^{-1} (Fig. 2.9).

Tabular summary

Group	Table	Page
Single bonds		
C—H	2.1 to 2.3	44–45
O—H	2.4	45
N—H	2.5, 2.6	46
S—H	2.7	46
P—H	2.7	46
Triple bonds		
C≡C	2.8	47
X≡Y	2.8	47
Cumulated double bonds		
C=C=C	2.9	47–48
N=C=O	2.9	47
X=Y=Z	2.9	47

Group	Table	Page
Double bonds		
C=O	2.10	48–51
C=N	2.11	52
N=N	2.12	52
C=C	2.13	52
N=O	2.14	52
Aromatics		
⬡—X$_n$	2.15	53
	2.16	53
Fingerprint region		
S-derivatives	2.17	54
P-derivatives	2.18	54
C–O single bonds	2.19	54
Halogen compounds	2.20	54
Inorganic ions	2.21	54

6. Overview of Characteristic Absorptions

In Fig. 2.9 the IR spectrum is divided into 4 regions, which are more fully described in the assignment summaries (Fig. 2.10 – 2.14). In the region 1800 – 1500 cm^{-1} carbonyl bands (Fig. 2.13) are listed separately from other absorptions (Fig. 2.12).

For each individual functional group wavenumber regions are given where an absorption might appear; typical intensities (which can be an aid to assignments) are also given. Intensities in IR spectra are not as easy to measure as in UV; they are generally described by the subjective terms strong (**s**), medium strong (**m**), weak (**w**), and variable (**v**).

In Tabs 2.1 to 2.21 typical bands of functional groups are listed in detail. The summary on page 39 serves as an index.

For the **interpretation** of the IR spectrum of an unknown compound the following procedure is recommended:

Firstly the three regions above 1500 cm^{-1} are checked using Figs. 2.10 – 2.13 to see if there are indications of particular structural elements or if structures can be excluded.

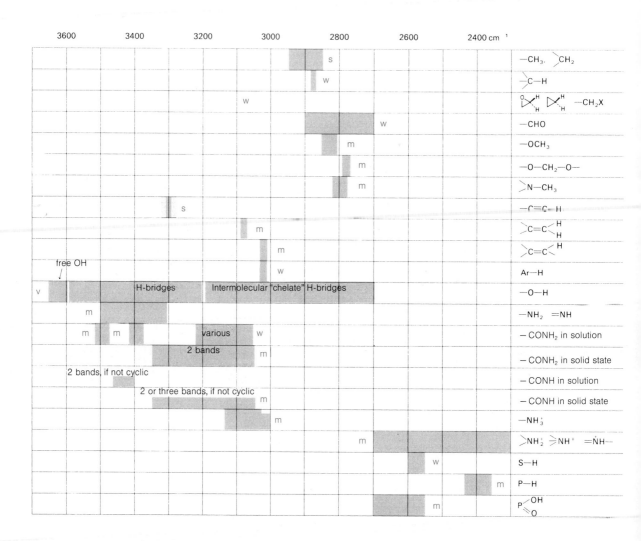

Fig. 2.10 Positions of the stretching vibrations of hydrogen (in the shaded areas the borders are less well defined); band intensity **s** strong, **m** medium, **w** weak, **v** variable.

A comparison of the *fingerprint* region with Fig. 2.14 will then show if typical bands are present, the presence or absence of which can support or disprove the structural hypothesis. Where there are uncertainties the Tables for individual functional groups (Tab. 2.1 to 2.21) can be consulted. If a firm structural hypothesis is indicated, it should always be checked by comparing the spectrum with one of an authentic sample (e.g. from a spectral catalogue); care must be taken to check whether the same measurement conditions (e.g. KBr/liquid film/Nujol) were used, since these can influence the spectrum (cf. Sec. 4.4, p. 35). If no comparison spectrum is available, the structure should be confirmed using other spectroscopic methods (e.g. NMR, MS).

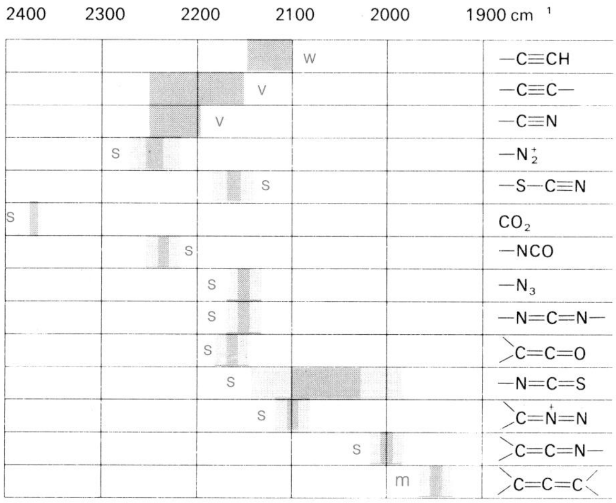

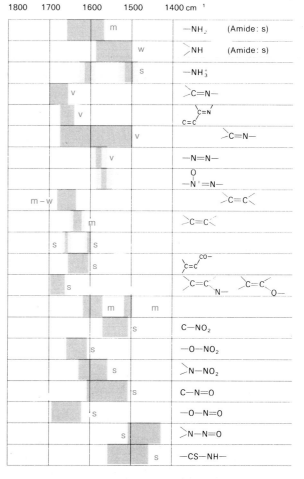

Fig. 2.11 Positions of the stretching vibrations of triple bonds and cumulated double bonds (**s** strong, **m** medium, **w** weak, **v** variable)

Fig. 2.12 Positions of the double bond stretching vibrations and (N–H) bending vibration (for carbonyl groups see Fig. 2.13); **s** strong, **m** medium, **w** weak, **v** variable

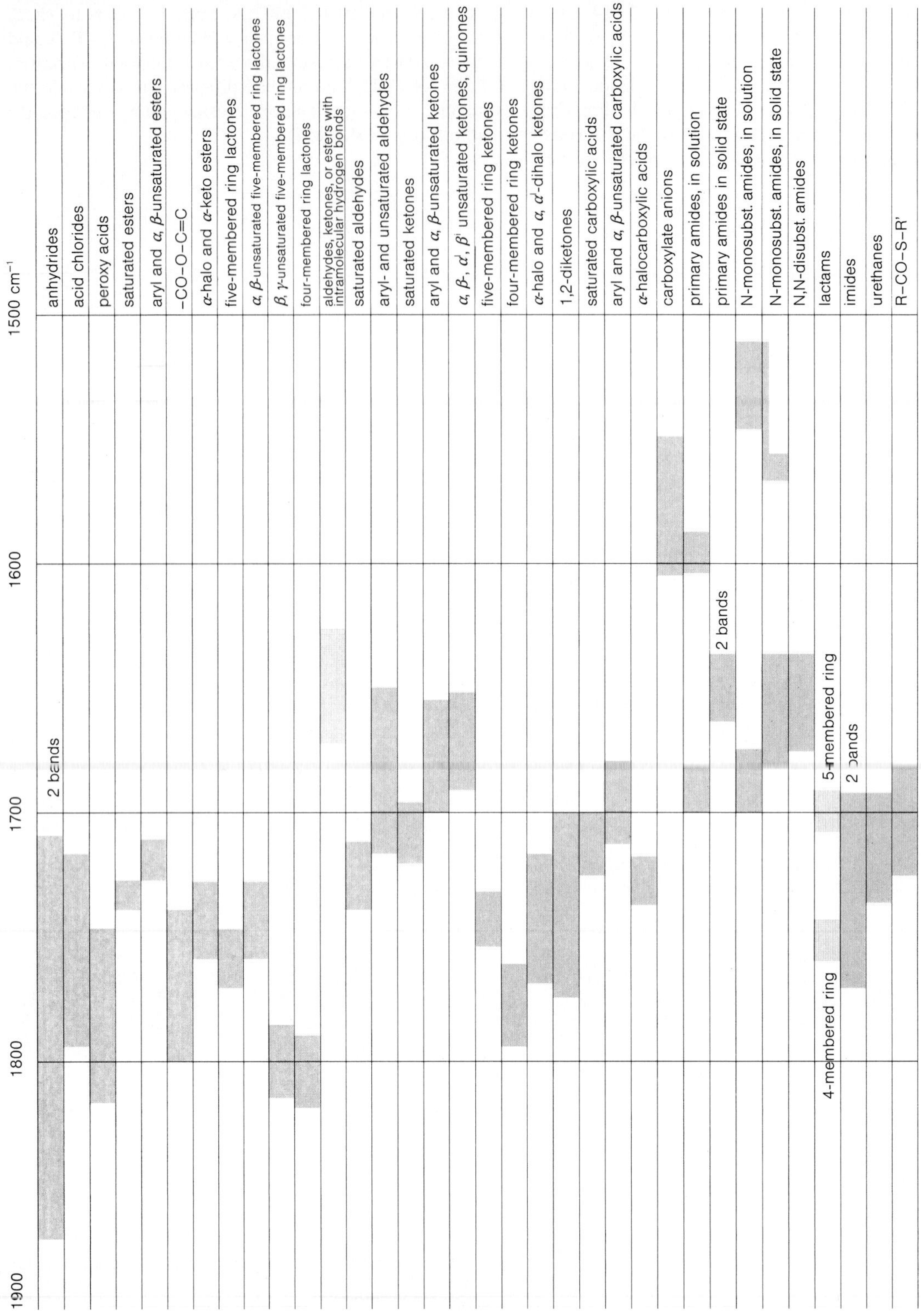

Fig. 2.13   Positions of carbonyl stretching vibrations (all bands are strong; wavenumbers see Tab. 2.10)

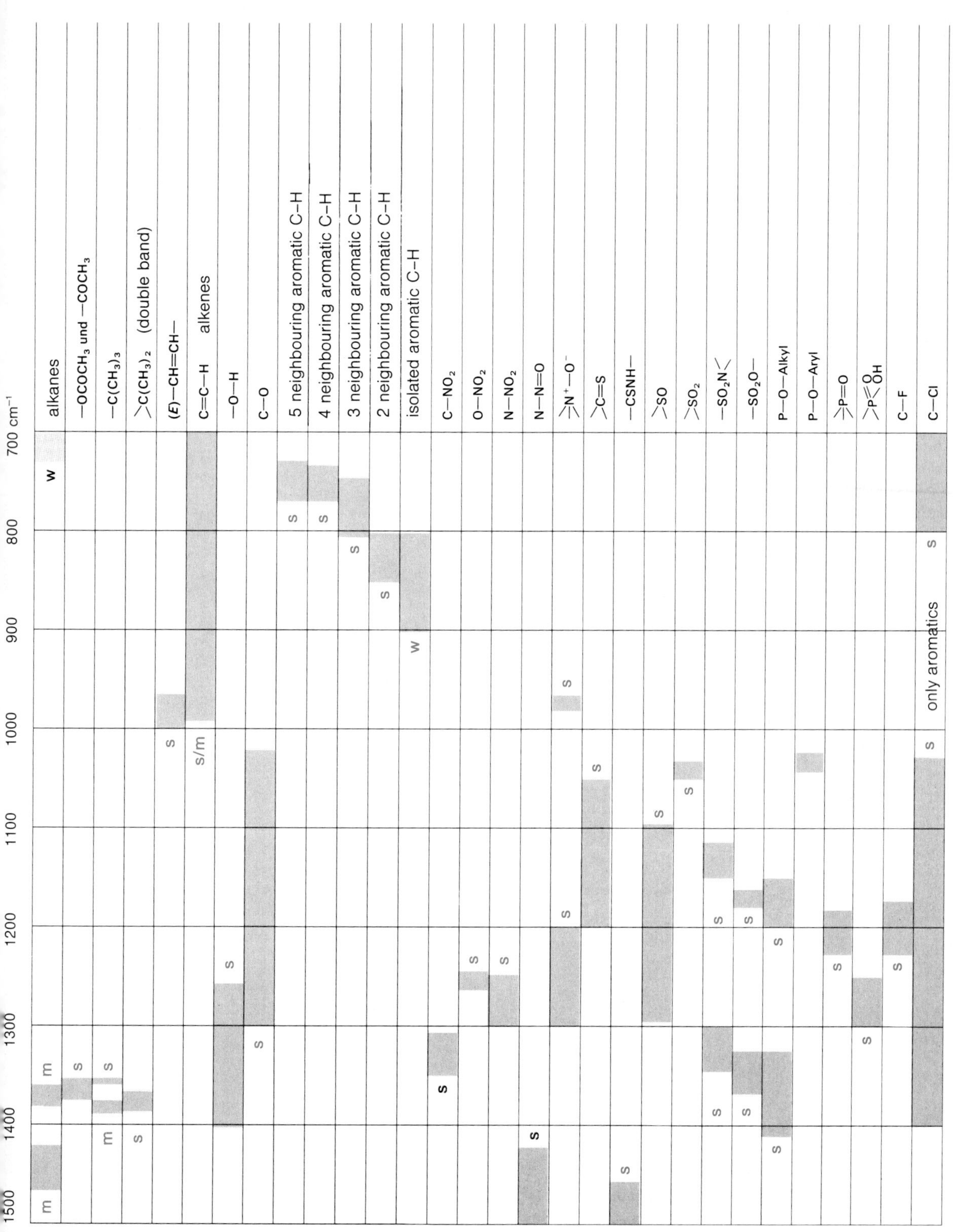

Fig. 2.14 — Characteristic absorptions in the fingerprint region. Wavenumber scale (cm⁻¹): 1500, 1400, 1300, 1200, 1100, 1000, 900, 800, 700.

Group	1500	1400	1300	1200	1100	1000	900	800	700
alkanes	m	m							
—OCOCH₃ und —COCH₃		s							
—C(CH₃)₃		m							
⟩C(CH₃)₂ (double band)		s							
(E)—CH=CH—						s			
C=C—H alkenes			s	s		s			bar
—O—H			s	s					
C—O			s		s	s			
5 neighbouring aromatic C—H								s	s
4 neighbouring aromatic C—H								s	
3 neighbouring aromatic C—H								s	
2 neighbouring aromatic C—H							s		
isolated aromatic C—H							w		
C—NO₂			s	s					
O—NO₂	s			s					
N—NO₂				s					
N—N=O	s				s				
⟩N⁺—O⁻				s		s			
⟩C=S					s				
—CSNH—					s				
⟩SO					s	s			
⟩SO₂					s	s			
—SO₂N⟨		s	s	s	s				
—SO₂O—		s	s	s					
P—O—Alkyl				s		s			
P—O—Aryl			s	s					
⟩P=O				s					
⟩P⟨O/OH		s	s	s					
C—F								s	
C—Cl (only aromatics)			s	s	s	s		s	s

Fig. 2.14 Characteristic absorptions in the fingerprint region (s = strong, m = medium, w = weak)

7. IR Absorptions of Single Bonds to Hydrogen

7.1 C–H Absorptions

The chemically simple alkanes (paraffins) also show simple IR spectra (see Fig. 2.8, p. 36). There are several reasons for this fact

— several absorptions are "symmetry forbidden"
— many absorptions coincide
— many absorptions are too weak in intensity

The localised vibrations of the CH_2 group have already been described on p. 37. The absorption regions of methyl, methylene, and methine groups are summarised in Tab. 2.1. Since these groups do not participate in hydrogen bonding, the positions of the bands are hardly affected by the chemical environment or the state in which the substance is measured.

Since most organic molecules contain (C–H) bonds of the alkane type, the absorptions arising from them are of little diagnostic value. The absence of a C–H bond in the spectrum is of course proof of the absence of this structural element in the compound being investigated. Unsaturated and aromatic (C–H) stretching vibrations can be easily distinguished from (C–H) absorptions in saturated structures:

saturated C – H: wavenumber $\tilde{v} < 3000$ cm^{-1}
C = C – H: wavenumber $\tilde{v} > 3000$ cm^{-1}

The absorptions of unsaturated and aromatic (C–H) stretching vibrations also appear with much lower intensity.

In the following Tables the positions of the bands of (C–H) vibrations are summarised.

Tab. 2.1 (C–H) absorption bands

Group	Band	Notes
$\diagdown$CH$_2\diagup$	2960–2850 (s)	normally 2-3 bands; (C–H) stretching vibration
—CH$_3$		
—ĊH	2890–2880 (s)	
$\diagdown$CH$_2\diagup$	1470–1430 (nm)	(C–H) deformation vibration
—CH$_3$		
—CH$_3$	1390–1370 (m)	symmetric deformation vibration
$\diagdown$CH$_2\diagup$	≈ 720 (w)	CH$_2$ rocking vibration

Tab. 2.2 Special (C–H) absorption bands

Group	Band	Notes
cyclopropane	≈ 3050 (w)	(C–H) stretching vibration, see alkenes
epoxide C–H —CH$_2$-Halogen		
–CHO	2900–2700 (w)	two bands, one near 2720 cm^{-1}; (C–H) stretching vibration of the aldehyde group has almost the same frequency as the first overtone of the (H–C=O) deformation; as a result of the Fermi resonance (see p. 38) two bands of similar intensity are observed; this double band can generally be used for the identification of aldehydes
—O—CH$_3$	2850–2810 (m)	
—O—CH$_2$—O—	2790–2770 (m)	
$\diagdown$N—CH$_3\diagup$	2820–2700 (m)	NCH$_2$ groups can also occur in this region
—C(CH$_3$)$_3$	1395–1385 (m) 1365 (s)	see Fig. 2.16
$\diagdown$C(CH$_3$)$_2\diagup$	≈ 1380 (m)	an almost symmetrical doublet
—O—CO—CH$_3$ —CO—CH$_3$	1385–1365 (s) 1360–1355 (s)	the high intensity of the bands often dominates this region of the spectrum

Tab. 2.3   C–H in alkenes, alkynes, and aromatics (see also the absorptions of the (C=C) bond in Tab. 2.13 and 2.14, p. 52 and 53).

Group	Band	Notes
−C≡C−H	≈ 3300 (s)	
C=C with H, H (cis/geminal)	3095–3075 (m)	(C–H) stretching vibration, sometimes obscured by the much stronger bands of the saturated (C–H) absorption, which lie below 3000 cm⁻¹
C=C with H	3040–3010 (w)	
Aryl-H	3100–3000 (w)	often obscured
C=C with R, H / H, R	970–960 (s)	(C–H) out-of-plane deformation vibration. If the double bond is conjugated with a C=O group, it is shifted to 990 cm⁻¹
R−CH=CH₂	995–985 (s) and 940–900 (s)	
R₂C=CH₂	895–885 (s)	
C=C with R, H / R, H	840–790 (m)	
C=C with H, H / R, R	730–675 (m)	

Tab. 2.4   Alcoholic and phenolic O–H

Group	Band	Notes
water in solution	3710	
free O–H	3650–3590 (v)	sharp; (O–H) stretching vibration
–OH in hydrogen bond to **sp³** –O or N (e.g. alcohols); not in gas-phase spectra	3600–3200 (s)	often broad, can be sharp for some intra-molecular H-bonds; the lower the fre-quency, the stronger the H-bond
–OH in hydrogen bond to **sp²** O or N (e.g. carboxylic acids, Tab. 2.10)	3200–2500 (v)	broad; the lower the frequency, the stron-ger the H-bond; the band can sometimes be so broad that it is overlooked
water of crystalli-sation (solid state spectra)	3600–3100 (w)	often also a weak band at 1640–1615 cm⁻¹; traces of water in KBr discs show a broad band at 3450 cm⁻¹
—O—H	1410–1260 (s)	(O–H) deformation vibration
—C—OH	1150–1040 (s)	(C–O) stretching vibration

7.2 (O–H) and (N–H) Absorptions

The position of the (O–H) stretching vibration has long been used as a criterion for and measure of the strength of hydro-gen bonds. The stronger the hydrogen bonding is, the longer the (O–H) bond, the lower the vibrational frequency, and the broader and more intense the absorption band. The sharp band of the free "monomer" in the region 3650 to 3590 cm⁻¹ can be observed in the gas phase or in dilute solution, or when such factors as steric hindrance make hydrogen bonding impossible. Pure liquids, crystals, and many solutions only show the broad "polymeric" band in the region 3600 to 3200 cm⁻¹. In the liquid phase both bands are often observed.

Intramolecular hydrogen bonds of the non-chelating type (e.g. in 1,2-diols) show a sharp band in the region 3570 to 3450 cm⁻¹, the precise position being a measure for the strength of the hydrogen bond. A similar, although much less sharp, band is observed when the hydrogen bond only causes dimerisation. The "polymeric" band is generally broader. The different possi-bilities can be distinguished by dilution experiments; **intramo-lecular** hydrogen bonds are not affected, and the absorption band therefore remains unchanged; **intermolecular** hydrogen bonds in contrast are broken with increasing dilution, that means that the absorption band of the relevant hydrogen-bonded (O–H) decreases while the absorption due to free (O–H) increases.

Tab. 2.5 Amine, imine, ammonium, and amide (N–H) stretching vibration

Group	Band	Notes
amines and imines $\backslash$N–H/	3500–3300 (m)	primary amines show two bands in this region, the symmetric and the antisymmetric stretching vibration, secondary amines show weaker absorption
=N–H		(N–H) bands of pyrrole and indole are sharp
$-\overset{+}{N}H_3$ in amino acids in ammonium salts	3130 3030 (m) ≈ 3300 (m)	values for the solid state; broad; bands also (but not always) at 2500 and 2000 cm^{-1} (see text p. 64, under Fig. 2.30)
$\backslash +$NH$_2$/ $|+$ –NH $|$ $\backslash\backslash$ +NH/	2700–2250 (m)	values for the solid state; broad, because of the presence of overtone bands
unsubstituted amides –CO–NH$_2$	≈ 3500 (m) ≈ 3400 (m)	lowered by ≈ 150 cm^{-1} in the solid state and if H-bonds are present; often several bands at 3200–3050 cm^{-1}
N-monosubstituted amides –CO–NH–	3460–3400 (m)	two bands; lowered by H-bonds and in the solid state (see Fig. 2.28a); only one band for lactams
	3100–3070 (m)	a weak extra band in the solid state and with H-bonds

Tab. 2.6 (N–H) bending vibration (cf. also Tab. 2.10 for amide absorptions in this region)

Group	Band	Notes
–NH$_2$	1650–1560 (m)	see Fig. 2.26
$\backslash$NH/	1580–1490 (w)	often too weak to be seen
$-\overset{+}{N}H_3$	1600 (s) 1500 (s)	secondary ammonium salts show the band at 1600 cm^{-1}

Tab. 2.7 Various R–H

Group	Band	Notes
–S–H	2600–2550 (w)	weaker than O–H; not much affected by H-bonds
$\backslash$P–H/	2440–2350 (m)	sharp
$\overset{O}{\overset{\|}{-P}}$–OH $|$	2700–2560 (m)	associated O–H
R–D	the corresponding R–H frequency must be divided by 1.37	useful for suspected (R–H) bands, since deuteriation leads to a known shift to lower frequency

Spectra measured on samples in the solid state show only a broad, strong band in the region 3400 to 3200 cm^{-1}.

The absorption bands of the (N–H) stretching vibration (Tab. 2.5) may sometimes be confused with those of hydrogen bonded O–H. However the N–H absorption is usually sharper because of the much weaker tendency of the N–H group to form hydrogen bonds; moreover the (N–H) band has less intensity, and in dilute solutions the frequency is never as high as that of the free O–H around 3600 cm^{-1}. Weak bands, which

arise from overtones of the strong carbonyl absorption at 1800 to 1600 cm^{-1}, also appear in the region 3600 to 3200 cm^{-1} as in the example of cyclohexanone (see Fig. 2.17, p. 56).

The influence of hydrogen bonds is also seen when a carbonyl group acts as an acceptor, since its stretching vibration frequency is also reduced (cf. Tab. 2.10).

The characteristic series of bands in the region 3000 to 2500 cm^{-1}, shown by most carboxylic acids, can be seen in Fig. 2.19 (see p. 57). The band with the highest frequency corresponds to an (O–H) stretching vibration, the other absorptions arise from combination vibrations. These bands usually appear as a tooth-like series below the (C–H) absorption. Together with a carbonyl absorption at the appropriate position (see Tab. 2.10) these series are very useful for the identification of carboxylic acids.

Two bands appear for the (N–H) absorption in amides, known as Type **1** and **2**. In the carbonyl region of many amides (see Tab. 2.10) there are often also two bands. Hydrogen bonding lowers and broadens the frequencies of the (N–H) vibration less than in the case of (O–H) groups. The intensity of the (N–H) absorption is generally less than that of the (O–H) absorption.

8. IR Absorptions of Triple Bonds and Cumulated Double Bonds

Tab. 2.8 Triple bonds X≡Y

Group	Band	Notes
—C≡C—H	3300 (s)	(C–H) stretching vibration
	2140–2100 (w)	(C≡C) stretching vibration
—C≡C—	2260–2150 (v)	in polyacetylene compounds there are often more bands than there are (C≡C) bonds present[a,b]
—C≡N	2260–2200 (v)	(C≡C) stretching vibration; strong and shifted to the lower end of the region if conjugated; occasionally very weak or absent, for example some cyanhydrins show no N-absorption
isocyanides $-\overset{+}{N}\equiv\overset{-}{C}$	2165–2110	
nitrile oxides —C≡N→O	2300–2290	
diazonium salts $R-\overset{+}{N}\equiv N$	≈ 2250 ±20	
thiocyanates R—S—C≡N	2175-2160	aromatic R
	2140 (s)	aliphatic R

[a] Conjugation with (C=C) bonds and (C≡C) bonds lowers the frequency and increases the intensity. Conjugation with carbonyl groups generally has only a limited affect on the position of the bands.

[b] Symmetrical or almost symmetrical substitution makes the (C≡C) stretching vibration IR-inactive; it does however appear in the Raman spectrum

Tab. 2.9 Cumulated double bonds X=Y=Z

Group	Band	Notes
carbon dioxide O=C=O	2349 (s)	shows incomplete compensation with the **background** measurement (FTIR, see Sec. 3.2), particularly against air
isocyanates —N=C=O	2275–2250 (s)	very high intensity; position not affected by conjugation
azides —N₃	2160–2120 (s)	
carbodiimides —N=C=N—	2155–2130 (s)	very high intensity; splits into an asymmetrical doublet on conjugation with aryl groups
ketenes $\overset{\backslash}{\underset{/}{C}}=C=O$	≈ 2150 (s)	
isothiocyanates —N=C=S	2140-1990 (s)	broad and very intense
diazoalkanes $R_2C=\overset{+}{N}=\overset{-}{N}$	≈ 2100 (s)	
diazoketones $-CO-CH=\overset{+}{N}=\overset{-}{N}$	3100–2090	
$-CO-CR=\overset{+}{N}=\overset{-}{N}$	2070–2060	
keteneimines $\overset{\backslash}{\underset{/}{C}}=C=N-$	≈ 2000 (s)	

Tab. 2.9 continued

Group	Band	Notes
allenes $\diagdown$C=C=C$\diagup$	$\approx$ 1950 (ml)	two bands, if terminal allenes or electron-with-drawing groups e.g. (–COOH) are present

The identification of triple bonds and cumulated double bonds with the aid of IR spectra is relatively easy, because they absorb in a region in which practically no other strong bands appear.

The unusually high double bond frequencies in systems X=Y=Z are assumed to be caused by strong coupling between two separate stretching vibrations, whereby the symmetric and asymmetric stretching vibrations become widely separated. This type of coupling only occurs when two groups with similar vibration frequencies lie close together in a molecule. Other examples where such a coupling is found are the amide group and the carboxylate anion (Tab. 2.10).

9. IR Absorptions of Double Bonds C=O, C=N, C=C, N=N, N=O

The carbonyl absorption leads to the strongest bands in the IR spectrum and lies in a region relatively free of other absorptions (1800 to 1650 cm^{-1}). The following relationship applies to the band **intensities**:

carboxylic acid > ester > ketone $\approx$ aldehyde $\approx$ amide

The amide group has complicated vibrational properties, and its bands show considerable variations in intensity.

The carbonyl group is particularly interesting because of its tendency to undergo **intra- and intermolecular interactions**. From the position of the carbonyl absorption in the spectrum many varied influences of the molecular environment can be detected. The following rules apply:

– The stronger electronically withdrawing a group X in the system R–CO–X is, the higher the wavenumber (frequency).

– In α,β-unsaturated compounds the (C=O) frequency is 15 to 40 cm^{-1} lower (except in amides, where the shifts are minimal).

– Further conjugation has relatively little effect.

– Ring strain in cyclic compounds causes a relatively large shift to higher frequency. This phenomenon serves as a remarkably reliable test for ring size, allowing a clear distinction between four-, five- and larger-membered ring ketones, lactones, and lactams. Six- and larger-membered ring ketones show a normal (C=O) frequency, as shown by the corresponding open-chain compounds.

– Hydrogen bonding to a carbonyl group causes a shift of 40 to 60 cm^{-1} to lower frequency. This effect is shown by carboxylic acids, amides, enolised β-oxocarbonyl compounds, and o-hydroxy- and o-aminophenylcarbonyl compounds.

– The spectra of all carbonyl compounds show slightly lower values for the stretching vibrational frequencies when measured in the solid state, compared to values for dilute solutions.

– If more than **one** structural effect on the carbonyl group is present, the total effect is in most cases approximately equal to the sum of the individual effects.

The most strongly substituted double bonds have a tendency to absorb at the higher end of the frequency range, the least substituted at the lower end. The absorption can be very weak, if the double bond is more or less **symmetrically** substituted. In these cases it is possible to determine the vibration frequency from the Raman spectrum. For the same reason (E)-double bonds generally absorb less strongly than (Z)-double bonds. Tab. 2.3 gives information about =C–H vibration frequencies, from which additional structural information can be gained.

The stretching vibration frequencies of double bonds are affected by ring strain. A double bond exocyclic to the ring shows the same effect as cyclic ketones; the frequency increases as the ring gets smaller. A double bond **in** the ring shows the opposite trend: the frequency decreases as the ring gets smaller. The (C–H) stretching vibration frequency increases slightly with increasing ring strain.

Tab. 2.10 Carbonyl absorption C=O (all bands indicated are strong)

Group	Band	Notes	Group	Band	Notes
carboxylic acid anhydrides			alkyl−CO−O−C=C		
				1800–1750	the (C=C) stretching vibration band is also shifted to higher frequency
			esters with electronegative substituents, e.g.		
saturated	1850–1800 1790–1740	two bands, usually separated by ca. 60 cm⁻¹; the band with the higher frequency is stronger in acyclic compounds, that with the lower frequency in cyclic compounds	−C−CO−O− Cl	1770–1745	
			α-keto esters	1755–1740	
			lactones (lactones without ring strain are like open-chain esters)		
aryl and α,β- unsaturated	1830–1780 1770–1710			1730	
saturated five-membered ring	1870–1820 1800–1750			1750	
all types	1300–1050	one or two bands resulting from the (C–O) stretching vibration		1720	
carboxylic acid chlorides (acyl chlorides)				1760	
saturated	1815–1790			1775	
aryl and α,β- unsaturated	1790–1750			1770–1740	
diacyl peroxides				≈1800	
saturated	1820–1810 1800–1780			1840	
aryl and α,β- unsaturated	1805–1780 1785–1755		β-keto esters in the enol form with hydrogen bonds	≈1650	keto form normal wavelengths; hydrogen bonding of the chelate type causes a shift to lower frequencies (cf. normal esters), the (C=C) band usually lies at 1630 cm⁻¹ (s)
esters and lactones					
saturated	1750–1735				
C=C−CO−O−	1725–1750		all types	1300–1050	usually two strong bonds from the (C–O) stretching vibration

Tab. 2.10 Carbonyl Absorption C=O

Group	Band	Notes
aldehydes		

(cf. also Tab. 2.2 for C–H). All values are ca. 10–20 cm^{-1} lower when the spectra are measured in liquid film or solid state. Measurements in the gas phase give values ca. 20 cm^{-1} higher.

Group	Band	Notes
saturated	1740–1720	
aryl–CHO	1715–1695	o-hydroxy or amino groups shift these values to 1655–1625 cm^{-1} as a result of intramolecular hydrogen bonding
α,β-unsaturated	1705–1680	
α,β-; γ,δ- unsaturated	1680–1660	
β-ketoaldehydes in the enol form	1670–1645	lowered by hydrogen bonds of the chelate type

ketones

All values are ca. 10–20 cm^{-1} lower when the spectra are measured in liquid film or in the solid state. Measurements in the gas phase give values ca. 20 cm^{-1} higher.

Group	Band	Notes
saturated	1725–1705	
aryl	1700–1680	
α,β-unsaturated	1685–1665	
α,β-; α',β'- unsaturated and diaryl	1670–1660	
cyclopropyl	1705–1685	
six-membered and higher ring ketones	similar values to the corresponding open-chain ketones	
five-membered ring ketones	1750–1740	conjugation with (C=C) bonds etc. affect these values in a similar way to open-chain ketones
four-membered ring ketones	≈1780	

Group	Band	Notes
α-halogenated ketones	1745–1725	is affected by the conformation; the highest values occur when both halogens are co-planar with the C=O
α,α'-dihalo- genated ketones	1765–1745	
1,2 diketones s-*trans* (e.g. open chain)	1730–1710	antisymmetric stretching vibration frequency of both C=O groups; the symmetric vibration is IR inactive, but Raman active
1,2 diketones, s-*cis*, six-membered ring	1760 and 1730	
1,2 diketones, s-*cis*, five-membered ring	1775 and 1760	
o-amino – or o-hydroxyaryl ketones	1655–1635	low because of intramolecular hydrogen bonds; other substituents, steric hindrance, etc. affect the position of the bands
quinones	1690–1660	C=C normally at 1600 cm^{-1} (s)
troponoo	1650	near 1600 cm^{-1}, when hydrogen bonds are present (as in tropolones)

carboxylic acids

Group	Band	Notes
all types	3000–2500	(O–H) stretching vibration; a characteristic group of sharp bands caused by combination vibrations etc.
saturated	1725–1700	the monomer absorbs near to 1760 cm^{-1} but is rarely observed; solution spectra occasionally show both bands: one from the free monomer and one from the hydrogen-bonded dimer; solutions in ether give one band at 1730 cm^{-1}.

Tab. 2.10 Carbonyl compounds

Group	Band	Notes
α,β-unsaturated carboxylic acids	1715–1690	
aryl carboxylic acids	1700–1680	
α-halo-carboxylic acids	1740–1720	

carboxylate anions

$$R-C\underset{\text{O}^-}{\overset{\text{O}}{\langle}}$$

for amino acids see text under Fig. 2.30

most types	1610–1550 1420–1300	antisymmetric and symmetric stretching vibration

amides

$$R-C\underset{\text{NR}_2}{\overset{\text{O}}{\diagdown}}$$

(cf. also Tab. 2.5 and 2.6 for (N–H) stretching and bending vibrations)

Primary amides
–CO–NH₂ (–CO–NH$_2$)

in solution in solid state	≈1690 ≈1650	amide I (C=O) stretching vibration
in solution in solid state	≈1600 ≈1640	amide II [mostly (N–H) bending vibration]; amide I is usually stronger than amide II; in solid state the two bands may overlap

N-monosubstituted amides
–CO–NH–

in solution in solid state	1700–1670 1680–1630	amide I (Fig. 2.28)
in solution in solid state	1550–1510 1570–1515	amide II; only observed in open-chain amides amide I is usually stronger than amide II
N,N-disubstituted amides	1670–1630	since there are no hydrogen bonds, the spectra in solution and solid state are very similar

Group	Band	Notes
$R-CO-N-\overset{\mid}{C}=C\diagup$		shifted by +15 cm⁻¹ by the extra double bond
$C=C-\overset{\mid}{C}-CO-N\diagup$		also shifted by +15 cm⁻¹; this is an unusual effect of the double bond; it is assumed that the usual conjugative affect is here less important than the –I-effect of the double bond

lactams

(7-membered lactam)	1669	shifted to lower frequency in solid state (Fig. 2.22)
(6-membered lactam)	1670	
(5-membered lactam)	1717	
(4-membered lactam)	1750	
(3-membered lactam, R)	1850	

imides

six-membered rings	≈1710 and ≈1700	shifted by +15 cm⁻¹ by conjugation to multiple bonds
five-membered rings	≈1770 and ≈1700	

Ureas

$$\diagup N-\overset{\overset{\text{O}}{\|}}{C}-N\diagdown$$

≈1660

Tab. 2.10 Carbonyl compounds

Group	Band	Notes
six-membered rings	≈1640	
five-membered rings	≈1720	

Urethanes

$$\underset{\underset{\displaystyle \diagdown}{R-O-\overset{\overset{\textstyle O}{\parallel}}{C}-N}}{}\diagup$$ 1740–1690 amide II band appears when at least one hydrogen is bonded to nitrogen

Thioesters and acids

R−CO−SH	1720	
R−CO−SR	1690	
R=CO−SAr	1710	
Ar−CO−SR	1665	
Ar−CO−SAr	1685	

Tab. 2.11 Imines, oximes, etc. $\diagup$C=N$\diagdown$

Group	Band	Notes
$\diagdown$C=N−H$\diagup$	3400–3300 (m)	(N–H) stretching vibration; lowered by hydrogen bonding
$\diagdown$C=N−$\diagup$	1690–1640 (v)	difficult to identify because of large intensity differences and proximity of the (C=C)
α,β-unsaturated	1660-1630 (v)	
conjugated cyclic systems	1660–1480 (v)	stretching vibration region; oximes generally give very weak bands

Tab. 2.12 Azo compounds −N=N−

Group	Band	Notes
−N=N−	≈1575 (v)	very weak or inactive in IR; occasionally visible in Raman
$\overset{+}{-}$N=N−$\underset{\underset{O^-}{\mid}}{}$	≈1570	

Tab. 2.13 Alkenes $\diagdown$C=C$\diagup$

Group	Band	Notes
non-conjugated $\diagdown$C=C$\diagup$	1680–1620 (v)	can be very weak, if more or less symmetrically substituted
conjugated with aromatic rings	≈1625 (m)	more intense than for non-conjugated double bonds
dienes, trienes, etc.	1650 (s) and 1600 (s)	the band with the lower frequency is usually more intense and can hide or overlap with the band with the higher frequency
α,β-unsaturated carbonyl compounds	1640–1590 (s)	usually much weaker than the (C=O) band
enol esters, enol ethers and enamines	1690–1650 (s)	

10. Typical IR Absorptions of Aromatic Compounds

Aromatics show characteristic absorptions in several regions, from which they can usually be unequivocally identified:

3100–3000 cm^{-1} Aryl-H stretching vibration (see Tab. 2.3, p. 45),

2000–1600 cm^{-1} several weak bands from overtones and combination vibrations,

1600–1500 cm^{-1} (C=C) stretching vibrations; two or three bands, which provide a valuable method of identification (Tab. 2.15); polycyclic compounds and pyridines also show these absorptions,

1225–950 cm^{-1} fingerprint bands, which are of little diagnostic value,

900–680 cm^{-1} (C–H) deformation vibrations (out of plane); the number and positions of the bands are dependent on the number of neighbouring hydrogen atoms in the ring and show the degree of substitution (Tab. 2.16).

The diagnostic value of the bands at 2000–1600 and below 900 cm^{-1} is often reduced by the fact that these bands are often neither the only nor the strongest bands in these regions. Thus above 1600 cm^{-1} carbonyl groups interfere, below 900 cm^{-1} halogens, so that assignments should be made with care. In doubtful cases NMR spectra are often of assistance.

Tab. 2.14 Nitro- and nitroso-groups, nitrates, nitrites (N=O stretching vibration)

Group	Band	Notes
$-\overset{\vert}{\underset{\vert}{C}}-NO_2$	≈1560 (s) ≈1350 (s)	asymmetric and symmetric stretching vibration of the NO bond; lowered by conjugation to multiple bonds by ca. 30 cm^{-1}
nitrates $R-O-NO_2$	1640–1620 (s) 1285–1270 (s)	asymmetric and symmetric stretching vibrations
nitramines $\overset{\backslash}{\underset{/}{N}}-NO_2$	1630–1550 (s) 1300–1250 (s)	asymmetric and symmetric stretching vibrations
nitrites $R-O-NO$	1680–1650 (s) 1625–1610 (s)	the two bands are assigned to the s-*trans* and s-*cis* forms of the nitrite group
monomer dimer *E* *Z*	1600–1500 (s) 1290–1190 1425–1370	
$-\overset{\vert}{\underset{\vert}{C}}-NO$		
nitrosamines $\overset{\backslash}{\underset{/}{N}}-NO$	1460–1430 (s)	
N-oxides aromatic aliphatic $R_3\overset{+}{N}-O^-$	1300–1200 (s) 970– 950 (s)	pyridine *N*-oxide absorbs at 1250 cm^{-1} in nonpolar solvents; electron withdrawing substituents in the ring increase the frequency and *vice versa*
NO_3^-	1410–1340 860– 800	

In Figs. 2.23 to 2.24c (see pp. 59 and 60) the spectra of toluene and the three isomeric xylenes are shown. The spectrum of tryptophan (p. 64) also shows the characteristic bands for 1,2-disubstitution.

The values in Tab. 2.16 are also valid to a reasonable approximation for condensed ring systems and pyridines (see Fig. 2.29, p. 64). Strongly electron-withdrawing substituents generally shift the values to higher frequencies.

Tab. 2.15 Aromatic compounds (C=C stretching vibrations)

Group	Band	Notes
aromatic rings	≈1600 (m) ≈1580 (m) ≈1500 (m)	stronger, when there is further conjugation to the aryl ring usually the strongest of the two or three bands

Tab. 2.16 Substitution patterns of the benzene ring

Group	Band	Notes
five neighbouring H	770–735 (s) 710–685 (s)	monosubstitution; usually two bands (see toluene, p. 57)
four neighbouring H	760–740 (s)	1,2-disubstitution (see 1,2-dimethylbenzene, p. 57)
three neighbouring H	800–770 (s)	1,3-disubstitution, 1,2,3-trisubstitution
two neighbouring H	840–800 (s)	1,4-disubstitution, 1,3,4-trisubstitution, etc.
isolated H	900–800 (w)	1,3-disubstitution etc.; usually too weak to be of use

11. **IR Absorptions in the *Fingerprint* Region**

Alongside the previously mentioned out-of-plane vibrations of aromatics (Tab. 2.16, above) vibrations of groups which contain elements from the third and higher rows of the periodic table (e.g. phosphorus and sulfur compounds) and of single bonds (e.g. C–O, C–Halogen) produce important bands in this region. The typical absorptions for halogenated aromatics above 1000 cm^{-1} do not arise from stretching, but from skeletal vibrations.

Tab. 2.17 Sulfur compounds S

Group	Band	Notes
−S−H	2600–2550 (w)	(S–H) stretching vibration; weaker than O–H and less influenced by hydrogen bonding. This absorption is strong in the Raman spectrum
$\diagdown$C=S$\diagup$	1200–1050 (s)	
$\diagdown$C−N$\diagup$ (with H and S)	≈3400	(N–H) stretching vibration; shifted in the solid state to as low as 3150 cm^{-1}
	1550–1460 (s)	amide II
	1300–1100 (s)	amide I
$\diagdown$S=O$\diagup$	1060–1040 (s)	
sulfones $\diagdown$SO$_2$$\diagup$	1350–1310 (s)	
	1160–1120 (s)	
sulfonamides R−SO$_2$−N$\diagdown^\diagup$	1370–1330 (s)	
	1180–1160 (s)	
sulfonates R−SO$_2$−OR'	1420–1330 (s)	
	1200–1145 (s)	
sulfates RO−SO$_2$−OR'	1440–1350	
	1200–1145	

Tab. 2.18 Phosphorus compounds P

Group	Band	Notes
P−H	2400–2350 (s)	sharp
P−Phenyl	1440 (s)	sharp
P−O-Alkyl	1050–1030 (s)	
P−O-Aryl	1240–1190 (s)	
P=O	1300–1250 (s)	
P−O−P	970– 910 (s)	broad
O‖ −P−OH	2700–2560	hydrogen bonded O–H
	1240–1180 (s)	(P=O) stretching vibration

Tab. 2.19 Functional groups with C−O single bonds
This region suffers from multiple overlap! The bands are only significant in conjunction with other structural indicators

Group	Band	Notes
Alcohols C−OH	1250–1000 (S)	primary alcohols at the lowest, tertiary and phenols at the highest end of the region often doublet
ethers C−O−C	1150–1070 (s)	sometimes split, see Fig. 2.18 (p. 56)
$\diagdown$C−O−C	1275–1200 (s)	
$\diagup$C−O−C	1075–1020 (S)	
epoxides (with O)	~1250, ~900, ~800	
esters C−O−C (with O)	1330–1050 (s)	2 bands $\delta_{asym.}$ stronger, and at lower wavenumbers
CH$_3$CO−O−C	~1240	asymmetric stretching vibration
RCO−O−CH$_3$	~1165	

Tab. 2.20 Halogen compounds C−Hal

Group	Alkyl−Hal	Aryl−Hal	
C−F	1365–1120 (s)	1270–1100	
C−Cl	830– 560 (s)	1100–1030	skeletal vibrations
C−Br	680– 515 (s)	1075–1030	
C−I	≈ 500 (s)	≈ 1060	

Tab. 2.21 Inorganic ions

Group	Band	Notes
ammonium	3300–3030	all bands are strong
cyanide, thiocyanate, cyanate	2200–2000	
carbonate	1450–1410	
sulfate	1130–1080	
nitrate	1380–1350	
nitrite	1250–1230	
phosphate	1100–1000	

12. Examples of IR Spectra

The following spectra show the position, appearance and relative intensity of the absorption bands for typical representatives of some classes of compounds. The variety in the **finger-** **print** region demonstrates the utility of this region of the IR spectrum for the identification of compounds.

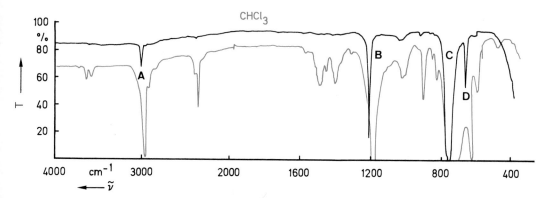

Fig. 2.15 Chloroform (as liquid film); black spectrum: 9 µm film thickness, blue spectrum: 100 µm. The figure shows the dependence of the strength of the bands on the thickness of the film for a commonly used solvent. In the regions of strong absorption the transparency using thick cells (>0.2 mm film thickness) is no longer sufficient for the detector to operate properly

A 3030 cm⁻¹ (C–H) stretching vibration v (CH)
B 1215 cm⁻¹ (C–H) bending vibration δ (CH)
C 760 cm⁻¹ asymmetric (C–Cl) stretching vibration
D 670 cm⁻¹ symmetric (C–Cl) stretching vibration

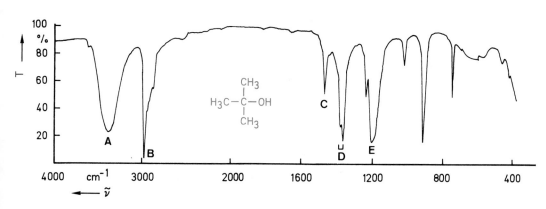

Fig. 2.16 *tert*-Butanol (as liquid film)
Alcohols are easily identified by the strong OH-band (A) and an intense and broad absorption between 1250–1000 cm⁻¹ (E)

A ≈ 3400 cm⁻¹ hydrogen bonded (O–H) stretching vibration; the preceding shoulder at 3605 cm⁻¹ is presumably due to non-associated O–H
B 2975 cm⁻¹ (C–H) stretching vibration $v_{as,s}(CH_3)$
C 1470 cm⁻¹ asymmetric (C–H) bending vibration
D 1380 cm⁻¹ characteristic double band for *t*-butyl groups
        1365 cm⁻¹ δ_s (C(CH$_3$)$_3$)
E 1200 cm⁻¹ (C–O) stretching vibration v (C–O)

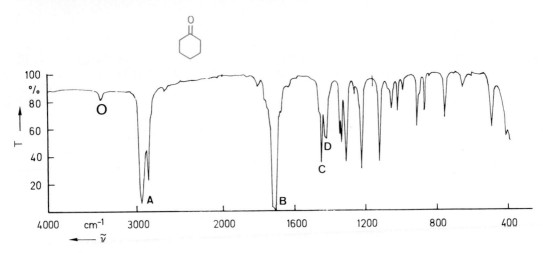

Fig. 2.17   Cyclohexanone (as liquid film)

O   3400 cm⁻¹ overtone vibration of the carbonyl group (see also Fig. 2.20 and 2.9, p. 57 and 38)
A   (C–H) stretching vibration $v_{as,s}(CH_3)$
B   1710 cm⁻¹ (C=O) stretching vibration v (C=O)
C   1450 cm⁻¹ (C–H) bending vibration δ (CH₂)
D   1420 cm⁻¹ (C–H) bending vibration next to C=O

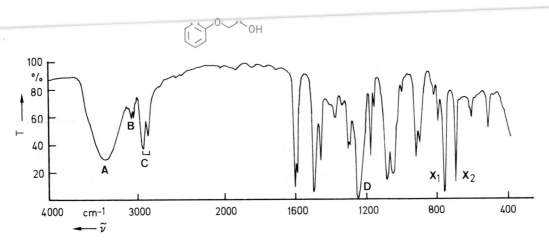

Fig. 2.18   2-Phenoxyethanol (as liquid film)
This example shows characteristic bands for an alcohol, an ether, and a monosubstituted aromatic ring

A   ≈ 3350 cm⁻¹ hydrogen bonded (O–H) stretching vibration
B   (C–H) stretching vibration of the benzene ring
C   (C–H) stretching vibration of the CH₂ groups
D   1250 cm⁻¹ (C–O) stretching vibration in aryl alkyl ethers; dialkyl ether bands lie at lower frequency (Tab. 2.19)
X₁   760 cm⁻¹ monosubstituted aromatic, i.e. five neighbouring H-atoms
X₂   695 cm⁻¹ (cf. toluene, Fig. 2.23)

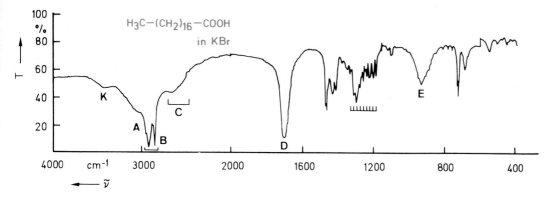

Fig. 2.19 Octadecanoic acid (stearic acid; in KBr)

Carboxylic acids associate by formation of hydrogen bonds (broad band for the dimeric form at 3000 cm⁻¹). The bending vibration of the hydrogen bonded complex at 930 cm⁻¹ is also characteristic. In longer chains ($> C_{12}$) **progression bands** are observed in the solid state; these are equidistant bands between 1350 and 1200 cm⁻¹, which arise from the (E)-oriented CH_2 groups (**twisting** and **rocking** vibrations).

A	≈3000 cm⁻¹	very broad OH band from hydrogen bonds
B		overlapping (C–H) stretching vibrations $\nu_{as,s}(CH_2, CH_3)$
C	2700 to 2500 cm⁻¹	characteristic shoulders, arising from overtones and combination vibrations
D	1700 cm⁻¹	(C=O) stretching vibration
E	930 cm⁻¹	(O–H) bending vibration from hydrogen bonds
		O–H from traces of water in the KBr

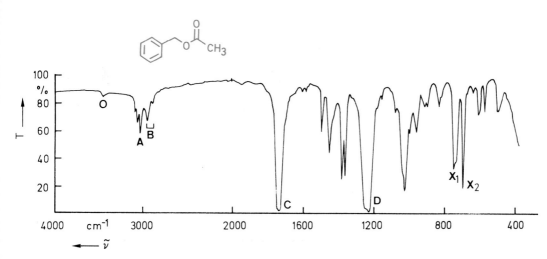

Fig. 2.20 Benzyl acetate (as liquid film)

O	3450 cm⁻¹	probably not traces of water, but overtone vibration of the carbonyl group (cf. Fig. 2.17)
A	3050 to 3020 cm⁻¹	(C–H) stretching vibration of the benzene ring
B	2960 to 2880 cm⁻¹	(C–H) stretching vibration of the CH_3 group
C	1740 cm⁻¹	(C=O) stretching vibration
D	1230 cm⁻¹	(C–O) stretching vibration; position is characteristic for the acetyl group
X₁	750 cm⁻¹	monosubstituted aromatic
X₂	700 cm⁻¹	(cf. toluene, Fig. 2.23)

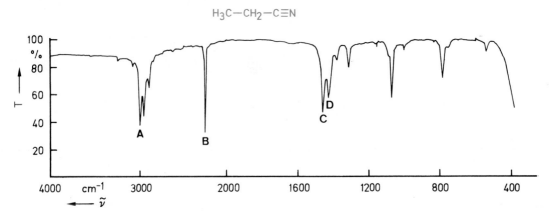

Fig. 2.21 Propionitrile (as liquid film)
Absorption bands in the region 2300–2000 cm^{-1} are almost certain indicators of triple bonds (see Tab. 2.8, p. 47)

A   (C–H) stretching vibrations $\nu_{as,s}$(CH$_2$, CH$_3$)
B   2250 cm^{-1} (C≡N) stretching vibration
C   1460 cm^{-1} (C–H) bending vibration
D   1430 cm^{-1} (C–H) bending vibration next to C≡N

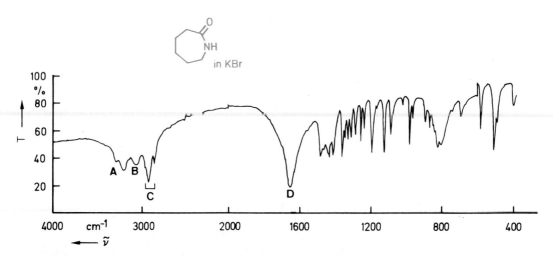

Fig. 2.22 ε-Caprolactam (hexahydro-2H-azepin-2-one) (in KBr)
Example of a cyclic carboxylic acid amide with absence of the amide II band (cf. Fig. 2.28a, p. 62). The many sharp bands in the fingerprint region are typical for aliphatic rings

A   3295 cm^{-1} (N–H) stretching vibrations in N-monosubstituted amides
        3210 cm^{-1}
B   3100 cm^{-1} combination band ν(C=O) + δ (N–H)
C   (C–H) stretching vibration $\nu_{as,s}$(CH$_2$)
D   1660 cm^{-1} (C=O) stretching vibration

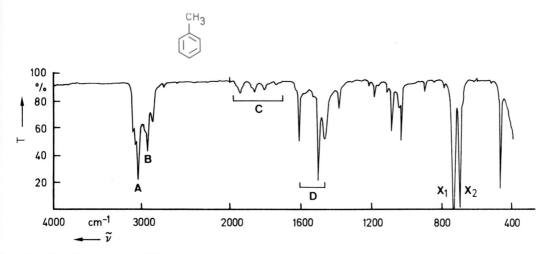

Fig. 2.23 Toluene (as liquid film)

A	aromatic (C–H) stretching vibrations
B	aliphatic (C–H) stretching vibrations
C	overtones and combination vibrations of aromatics
D	(C=C) stretching vibrations, typical of aromatics
X_1 730 cm^{-1}	monosubstituted aromatic (five neighbouring H-atoms); H bending vibration (out of plane); (see Tab. 2.16, p. 53)
X_2 695 cm^{-1}	ring bending vibration, also characteristic for a monosubstituted benzene

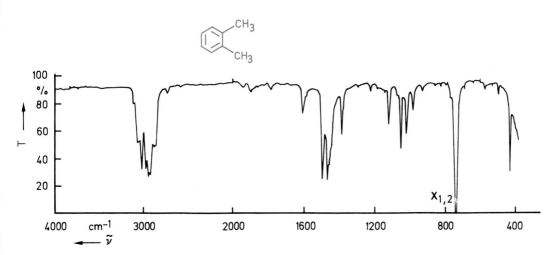

Fig. 2.24a 1,2-Dimethylbenzene (o-xylene) (as liquid film)

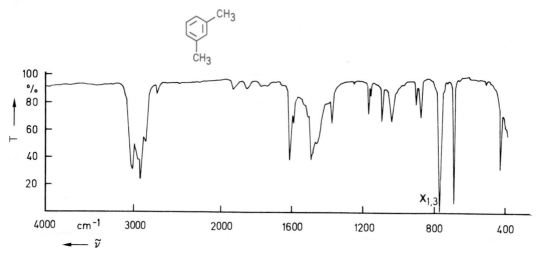

Fig. 2.24b 1,3-Dimethylbenzene (*m*-xylene) (as liquid film)

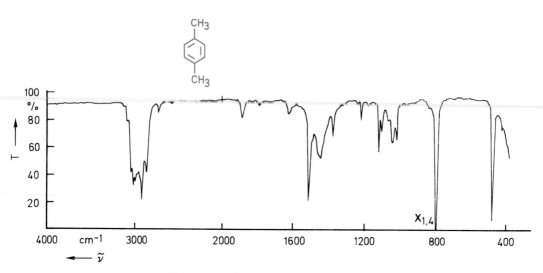

Fig. 2.24c 1,4-Dimethylbenzene (*p*-xylene) (as liquid film)

Determination of the substitution pattern from Tab. 2.16 (p. 53)

X$_{1,2}$ 740 cm^{-1} typical of four neighbouring H on an aromatic (1,2-disubstitution)
X$_{1,3}$ 770 cm^{-1} three neighbouring H (1,3-disubstitution)
X$_{1,4}$ 770 cm^{-1} two neighbouring H (1,4-disubstitution)
For the assignment of the remaining bands compare with the spectrum of toluene (see Fig. 2.23)

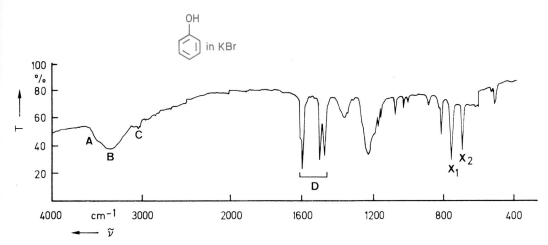

Fig. 2.25   Phenol (in KBr)

A   3500 cm^{-1} (O–H) stretching vibration in hydrogen bonded dimer
B   3360 cm^{-1} (O–H) stretching vibration in hydrogen bonded polymer
C   3040 cm^{-1} (C–H) stretching vibration in aromatic
D   (C=C) stretching vibrations, typical of aromatics (cf. Fig. 2.33)
X$_1$ 755 cm^{-1}
X$_2$ 690 cm^{-1} monosubstituted aromatic (see Tab. 2.16, p. 53)

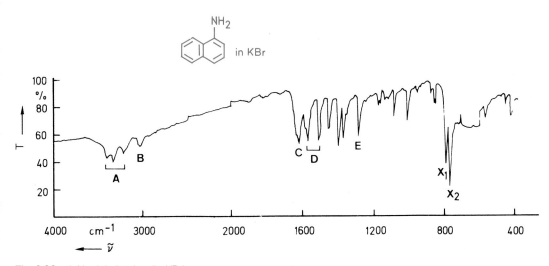

Fig. 2.26   1-Naphthylamine (in KBr)

A   (N–H) stretching vibration (various degrees of association)
B   3040 cm^{-1} (C–H) stretching vibration in aromatic
C   1620 cm^{-1} (N–H) bending vibration
D   1570 cm^{-1} aromatic (C=C) stretching vibrations
        1510 cm^{-1}
E   1290 cm^{-1} (C–N) stretching vibration
X$_1$ 795 cm^{-1} monosubstituted aromatic (the values in Tab. 2.16 apply approximately to naphthalenes)
X$_2$ 770 cm^{-1}

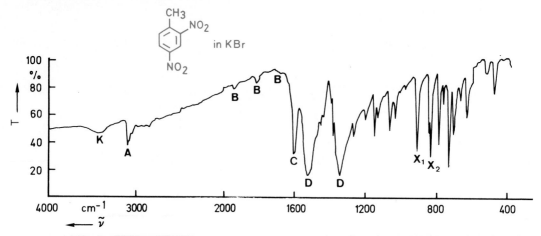

Fig. 2.27 2,4-Dinitrotoluene (in KBr)

A 3100 cm^{-1} aromatic (C–H) stretching vibration
B aromatic overtone vibrations
C 1600 cm^{-1} aromatic (C=C) stretching vibrations
D 1520 cm^{-1} asymmetric and symmetric N=O stretching vibration
  1340 cm^{-1} (conjugated to aromatic system)
X$_1$ 915 cm^{-1} probably the two bands of the out-of-plane (C–H) bending vibrations
X$_2$ 840 cm^{-1} indicative of 1,2,4-substitution
K traces of water in the KBr disc

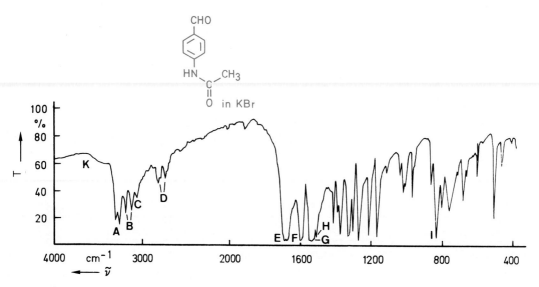

Fig. 2.28a 4-Acetylaminobenzaldehyde (in KBr)

A 3300 and N–H in N-monosubstituted amides
  3260 cm^{-1}
B 3190 and amide bands of unknown origin
  3110 cm^{-1}
C 3060 cm^{-1} aromatic C–H
D 2810 and C–H in aldehydes
  2730 cm^{-1}

E 1690 and aldehyde-carbonyl and amide I
  1670 cm^{-1}
F 1600 cm^{-1} benzene ring
G 1535 cm^{-1} amide II
H 1515 cm^{-1} benzene ring
I 835 cm^{-1} p-disubstituted benzene ring
K shoulder of an OH band from traces of water in the KBr disc

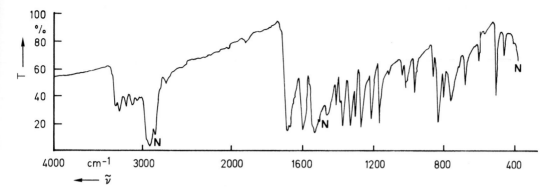

Fig. 2.28b   4-Acetylaminobenzaldehyde (in Nujol)

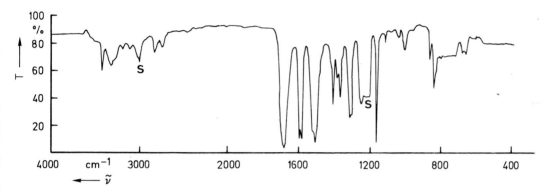

Fig. 2.28c   4-Acetylaminobenzaldehyde (in CHCl₃)

The spectrum in Fig. 2.28a shows a multiplicity of bands which is typical for N-monosubstituted amides. The appearance of so many bands in the spectra of amides is presumably due to the many possibilities for association, of which only one is shown on p. 47.

In contrast to Fig. 2.28a, the spectrum in Fig. 2.28b was measured in Nujol. As a result, the aldehyde (C–H) band (**D** in Fig. 2.28a) is obscured by the strong Nujol band (marked with **N**). On the other hand the absorption marked **K** has disappeared, since it was due to the traces of moisture in the KBr disc.

The spectrum in Fig. 2.28c shows the same compound in solution. This causes several changes: the region of the (N–H) vibration is much different, and the amide I band is somewhat shifted to higher frequency. This leads to overlap with the aldehyde (C=O) absorption. Such differences are generally to be expected on going from the crystalline state to solution, since the latter leads to a reduction in the **inter**molecular interactions. The change is most apparent for the vibration frequen-

cies of those functional groups which are most strongly involved in the association.

The benzene absorption at 1600 cm^{-1} is now split into two separate bands, i.e. spectra in solution are often better resolved than solid state spectra. On the other hand solid state spectra show considerably more bands in the fingerprint region.

The bands marked **S** originate from the solvent, which is not totally compensated for by the double beam technique.

Amino acids (Fig. 2.30) show spectra of zwitterionic groups. The (N–H) absorption of the primary ammonium group (NH$_3^+$) is overlayed by the bands of the saturated C–H bonds. The two bands at 2500 and 2000 cm^{-1} are often seen when an $-$NH$_3^+$ group is present, and are due to overtones and combination vibrations. In the double bond region there are several bands of which at least one arises from the ionised carboxyl group.

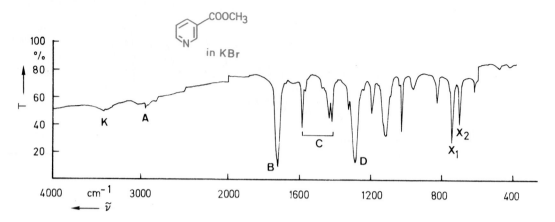

Fig. 2.29 Nicotinic acid methyl ester (in KBr)

A 2950 cm^{-1} (C–H) stretching vibration v (CH$_3$); the aromatic (C–H) stretching vibration is only weakly visible (above 3000 cm^{-1})
B 1725 cm^{-1} (C=O) stretching vibration; (C=C) and (C=N) stretching vibration
D 1290 cm^{-1} (C–O) stretching vibration
X$_1$ 745 cm^{-1} monosubstituted aromatic (the values in Tab. 2.16 apply approximately to pyridines)
X$_2$ 705 cm^{-1}
K traces of water in the KBr disc

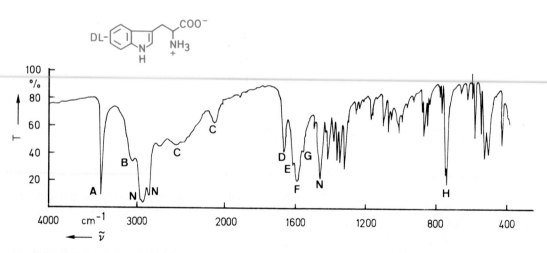

Fig. 2.30 D,L-Tryptophan (in Nujol)

A 3400 cm^{-1} indole (N–H) stretching vibration
B 3030 cm^{-1} broad "ammonium" band from –NH$_3^+$
C ≈2500 and two bands, very common in amino acids, also appear in primary ammonium salts
     ≈2100 cm^{-1}
D 1665 cm^{-1} amino acid I; unusually strong
E 1610 cm^{-1} probably aryl group
F 1585 cm^{-1} amino acid II; ionised carboxyl group –COO$^-$
G 1555 cm^{-1} –NH$_3^+$ bending vibration
H 755 or 745 cm^{-1} (C–H) out-of-plane bending vibrations of a 1,2-disubstituted benzene ring
N Nujol bands

13. Information Technology as an Aid to IR Spectroscopy

All modern IR spectrometers operate in on-line mode; the actual measurement device forms a unit with the computer and storage media.

IR software currently available can be divided into five categories:

1. Software for spectrometer operation (setting measurement parameters, etc.)

2. Software for spectral processing (peak recognition, expansions of sections of the spectrum, overlaying spectra for comparison purposes, etc.)

3. On-line catalogues of spectra: both general (all chemicals of a fine chemicals supplier) and specialised catalogues (e.g. spectra of active species, drugs) are available. Such software is often capable of comparing a measured spectrum with spectra from the catalogue and determine a correlation factor based on positions and intensities of the bands. The quality of the results is dependent on the quality of the digitisation and the coverage of the catalogue. Often measured spectra can be added to the catalogue, which improves the applicability of the system.

4. Software for spectral interpretation: this extends from simple systems, which make suggestions based on peak positions or list typical absorptions for a functional group, to systems employing pattern recognition and similar chemometric methods.

5. Software for the combination of spectroscopic methods (IR, NMR, MS, UV) with one another and with chemical structures, which are held in relational databases so that for a specific structural element recognised in the IR spectrum the corresponding NMR signals, fragmentations in the MS or UV bands can be shown, right through to the co-ordinated interpretation of the various spectra, using artificial intelligence methods to reproduce human skills of reasoning.

Software of types 1 and 2 is an integral component of the IR spectrometer. It is delivered by the manufacturer, as are usually spectral catalogues and simple interpretation software. More powerful software of types 3 to 5 can be purchased as accessories from the instrument manufacturers or from scientific software companies. The capabilities of such programs are constantly increasing in parallel with the increasing performance of computers; a comprehensive treatment is outside the range of this chapter.

14. Quantitative IR Spectroscopy

The quantitative determination of the concentration of a substance in a solution or mixture can be made with the assistance of IR spectroscopy. As in UV spectroscopy the Beer-Lambert law describes the relationship between absorbed light and concentration:

$$\lg \frac{I_0}{I} = \varepsilon \cdot c \cdot d = E_\lambda$$

The absorption at a specific wavelength is proportional to the concentration c and the transmission distance (cell thickness) d. The measured quantity is the ratio I_0/I of the light **before** and **after** its passage through the sample. The quantity $\lg I_0/I$ is the absorbance E_λ and ε the extinction coefficient. In the formula above there are three variables, c, d, and E_λ while ε is a constant for the substance. The aim of quantitative IR analysis is therefore to determine the concentration c from E_λ for a characteristic absorption band; d is the thickness of the cell.

The Beer-Lambert law is however only strictly valid at **low** concentration, e.g. for the very dilute solutions used for UV spectroscopy. Reflections and diffraction of the incident light also

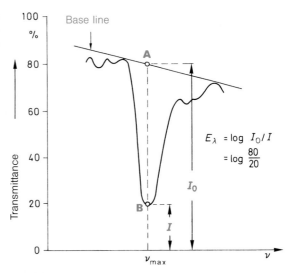

Fig. 2.31   Base line method for the determination of the absorbance E_λ

affect the determination of I_0/I. Thus KBr discs are only suitable for **semi**-quantitative IR measurements.

The quantitative determination of a sample using IR spectroscopy essentially requires the establishment of an empirical **calibration curve**. Several solutions of different concentrations are prepared and the resulting absorbances of a characteristic absorption plotted against the concentration. Since there is initially no reference point against which to measure the absorbance, the base line method is often used. This consists of using as the base line (i.e. the line of no absorbance) a straight line, often drawn as a tangent to the absorption curve as in Fig. 2.31. The ratio I_0/I can then be easily determined for each concentration.

The calibration curve is then generated by plotting the absorbances against the concentration (Fig. 2.32). From the calibration curve an unknown concentration c_x can be determined from a measurement of E_λ.

Quantitative IR analysis nowadays finds practical application in the plastics industry and in quality control of pharmaceuticals and agricultural pesticides.

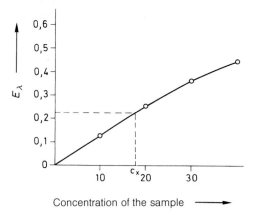

Fig. 2.32   Calibration curve

15. Raman Spectroscopy

The Raman effect was predicted theoretically by **A. Smekal** in 1923 and observed experimentally five years later by **C. V. Raman**.

Raman spectra are not normally recorded routinely, and the organic-oriented chemist seldom uses Raman spectroscopy for structural determination. Nevertheless a Raman spectrum can for certain specialised problems be a useful complement to IR spectroscopy, for example for the measurement of aqueous solutions, single crystals, and polymers. The applicability of Raman spectroscopy has also become much easier and quicker by the application of **laser technology**.

15.1 The Raman Effect

If a liquid or a concentrated solution of a substance is irradiated with monochromatic light (e.g. with an argon laser, which has a wavelength of 488 nm = 20492 cm^{-1}) then:

–   the majority of the light passes through the sample unaffected (transmission)

–   a small portion of the light (factor 10^{-4}) is scattered in all

directions, but retains the frequency of the irradiating light; this is called **Rayleigh scattering** and can be thought of as arising from elastic impacts of the light quanta with the molecules.

–   an even smaller portion of the light (factor 10^{-8}) is also scattered in all directions, but has a frequency distribution; it arises from absorption and re-emission combined with a vibrational excitation or extinction. This scattered radiation can be spectrally analysed and recorded with a photoelectric detector. The difference between the frequency of the irradiating line and a Raman line is the frequency of the relevant vibration.

The Raman effect is thus a consequence of the interaction between matter and electromagnetic radiation. The Raman spectrum is an **emission spectrum**. The frequency of the Raman lines or bands can be larger or smaller than the excitation frequency v_0 (Rayleigh line). Characteristic for a molecule are the **differences** between the Raman frequencies and the excitation frequency v_0. They are independent of v_0 and can also be found in the IR spectrum as absorption bands (see selection rules).

The origin of the Raman effect can be explained as follows: when the laser beam hits molecules of the sample (and the excitation energy is insufficient for an electronic transition) the interaction either causes elastic scattering (Rayleigh scattering) or a part of the light energy is taken up by increasing the vibrational energy of the molecule, i.e. the scattered light is of **lower energy** (longer wavelength). If the excitation beam hits a molecule in an excited vibrational state, the same interaction results in the emission of light of **higher energy** (shorter wavelength). The Raman lines on the longer-wavelength side of the Rayleigh frequency are called **Stokes** lines, those on the shorter-wavelength side **anti-Stokes** lines.

15.2 Selection Rules

As explained in Sec. 2 (see p. 31) the generation of an IR absorption requires that a change in the **dipole moment** of the molecule is caused by the vibration. For the appearance of a Raman line however it is necessary that there is a change in the **polarisability** of the molecule. The polarisability is a measure of the ease of deformation of the electron cloud around an atom or molecule, e.g. greater for I^- than for Cl^- or Br^-.

These selection rules have an important consequence: in symmetrical molecules vibrations which are symmetric about the symmetry centre are IR inactive (no change of dipole moment) but Raman active. Conversely, vibrations which are not symmetric about the symmetry centre are Raman inactive (forbidden) and in the IR spectrum generally active (allowed). This can be demonstrated by the simple example of the carbon dioxide molecule (Fig. 2.33). In the symmetric stretching vibration with amplitudes **a** and **b** it is apparent that there is no change in the dipole moment. This vibration is therefore IR inactive, and causes no absorption band in the spectrum. The polarisability in the compressed state **a** is however different from that in the extended state **b**, therefore this vibration is Raman active. This underlines the importance of Raman spectroscopy for symmetrical molecules. For the antisymmetric stretching vibration (**c,d**) however the situation is reversed. The polarisability remains the same, whereas the dipole moment changes. Thus this vibration does not appear in the Raman spectrum. The changes of polarisability α and dipole moment μ for the stretching vibrations of the CO_2 molecule are also shown graphically in Fig. 2.33.

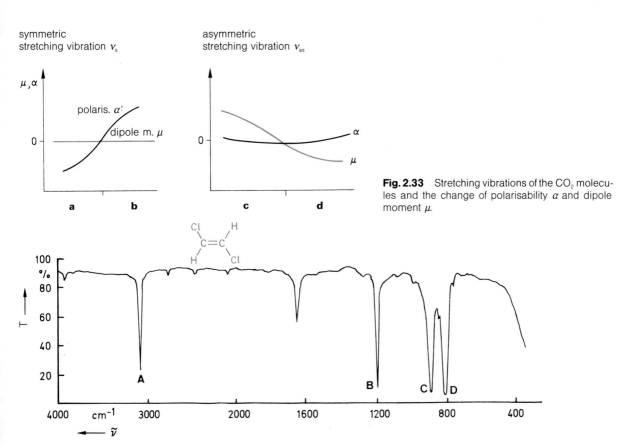

symmetric stretching vibration v_s

asymmetric stretching vibration v_{as}

Fig. 2.33 Stretching vibrations of the CO_2 molecules and the change of polarisability α and dipole moment μ.

Fig. 2.34 IR spectrum of dichloroethylene

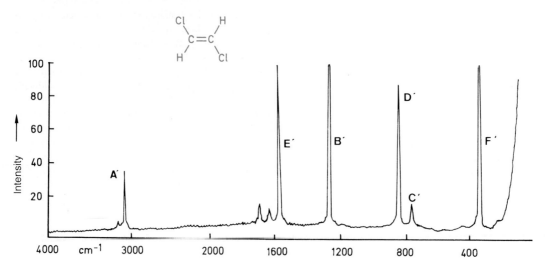

Fig. 2.35　Laser Raman spectrum of (*E*)-dichloroethylene

As an illustration the IR and Raman spectra of (*E*)-dichloro-ethylene (i.e. a symmetrical molecule) are shown in Figs. 2.34 and 2.35. These show clearly how IR and Raman spectroscopy yield complementary pictures of the vibrations in a molecule: in the IR spectrum the absorptions from the **antisymmetric** vibrations appear, whereas the Raman spectrum shows the emission bands of the **symmetric** vibrations. In Tab. 2.22 the individual vibrations are assigned to their respective bands.

Tab. 2.22　Assignment of the bands in Figs. 2.34 and 2.35

vibration type	anti-symmetric vibration (IR active)	IR band Fig. 2.34 (cm^{-1})	symmetric vibration (Raman active)	Raman band Fig. 2.35 (cm^{-1})
ν (C—H)		3090 (**A**)		3070 (**A′**)
ν (C—Cl)		817 (**D**)		844 (**D′**)
δ (C—H)		1200 (**B**)		1270 (**B′**)
γ (C—H)		895 (**C**)		760 (**C′**)
ν (C=C)	–	–		1576 (**E′**)
δ (C—Cl)	below 300 cm^{-1} in IR	–		350 (**F′**)

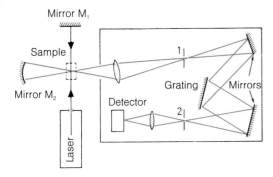

Fig. 2.36 Construction scheme of a Raman spectrometer

15.3 Raman Spectrometers

For the measurement of a Raman spectrum a very intense source of monochromatic light is required with a wavelength between the UV and the IR regions, since in that region there are very few interfering absorptions. However fluorescence radiation from impurities in the sample can completely obscure the Raman spectrum and make the measurement of a spectrum impossible.

The introduction of lasers in the 1960's reduced the amount of sample required to a few milligrams, the time of measurement from hours to minutes and improved the signal-to-noise ratio. These improvements are due to the enormous increase in the radiation density (some 10 orders of magnitude) over the low pressure mercury lamp.

In Fig. 2.36 the construction of a Raman spectrometer is shown schematically. The monochromatic light emitted by the laser passes through the sample and is reflected from the mirror M_1 in order to double the intensity. The Raman scattering is usually observed **perpendicular** to the direction of the transmitted radiation and focused onto the entry slit 1 by a lens. The mirror M_2 doubles the intensity of the scattered light. The light is split up into its spectral components by a grating and focused onto a photoelectric detector after passing through the exit slit 2.

15.4 Applications

Raman spectroscopy is particularly well suited for the study of non-polar or weakly polar bonds such as $C{\equiv}C$, $C{=}C$, $N{=}N$, $C{-}C$, $O{-}O$, $S{-}S$, or rings. The skeletal vibrations of $(C{-}C)$ bonds in rings are usually much stronger in the Raman spectrum than in the IR spectrum. This allows the assignment of the structures of molecular frameworks. On the other hand the strong and characteristic IR bands of polar groups like $C{=}O$ and $O{-}H$ are only weakly visible in the Raman spectrum.

An advantage of Raman spectroscopy is the possibility of using water as a solvent. In IR spectroscopy water is unsuitable because of its own strong absorptions and the use of sodium chloride cells. In contrast Raman spectra of aqueous solutions can be easily obtained since glass cells are used and water has a Raman spectrum consisting of only a few, weak lines.

The greatest area of application of Raman spectroscopy is not however for structural assignment, but in problems of interpretation and band assignment. The complete analysis of a vibrational spectrum requires the assignment of each IR or Raman **band** to the corresponding **vibration**, whereas for structural analysis it is sufficient to assign **bands** to **structural elements**.

For an exact assignment it is often necessary to consider further properties of a band beyond its position, intensity, and shape. One such property is the **degree of depolarisation** ϱ of a Raman band. This allows the assignment of Raman bands to vibrations of specific symmetry types, not however to specific structural elements.

The polarisability of a molecule is – like its dipole moment – a vector quantity. This means that energy can only be taken up from electromagnetic radiation if the directions of the electric vector and components of the polarisability are the same. With the introduction of lasers as the light source it has also become possible to use linearly polarised light. Consequently the degree of depolarisation can be determined. It is defined as the quotient of two intensities of different polarisation

$$\varrho = \frac{I_\perp}{I_\parallel}$$

In this equation $I_\perp$ represents the intensity of the scattered radiation perpendicular to the polarisation of the irradiating laser light, and $I_\parallel$ parallel to it. In practice is determined by measuring the Raman spectrum twice, with the polarisation plane of the irradiating laser rotated by 90°. For each band, ϱ depends on the symmetry properties of the relevant vibration.

Literature

Text Books

Bellamy, L. J. (1966), Ultrarot-Spektrum und chemische Konstitution, D. Steinkopff Verlag, Darmstadt.

Brügel, W. (1969), Einführung in die Ultrarotspektroskopie, D. Steinkopff Verlag, Darmstadt.

Cross, A. D. (1964), Introduction to Practical Infrared Spectroscopy, Butterworth, London.

Fadini, A., Schnepel, F. M. (1985), Schwingungsspektroskopie, Georg Thieme Verlag, Stuttgart.

Günzler, H., Böck, H. (1983), IR-Spektroskopie, Verlag Chemie, Weinheim.

Hediger, H. J. (1971), Infrarotspektroskopie, Akademische Verlagsgesellschaft, Frankfurt/M.

Kemmner, G. (1968), Infrarot-Spektroskopie, Grundlagen, Anwendung, Methoden, Franckh'sche Verlagsbuchhandlung, Stuttgart.

Nakamoto, K. (1963), Infrared Spectra of Inorganic and Coordination Compounds, Wiley, New York.

Nakanishi, K. (1962), Infrared Absorption Spectroscopy, Holden Day, San Francisco.

Parker, F. S. (1971), Applications of Infrared Spectroscopy in Biochemistry, Biology and Medicine, Hilger, London.

Rao, C. (1963), Chemical Applications of Infrared Spectroscopy, Academic Press, New York.

Van der Maas, J. H. (1972), Basic Infrared Spectroscopy, Heyden, London.

Volkmann, H. (1972), Handbuch der Infrarot-Spektroskopie, Verlag Chemie, Weinheim.

Weidlein, J., Müller, U., Dehnicke, K. (1982), Schwingungsspektroskopie, Georg Thieme Verlag, Stuttgart, New York.

Weitkamp, H. (1973), IR-Spektroskopie, in Methodicum Chimicum (Korte, F.), Vol. 1, Georg Thieme Verlag, Stuttgart.

Williams, D. H., Fleming, I. (1991), Spektroskopische Methoden in der organischen Chemie, Georg Thieme Verlag, Stuttgart.

Spectrum Catalogues

Hershenson, H. M., Infrared Absorption Spectra, Academic Press, New York. Index for 1945-1957 (1959); Index for 1958-1962 (1964).

An Index of Published Infrared Spectra (1960), Vol. 1 and 2, H. M. S. O., London.

Dokumentation der Molekülspektroskopie, Verlag Chemie, Weinheim. (A Collection of Assigned Coded and Indexed Spectra).

Sadtler Standard Spectra (1970), Heyden, London. (A collection of ca. 60 000 spectra with annual supplements).

The Aldrich Library of Infrared Spectra (1975), Aldrich Chemical Company, Milwaukee. (Ca. 10 000 spectra).

Schrader, B., Raman/Infrared Atlas of Organic Compounds (1989), VCH Verlagsgesellschaft, Weinheim.

Quantitative IR Analysis

Kössler, I. (1961), Methoden der Infrarot-Spektroskopie in der chemischen Analyse, Akademische Verlagsgesellschaft, Leipzig.

Weitkamp, H., Barth, R. (1976), Einführung in die quantitative Infrarot-Spektrophotometrie, Georg Thieme Verlag, Stuttgart.

Coupled Techniques

Herres, W. (1987), Capillary Gas Chromatography – Fourier Transform Infrared Spectroscopy, Hütig Verlag, Heidelberg, Basel, New York.

Raman Spectroscopy

Brandmüller, J., Moser, H. (1962), Einführung in die Raman-Spektroskopie, D. Steinkopff Verlag, Darmstadt.

Freeman, S. K. (1974), Applications of Laser Raman Spectroscopy, Wiley, New York.

Gilson, T. R., Hendra, P. J. (1970), Laser Raman Spectroscopy, Wiley, London.

Loader, J. (1970), Basic Laser Raman Spectroscopy, Heyden, London.

Schrader, B. (1973), Raman-Spektroskopie, in Methodicum Chimicum (Korte, F.), Vol. 1, Georg Thieme Verlag, Stuttgart.

3 Nuclear Magnetic Resonance Spectroscopy

1. Fundamental Physical Principles

1.1 The Resonance Phenomenon

Most atomic nuclei have angular momentum p **(nuclear spin)** and therefore a **magnetic moment** $\mu = \gamma \, p$. The **gyromagnetic ratio** γ is a characteristic constant for each individual nuclear type. From quantum theory

$$p = \sqrt{I(I+1)} \cdot \frac{h}{2\pi}$$

and

$$\mu = \gamma \cdot \sqrt{I(I+1)} \cdot \frac{h}{2\pi}.$$

I is the **nuclear angular momentum quantum number** or **nuclear spin quantum number** of the particular atomic nucleus and can have integer or half-integer values (Tab. 3.1).

$$I = 0, 1/2, 1, 3/2, 2, 5/2, 3, \ldots$$

In a **homogeneous, static magnetic field** B_0 the angular momentum vector P can take up specific selected angles to the B_0 vector **(quantisation of direction)**. In these positions the components of p in the direction of the field are given by

$$p_B = m \cdot \frac{h}{2\pi}.$$

For the **orientational** or **magnetic quantum number** m the allowed values are

$$m = +I, I-1, I-2, \ldots, -I+1, -I$$

The total of $(2I+1)$ eigen states, the so-called nuclear Zeeman levels, have different energies, given by:

$$E_m = -\mu_B \cdot B_0 = -\gamma \cdot p_B \cdot B_0 =$$
$$= -\gamma \cdot m \cdot \frac{h}{2\pi} \cdot B_0$$
$$(m = +I, \ldots, -I)$$

For the hydrogen nucleus, the proton, $I = 1/2$, and therefore $m = \pm 1/2$.

The resultant energy level scheme is shown in Fig. 3.1. In the lower energy state m precesses with the Larmor frequency $v_0 = |\gamma| \cdot B_0/2\pi$ about B_0, in the higher energy state conversely about $-B_0$. (If E_m is defined with positive sign, then the magnetic quantum numbers m in Fig. 3.1 must be exchanged).

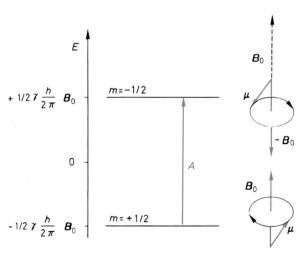

Fig. 3.1 Energy levels of protons in a magnetic field B_0

In thermal equilibrium the ^1H nuclei are subject to the **Boltzmann distribution**. Since the energy difference

$$\Delta E = \gamma \cdot \frac{h}{2\pi} \, \boldsymbol{B}_0$$

is very small compared to the average thermal energy, the lower energy state is only very slightly more populated. The relationship of the populations is given by

$$\frac{N_{(m=-1/2)}}{N_{(m=+1/2)}} = e^{-\frac{\Delta E}{kT}}$$

Irradiation with quanta of energy of ΔE induces spin inversion. Because of the difference in populations the absorption A dominates*. The **resonance condition** is defined by the relationship:

$$h\nu = \Delta E = \gamma \frac{h}{2\pi} \cdot \boldsymbol{B}_0$$

---

*Nuclear magnetic emission is also possible, as in the CIDNP effect (chemically induced nuclear polarisation). For details see NMR texts quoted in the bibliography.

For a field strength of 1.4T the resonance frequency for protons $\nu = f(\boldsymbol{B}_0)$ is near to 60 MHz, corresponding to radio waves with $\lambda = 5$ m. If the resonance condition is fulfilled the absorption very rapidly eliminates the population difference between the two nuclear Zeeman levels; the system is then described as being saturated, unless the reverse process, **relaxation**, takes place to a sufficient extent.

The energy released by the transition of a nucleus from the higher into the lower energy state can be absorbed by the environment in the form of heat (**spin-lattice relaxation**). This process takes place with a rate constant $1/T_1$. T_1 is called the **longitudinal relaxation time** because the process alters the magnetisation in the direction of the field. The transverse magnetisation also varies with time because of the interaction of the magnetic moments with each other (**spin-spin relaxation**). The rate of this process is known as the **transverse relaxation time** T_2.

As shown above, a magnetic moment $\mu \neq 0$ is a precondition for the nuclear magnetic resonance experiment. (The only nuclei with magnetic moment $\mu = 0$ are the e,e-nuclei with even mass and atomic numbers.) Furthermore it is of advantage if $I = 1/2$, since nuclei with larger spin quantum numbers also possess an electric **nuclear quadrupole moment**, which produces complications in the spectra (signal broadening).

Tab 3.1 Properties of nuclei of relevance to NMR spectroscopy of organic compounds

Isotope	Spin quantum number	gyromagnetic ratio γ [rad/Ts]	magnetic moment μ (in units of μ_N)	natural abundance (%)	relative sensitivity per nucleus	absolute sensitivity taking account of natural abundance	Resonance frequency ν_0 (MHz) at a field of 2.3488 T
^1H	1/2	26.752	2.793	99.985	1.000	1.000	100.000
$^2H \equiv D$	1	4.107	0.857[b]	0.015	0.010	$1.45 \cdot 10^{-6}$	15.351
^6Li	1	3.937	0.822[b]	7.42	0.009	$6.31 \cdot 10^{-4}$	14.716
^7Li	3/2	10.396	3.256[b]	92.58	0.294	0.27	38.862
^{10}B	3	2.875	1.801[b]	19.6	0.020	$3.90 \cdot 10^{-3}$	10.747
^{11}B	3/2	8.584	2.688[b]	80.4	0.165	0.13	32.084
^{13}C	1/2	6.728	0.702	1.10	0.016	$1.76 \cdot 10^{-4}$	25.144
^{14}N	1	1.934	0.404[b]	99.634	0.001	$1.01 \cdot 10^{-3}$	7.224
^{15}N	1/2[a]	-2.712	0.283	0.366	0.001	$3.85 \cdot 10^{-6}$	10.133
^{17}O	5/2[a]	-3.628	1.893[b]	0.038	0.029	$1.08 \cdot 10^{-5}$	13.557
^{19}F	1/2	25.181	2.627	100.0	0.833	0.833	94.077
^{29}Si	1/2[a]	-5.319	0.555	4.67	0.008	$3.69 \cdot 10^{-4}$	19.865
^{31}P	1/2	10.841	1.132	100.0	0.066	0.066	40.481
^{33}S	3/2	2.053	0.643[b]	0.76	0.003	$1.72 \cdot 10^{-5}$	7.670
^{77}Se	1/2	5.101	0.532	7.6	0.007	$5.25 \cdot 10^{-4}$	19.067

[a] In these cases $\gamma < 0$, i.e. the magnetic moment and nuclear spin point in opposite directions
[b] These nuclei also have an electric quadrupole moment

In organic chemistry the most important nuclei are ^1H, ^7Li, ^{11}B, ^{13}C, ^{15}N, ^{17}O, ^{19}F, ^{29}Si, ^{31}P, and ^{77}Se. For ^{13}C, ^{15}N, ^{29}Si, and ^{77}Se the low natural abundance is disadvantageous.

A large magnetic moment for the nucleus to be studied is advantageous, since the signal intensity is proportional to the third power of the magnetic moment. Table 3.1 contains the important properties of those nuclei which are of relevance to organic chemistry.

The energy level scheme (Fig. 3.1) needs to be changed for nuclei with $I > 1/2$ in accordance with the equation for E_m. Since $\Delta m = 1$ however the resonance condition remains the same. At a constant field B_0 the resonance frequencies v of the various nuclei are directly related to the γ-values. Since $\gamma(^1H)/(\gamma(^{13}C)) = 26.75/6.73 = 3.975$ it follows that a ^1H resonance frequency of e.g. 90 MHz corresponds to a ^{13}C resonance frequency of $90/3.975 \approx 22.6$ MHz.

In the following sections 1.2 to 1.5 the important properties of the resonance signals will be treated:

– their **position** (resonance frequency) in the chapter on chemical shifts, p. 73,
– their **fine structure** in the section on spin-spin coupling, p.74,
– their **line width**, p. 83, and
– their **intensity**, p. 84.

1.2 Chemical Shift

The exact **resonance frequency** of a particular nuclear type depends in characteristic fashion on the environment of the nucleus. The **effective field strength** at the nucleus differs from B_0 by the induced field

$$B_{\text{eff}} = B_0 - \sigma B_0.$$

Including the dimensionless **shielding constant** σ in the resonance condition leads to

$$v = \frac{\gamma}{2\pi} B_0 (1 - \sigma).$$

The stronger a nucleus is shielded, i.e. the larger σ is, the smaller B_{eff} becomes; this means that, for a constant frequency, the applied field B_0 must be stronger in order to bring the nucleus to resonance. Similar considerations show that at constant B_0 field v decreases as the shielding increases.

Because of the relationship $v = f(B_0)$ the position of the NMR absorption cannot be given by an absolute scale of values of v or B_0. Instead the signal position is related to that of a **reference compound**. For ^1H and ^{13}C NMR spectroscopy **tetramethylsilane** [TMS, $Si(CH_3)_4$] is the usual standard. At an observation frequency v the difference of the positions of the signals of the observed nucleus X and TMS is given by

$$\Delta B = B(X) - B(\text{TMS})$$

and similarly for a frequency scale in Hz

$$\Delta v = v(X) - v(\text{TMS}) = \frac{\gamma}{2\pi} \cdot \Delta B$$

To specify the position of the signal the **chemical shift** δ of the nucleus X is defined by

$$\delta(X) = 10^6 \frac{\Delta v}{v} \quad \text{with} \quad \delta(\text{TMS}) = 0.$$

δ is a dimensionless quantity, independent of the measurement frequency or the magnetic field strength, characteristic of the observed nucleus in its environment. (In ^1H NMR the τ-scale was previously popular; τ was defined by $\tau = 10 - \delta$.) Since Δv is very small compared to v, the factor 10^6 has been introduced and δ is quoted in **ppm (parts per million)**. The range of the δ-scale for ^1H NMR is about 10, for ^{13}C NMR about 200 ppm. If extreme values are considered, the ranges increase to 40 or 350 ppm respectively. The exact resonance frequencies vary about v_0 (Table 3.1) in the ppm range.

Fig. 3.2 shows as an example the ^1H NMR and the ^{13}C NMR spectrum of acetic acid (**1**). In each case two absorptions are observed, the H- and C-atoms of the carboxy group appearing at lower field. These atoms (nuclei) are therefore less shielded than the methyl protons or the methyl carbon atom respectively. The calculation of the δ values follows from the equation given above for the chemical shifts.

Let us take the methyl signal in the ^1H NMR as an example. At an observation frequency of 60 MHz it appears 126 Hz to low field of the TMS signal. This corresponds to 2.10 ppm. Following current convention one writes $\delta = 2.10$.

$$\delta_H(CH_3) = 10^6 \frac{126}{60 \cdot 10^6} = 2.10$$

δ values are positive in the direction of increasing resonance frequency.

The sensitivity of the **chemical shift** to changes in the environment of the measured nucleus is of great importance for the determination of the structures of organic compounds. The **shielding constant** σ, which determines the resonance position, is made up of three terms

$$\sigma = \sigma_{\text{dia}} + \sigma_{\text{para}} + \sigma'$$

The diamagnetic term σ_{dia} corresponds to the opposing field induced by the external field in the electron cloud surrounding the nucleus. Electrons near to the nucleus shield more than distant ones. The paramagnetic term σ_{para} corresponds to the excitation of p-electrons in the field and has an opposite effect to the diamagnetic shielding. Since for hydrogen only

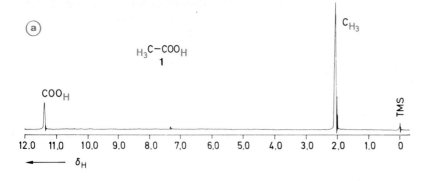

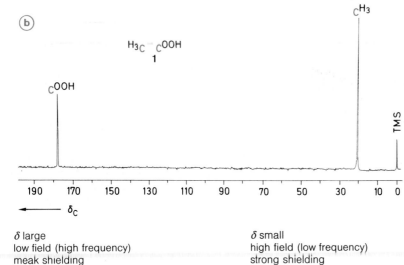

Fig. 3.2 **a** ^{1}H NMR spectrum of acetic acid **1** in CDCl$_3$

δ large	δ small
low field (high frequency)	high field (low frequency)
meak shielding	strong shielding

Fig. 3.2 **b** ^{13}C NMR spectrum of acetic acid in CDCl$_3$ (^{1}H broad band decoupled, i.e. without the effects of the ^{13}C, ^{1}H coupling, see Sect. 1.3)

s-orbitals are present only σ_{dia} is important. For higher nuclei like ^{13}C the paramagnetic term dominates. The term σ^i describes the influence of neighbouring groups, which can decrease or increase the field at the nucleus. Finally σ depends on intermolecular effects, which can be included by the addition of an extra term σ_{Medium}.

1.3 Spin-Spin Coupling

The signals observed in NMR spectra often show fine structure. Depending on the number of components of each signal they are referred to as singlets, doublets, triplets, quartets, etc., or in general multiplets. The cause of this fine structure is the interaction with neighbouring nuclei which possess a magnetic moment. This **spin-spin coupling** occurs between nuclei of the same type (**homonuclear**) and between nuclei of different type (**heteronuclear**) and means that the orientation of the spin of nucleus A influences the local field at the coupling nu-

cleus X and vice versa. For two nuclei A and X, which both have nuclear spin 1/2, there exist in principle four energy levels, corresponding to the four possible orientations of the two nuclei. Without spin-spin interaction ($J=0$) each of A and X shows two absorptions of the same energy (Fig. 3.3, middle). This degeneracy is lifted by the **coupling** J. J is defined as having a positive sign if the energy levels of spin states where both nuclei have the same spin orientation with respect to the external field B_0 are raised. Spin states with opposite orientations are decreased in energy. The reverse situation applies when $J < 0$. Both cases cause the splitting of the A and X signals into doublets (Fig. 3.3).

The magnitude of the coupling is given by the **coupling constant** J, which in this case can be measured directly from the separation of the X lines or from the identical separation of the A lines. For proton-proton coupling the values lie between about –20 and +20 Hz. With other nuclei much larger values can occur. Thus in acetylene the ^{13}C–^{13}C coupling is 171.5 Hz and the (C–H)-coupling is 250 Hz.

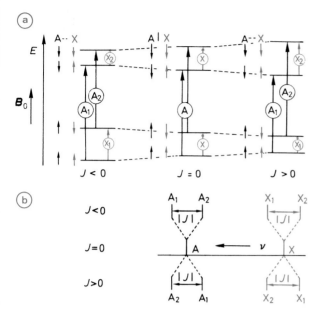

$J < 0$ $J = 0$ $J > 0$

Fig. 3.3 **a** The four possible spin combinations and energy levels of a two spin system ($m = \pm 1/2$) and the corresponding NMR transitions for $J \neq 0$: ↑ A resonances, ↑ X resonances **b** Stick spectra of the cases where $J < 0, J = 0, J > 0$

Particularly important is the fact that the coupling constant is independent of the external field B_0. Two lines in a spectrum can either be singlet signals of two uncoupled nuclei with different chemical shifts or a doublet, arising from one nucleus, which forms an AX system with another, coupling, nucleus. A distinction between these two cases is easily made by measuring two spectra at different frequencies (see Sect. 3.8, p. 124). If the separation remains the same, the cause is coupling; if the separation (in Hz) increases with increasing observation frequency, the lines are two singlets of different chemical shift. (On the δ scale the separation is independent of the observation frequency.)

The coupling between two nuclei A and X, in an isotropic liquid phase, takes place in general through the bonds within the molecule (**scalar coupling**).

If there is **one** intermediate bond the coupling is referred to as a **direct coupling** ^1J, e.g. $^1H-^{19}F$, $^1H-^{13}C$ etc. This term should not be confused with the **direct dipole-dipole coupling**, a through space effect which appears in orientated phases (liquid crystalline states, solids). In isotropic fluids these non-scalar coupling are averaged to zero by the thermal motion of the molecules.

With two or three intermediate bonds the coupling is referred to as a **geminal coupling** ^2J or a **vicinal coupling** ^3J respectively, e.g.

^2J :

^3J :

With increasing numbers of bonds between A and X J_{H-H} generally decreases. In the case of couplings with heavier nuclei there is often a non-uniform decrease of $|J|$ with the number of bonds, instead there may be an intermediate maximum. For the detection of **long range couplings** ^nJ the resolving power of the spectrometer is of critical importance.

The complexity of the coupling pattern increases with the number of coupling nuclei. If the chemical shifts δ_i of all nuclei i in a molecule are known, as well as the coupling constants $^nJ_{i,j}$ for all the possible pairs of nuclei, the NMR spectrum can be calculated. Conversely the δ and J values can be directly determined from simple spectra.

In order to treat the coupling phenomena in more complicated systems than the AX case described above, it is necessary to have a general systematic **nomenclature for spin systems**.

n isochronous nuclei, i.e. n nuclei, which have the same chemical shift, either by coincidence or as a result of their chemical equivalence (cf. Sect. 2.1, p. 86), form an A_n system. If there is in addition a set of m nuclei, again isochronous within the set, the spin system is described as A_nB_m, A_nM_m, or A_nX_m, depending on whether the resonance frequency n of the second set differs slightly, moderately, or strongly from v_A. This notation has an important restriction, namely that for any nuclear combination A_iB_j (or A_iM_j or A_iX_j as appropriate) ($i = 1, ..., n; j = 1, ..., m$) the coupling is the same.

Isochronous nuclei A_i, which only have **one** spin-spin interaction with nuclei of a neighbouring group are said to be **magnetically equivalent**. The same applies to the nuclei B, M, or X. (Since the spin-spin coupling is a reciprocating property, it is not possible for the A spins of an A_nB_m system to be magnetically equivalent and the B spins to not be.) Where there are more than two sets of spins, e.g. in the system $A_nB_mX_l$, the definition of magnetic equivalence requires that there is only a single value of J_{AB}, J_{AX}, and J_{BX}. **Isochronosity** is a necessary, but not sufficient, condition for magnetic equivalence, while magnetic equivalence is a sufficient, but not necessary, condition for isochronosity. For a better understanding this can be demonstrated by two examples. Difluoromethane (**2**) has two isochronous H nuclei and two isochronous F nuclei. Each coupling $^2J(H,F)$ is the same. The two H nuclei and the two F nuclei are therefore as pairs chemically and magnetically equivalent. CH_2F_2 has an A_2X_2 spin system. In 1,1-difluoroethylene (**3**) the H nuclei are also chemically equivalent, as are the F nuclei, but there are two different couplings $^3J(H,F)$ (as seen from a single hydrogen, (*Z*)- and (*E*)-fluorines couple differently). Isochronous nuclei which are not magnetically equivalent are indicated by a dash. 1,1-Difluoroethylene (**3**) has an AA'XX' spin system.

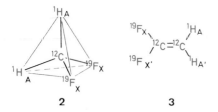

In each molecule the ^1H and ^{19}F spectra show one half of the total spectrum; in the ^1H spectrum the A part and in the ^{19}F spectrum the X part. If one considers the case where the spin-free ^{12}C nuclei are replaced by ^{13}C nuclei, one would obtain in the coupled ^{13}C spectrum for difluoromethane (**2**) the M part of an A_2MX_2 system and for 1,1-difluoroethylene (**3**) the MN part of an AA'MNXX' system.

For the interpretation of the coupling patterns it is important to note that **spin-spin coupling between magnetically equivalent nuclei has no effect on the spectrum**, although such nuclei do in fact have a coupling.

A single set of nuclei, as appears in for example the ^1H spectrum of methane, ethane, ethylene, acetylene, or benzene, gives a singlet absorption. (The coupling between ^1H and ^{13}C is not usually observed in routine spectra, since the natural ^{13}C content is small: 1.1%.) In these examples the protons are isochronous on account of their chemical equivalence; but even if the isochronosity is coincidental no splitting is observed. An example is given in Fig. 3.4. Methyl 3-cyanopropionate (**4**) shows a single, unsplit signal at $\delta = 2.68$ for the two chemically inequivalent methylene groups.

Spin systems of the A_nX_m or A_nM_m type with two sets of magnetically equivalent nuclei are easy to interpret, as long as $|\nu_A - \nu_M|$ is at least a factor of ≈ 10 greater than $J_{A,M}$. The number of lines, the so-called **multiplicity of the signal**, is then

for A: $m \cdot 2I_X + 1$
for X: $n \cdot 2I_A + 1$

I_X and I_A are the spins of the nuclei X and A. When $I_X = I_A = 1/2$ a **first order spectrum** is obtained with

($m + 1$) lines in the A part and
($n + 1$) lines in the X part.

Consider as an example the ^1H spectrum of ethyl bromide (**5**; Fig. 3.5). It has an A_3X_2 spin system. The local field at the position of the three chemically and magnetically equivalent methyl protons is affected by the nuclear spin of the two methylene protons. These can be both parallel, both antiparallel, or one parallel, one antiparallel to the external field. There are four resultant energy levels, of which the two with opposing spins are degenerate. Because of the equal population probabilities for the individual spin states the coupling from the methylene protons results in a triplet signal for the methyl

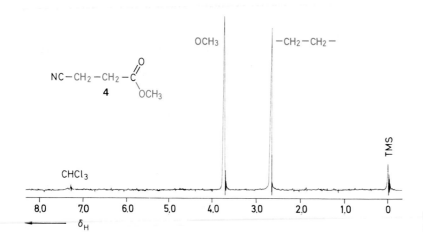

Fig. 3.4 ^1H NMR spectrum (60 MHz) of methyl 3-cyanopropionate (**4**) in $CDCl_3$

protons, with an intensity distribution of 1:2:1. The chemical shift is given by the **centre of gravity** of the signal: 1.67 ppm. In an exactly analogous way the local field at the position of the methylene protons is influenced by spin-spin coupling from the methyl protons. For the three protons of the CH_3 group there are eight spin combinations. The one of lowest energy has a total spin $m=3/2$, the one of highest energy $m=-3/2$; in between are three degenerate states with $m=+1/2$ and three degenerate states with $m=-1/2$. The methylene group therefore produces from the coupling with the methyl group a quartet with the intensity distribution 1:3:3:1 (Fig.3.5). The centre is at $\delta=3.43$ ppm. The separation of the lines in the triplet and the quartet is equal to the coupling constant J.

Tab. 3.2 shows the splitting patterns for the signal of a nucleus (or group of magnetically equivalent nuclei) in first order spectra, as they vary with the number of coupling partner nuclei.

The **chemical shifts** of the nuclei in an A_nX_m spin system are given by the centres of the multiplets. The coupling constant J_{AM} can be directly measured in Hz from the separation of any two neighbouring lines in either the A part or the X part of the spectrum. (cf. Fig. 3.5). As a further example the ^{13}C spectra of chloroform compared to deuteriochloroform and of dichloromethane compared to $CDHCl_2$ and CD_2Cl_2 will be considered. For $CHCl_3$ a doublet is obtained, for $CDCl_3$ a triplet (Fig. 3.6). Because of the isotope effect there is a small difference in

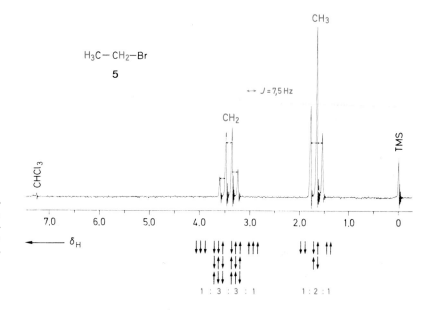

Fig. 3.5 ^{1}H NMR spectrum of ethyl bromide (**5**) in $CDCl_3$. (The splitting of the methyl signal into a triplet and the methylene signal into a quartet is explained by the spin orientations of the coupling protons in the neighbouring groups)

Tab. 3.2 Coupling patterns in first order spectra caused by spin-spin interactions

Number of coupling neighbouring nuclei with spin $I=1/2$	$I=1$	Number of lines (Signal multiplicity)	Relative intensities[a]
0		1 (Singlet)	1
1		2 (Doublet)	1 : 1
2		3 (Triplet)	1 : 2 : 1
3		4 (Quadruplet, Quartet)	1 : 3 : 3 : 1
4		5 (Quintublet, Quintet)	1 : 4 : 6 : 4 : 1
5		6 (Sextet)	1 : 5 : 10 : 10 : 5 : 1
6		7 (Septet)	1 : 6 : 15 : 20 : 15 : 6 : 1
	0	1 (Singlet)	1
	1	3 (Triplet)	1 : 1 : 1
	2	5 (Quintublet, Quintet)	1 : 2 : 3 : 2 : 1
	3	7 (Septet)	1 : 3 : 6 : 7 : 6 : 3 : 1

[a] When $I=1/2$ the relative intensities are the binomial coefficients which can be calculated from Pascal's triangle

the chemical shifts. The intensity ratios are 1:1 for $CHCl_3$ and 1:1:1 for $CDCl_3$. These ratios can be directly taken from Tab. 3.2, bearing in mind that protons have a spin of 1/2 and deuterons a spin of 1. Very noticeable is the difference in the coupling constants. The (C–H) coupling is larger by a factor which is very nearly equal to the ratio of the gyromagnetic ratios: $\gamma_H/\gamma_D \approx 6.5$.

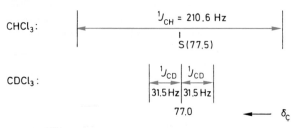

$CHCl_3$:

$^1J_{CH} = 210.6$ Hz

$S(77.5)$

$CDCl_3$:

$^1J_{CD}$ | $^1J_{CD}$

31.5 Hz | 31.5 Hz

77.0 $\longleftarrow \delta_C$

Fig. 3.6 ^{13}C stick spectra of $CHCl_3$ and $CDCl_3$ (taking into account the (C–H) and (C–D) coupling)

In the coupled ^{13}C spectrum CH_2Cl_2 gives a 1:2:1 triplet ($\delta_C = 53.8$, $^1J_{CH} = 177.6$ Hz), CD_2Cl_2 a 1:2:3:2:1 quintet and $CHDCl_2$ a doublet of triplets with six lines of equal intensity. The rule is:

If a nucleus A or a set of magnetically equivalent nuclei A_n couples with two sets of neighbouring nuclei M_m and X_l, the multiplicity of the signal of A is the product of the multiplicities caused by M and X, i.e. for spin 1/2 nuclei $(m+1) \cdot (l+1)$.

The doublet in the A part of an AM spectrum for example becomes a doublet of doublets from additional AX coupling. If by coincidence $J_{AM} = J_{AX}$ then two of the four lines fall together, and a 1:2:1 triplet is obtained (Fig. 3.7).

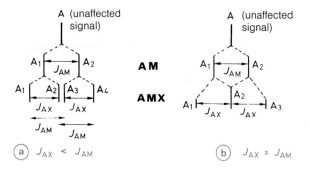

A (unaffected signal)

A_1 ⎍ A_2
J_{AM}

A M

A_1 ⎍ A_2 A_3 ⎍ A_4
J_{AX} J_{AX}
J_{AM} J_{AM}

A M X

(a) $J_{AX} < J_{AM}$

A (unaffected signal)

A_1 ⎍ A_2
J_{AM}

A_1 ⊢ A_2 A_3
J_{AX} J_{AX}

(b) $J_{AX} = J_{AM}$

Fig. 3.7 Coupling pattern of the A part of an AMX spectrum

Generally the $(m+1) \cdot (l+1)$ lines of the A part of an $A_nM_mX_l$ spin system form an $(m+l+1)$ multiplet if $J_{AM} = J_{AX}$. The A nuclei behave as if they were "seeing" $(m+l)$ magnetically equivalent neighbouring nuclei. Consider for example the iso-

propyl and *n*-propyl groups in the structurally isomeric nitropropanes (**6**) and (**7**).

$H_3C-CH-CH_3$
⎮
NO_2

6

$H_3C-CH_2-CH_2-NO_2$
γ β α

7

2-Nitropropane (**6**) forms an A_6X spin system. The six methyl protons are chemically and magnetically equivalent. The coupling with the methine proton splits their signal into a doublet. The methine proton itself appears as a septet at lower field (Fig. 3.8).

The protons in 1-nitropropane (**7**) form an $A_3M_2X_2$ spin system, assuming magnetic equivalence of the protons within each methylene group (but see Sect. 2.2, p. 88). The methyl protons A and the α-CH_2 protons X each have the two protons of the β-methylene group as neighbours. For the A and X signals a triplet is therefore expected, and for the M protons a dodecuplet $(12 = (3+1) \cdot (2+1))$. Since the coupling constants $^3J_{AM}$ and $^3J_{AX}$ however are in practice of equal magnitude, a sextet is observed for the β-methylene group of (**7**) (see Fig. 3.9).

If a molecule has two homonuclear sets of nuclei A_nB_m, where the quotient $|\nu_A - \nu_B|/J_{AB}$ is less than 10, then the rules for **first order** spectra lose their validity. In Fig. 3.9 one can already see that the intensities of the lines in the two triplets are no longer exactly in the ratio 1:2:1. The lines which lie nearer to the signal of the coupling partners (βCH_2 group) are more intense than those that lie further away. This phenomenon is referred to as the **roof effect**. Since

$$\frac{|\nu_A - \nu_M|}{J_{AM}} < \frac{|\nu_X - \nu_M|}{J_{XM}}$$

the roof effect is more marked in the methyl triplet than in the methylene triplet. For the same reason the overriding roof effect in the β-methylene group is that towards the methyl triplet. In complicated spectra the roof effect can help to identify the coupling partners. **Second order** spectra can also occur with $\Delta\nu/J > 10$ in those cases where sets of nuclei are present which are chemically equivalent, but magnetically non-equivalent (e.g. $AA'XX'$). In general spectra can be classified as **zero order** (only singlet signals), **first order**, and **second order**.

The simplest second order spectrum is produced by the AB system. As in the AX case there are four lines (Fig. 3.10).

The spectrum shows the same symmetry about the centre $\frac{1}{2}(\nu_A + \nu_B)$ as the AX system, and the separation of the two A or B lines is also equal to the coupling constant J_{AB}. However the intensities of the lines are fundamentally different for AX

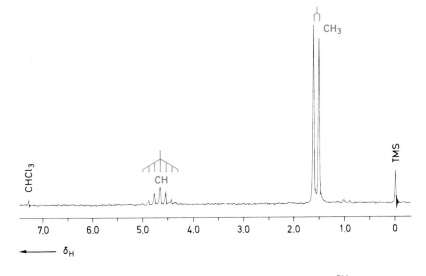

Fig. 3.8 ¹H NMR spectrum of 2-nitropropane (**6**) in CDCl₃

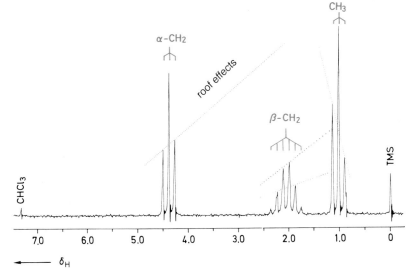

Fig. 3.9 ¹H NMR spectrum of 1-nitropropane (**7**) in CDCl₃

and AB systems. As Fig. 3.10 shows, the appearance of the spectrum depends on the ration $\Delta v/J$. For Δv near 0 the appearance of the spectrum approaches the singlet of an A_2 system. The opposite limit, when Δv is large, approaches an AX spectrum.

The analysis of an AB spectrum will be demonstrated with the example of (*E*)-cinnamic acid (**8**) (Fig. 3.11). The olefinic protons form an AB system (long range coupling to the carboxyl proton or the phenyl protons is not observed at this resolution).

Since
$v_1 - v_2 = v_3 - v_4 = 16\ \text{Hz}$ it follows immediately that $J_{AB} = 16\ \text{Hz}$.

Since the separation
$v_1 - v_3 = v_2 - v_4 = \sqrt{(v_A - v_B)^2 + J^2} = 82\ \text{Hz}$ it follows that $v_A - v_B$
$= 80\ \text{Hz}$ (at 60 MHz).

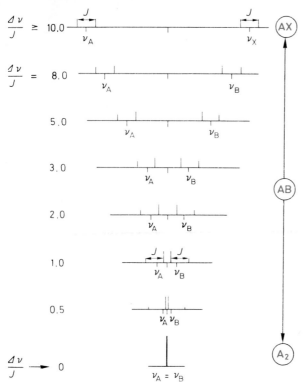

Fig. 3.10 Stick spectra of an AB system with constant coupling J_{AB} and varying ratio $\Delta v/J$

From the known position of the centre point of the spectrum
$$\tfrac{1}{2}(v_A + v_B) = \tfrac{1}{2}(v_1 + v_4) = \tfrac{1}{2}(v_2 + v_3)$$
the positions of v_A and v_B can be calculated. On the δ scale the values obtained are $\delta_A = 7.82$ and $\delta_B = 6.47$. The assignment of these two signals to the two different olefinic protons cannot be made from the spectrum alone; it requires the use of a system of chemical shift increments or comparison with similar compounds.

Of the possible **three-spin systems** A_3, A_2X, A_2M, A_2B, AMX, ABX, and ABC the A_2B, ABX, and ABC cases cannot be treated according to first order rules. The maximum possible number of lines in these three systems is 9, 14, and 15 respectively, though some of these are often weak transitions which do not appear in routine spectra. (Spin systems AXX' or ABB' only occur when two chemically non-equivalent nuclei X or B are coincidentally isochronous, i.e. have the same chemical shift).

For the exact analysis of three- or more-spin systems the reader should refer to NMR textbooks. In cases which are only weakly second-order it is often possible to make an approximate analysis on the basis of first order rules. An example is the 1H NMR spectrum of phenyloxirane (styrene oxide **9**, Fig. 3.12). It has the ABM system. If it is treated as an AMX case, the parameters obtained (chemical shifts and coupling constants) deviate only insignificantly from those obtained by an exact analysis as an ABX case. (The analysis of an ABX case will be described with reference to a heteronuclear example in Sect. 6.2, p. 204 ff.)

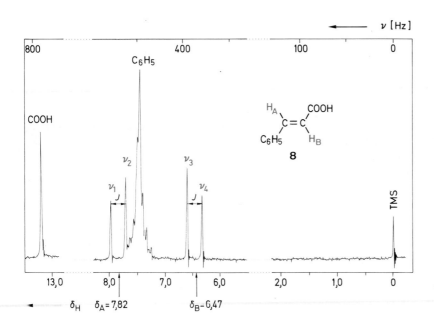

Fig. 3.11 60 MHz ^{1}H NMR spectrum of (E)-cinnamic acid (**8**) in CDCl$_3$ ($v_1 = 478$ Hz, $v_2 = 462$ Hz, $v_3 = 396$ Hz, $v_4 = 380$ Hz, $v_A = 469$ Hz, $v_B = 388$ Hz)

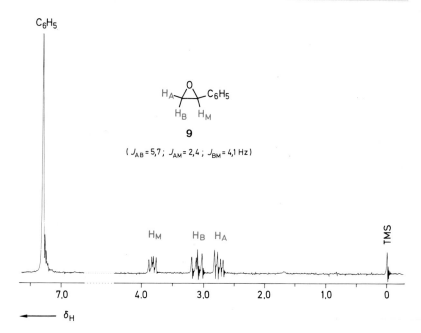

Fig. 3.12 ^{1}H NMR spectrum of phenyloxirane (**9**) in CDCl$_3$

As shown on p. 78 an AMX system with three different coupling constants has four lines for each nucleus A, M, and X. (In an ideal case all twelve lines should have the same intensity.) The coupling constants can be obtained directly from the frequencies of the individual lines (see Sect. 3.7).

Of the **higher spin systems** only a few symmetrical **four-spin systems** will be considered further (Tab. 3.3).

Tab. 3.3 Symmetrical spin systems with two sets of nuclei

System	chemical shifts		couplings	transitions, max. number of lines[a]
1. order				
A$_2$X$_2$	v_A	v_X	J_{AX}	6
higher order				
AA'XX'	v_A	v_X	$J_{AA'}$, J_{AX}, $J_{AX'}$, $J_{XX'}$	20
A$_2$B$_2$	v_A	v_B	J_{AB}	16 (18)
AA'BB'	v_A	v_B	$J_{AA'}$, J_{AB}, $J_{AB'}$, $J_{BB'}$	24 (28)

[a] for spins with $I = 1/2$

For a better understanding of these spin systems the following examples have been selected, and arranged so that the chemical shift difference decreases from left to right.

1,2-Disubstituted ethanes like 3-nitropropionitrile (**13**) deserve special attention. While the examples cyclopropene (**10**), methane (**12**), thiophene (**14**), and ethylene (**15**) are "rigid" molecules, rotation about the C—C bond must be considered for 1,2-disubstituted ethanes. This results in the chemical equivalence of the two protons labelled A as well as the two protons labelled X. The rotation does not however necessarily lead to magnetic equivalence, i.e. $^3J_{AX}$ and $^3J_{AX'}$ can be different (see Sect. 2.2, p. 88). Whether (**13**) is an AA'XX' or an AA'BB' system then depends on the difference $|v_A - v_X|$.

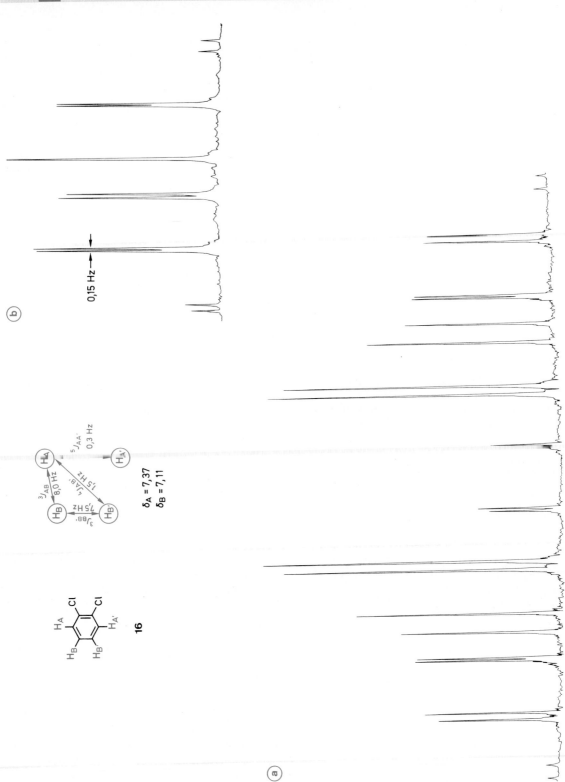

Fig. 3.13 High resolution ^{1}H NMR spectrum of o-dichlorobenzene (**16**) **a** AA'BB' system with 24 lines symmetrically spaced about the centre, measured at 90 MHz; **b** MM' part of an AA'MM' spin system, measured at 400 MHz

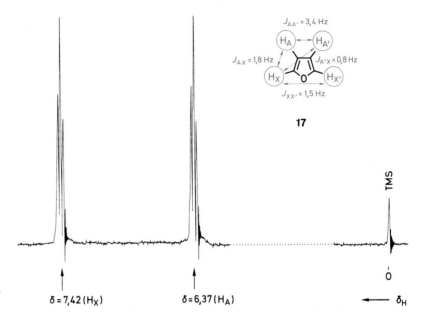

Fig. 3.14 ^{1}H NMR spectrum of furan (**17**) in CDCl$_3$ (AA'XX' system, apparently consisting of two triplets)

$\delta = 7.42\,(H_X)$ $\delta = 6.37\,(H_A)$ δ_H

Fig. 3.13 shows the AA'BB' spectrum of *o*-dichlorobenzene under conditions of high resolution. 24 lines can be identified, thus allowing all the parameters to be determined. In many cases of four-spin systems the spectra obtained under routine conditions show remarkably few lines. An example is the ^{1}H spectrum of furan (**17**; Fig.3.14), an AA'XX' spin system. The interpretation of the two triplets as arising from an A$_2$X$_2$ system with a single value for J_{AX} would be incorrect. An exact analysis gives $J_{AX} = J_{A'X'} = 1.8$ and $J_{A'X} = J_{AX'} = 0.8$ Hz.

To conclude this section it should be realised that the splitting of an NMR signal depends not only on the resolution of the instrument, but also on the measurement frequency (proportional to the field strength). For example, the quotient $\Delta v/J$, which determines whether or not spectra are first order, increases on going from 60 MHz (1.41 T) to 360 MHz (8.45 T) by a factor of 6; that means that a spectrum which is first order at 360 MHz can be second order at 60 MHz and therefore look quite different! A spin system with chemically equivalent, but magnetically non-equivalent nuclei will give a second order spectrum irrespective of the field (see Sect. 3.13), although the appearance of the spectrum can change considerably as the field is changed.

1.4 Line Widths

Fig. 3.15 shows the typical shape of an NMR signal. The **line width b** measured at half-height is considerably greater than the "natural linewidth" predicted by the Heisenberg uncertainty principle. b depends not only on the inhomogeneity of the magnetic field but also on long range couplings and on the **relaxation times** T_1 and T_2 of the nucleus giving rise to the signal.

Nuclei with an electric quadrupole moment, like ^{14}N, or paramagnetic compounds present in the sample reduce the spin-lattice relaxation time T_1 and therefore broaden the lines. (Paramagnetic compounds themselves are poorly suited to NMR measurements for this reason.) In an analogous fashion a reduction of the spin-spin relaxation time T_2, caused perhaps by an increase in the viscosity of the solution, also increases the line width. A special effect is the **line broadening** caused by **exchange phenomena**. In these cases a distinction must be made between **intermolecular** and **intramolecular processes**. An example of the former is **proton transfer** in carboxylic acids, alcohols, or amines. If the ^{1}H spectrum of aqueous methanol is measured at 0°C two OH signals are observed if the water concentration is less than 5%. At higher water concentrations the proton transfer is accelerated. At first a broadening of the OH signals is observed, and finally in the region where exchange is fast a single OH signal is observed at an averaged chemical shift. As the lines broaden so the coupling of the OH protons with the protons of the methyl groups disappears. In

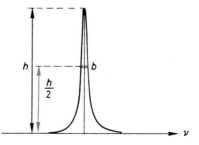

Fig. 3.15 Shape of an NMR signal; Lorentzian curve, h height, b linewidth

pure methanol this latter effect can be produced by raising the temperature from 0°C to +10°C. (Acceleration of proton exchange on warming.)

Intramolecular exchange phenomena arise from **fluxionality of molecules** (rotation, inversion etc.) or chemical reactions (rapid rearrangements, valence isomerism etc.). Examples will be discussed in Sect. 2.2 (p. 88). As a general rule two nuclei, which exchange their chemical environments, will give two separate signals, v_A and v_B if the exchange is slow on the NMR time scale. This is the case if the **lifetime** τ in the two states satisfies the condition $\tau \cdot | v_A - v_B | \gg 1$.

Conversely, if $\tau \cdot | v_A - v_B | \ll 1$ a single, averaged signal is obtained. In the intermediate region, where $\tau \cdot | v_A - v_B | \cong 1/2\,\pi$, the **coalescence** of the two signals is observed. In this region the line shape is very dependent on τ. Since τ is a function of the temperature, the spectra in this region are very temperature dependent. Determination of the coalescence point or more precisely a **lineshape analysis** allows the thermodynamic parameters of such processes to be calculated.

1.5 Intensity

The **area under the absorption curve** of an NMR signal is a measure of the intensity of the transition. The integration is drawn by the spectrometer as a step curve.

In **¹H spectra** the intensity measured by the height of the steps is proportional to the number of ¹H nuclei in the molecule which are responsible for the signal (see Fig. 3.16, ethyl p-toluene-sulfonate, **18**).

In the quantitative analysis of mixtures the number of chemically equivalent protons responsible for each signal must be taken into account. If the area F_A corresponds to n_A protons of substance A (as determined by the structural formula) and the area F_B similarly corresponds to n_B protons of substance B, then the molar concentrations c in the solutions being measured are given by

$$\frac{c_A}{c_B} = \frac{F_A \cdot n_B}{F_B \cdot n_A}$$

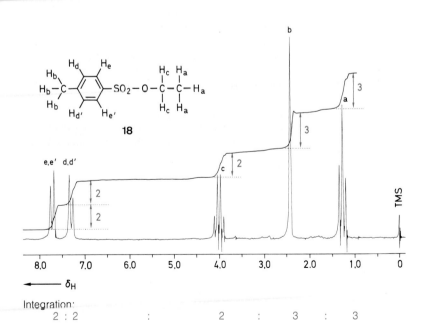

18

Integration:
2 : 2 : 2 : 3 : 3

Fig. 3.16 ¹H NMR spectrum of ethyl p-toluenesulfonate (**18**) (60 MHz, CCl₄) with integration curve

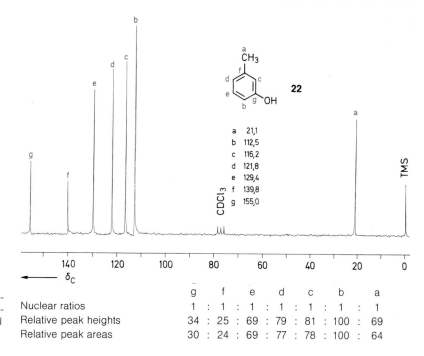

Fig. 3.17 ¹³C NMR spectrum of *m*-cresol (**22**) (in CDCl₃, ¹H broadband decoupled) with an evaluation of the signal intensities

	g		f		e		d		c		b		a
Nuclear ratios	1	:	1	:	1	:	1	:	1	:	1	:	1
Relative peak heights	34	:	25	:	69	:	79	:	81	:	100	:	69
Relative peak areas	30	:	24	:	69	:	77	:	78	:	100	:	64

In routine **¹³C spectra** exact quantitative conclusions cannot be drawn from the intensities of the signals. The intensity of a signal is proportional to the effective difference in populations between the energy levels giving rise to the signals, and therefore depends critically on the relaxation times. The T_1 relaxation times of ¹³C nuclei lie in the region from 10^{-1} to $3 \cdot 10^2$ s. Protons bound directly to the relevant ¹³C nucleus are most effective at reducing T_1. Quaternary ¹³C atoms show the longest T_1's, and consequently their signals are the weakest. Furthermore, molecular size and molecular motions have considerable influence on the longitudinal relaxation. This is clearly shown by the examples of toluene (**19**), styrene (**20**), and polystyrene (**21**).

The time interval between two successive pulses (cf. p. 103) is generally too short in ¹³C NMR to allow relaxation of the spin system to the equilibrium state.

Fig. 3.17 shows the ¹³C spectrum of 3-methylphenol (*m*-cresol **22**). The seven ¹³C signals have quite different intensities.

Irradiation at or near the NMR frequency of a nucleus leads to perturbation of the relaxation of neighbouring nuclei. This leads to a change in the intensities of the signals of these neighbouring nuclei (**nuclear Overhauser effect**). In ¹H NMR this only occurs in double resonance experiments (see p. 131 ff). In ¹³C NMR however routine spectra are measured with ¹H broad-band decoupling. This removes the splitting caused by the coupling to the protons – the spectrum consists of individual singlet peaks (see Fig. 3.2b, p. 74, or 3.17). The heteronuclear nuclear Overhauser effect caused by the decoupling produces increases in intensity of up to 200%. **Differential relaxation and differential nuclear Overhauser effects are therefore the reason for the deviations of measured from theoretical intensity ratios of ¹³C NMR signals.** Furthermore, the large influence of the measurement conditions on peak intensities must also be taken into account. It is however possible to avoid these disadvantages of ¹³C NMR spectroscopy by removing the effects and thus obtaining spectra which can be integrated if necessary (see Sect. "Integration of NMR spectra", p. 169 ff.).

The T_1 values, given in seconds, show marked differences between large and small molecules and between protonated and quaternary C-atoms. The anisotropy of the molecular motions also plays a role. Thus the *para*-carbons, which lie on the preferred axis of rotation, have the lowest T_1 values. (In special cases this effect can be used for signal assignments.)

2. NMR Spectra and Molecular Structure

2.1 Molecules with "Rigid" Atomic Positions

The number of NMR signals which appear in a spectrum is dependant on the symmetry of the molecule under investigation. **Two nuclei in a molecule are chemically equivalent if they can be transposed by a symmetry element present in the molecule or if they become identical in the time average as a result of a fast intramolecular exchange process.** For a thorough understanding a few examples with a rigid carbon skeleton will be considered. Tab. 3.4 gives the number of expected ^{13}C and ^{1}H NMR signals of a few selected structures of various symmetry (point groups). The higher the symmetry of the molecule, the fewer the number of signals. In buckminsterfullerene C_{60} for example all 60 carbon atoms have the same chemical shift ($\delta = 143.2$).

The ways in which ^{13}C and ^{1}H NMR spectroscopy usefully complement each other is demonstrated through the example of some substituted benzenes in Tab. 3.5.

Tab. 3.4 Chemical equivalence of ^{1}H and ^{13}C nuclei in selected structures of various symmetries (point groups)

Structure	Point group	applicable symmetry elements	Groups of chemically equivalent carbon nuclei	Groups of chemically equivalent hydrogen nuclei
1,1,2-Trichlorocyclopropane	C_1	–	C–1 C–2 C–3	H_A H_B H_M (ABM)
trans-1,2-Dichlorocyclopropane	C_2	C_2	C–1, C–2 C–3	H_A, $H_{A'}$ H_M, $H_{M'}$ (AA'MM')
1,2-Dichlorocyclopropane	C_{2v}	2σ (σ, C_2)	C–1 C–2, C–3	H_A, $H_{A'}$, $H_{A''}$, $H_{A'''}$ Singlet
all-cis-1,2,3-Trichlorocyclopropane	C_{3v}	C_3 (3σ)	C–1, C–2, C–3	H_A, H_A, H_A Singlet

Tab. 3.4 Chemical equivalence of ^1H and ^{13}C nuclei in selected structures of various symmetries (point groups)

Structure	Point group	applicable symmetry elements	Groups of chemically equivalent carbon nuclei	Groups of chemically equivalent hydrogen nuclei
Fumaric acid dinitrile	C_{2h}	$i \equiv S_2$ (C_2)	C–1, C–4 C–2, C–3	H_A, H_A Singlet
r-1,t-3-Dibromo-c-2,t-4-dichlorocyclobutane	C_i	$i \equiv S_2$	C–1, C–3 C–2, C–4	H_A, $H_{A'}$ H_B, $H_{B'}$ (AA'BB')
1,2,6-Trichlorobicyclo-[2,2,2]octa-2,5-diene-8,8-dicarboxylic acid	C_s	σ	C–1 C–2, C–6 C–3, C–5 C–4 C–7 C–8 C–9, C–10	H_A H_x, H_x (AX_2) H_B, H_B Singlet H_y, H_y Singlet Long range couplings ignored
Allene	D_{2d}	2σ, S_4	C–1, C–3 C–2	H_A, $H_{A'}$, $H_{A''}$, $H_{A'''}$ Singlet
Naphthalene	D_{2h}	2σ (3 C_2)	C–1, C–4, C–5, C–8 C–2, C–3, C–6, C–7 C–4a, C–8a	H_A, $H_{A'}$, $H_{A''}$, $H_{A'''}$ H_B, $H_{B'}$, $H_{B''}$, $H_{B'''}$ (AA'A''A'''–BB'B''B''')
1,3,5,-Trifluorobenzene	D_{3h}	C_3 (3σ, 3C_2)	C–1, C–3, C–5 C–2, C–4, C–6	H_A, $H_{A'}$, $H_{A''}$ (A-part of AA'A''XX'X'')

Tab. 3.5 Number of ^{13}C signals and spin systems in the ^1H spectra of disubstituted benzenes

Arrangement	^{13}C signals (no coupling)		^1H signals (spin system)
Same substituents			
o	3 $\begin{cases} C_1 = C_2 \\ C_3 = C_6 \\ C_4 = C_5 \end{cases}$		AA'BB'
m	4 $\begin{cases} C_1 = C_3 \\ C_2 \\ C_4 = C_6 \\ C_5 \end{cases}$		AB$_2$C
p	2 $\begin{cases} C_1 = C_4 \\ C_2 = C_3 = \\ C_5 = C_6 \end{cases}$		AA'A''A''' Singlet
Different substituents			
o	6 different carbons		ABCD
m	6 different carbons		ABCD
p	4 $\begin{cases} C_1 \\ C_2 = C_6 \\ C_3 = C_5 \\ C_4 \end{cases}$		AA'BB'

2.2 Intramolecular Motion

As was stated at the beginning of this section, nuclei which are not related by a symmetry element can be chemically equivalent if they become identical as a result of an intramolecular motion. For example the three protons of a freely rotating methyl group are chemically equivalent. Alanine (**23**) possesses no symmetry element, and therefore certainly no C_3 axis through the methyl carbon atom; nevertheless the three protons H_A are identical because of the rotation of the methyl group. The A_3X system of the four protons bound to carbon atoms gives in D_2O a doublet at $\delta = 1.48$ and a quartet at $\delta = 3.78$.

23

In the same way the tertiary butyl group of a compound (**24**) shows a single ^1H signal (singlet) and two ^{13}C signals for the quaternary and the three primary C atoms. Deviations from this behaviour occur when the rotation of the t-butyl group is restricted. There are only a few examples of the freezing out of a methyl rotation; thus 9-methyltrypticene-1,4-quinone (**25**) shows anisochronous methyl protons at $-141°C$.

24   **25**

For CX_2 groups ($X = H, CH_3$ etc.) the situation is more complicated. Rotation does not always average the differences in the chemical environments of the two protons of a methylene group. Consider for example the compounds (**26**).

26

For $Y = a$ or $Y = b$ there are conformations with a plane of symmetry.

If the CH_2R group rotates, the chemical environments of the two methylene protons become different. The change for H^1 on rotation in a particular direction is exactly the same as the change for H^2 on rotation in the opposite direction. Independent of the population of the various rotational conformers H^1 and H^2 are therefore chemically equivalent, and give a singlet if there is no coupling. If $Y \neq a,b$ this is not possible. H^1 and H^2 are not identical in any of the conformers and form an AB system. Such protons are described as being **diastereotopic**. This is also true for the case where $Y = CH_2R$. The presence of neighbouring **chiral** or **prochiral** groups can therefore lead to the non-equivalence of the protons of a methylene group. (If however two methylene protons can be interchanged by a symmetry operation which applies to the whole molecule, then chiral or prochiral groups do not effect the chemical equivalence!) The same considerations apply to the X signals of a CX_2 group, e.g. for the ^{13}C and ^1H signals of the two methyl groups of an isopropyl group.

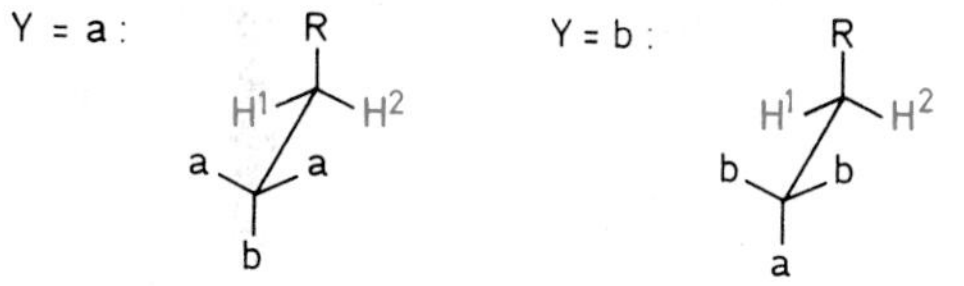

Y = a:

Y = b:

27

28

In general there are the following possibilities for two X groups in a molecule; **homotopic, enantiotopic,** and **diastereotopic.** The three cases are distinguished by considering the substitution of each of the protons in turn by a hypothetical achiral group T, not already present in the molecule, and comparing the molecules X|T and T|X so formed

29

30

	X / X	
homotopic	**enantiotopic**	**diastereotopic**
if X|T ≡ T|X	if X|T and T|X are optical isomers (enantiomers)	if X|T and T|X are diastereomers

Since enantiotopic groups are interconverted by rotation-reflection axes S_n (including $S_1 \equiv \sigma$ and $S_2 \equiv i$) they can only occur in achiral molecules. Chiral molecules can contain diastereotopic groups and, if rotation axes are present, homotopic groups.

- **Homotopic groups are chemically equivalent and always lead to a single signal for each nuclear type.**

- **Enantiotopic groups give isochronous signals in achiral or racemic media,** anisochronicity may be observed in a chiral medium

- **Diastereotopic groups are chemically non-equivalent, and can only be isochronous by coincidence.**

For a better understanding of these situations a few examples will be considered. Whereas the three methyl protons in alanine (**23**) are chemically equivalent, the two diastereotopic protons in phenylalanine (**27**) form the AB part of an ABC system.

In valine (**28**) and leucine (**29**) the asymmetric carbon atom, a chiral centre, causes the methyl groups to be non-equivalent in the ^1H and ^{13}C spectra. As the distance between the isopropyl group and the chiral centre increases, so the differences in chemical shift decrease. Thus for cholesterol (**30**) the ^{13}C spectrum does indeed show two separate methyl signals for the isopropyl group, but the ^1H spectrum does not distinguish between them.

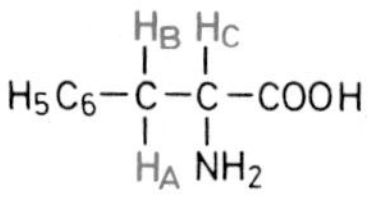

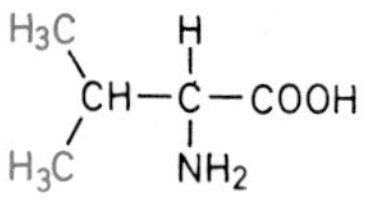

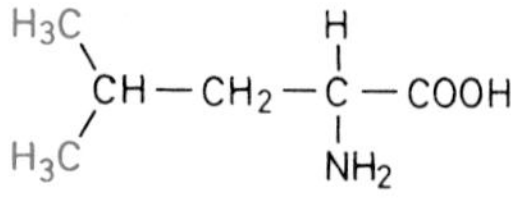

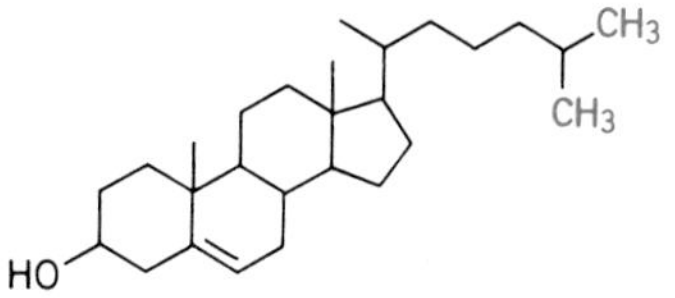

2,4-Diaminoglutaric acid (**31**) exists as two enantiomers and an achiral *meso*-form. The two chiral molecules possess a C_2 axis in the conformation drawn, and the two methylene protons are therefore chemically equivalent.

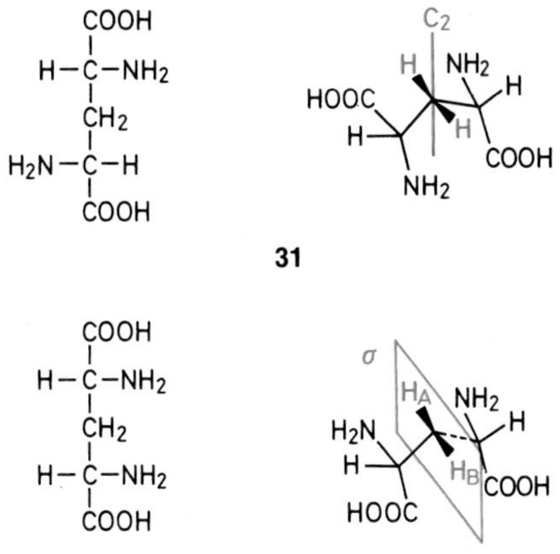

31

In contrast in the *meso*-form the CH_2 group forms the AB part of an ABC_2 system.

Finally a prochiral case will be considered. In glycerine (**32a**) or in citric acid (**32b**) for example the two CH_2 groups are enantiotopic. The two H atoms within a methylene group are however diastereotopic ($AA'BB'C$ or AB system, respectively).

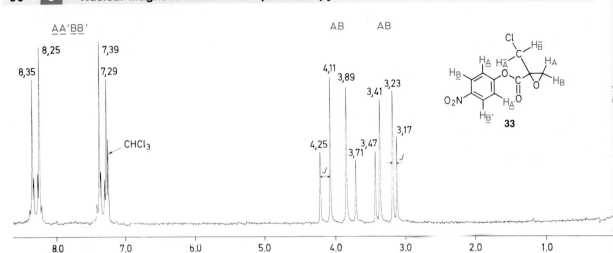

Fig. 3.18 ¹H NMR spectrum of the *p*-nitrophenyl ester of 1-chloromethyloxirane-1-carboxylic acid (**33**) in CDCl₃

The unhindered rotation of phenyl groups fundamentally leads to the equivalence of the two *o*- and of the two *m*-protons. This also applies in the presence of chiral centres. The spectrum of the compound (**33**) is shown as an example.

The two rigidly bound protons on the oxirane ring form an AB system with its centre at $\delta = 3.32$. The rotating CH₂Cl group bound to the chiral centre similarly gives an AB system (centre at $\delta = 4.00$). Rotation of the *p*-nitrophenyl group by 180° reproduces the same structure, the relevant protons and C-atoms are identical in the time average. In the ¹³C spectrum there are four signals for aromatic carbons and in the ¹H spectrum an AA'BB' pattern (Fig. 3.18).

The non-equivalence of the two protons in a methylene group attached to a chiral or suitable prochiral centre is also of relevance for amines. In a "rigid" compound of type (**34**) the two protons of the methylene group are diastereotopic. This is also true for the case where R² = CH₂R¹. As a result of the rapid **inversion at the nitrogen atom** they become chemically equivalent (enantiotopic). The inversion can be slowed down by including the nitrogen atom in a ring. A typical example is 1-ethylaziridine (**35**), the ring protons of which give an AA'BB' system at room temperature. Only above 100°C does the inversion at nitrogen lead to chemical equivalence (AA'A"A'''). (The two methylene protons of the ethyl group are equivalent even without the inversion!)

A further way of slowing the inversion at nitrogen atoms in amines is by protonation.

$$R^1-\overset{\displaystyle H}{\underset{\displaystyle H}{C}}-\overset{\displaystyle R^2}{\underset{\displaystyle R^3}{N}} \qquad (R^1,R^2,R^3 \neq H; \; R^2 \neq R^3)$$

34

35

In the head-to-tail polymerisation of vinyl monomers RCH=CH₂ or RR'C=CH₂ a chain is formed in which every second carbon atom is a chiral centre. There are three distinguishable types of **tacticity**.

isotactic: sequence of identical configurations
syndiotactic: regular alternation of configuration
atactic: random (statistical) sequence

Shown below is an atactic polymer chain (36) which contains the isotactic and syndiotactic **diads D** and **triads T** and the heterotactic triads **T**. In the syndiotactic triads the methylene protons are homotopic from a consideration of the local symmetry. They are therefore observed in a syndiotactic polymer as a singlet. In an isotactic polymer on the other hand they form an AB system. In both cases the groups R and R' are chemically equivalent. For atactic polymers it is usually sufficient to consider triads for R and R', whereas for the methylene protons the local symmetry must be determined at the tetrad level. The spectrum is therefore composed of elements arising from iso- and syndiotactic triads and their "connecting links" the heterotactic triads. The region of the *geminal* protons in particular can become very complex.

Let us consider the three energetically preferred "staggered" conformations (38a), (38b), and (38c) with relative populations p_1, p_2, and p_3 ($p_1 + p_2 + p_3 = 1$).

In (38a) the substituents X and Y are *anti* to each other. The molecule has a plane of symmetry. The protons form an AA'BB' system with two chemical shifts ν_A and ν_B and four coupling constants. The two conformations (38b) and (38c) form a pair of enantiomers with identical spectra (in an achiral medium) corresponding to the ABCD system (four chemical shifts, six coupling constants). Moreover $p_2 = p_3$, since (38b) and (38c) have the same energy.

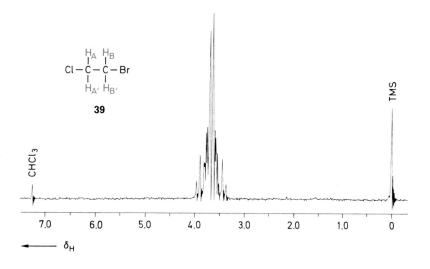

| | 38a (p_1) | 38b (p_2) | 38c ($p_3 = p_2$) |

If rotation is fast (38) shows a spectrum with averaged shift and coupling parameters (the blue numbers identifying the H atoms are retained during the rotation). The averaged chemical shift of H^1 for example is given by

$$\nu_1 = p_1 \nu_A + p_2 \nu_{\bar{B}} + p_2 \nu_{\bar{A}}$$

and for H^2 by

$$\nu_2 = p_1 \nu_{A'} + p_2 \nu_{\bar{A}} + p_2 \nu_{\bar{B}}.$$

Since $\nu_A = \nu_{A'}$ (plane of symmetry) $\nu_1 = \nu_2$. Similarly $\nu_3 = \nu_4$. The question now is whether (38) should be classified as an A_2B_2 spin system with **one** coupling constant $^3J_{AB}$ or as an AA'BB' system with $^3J_{AB} + ^3J_{AB'}$. To decide this question we must com-

Following the description of **chemical equivalence** of the two protons of a methylene group we will now consider their **magnetic equivalence**. While ethyl compounds (37) without steric hindrance form A_3B_2, A_3M_2, or A_3X_2 systems depending on the chemical shift difference between the CH_3 and CH_2 groups, it is at first sight surprising that 1,2-disubstituted ethanes (38) form AA'BB', AA'MM', or AA'XX' systems.

X–CH₂–CH₃

37 (X ≠ H)

X–CH₂–CH₂–Y

38 (X ≠ Y ≠ H)

Cl–C–C–Br with H_A, H_B and $H_{A'}$, $H_{B'}$

39

CHCl₃   TMS

7.0   6.0   5.0   4.0   3.0   2.0   1.0   0

Fig. 3.19 ¹H NMR spectrum of 1-bromo-2-chloroethane (**39**) in CDCl₃

δ_H

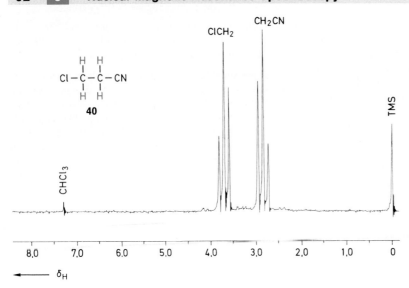

Fig. 3.20 ¹H NMR spectrum of 3-chloro-propiononitrile (**40**) in CDCl₃

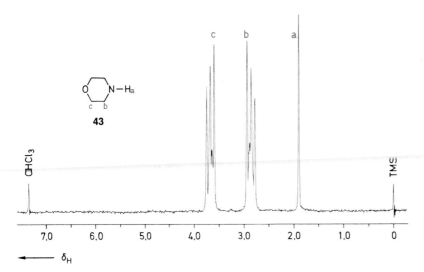

Fig. 3.21 ¹H NMR spectrum of morpholine (**43**) in CDCl₃ at room temperature

pare the coupling between H¹ and H³ with that between H¹ and H⁴.

$$^3J_{1,3} = p_1 J_{AB} + p_2 J_{BC} + p_2 J_{\bar{A}D}$$
$$^3J_{1,4} = p_1 J_{AB'} + p_2 J_{\bar{B}D} + p_2 J_{\bar{A}C}$$

In contrast to the chemical shifts the *vicinal* coupling constants do not necessarily average to the same value on rotation. $^3J_{1,3}$ can be different from $^3J_{1,4}$; that means that (**38**) must be described as an AA'BB' system. A nice example is the spectrum of 1-bromo-2-chloroethane (**39**) reproduced in Fig. 3.19. $^3J_{1,3}$ and $^3J_{1,4}$ can however be so similar, that a simpler ¹H spectrum is obtained. An example is 3-chloropropionitrile (**40**)

(Fig. 3.20). One should however take care not to assume from the 60 MHz spectrum that (**40**) has an A₂M₂ spin system.

In ring systems it is necessary to consider the temperature dependent **ring inversion** before deciding on the chemical equivalence of the various nuclei. The ¹H NMR spectrum of cyclohexane for example shows a singlet signal at $\delta = 1.43$ at room temperature. *Axial* and *equatorial* protons are equivalent as a result of the rapid inversion. On cooling this process becomes slower and is eventually frozen out. In monosubstituted cyclohexanes (**41**) this leads to the appearance of isomers with the substituent in *axial* and *equatorial* positions. Similar considerations apply to the ¹³C spectrum of 7,7-dimethylcycloheptatriene (**42**). The methyl C atoms are only chemically equivalent at temperatures where the ring inversion is rapid.

41

42

Chemical **and** magnetic equivalence can be discussed with reference to morpholine (**43**) as an example.

43

This system can be considered as a further example of the systems found in 1,2-disubstituted ethanes (**38**) above. In the chair form (**43**) has an ABCD spin system, if transannular couplings are ignored. The rapid ring inversion at room temperature converts this into an AA'MM' system, i.e. the two protons of each methylene group become chemically, but not magnetically equivalent.

In the cases discussed so far the rapid ring inversion at room temperature is slowed down by cooling, which can be studied by the measurement of low temperature spectra. 4,3,7,8-Dibenzocyclooctyne (**44**) exists at room temperature in the chiral C_2 configuration. The aliphatic protons form an AA'BB' system (Fig. 3.22). On warming the ring inversion becomes apparent. The signals first become broader, combine at the **coalescence temperature** (112°C), and at 145°C in the region of rapid ring inversion form a singlet. H_A and H_B exchange so quickly, that a single, averaged signal is obtained in the NMR spectrum. The coupling of the protons no longer affects the spectrum. The ring inversion in this case is equivalent to the racemisation of the compound.

Further, very interesting examples where NMR is of use in the study of the flexibility of rings are the annulenes (see p. 105).

Alongside "sterically" hindered rotation about σ-bonds and "pseudorotation" in rings another important example is the **rotation about bonds with partial double bond character**.

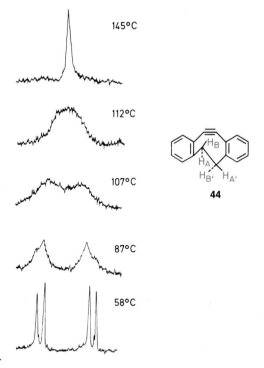

44

Fig. 3.22 Temperature dependent 90 MHz ¹H NMR spectra of (**44**) in deuteriobromoform (from Meier H., Gugel, H., Kolshorn, H. (1976), Z. Naturforsch. B, **31**, 1270)

Typical cases are groups

$$D-Z\overset{A}{\underset{\diagdown}{\parallel}} \quad \longleftrightarrow \quad \overset{+}{D}=Z\overset{A^-}{\diagup}$$

An electron acceptor (A = O,S) and an electron donor D are bound to a central atom (Z = C,N). As a result of the participation of the dipolar resonance form the D⋯Z bond (N⋯C, C⋯C, O⋯C, C⋯N, N⋯N, O⋯N) shows restricted rotation.

At room temperature the two methyl groups of dimethylformamide (**45**) and dimethylnitrosamine (**46**) for example are already chemically non-equivalent.

46

¹H-Resonance
a: $\delta = 3.0$
b: $\delta = 3.8$

¹³C-Resonance
a: $\delta = 32.6$
b: $\delta = 40.5$

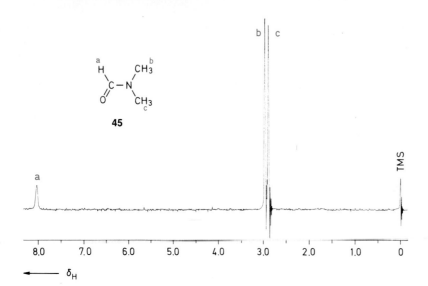

Fig. 3.23 ¹H NMR spectrum of dimethyl-formamide (**45**) at room temperature (above 120°C a singlet is obtained for the methyl groups)

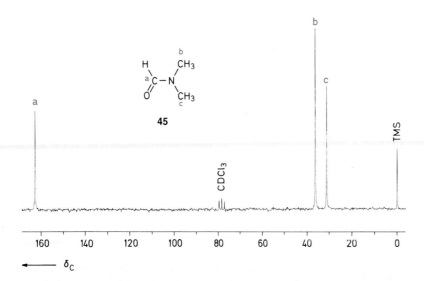

Fig. 3.24 ¹³C NMR spectrum of dime-thylformamide (**45**) in CDCl₃ (¹H broad band decoupled)

Amides with only one substituent on the N atom (e.g. *N*-ethylacetamide, **47**) exist exclusively or predominately in the form shown, with the substituent and the carbonyl O atom mutually (*Z*).

In enamines too the (C–N) rotation can be frozen out at room temperature, as the example of 3-dimethylamino-1,2-dihydropentalene (**48**) shows:

	¹H-Resonance	¹³C-Resonance
a:	$\delta = 1.1$	a: $\delta = 14.6$
b:	$\delta = 3.2$	b: $\delta = 33.7$
c:	$\delta = 8.2$	CO: $\delta = 169.5$
d:	$\delta = 2.0$	d: $\delta = 22.5$

47

48

Position	δ (^{1}H-Resonance)	δ (^{13}C-Resonance)
1	2.91	22.9
2	3.13	38.7
3	–	163.6
3a	–	122.4
4	6.21	107.3
5	6.68	129.4
6	5.91	106.3
6a	–	147.5
CH_3	3.20/3.32	41.1/41,5

2.3 Chemical Exchange Processes

Alongside internal molecular motions inter- and intramolecular chemical processes are also important for the determination of equivalence of nuclei.

Rearrangement reactions are generally so slow on the NMR time scale that both isomers can be observed in the spectrum. The same applies to **tautomerism**, as shown by the example of acetylacetone (**49**) (Fig. 3.25a and b).

At room temperature in $CDCl_3$ the keto form (**49a**) is present in about 14% concentration and the enol form (**49b**) in about 86%. Only on raising the temperature does the reversible prototropy between C and O become rapid enough to result in the observation of averaged signals. From the observed spectra it can further be seen that (**49a**) and (**49b**) both possess a similar symmetry element which causes the chemical equivalence of the methyl groups and the carbonyl C atoms. There are two possible explanations for this. The acidic proton could shift between the two O atoms (with simultaneous shift of the double bonds) so quickly, that the enol form *de facto* appears to be symmetrical. The other possibility is a change of the coordina-

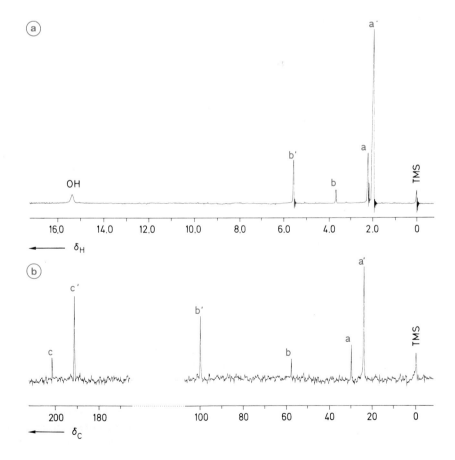

Fig. 3.25 NMR spectra of acetylacetone (**49**) in $CDCl_3$ at room temperature
a ^{1}H NMR spectrum
b ^{13}C NMR spectrum

tion position of the proton on the mesomeric β-diketonate. The temperature dependence of the ^1H, ^{13}C, and ^{17}O spectra of the enol form (even in unsymmetrical cases) supports the resonance model, in which the dynamic phenomenon relates only to the proton and not to the chain.

A similar question arises with tropolone (**50**). In the ^1H spectrum it shows an AA'BB'C spin pattern. If the mobile proton is replaced by a methyl group (**51**) which cannot migrate, an ABCDE spectrum is obtained.

50   **51**

The measurement of tautomers is naturally not confined to equilibrium situations. The proportion of vinyl alcohol (**52b**) in acetaldehyde (**52a**) for example is under the detectable limit for NMR spectroscopy. It can be measured after selective formation of the meta-stable species if the rearrangement is sufficiently slow [$\delta(^{13}C)$ and $\delta(^1H)$]:

52 a   **52 b**

Besides reversible proton shifts, **tautomerism**, reversible electron shifts, **valence tautomerism**, are of especial interest. The "growth" of this type of reaction in recent years would have been unthinkable without NMR spectroscopy. Bullvalene (**53**) will be described as a classic example. It undergoes a degenerate Cope rearrangement between 10!/3 identical isomers. At 120°C this process is so fast that a sharp singlet is obtained for all 10 protons and for all 10 carbons. The term **fluxional structure** is used to describe such situations. Below −60°C the region of slow exchange is reached. Four distinct ^{13}C signals are observed, in line with the symmetry of the rigid structure. (In the ^1H spectrum some signals are coincidentally isochronous.) The **coalescence region** is at room temperature. At 15°C the ^1H spectrum shows a very broad band; the ^{13}C signals are lost in the noise.

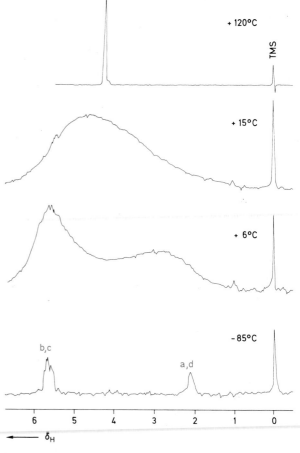

Fig. 3.26 Temperature dependent ^1H NMR spectra of bullvalene (**53**) in carbon disulfide.
(At −85°C the signal of the olefinic protons b and c are observed at low field (6H) and the signal of the three protons a on the three membered ring and the bridgehead proton d at high field (from Schröder, G. et al. (1965) Angew. Chem. **77**, 774)

53

The activation energy for this valence tautomerism is ca 49 kJ·mol^{-1}. (The rearrangement also takes place in the solid state. There however a reorientation of the molecule is also necessary, in order to retain the position of the molecule in the lattice; the activation energy in the solid is 63 kJ·mol^{-1}.)

Fluxional systems must be clearly distinguished from **resonance systems**. If we compare a monosubstituted cyclooctatetraene (**54**) and a monosubstituted benzene (**55**):

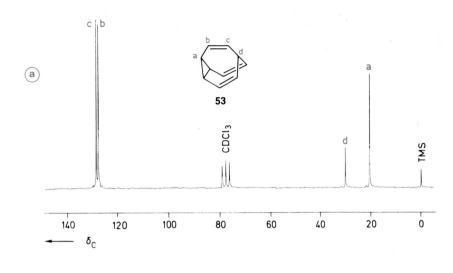

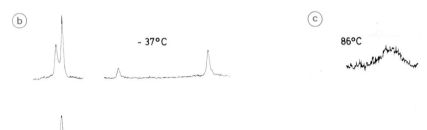

Fig. 3.27 Temperature dependent
^{13}C NMR spectra of bullvalene (**53**)
from Günther, H., Ullmen, J. (1974),
Tetrahedron **30**, 3781)
a Spectrum at –62°C in CDCl₃,
   broad band decoupled
b Spectra in the region of slow
   exchange (–27 .... –10°C)
c Spectra in the region of fast
   exchange (+86 .... +128°C)

At room temperature the double bonds in cyclooctatetraene oscillate. There is a simultaneous inversion of the tub-shaped ring. Both processes are fast on the NMR time scale, so that instead of eight different ring C-atoms and seven different H-atoms only five ^{13}C signals and four ^1H signals are observed ($H_A = H_{\bar{A}}$, $H_B = H_{\bar{B}}$, $H_C = H_{\bar{C}}$).

In the benzene derivative (**55**) the protons *ortho*- to the substituent R are chemically equivalent, as are the *meta*- protons. The same applies to the corresponding C atoms. This is an example of the **static** phenomenon of resonance in these planar molecules and not a **dynamic** process as in the case of cyclooctatetraene. On cooling (**54**) the movement of the double bonds, and independently also the ring inversion, become slower and eventually freeze out.

In (**55**) there is no such temperature dependence of the NMR spectra. This difference is clarified by the energy diagrams in Fig. 3.28. As well as H and C, heteroatoms can of course also be involved in exchange processes. An example are the furoxanes (**56**).

In ¹H NMR the ABCD spin system of benzofuroxan changes to an AA'BB' spin system on warming; in the ¹³C NMR the six signals reduce to three.

Intra- and intermolecular exchange processes also play an important role in the NMR spectroscopy of organometallic compounds and metal complexes.

2,4-Dimethyl-2,3-pentadiene (tetramethylallene) for example forms an $Fe(CO)_4$ complex (**57**), in which all four methyl groups are equivalent at room temperature. In the η^2-complex the iron atoms changes its π ligand rapidly. At low temperature however a "frozen out" structure with reduced symmetry

is observed. The 2:1:1 distribution of the methyl groups is a consequence of the iron atoms being attached to only **one** double bond.

For many organometallic compounds the temperature dependence of the NMR spectra is a consequence of association and dissociation processes.

Trimethylaluminium (**58**) for example shows a single methyl signal at room temperature, which splits below −40°C into two signals. This is an example of a monomer–dimer equilibrium, which is so slow at low temperatures that a distinction can be made between terminal and bridging methyl groups. The monomer concentration is below the limit of sensitivity.

At the end of this section some mention will be made of the application of the temperature dependence of NMR spectra in **kinetic studies**. In the processes mentioned the intramolecular motion (rotations, inversions, etc.) and the intermolecular reactions are **equilibria** between two or more **conformers**, **tautomers**, or **valence isomers**. If we take the simple case of a reversible exchange process between A and B with first order kinetics:

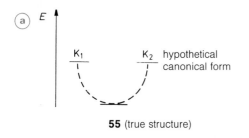

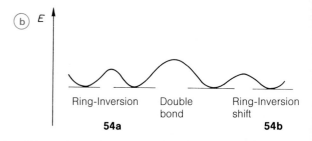

Fig. 3.28 **a** Schematic energy diagram for the benzene derivative (**55**) and its two hypothetical Kekulé structures **b** Schematic energy diagram for the cyclooctatetraene derivative (**54**) with valence isomerism between the structures (**54a**) and (**54b**) and ring inversion. (From experiments with the six-fold deuteriated compound with R = C(CH₃)₂OH values of $\Delta G^* = 61.5$ kJ · mol⁻¹ for the ring inversion and $\Delta G^* = 71.6$ kJ · mol⁻¹ for the double bond shift (both at −2°C) were determined)

$$A \underset{k'}{\overset{k}{\rightleftarrows}} B$$

The relative populations are n_A and n_B ($n_A + n_B = 1$). If the energy contents of A and B are different, then the distribution will not be 1:1, but a temperature dependent equilibrium

$$\frac{n_B}{n_A} = e^{-\frac{\Delta G}{RT}}$$

ΔG difference in Gibb's free energy
R gas constant
T absolute temperature

The **Eyring equation** gives the rate constant k

$$k = \frac{RT}{N_A \cdot h} e^{-\frac{\Delta G^{\#}}{RT}}.$$

$\Delta G^{\#}$ free energy of activation
N_A Avogadro's number
h Planck's constant

If the exchange $A \rightleftarrows B$ is slow the signals of A and B are observed separately in the NMR spectrum, if it is fast however only an averaged signal is observed for the exchanging nuclei. (The terms "slow" and "fast" are defined by comparison to k or $\tau = 1/k$ as described in Sec. 1.4, p. 84).

Fig. 3.29 shows the simple case where A and B have singlet signals of equal intensity which then merge on fast exchange to a twice as intensive signal ν_m. If the effect of temperature on the shifts ν_a and ν_b can be ignored then

$$\nu_m = \frac{\nu_A + \nu_B}{2}.$$

Between the regions of fast and slow exchange broad signals appear. The curve labelled T_c is the coalescence situation. T_c is called the **coalescence temperature**. k at coalescence is given approximately by

$$k_{T_c} = \frac{\pi}{\sqrt{2}} |\nu_A - \nu_B| \quad \text{also } k_{T_c} \approx 2.22 \, \Delta\nu$$

Inserting this relation into the Eyring equation results in

$$\frac{\pi}{\sqrt{2}} |\nu_A - \nu_B| = \frac{RT_c}{N_A \cdot h} e^{-\frac{\Delta G^{\#}}{RT_c}}$$

or

$$\Delta G^{\#} = RT_c \cdot \ln \frac{RT_c \sqrt{2}}{\pi \cdot N_A \cdot h |\nu_A - \nu_B|}.$$

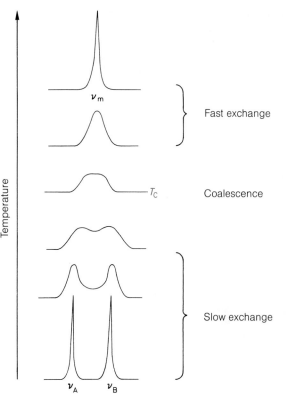

Fig. 3.29 Schematic representation of the temperature-dependent NMR spectra for a process $A \rightleftarrows B$ with no coupling between the exchanging nuclei

If T_c is measured in K and the shifts n in Hz, then the free energy of activation is given (in kJ) by

$$\Delta G^{\#} = 19.1 \cdot 10^{-3} \cdot T_c (9.97 + \log T_c - \log |\nu_A - \nu_B|).$$

Considerably more accurate than this approximate solution based on the coalescence temperature is **line shape analysis** (for details see the NMR literature quoted in the bibliography).

If one assumes that

$$|\nu_A - \nu_B| \leq \begin{array}{l} 150 \text{ Hz (in } {}^1\text{H NMR)} \\ 300 \text{ Hz (in } {}^{13}\text{C NMR)} \end{array}$$

then this method can be used to study the reversible interconversion of states which have a **lifetime** $\tau = 1/k$ between 10^1 and 10^3 s; in certain cases τ values one or two orders of magnitude smaller can be determined. This method can be extended to the case of coupling nuclei (see for example Fig. 3.22, p. 93).

In such cases

$$k_{T_c} \approx 2.22 \sqrt{\Delta\nu^2 + 6J_{AB}^2}$$

Since Δv increases with the measurement frequency, the coalescence temperature T_C must also increase with B_0. If in one molecule there are several pairs of nuclei with different Δv, then their T_c values must also vary. Any specification of the **coalescence temperature** must therefore always be accompanied by the measurement frequency and the relevant pair of nuclei A/B.

The kinetics of slower reactions can naturally also be determined by NMR. Concentrations of components being formed or disappearing can be determined by integration of the signals. By application of the Gibbs-Helmholtz equation

$\Delta G^{\neq} = \Delta H^{\neq} - T \Delta S^{\neq}$ the enthalpy $\Delta H^{\neq}$ and entropy of activation $\Delta S^{\neq}$ can be determined. By taking logarithms of both sides of the Eyring equation

$$\log \frac{k}{T} = 10.32 - \frac{\Delta H^{\neq}}{19.1} \cdot \frac{1}{T} + \frac{\Delta S^{\neq}}{19.1} \quad .$$

and plotting $\log k/T$ against $1/T$ $\Delta H^{\neq}$ is derived from the slope and $\Delta S^{\neq}$ from the intercept of the straight line obtained. Obviously it is advantageous to have as many values of k and T as possible.

3. ¹H NMR Spectroscopy

3.1 Sample Preparation and Measurement of Spectra (CW and PFT Techniques)

NMR spectra for analytical purposes are normally measured in solution. A concentrated, but not viscous solution is made in a proton-free solvent. Alongside CCl_4 and CS_2 a variety of deuterated solvents are commercially available (Tab. 3.6). The most commonly used is $CDCl_3$. Since the degree of deuteration is always somewhat less than 100%, weak signals from the solvent must always be expected. The δ values of these solvent signals are given in Tab. 3.6. Whereas the $CHCl_3$ impurity in $CDCl_3$ (typically 0.2%) gives a singlet signal at $\delta = 7.24$, solvents with CD_3 groups give a quintet because of the coupling to deuterium ($I = 1$) in CHD_2 groups (cf. p. 77).

Polar solvents also often have a detectable concentration of water, which gives rise to an H_2O or HOD signal.

The choice of solvent has some affect on the measured shifts. In the case of overlapping signals the **solvent shift** can be useful. C_6D_6 is particularly suitable because of its high magnetic anisotropy (see for example Fig. 3.39, p. 126).

[D6]-Dimethyl sulfoxide ([D6]-DMSO) slows the proton exchange of OH groups and is recommended as solvent if the coupling patterns of OH protons are to be observed (see p. 109).

As **reference substance** for establishing the zero point of the δ scale tetramethylsilane (TMS) is used, either added directly to the solution (**internal standard**) or in a capillary tube inside the

measurement tube (**external standard**). If an external standard is used the δ values for the chemical shifts must be corrected:

$$\delta_{\text{corr.}} = \delta_{\text{meas.}} + 6.67 \cdot 10^5 \cdot \pi \cdot [\chi_v(\text{Standard}) - \chi_v(\text{Sample})]$$

χ_v volume susceptibility

The use of TMS is unnecessary, if, for example, the residual $CHCl_3$ signal in $CDCl_3$ can be directly used.

Measurements are normally made at room temperature. **Low** or **high temperature spectra** are important for the study of intramolecular mobility (rotations, inversions, etc.) and for kinetic studies of reactions.

Fig. 3.30 gives a schematic representation of the construction of an NMR spectrometer. The magnetic field, produced by a permanent or electro-magnet, should be as homogeneous as possible. It splits the nuclear Zeeman levels (see Sec. 1.1, p. 71) by an amount proportional to B_0. The sample is contained in a tube, which is rotated about its vertical axis. This averages out horizontal field inhomogeneities.

The nuclei are excited by a radio frequency transmitter of high stability. At resonance the induced magnetisation created by the spin inversion results in a current in the receiver coil, which is perpendicular to both the magnetic field and the transmitter coil. Instead of the second coil a bridge circuit can be used. The amplified signal is recorded on an x,y-plotter, which draws the spectrum.

Tab. 3.6 Solvents for ¹H NMR spectroscopy

Solvent	¹H NMR shift δ	Mpt.* (°C)	Bpt-$_{760}$* (°C)
Carbon tetrachloride (CCl$_4$)	–	– 23	77
Carbon disulfide (CS$_2$)	–	– 112 T	46
Hexachloro-1,3-butadiene (C$_4$Cl$_6$)	–	– 21	215 H
Dichlorofluoromethane (CCl$_2$F$_2$)	–	– 160 T	– 30
[D$_1$] Chloroform (CDCl$_3$)	7.24	– 64	61
[D$_4$] Methanol (CD$_3$OD)	3.35	– 98 T	64
	4.78		
[D$_6$] Acetone (CD$_3$COCD$_3$)	2.04	– 95 T	56
[D$_6$] Benzene (C$_6$D$_6$)	7.27	6	80
[D$_{12}$] Cyclohexane (C$_6$D$_{12}$)	1.42	7	81
[D$_8$] Toluene (C$_6$D$_5$CD$_3$)	2.30	– 95	111
	7.19		
[D$_5$] Nitrobenzene (C$_6$D$_5$NO$_2$)	7.50	6	211 H
	7.67		
	8.11		
[D$_2$] Dichloromethane (CD$_2$Cl$_2$)	5.32	– 97 T	40
[D$_1$] Bromoform (CDBr$_3$)	6.83	8	150 H
[D$_2$] 1,1,2,2-Tetrachloroethane (C$_2$D$_2$Cl$_4$)	6.00	– 44	146 H
[D$_3$] Acetonitrile (CD$_3$CN)	1.93	– 45	82
[D$_{10}$] Diethyl ether (C$_4$D$_{10}$O)	1.07	– 116 T	35
	3.34		
[D$_8$] Tetrahydrofuran (C$_4$D$_8$O)	1.73	– 108 T	66
	3.58		
[D$_8$] Dioxane (C$_4$D$_8$O$_2$)	3.58	12	102
[D$_6$] Dimethyl-sulfoxide (CD$_3$SOCD$_3$)	2.49	19	189 H
[D$_5$] Pyridine (C$_5$D$_5$N)	7.19	– 42	115
	7.55		
	8.71		
[D$_2$] Water (D$_2$O)	4.65	0	100
[D$_4$] Acetic acid (CD$_3$COOD)	2.03	17	118
	11.53		
[D$_1$] Trifluoroacetic acid (CF$_3$COOD)	11.5	– 15	72
[D$_{18}$] Hexamethylphosphoric triamide (HMPT) [(CD$_3$)$_2$N]$_3$PO	2.53	7	233 H

* Refers to the undeuteriated compound
T suitable for low temperature measurements
H suitable for high temperature measurements

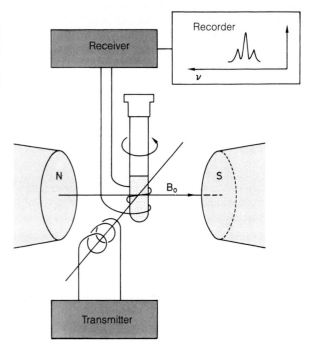

Fig. 3.30 Schematic diagram of an NMR spectrometer

The rotation of the sample tube can lead to spinning side bands, which lie symmetrically either side of the main signal. Their distance from the main signal increases with increasing rotational frequency, and their intensity becomes so small that confusion with real signals can be eliminated.

Further signals symmetrically placed about the main signal are the so-called ¹³C satellites. They result from coupling to ¹³C nuclei. Since the natural abundance of ¹³C is only 1.1% they are only observed for the strongest signals in routine ¹H NMR spectra (but see p. 129).

Fig. 3.31 shows all the expected "spurious" signals in the case of chloroform dissolved in hexadeuterioacetone.

For the measurement of routine spectra spectrometers with a ¹H frequency of 60, 80, 90, 100 or 200 MHz (1.41; 1.88; 2.11; 2.35; 4.70 T) are normally sufficient. If higher spectral dispersion or signal-noise ratio (sensitivity) are required instruments with a ¹H frequency of 250, 270, 300, 360, 400, 500, 600, or even 750 MHz (5.87; 6.34; 7.05; 8.45; 9.39; 11.74; 14.10; 17.63 T) are available. It should be emphasised, however, that multiplet signals which arise from spin-spin coupling cannot be better resolved at higher field; on the contrary, line widths generally increase with field strength. Field strengths corresponding to a ¹H frequency above 100 MHz require super-conducting magnets which must be cooled with liquid helium. (For the use of high field instruments see p. 124.)

There are two possible ways of achieving resonance (see Sec. 1.1). Either the frequency ν can be varied at constant field B_0 (**frequency sweep**) or the field B can be varied at constant frequency (**field sweep**). In both methods the individual signals are observed consecutively by constant variation of ν or B; this is described as a **CW technique (continuous wave)**.

If less than ca. 20 mg of the substance to be measured are available, or if less than 20 mg are soluble in ca. 0.5 ml of solvent, alternative methods of measurement must be employed. To improve the signal-noise ratio when measuring dilute solutions many spectra can be measured consecutively, stored in a small computer and the resultant spectra averaged (so-called time-averaging, CAT-method, computer averaged transients). The random noise tends to cancel out, so the **signal-noise ratio** S/N improves by a factor of root $\sqrt{n}$, where n is the number of individual scans. The time required for this procedure limits its usefulness.

Nowadays the **pulsed Fourier transform technique** (PFT-NMR spectroscopy) is a superior solution to the problem. This allows the measurement of ^{1}H NMR spectra on samples of 1 mg (or even less).

In contrast to the **CW technique** all nuclei of a particular type, for example all protons in a sample, are excited simultane-ously by an intense radiofrequency pulse. As explained in Sec. 1.1 (p. 71) a consequence of the Boltzmann distribution is that in an external field $\boldsymbol{B}_0$ more nuclei precess about the direction of $\boldsymbol{B}_0$ than about the opposite direction. The vector sum of the magnetic moments of these excess nuclei results in an **equilibrium magnetisation** M_0 in the direction of $\boldsymbol{B}_0$ (**longitudinal magnetisation**). The high-frequency pulse rotates $\boldsymbol{M}$ by an angle α from its equilibrium direction, creating a **transverse magnetisation**. Thus if $\alpha = 90°$ then $\boldsymbol{M}$ is perpendicular to $\boldsymbol{B}_0$.

The duration of the high-frequency pulse (**pulse width**) is of the order of a few µs. After this short perturbation the nuclei return to their equilibrium state. The transverse magnetisation decays at a rate corresponding to the effective **relaxation time** T_2^*. (Unavoidable inhomogeneities in the field lead to the effective relaxation time T_2^* being less than T_2). The longitudinal magnetisation also returns to its (non-zero) equilibrium value. The longitudinal relaxation time T_1 is $\geqq T_2$. (See Sec. 1.1, p. 72 for further details on relaxation times.)

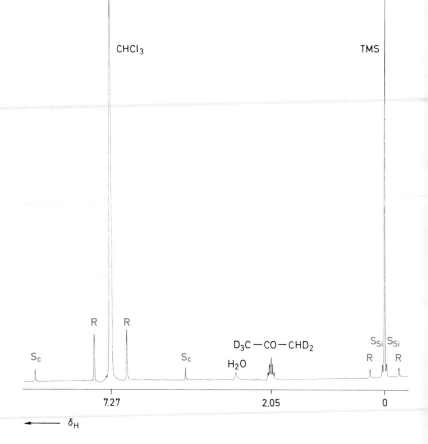

Fig. 3.31 ^{1}H NMR spectrum of chloroform in hexadeuterioacetone with "spurious" signals: CD_3–CO–CHD_2 (incomplete deuteriation), H_2O (water content)

R   Spinning sidebands (because of too slow rotation)

S_c ^{13}C satellites (^{13}C, ^{1}H coupling in $CHCl_3$)

S_{si} ^{29}Si satellites (^{29}Si, ^{1}H coupling in TMS)

After the end of the pulse the decay of the transverse magnetisation, known as the **FID (free induction decay)** is measured. The FID takes the form of a complex interferogram of superimposed damped vibrations. A mathematical operation, the **Fourier transformation**, produces a normal NMR spectrum from the FID. The signal is transformed from a scale of time (time domain) to a scale of frequency (frequency domain).

The time taken for a pulse and the subsequent measurement of the FID is so short, that a large number of FID's can be accumulated before the Fourier transformation is carried out. The increase in the number of scans allows an enormous increase in sensitivity. If a pulse angle of 90° is used a long delay (pulse delay) of ca. 5 T_1 must be left before the next pulse to allow the equilibrium situation to be achieved. In practice **pulse widths** are chosen which result in $\alpha < 90°$.

An additional improvement in the signal-noise ration is achieved by a mathematical manipulation of the FID before Fourier transformation. The FID is multiplied by an exponential function e^{-ct}. This improvement in signal-noise is achieved at the cost of resolution, since the damping of the FID results in an artificial broadening of the lines. The opposite effect is achieved by the use of a function e^{+ct}. A better method for improving the resolution is multiplication by a function $e^{c_1t - c_2t^2}$ (c_1, $c_2 > 0$).

The quadratic term in the exponent corresponds to a Lorentz-Gauss transformation. The resolution is improved at the cost of poorer signal-noise ratio (**resolution enhancement**). The individual lines become sharper (lower line width) and often more components of a complex coupling pattern can be recognised. Excessive "Gaussing" however leads to negative components of the signals.

Very high demands are placed on the constancy of the magnetic field. This is achieved by correcting the field based on measuring it by means of deuterium resonance frequencies (**lock**). Deuteriated solvents therefore have a secondary function of providing the lock signal. (If undeuteriated solvents are used, an external lock may be provided). A third function is to provide a reference for ¹H spectra from the ¹H signals arising from the small amount of protonated solvent resulting from the inevitably incomplete deuteration. In such cases the addition of TMS may be unnecessary.

At the end of this section it should be pointed out that the terms NMR absorption and NMR signal are often interchangeably used, despite the fact that the PFT technique is no longer a direct measure of NMR absorbance.

3.2 ¹H Chemical Shifts

Ignoring media effects the chemical shift of a proton is determined by three factors:

the distribution of electronic charge;

anisotropy effects;

steric effects.

Electronic Effects

The electron shells of a nucleus and its direct neighbours shield the nucleus from the external field B_0 (see Sec. 1.2, p. 73). The electron density and therefore the shielding constant σ are affected by inductive and mesomeric effects. Consider first element-hydrogen bonds X–H. With increasing electronegativity of X the shielding of the proton is reduced and the signal is shifted to lower field. The ¹H shifts of ethanol (**59**) and ethanethiol (**60**) are given as examples. The OH protons absorb under comparable measurement conditions at lower field than the SH protons.

$$H_3C - CH_2 - OH \qquad\qquad H_3C - CH_2 - SH$$

$$\delta : \quad 1.24 \quad 3.71 \quad 2.56 \qquad\qquad 1.30 \quad 2.44 \quad 1.46$$

59 (in CCl₄) **60** (in CCl₄)

Whereas a comparison of the shifts of the α- protons of ethanol and ethanethiol shows a similar effect, the electronegativity of X is no longer dominant for the shifts of the β- protons.

As Tab. 3.7 shows, the low field shift is markedly enhanced by the presence of additional electronegative substituents.

In contrast, protons bound to electropositive central atoms absorb at very high field. In metal complexes, for example iron tetracarbonyl hydride (**61**), other influences must also be considered. In the literature δ-values up to −30 are found for such compounds.

$$HFe(CO)_4 \qquad \delta = -10.5$$

61

Changes of charge density caused by formation of **positively** or **negatively charged ions** often have marked effects on shifts.

Tab. 3.7 ¹H shifts of the halogenated methanes (δ values)

	CH_3X	CH_2X_2	CHX_3
X = F	4.27	5.45	6.49
X = Cl	3.06	5.30	7.24
X = Br	2.69	4.94	6.83
X = I	2.15	3.90	4.91

(H₃C)₂N—⟨benzene⟩—H 6.59

δ: H 6.60 H 7.08 **62**

(H₃C)₃N⁺—⟨benzene⟩—H 7.60

H 7.98 H 7.66 **63**

H₃C—CH₂—CH₃ **64**
δ: 0.91 1.33 0.91

H₃C—ĊH—CH₃ **65**
5.06 13.50 5.06

[CH₂≈CH≈CH₂]⁺ **66**
δ: 8.97 9.64 8.97

[CH₂≈CH≈CH₂]⁻ **67**
2.46 6.28 2.46

Anisotropy Effects

Chemical bonds are in general magnetically anisotropic, that means that the susceptibility χ depends on the spatial orientation. Double and triple bonds, three-membered rings and cyclic conjugated systems show particularly strong magnetic anisotropy. Fig. 3.32 demonstrates the effect using cones of anisotropy. In the positive areas, marked in blue, the shielding is large. Protons in these regions are shifted to higher field (smaller δ-values). Protons in the negative regions absorb at low field (large δ-values) because of the reduced shielding.

Protons bonded to olefinic C atoms have δ-values between ca. 4 and 8, i.e. at much lower field than similar protons bonded to saturated C atoms. This is due to the anisotropy as well as the change in hybridisation. Aldehyde protons absorb between about 9.3 and 10.7. Here the effect of the electronegativity of the O atom adds to those of hybridisation and anisotropy. Acetylinic H atoms should be more deshielded than olefinic H atoms because of the polarity of the CH bond. The anisotropy effect of the C≡C bond however causes an opposite effect: the signals of protons on *sp* carbons are found between $\delta = 1.8$ and 3.2. If methyl groups are considered instead of H atoms, then bonding to C=C or C=O bonds also causes a low-field shift.

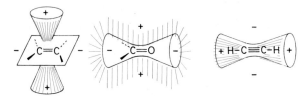

Fig. 3.32 Magnetic anisotropy of (C=C)–, and (C≡C)– bonds

The methyl protons therefore lie in the negative region of the cone of anisotropy (Fig. 3.32). Methyl groups bound to *sp* carbons also show low field shifts because of the small angle of the cone of anisotropy. These shift effects are well illustrated by the δ-values of the following compounds.

H₃C—CH₂—CH₃
0.91 1.33 0.91

propane **64**

1.66 H₃C   H 4.96
       C=C
5.73 H   H 4.88

propene **68**

2.20 H₃C
        C=O
9.80 H

acetaldehyde **69**

1.8 H₃C—C≡C—H 1.8

propyne **70**

The bicyclic compounds (**71**), (**72**), and (**73**) show further examples of these effects.

H H 1.21

H 1.49
H H 1.18
2.20

norbornane **71**

1.07 H H 1.32
   anti syn
exo
1.57 H

0.94 H H
endo 2.83 H 5.93

norbornene **72**

H H 1.95

H
3.53 H 6.66

norbornadiene **73**

In norbornane (**71**) the chemical shift of the bridgehead protons lies as expected at relatively high field. The introduction of one or two (C=C) bonds shifts them to successively larger δ-values. A similar effect is seen for the *syn*-protons of the methylene bridge; the signal of the *anti*-proton in (**72**) however is shifted to higher field. Similar differences are seen for the shifts of the *exo*- and *endo*-protons. Finally it is noteworthy that the olefinic protons in norbornadiene (**73**) absorb at unusually high δ-values. From these examples it can be appreciated that the anisotropy effects of individual bonds or structural elements often combine in quite complicated ways. The schematic representation of anisotropy effects in Fig. 3.32 can only be taken as a rough guide.

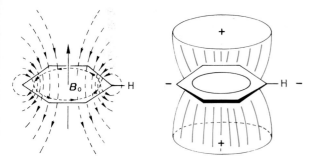

Fig. 3.33 Ring current model for aromatic systems (example: benzene)

The **ring current model** for the explanation of anisotropy effects in cyclic conjugated π-systems has proved to be particularly fruitful, despite certain theoretical problems. As shown in Fig. 3.33 for a benzene ring, one can imagine that the magnetic field causes a "ring current" of the electrons in an aromatic ring. The opposing field thus induced is described by the lines of force represented by the dashed lines. In the positive zone above, below, and inside the benzene ring the B_0 field is reduced, i.e. the shielding increased. The signals of protons lying in these zones experience a high-field shift. The signals of protons in the negative zone outside the ring are shifted to low field.

The ¹H shifts of the following examples show these effects very clearly.

If the ring current model is extended to condensed aromatics, the estimates of the chemical shifts must be summed over all the rings involved.

In the annulene series the ring current is often used as a qualitative criterion for aromaticity. The model proposed for benzene in Fig. 3.33 applies to planar, cyclic conjugated $(4n+2)\,\pi$ electron systems. The "external" protons are deshielded, the "internal" protons shielded. In contrast to this **"diamagnetic" ring current** of the **"diatropic"** compounds (aromatics) cyclic $(4n)\pi$ electron systems, **"paratropic"** compounds (antiaromatics) show a **"paramagnetic" ring current**. Shielding and deshielding are then exactly reversed. This does *not* mean that the direction of the induced field is reversed; instead, antiaromatics have a small HOMO-LUMO energy difference based on the Jahn-Teller splitting. In the magnetic field B_0 this leads to a mixing of wave functions of the excited states with the ground state, increasing the paramagnetic part of the shielding constant (cf. p. 73). The ring current effect is an easily determined physical criterion of **aromaticity**. An illustration is given by the aromatic (diatropic) [18]annulene (**80**) and the antiaromatic (paratropic) [16]annulene (**81**).

80

	12 H$_a$: $\delta =$	9.28
$-70\,°C$	6 H$_i$: $\delta =$	-2.99
$+110\,°C$	18 H: $\delta =$	5.45

81

	12 H$_a$: $\delta =$	5.2
$-120\,°C$	4 H$_i$: $\delta =$	10.32
$+30\,°C$	16 H: $\delta =$	6.70

As the temperature increases the mobility of these annulenes increases to such an extent that the internal and external protons exchange their positions and only an averaged signal is observed. While [18]annulene shows separate signals for the two kinds of protons at room temperature, the much reduced activation energy in [16]annulene results in a singlet.

The ring current concept as a model for the anisotropy effect can be extended to heteroaromatics [cf. for example tetrahydropyridine (**82**) and pyridine (**83**)].

74

75

76

77

78

79

H 5.77
H 5.72

82

H 7.75
H 7.38
H 8.59

83

Another good example is coproporphyrin (**84**).

COOCH3
CH2
CH2 Ha   CH3
H3C   CH2—CH2—COOCH3
NHi   N
Ha   Hd
N   HiN
H3COOC—CH2—CH2   CH3
H3C   Ha   CH2
CH2
COOCH3

84

The NH protons exchange their positions on the four nitrogen atoms so quickly that all the pyrrole rings are equivalent. Only two methyl signals are observed, one being due to the ester groups. The signal of the NH protons is at $\delta = -4$ and is shifted by 11 ppm compared to the NH signal of pyrrole. The protons of the methine bridges absorb at $\delta = 10$, i.e. at very low field.

The ring current model applied to the periphery of the porphyrin framework gives a simple explanation for these effects, the NH protons being internal and the methine protons external protons.

The shifts of the quasiaromatic ions **85** to **89** show, in addition to the effects of anisotropy, a marked influence of the electron density caused by the charge.

δ :   11.1   4.0   5.5   7.7   5.7
        85   **86**   **87**   **88**   **89**

The dianion of [16]annulene shows a signal for the four internal protons at very high field ($\delta = -8.17$) and for the twelve external protons at low field. This contrast to the uncharged compound is typical for the transition from $4n$ to $(4n+2)\pi$ electron systems.

Anisotropy effects are important for a number of other, not necessarily conjugated ring systems. Mention will be made here of the cyclopropane ring, protons on which are strongly

Tab. 3.8 ^{1}H chemical shifts of cycloalkanes and cyclic ethers δ values in CDCl$_3$ or CCl$_4$)

	δ values		δ values		
			α	β	γ
Cyclopropane	0.22	Oxirane	2.54	–	–
Cyclobutane	1.96	Oxetane	4.73	2.72	–
Cyclopentane	1.61	Tetrahydrofuran	3.75	1.85	–
Cyclohexane	1.44	Tetrahydropyran	3.56	1.58	1.58
Cyclodecane	1.51				
Cyclododecane	1.34				

shielded (Tab. 3.8). The effect is also shown by heterocyclic three-membered rings.

Cyclic π-electron systems, which are neither aromatic nor antiaromatic because of their strong deviation from planarity, show shifts which confirm their analogy to open chain alkenes. Thus cyclooctatetraene (**90**) shows a singlet at $\delta = 5.69$. Here it must be realised that this signal is averaged by two processes which are both fast on the NMR time scale: ring inversion and double bond shifts (degenerate valence isomerism, automerisation, see p. 96).

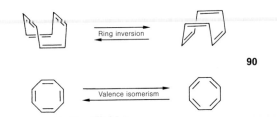

Ring inversion

Valence isomerism

90

3.3 ^{1}H, ^{1}H Coupling

The magnetic coupling between two nuclei in a molecule is generally transmitted by the intervening bonds. (Scalar coupling through space is however also known. This through-space coupling occurs when two nuclei are brought so close together by steric compression that their orbitals overlap). The quantitative measure for the coupling is the **coupling constant** $^n J$. n indicates the number of bonds. The most important ^{1}H,^{1}H couplings are summarised in Tab 3.9.

Tab. 3.9 ¹H, ¹H couplings

Type of coupling	Coupling constant ⁿJ (order of magnitude)	Structural elements		
direct	¹J(276 Hz)	H–H		
geminal	²J (0...30 Hz) usually negative			
vicinal	³J (0...20 Hz) positive			
long-range	⁴J (0...3 Hz) positive or negative			
	⁵J (0...2 Hz) positive			

The **sign of the coupling constant** cannot be determined from first order spectra. Relative, but not absolute, signs can sometimes be determined from second order spectra. Absolute signs are based on the assumption that ¹J(¹³C,¹H) is positive. ²J(H,H) couplings are usually negative, ³J(H,H) couplings positive, and long-range couplings either positive or negative.

The coupling between magnetically equivalent protons has no effect on ¹H NMR spectra. They can be determined by deuteriation or from ¹³C satellites (see p. 127 and p. 129).

As n increases the coupling generally gets weaker, i.e. |J| decreases.

H–H	H–CH₂–H	H–CH₂–CH₂–H	H–(CH₂)₃–H
91	**12**	**92**	**64**
276 Hz	12.4 Hz	8.0 Hz	< 1 Hz

Geminal Coupling

The geminal coupling increases with the s-character of the hybrid orbitals.

12	**93**	**15**
–12.4 Hz	–4.5 Hz	+2.5 Hz

The following examples show the effect of various substituents.

94	**77**	**95**	**96**	**97**
–22.3 Hz	–14.5 Hz	–10.8 Hz	±0 Hz	+5.5 Hz

98	**99**	**100**	**101**
–3.2 Hz	–1.3 Hz	+7.1 Hz	+42.2 Hz

More complex effects are shown by the examples (**102**) and (**103**). The effect of a neighbouring oxygen atom (more electronegative) on the *geminal* coupling of a methylene group is an increase (less negative) compared to methane (**12**). A neighbouring π-bond, here CO, reduces the coupling constant (more negative).

102	**103**

Vicinal Coupling

The generally positive $^3J(H,H)$ coupling depends, apart from substituent effects, to a large degree on the molecular structure. The bond length l, the bond angle α, and the dihedral angle Φ are all important.

If, as in the cases of olefins or aromatics, there is no possibility of the dihedral angle varying (rotation about the C–C bond) then ^3J reduces with increasing bond length l and increasing bond angle α.

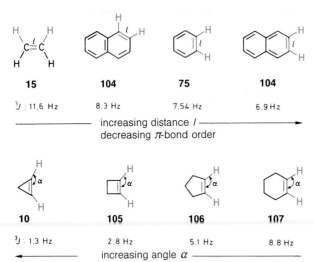

15	**104**	**75**	**104**
^3J : 11.6 Hz	8.3 Hz	7.54 Hz	6.9 Hz

increasing distance l —→
decreasing π-bond order

10	**105**	**106**	**107**
^3J : 1.3 Hz	2.8 Hz	5.1 Hz	8.8 Hz

←— increasing angle α

$^3J(E)(\Phi = 180°)$ is always larger than $^3J(Z)(\Phi = 0°)$;

	$^3J(Z)$	$^3J(E)$	$^2J(gem)$
R = H	11.6	19.1	2.5
R = C_6H_5	11.5	18.6	1.1
R = OCH_3	6.7	14.0	−2.0
R = F	4.7	12.8	−3.2

108

109
$^3J(Z)$ = 12.3 Hz

8
$^3J(E)$ = 15.8 Hz

In small rings or rigid bicyclic systems like norbornane the ^3J coupling of cis-protons can however be larger than that of *trans*-protons. For cyclopropane $^3J_{cis}$ is about 8 and $^3J_{trans}$ about 5 Hz.

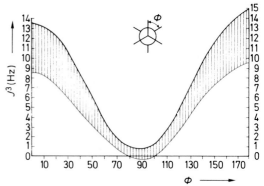

Fig. 3.34 Dependance of the *vicinal* coupling constant ^3J on the dihedral angle Φ

——— Karplus curve

$$^3J = \begin{cases} 8{,}5\,\cos^2\phi - 0{,}28 & \text{für} \quad 0° \le \phi \le 90° \\ 9{,}5\,\cos^2\phi - 0{,}28 & \text{für} \quad 90° \le \phi \le 180° \end{cases}$$

For ^3J-couplings across "freely" rotating C–C bonds the size of the coupling constant varies with the dihedral angle Φ. A quantitative estimate is given by the **Karplus curve** and related equations (Fig. 3.34). The measured couplings are particularly for $\Phi = 0°$ and $\Phi = 180°$ somewhat larger than predicted (shaded area).

When there is rapid rotation an averaged value is obtained for ^3J. To a first approximation it can be assumed that the three staggered conformations are equally populated. The average value is then given by:

$$^3J = \frac{^3J(60°) + {}^3J(180°) + {}^3J(300°)}{3}$$

$$\approx \frac{3.5 + 14 + 3.5}{3} \approx 7 \text{ Hz}$$

In practice the coupling constants ^3J in ethyl groups vary about this value, the effect of the electronegativity of R being noticeable.

H_2C–CH–R

37

R = Li	$^3J = 8.4$ Hz
R = H	8.0 Hz
R = C_6H_5	7.6 Hz
R = CH_3	7.3 Hz
R = OC_2H_5	7.0 Hz

Vicinal couplings from aldehyde protons to neighbouring protons on saturated carbon atoms are conspicuously small:

$^3J \approx 1.....3$ Hz

$$=\overset{\displaystyle H}{\underset{}{C}}-\overset{\displaystyle H}{\underset{\overset{\displaystyle \|}{O}}{C}}\qquad ^{3}\!J \approx 5\ldots 8\,Hz$$

In the chair conformation of cyclohexane (**110**) three *vicinal* couplings can be distinguished:

$$^{3}J_{aa} \approx 7 \ldots 12\ Hz\ (\phi = 180°)$$
$$^{3}J_{ee} \approx 2 \ldots 5\ Hz\ (\phi = 60°)$$
$$^{3}J_{ae} \approx 2 \ldots 5\ Hz\ (\phi = 60°)$$

110

D-Glucose (**111** ⇄ **112**) may be considered as a specific example. An aqueous solution shows two doublets for the proton on C-1 at $\delta = 5.22$ and 4.63, that at lower field being assigned to α-D-glucose (**111**), because it has the smaller coupling to the proton at C-2 ($^{3}J_{ae} < {}^{3}J_{aa}$).

α-D-Glucose **111**
$(\,{}^{3}J_{ae} = 3.5\ Hz\,)$

β-D-Glucose **112**
$(\,{}^{3}J_{aa} = 7.7\ Hz\,)$

If the relevant C—C bond carries electronegative substituents, the coupling constant is reduced. For organometallic compounds it is correspondingly increased. This **substituent effect** is observed for saturated, unsaturated, and heteroaromatic compounds.

In conformational analysis the ^{3}J couplings of a proton H_X with the two non-equivalent protons H_A and H_M of an α-CH2 group are often of interest. In an ideal case the two dihedral angles Φ_1 and Φ_2 differ by 120°. This leads to the coupling constants and splitting patterns shown in Fig. 3.35.

Vicinal Couplings to Exchanging Protons

The coupling of OH protons with *vicinal* CH protons can only be observed for very pure, water- and acid-free alcohols. Dimethyl sulphoxide (DMSO) has established itself as the favoured solvent, since **proton exchange** in DMSO at room temperature is sufficiently slow (cf. p. 100).

Exactly similar circumstances apply to NH protons. The *vicinal coupling* ^{3}J(CH—NH) is only observable if the (base catalysed) proton transfer is slow. This is often the case for the structural element $=\overset{|}{C}-NH-\overset{|}{CH}-$ (aromatic amines, enamines, amides).

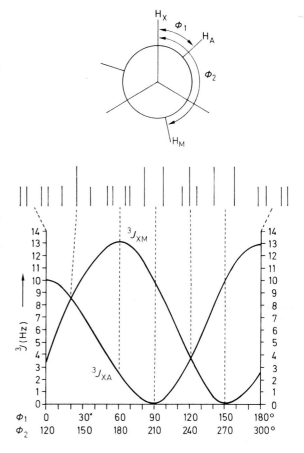

Fig. 3.35 *Vicinal* coupling constants and splitting patterns for the H_x signals in an AMX system $>CH_X$—CH_AH_M—

In trifluoracetic acid the proton exchange of ammonium cations is so slow that the coupling ^{3}J(CH—$\overset{+}{N}$H) is observable as a splitting of the (C—H) signal.

$$^{3}\!J(-NH-\overset{|}{CH}-) \approx {}^{3}\!J(-\overset{+}{N}H-\overset{|}{CH}-)$$

The conformational dependance of the ^{3}J coupling constant in alcohols, amines, and amides is similar to that of the vicinal CH—CH coupling. For free rotation:

$$^{3}\!J\,(CH-OH) \approx 4\text{-}5\ Hz$$
$$^{3}\!J\,(CH-NH) \approx 5\text{-}6\ Hz$$
$$^{3}\!J\,(CH-NH-\underset{\overset{\|}{O}}{C}) \approx 7\ Hz$$

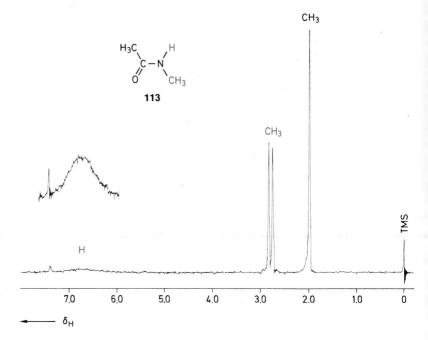

Fig. 3.36 ^{1}H NMR spectrum of N-methyl acetamide (**113**) in CDCl$_3$ (the compound exist exclusively in the conformation with s-*trans*-oriented methyl groups)

It is worth mentioning that in amides, for example, the splitting of the CH signal can appear even when the NH proton shows a broad signal (cf. the ^{1}H NMR spectrum of N-methylacetamide (**113**), Fig. 3.36).

To conclude this section a few comments about the direct ^{14}N,^{1}H coupling are appropriate. In the NH$_4^+$ ion it is 52.8 Hz. ^{14}N with a nuclear spin of 1 has an electric quadrupole moment (see Sec. 1.1, p. 71). In compounds other than the highly symmetrical NR$_4^+$ ions the contribution of the quadrupole to the relaxation is so large that ^{14}N,^{1}H coupling is not observed. Nonetheless this is a further cause, apart from proton exchange, for broad NH signals. It can however be removed by irradiation at the resonance frequency of the ^{14}N nucleus (see Fig. 3.50, p. 134).

Long Range Coupling

Long range couplings extend over four or more bonds. In open chain, saturated compounds they are usually smaller than 1 Hz and are not observed. In bicyclic and polycyclic systems the situation changes when a fixed W-orientation of the four bonds in a ^4J coupling is caused by the molecular structure.

$$^4J_{1,4} = +7 \text{ Hz}$$
$$^4J_{1,3} = {}^4J_{2,4} \approx {}^4J_{2,3} \approx 0 \text{ Hz}$$

114

An extreme example is bicyclo[1.1.1]pentane (**115**), in which the bridgehead protons have a ^4J coupling of 18 Hz.

$$^4J = 18 \text{ Hz}$$

115

In unsaturated compounds **allylic** ^4J and **homoallylic** ^5J **couplings** are observed:

$$^4J(Z) \approx -3 \text{ to } +2 \text{ Hz}$$
$$^4J(E) \approx -3.5 \text{ to } +2.5 \text{ Hz}$$

$$^5J(Z) \approx {}^5J(E) \approx 0 \text{ to } 2.5 \text{ Hz}$$

Larger magnitudes of ^4J and ^5J couplings can occur in alkynes and allenes.

$$^5J \approx 1 \text{ to } 3.0 \text{ Hz}$$

In carbocyclic and heterocyclic compounds considerably larger ^4J and ^5J values can be observed:

116

Y = O, NH

$J \approx 0\text{-}7$ Hz

117

Z = CH, N

$^5J \approx 5\text{-}11$ Hz

For a better understanding the ¹H NMR spectrum of 1,3-butadiene (**118**) with all the couplings may be discussed. The AA'BB'CC' system with three chemically non-equivalent kinds of protons shows three signals and nine different couplings.

118

$\delta_A = \delta_{A'} = 5.06$ $\delta_B = \delta_{B'} = 5.16$ $\delta_C = \delta_{C'} = 6.27$

$^2J_{AB} = {}^2J_{A'B'} = 1.8$ Hz $^4J_{BC'} = {}^4J_{B'C} = -0.8$ Hz

$^3J_{AC} = {}^3J_{A'C'} = 10.2$ Hz $^5J_{AA'} = 1.3$ Hz

$^3J_{BC} = {}^3J_{B'C'} = 17.1$ Hz $^5J_{AB'} = {}^5J_{A'B} = 0.6$ Hz

$^3J_{CC'} = 10,4$ Hz $^5J_{BB'} = 0.6$ Hz

$^4J_{AC'} = {}^4J_{A'C} = -0.9$ Hz

Long range couplings are of particular importance in the spectra of aromatic and heteroaromatic rings. A comparison of benzene (**75**) and 1,3-cyclohexadiene (**74**) shows the influence of the different bond orders.

75

H 7.54 Hz
0.66
1.37 Hz

74

H 9,4 Hz
1,1
0,9 Hz
5.1 Hz

In substituted benzenes:

$^3J_{\text{ortho}} = 6.9 \dots 9.0$ Hz
$^4J_{\text{meta}} = 0.9 \dots 3.0$ Hz
$^4J_{\text{para}} = 0 \dots 1.0$ Hz

For benzene derivatives the substitution is often recognisable from the splitting patterns. This even applies to identical spin systems with different relationships of the coupling constants; thus *o*-substituted benzenes with two identical substituents have the same AA'BB' or AA'XX' spin system as *p*-substituted benzenes with two different substituents (cf. Tab 3.5). Nevertheless, as Fig. 3.37 shows, a distinction between the two cases is easily made from the patterns of lines observed.

Condensed benzenoid arenes, e.g. naphthalene (**104**) have similar ^3J, ^4J, and ^5J couplings to benzene.

104

$\delta(H_A) = 7.66$
$\delta(H_B) = 7.30$

$^3J_{AB} = 8.3$ Hz $^5J_{AA'} = 0.7$ Hz
$^3J_{BB'} = 6.9$ Hz $^5J_{AA''} = 0.9$ Hz
   $^5J_{AB'''} = 0.2$ Hz
$^4J_{AB'} = 1.3$ Hz $^6J_{AB''} = -0.1$ Hz
$^4J_{AA'''} = -0.5$ Hz $^6J_{BB''} = 0.1$ Hz
   $^7J_{BB'''} = 0.3$ Hz

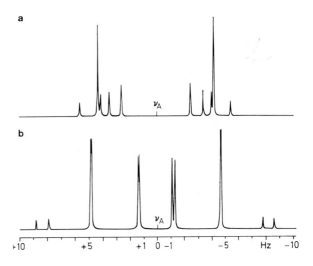

a

b

⊢10   +5   +1 0 −1   −5   Hz   −10

Fig. 3.37 A parts of the AA'XX' systems of disubstituted benzenes with following parameters:

$|\nu_A - \nu_X| = 240$ Hz at 400 MHz

a)

119

$^3J_{AX} = {}^3J_{A'X'} = 8.0$ Hz
$^4J_{AA'} = 1.2$ Hz
$^4J_{XX'} = 1.8$ Hz
$^5J_{AX'} = {}^5J_{A'X} = 0.5$ Hz

b)

120

$^3J_{AX} = {}^3J_{A'X'} = 8.0$ Hz
$^3J_{XX'} = 7.0$ Hz
$^4J_{AX'} = {}^4J_{A'X} = 1.5$ Hz
$^5J_{AA'} = 0.5$ Hz

In heterocyclic systems the deviations are often greater, in particular certain ^3J coupling constants can become quite small:

83 Pyridine   **121** Pyridinium ion

	83 Pyridine	121 Pyridinium ion
$\delta(H_A)$	8.60	9.23
$\delta(H_B)$	7.25	8.50
$\delta(H_C)$	7.64	9.04
$^3J_{AB} = {}^3J_{A'B'}$	5.5 Hz	6.0 Hz
$^3J_{BC} = {}^3J_{B'C}$	7.6 Hz	8.0 Hz
$^4J_{AC} = {}^4J_{A'C}$	1.9 Hz	1.5 Hz
$^4J_{AA'}$	0.4 Hz	1.0 Hz
$^4J_{BB'}$	1.6 Hz	1.4 Hz
$^5J_{AB'} = {}^5J_{A'B}$	0.9 Hz	0.8 Hz

17 Furan   **122** Pyrrole   **14** Thiophene

	17 Furan	122 Pyrrole	14 Thiophene
$\delta(H_A) = \delta(H_{A'})$	7.38	6.62	7.20
$\delta(H_B) = \delta(H_{B'})$	6.30	6.05	6.96
$^3J_{AB} = {}^3J_{A'B'}$	1.8	2.6	4.8
$^3J_{BB'}$	3.4	3.5	3.5
$^4J_{AB'} = {}^4J_{A'B}$	0.9	1.3	1.0
$^4J_{AA'}$	1.5	2.1	2.8

3.4 Coupling to Other Nuclei

For coupling between ^1H and other nuclei the natural abundance of these nuclei must be considered. The isotope ^{13}C ($I = 1/2$) makes up 1.1% of natural carbon, alongside 98.9% of the spin free ^{12}C. The $^1H, {}^{13}C$ coupling therefore leads to signals with only about 1% of the intensity of the corresponding ^1H signals without ^{13}C coupling. The measurement of these so-called ^{13}C **satellites** is described on p. 129. In routine ^1H spectra they can usually be ignored.

Tab. 3.10 $^1H, {}^{19}F$ coupling constants of selected fluorine compounds

Compound	type of coupling	$|J|$ in Hz
H—CH$_2$—F Fluoromethane	$^2J(H, F)$	46
H—CF$_2$—F Trifluoromethane	$^2J(H, F)$	80

H—CH$_2$—CH—F
           |
           H
 Fluoroethane
	$^2J(H, F)$
	 $^3J(H, F)$
		47
		 25

H—CH$_2$—CF$_2$—F 1,1,1-Trifluoroethane	$^3J(H, F)$	13
Fluoroethylene	$^2J(H, F)$ $^3J(Z\text{-}H, F)$ $^3J(E\text{-}H, F)$	85 20 52
1,1-Difluoroethylene	$^3J(Z\text{-}H, F)$ $^3J(E\text{-}H, F)$	≈ 1 34
(E)-1-Fluoropropene	$^2J(H, F)$ $^3J(H, F)$ $^4J(H, F)$	85 20 3
(Z)-1-Fluoropropene	$^2J(H, F)$ $^3J(H, F)$ $^4J(H, F)$	85 42 2

H—CH$_2$—C—F
         ||
         O
 Acetyl fluoride
	$^3J(H, F)$	7

H—C≡C—F Fluoroacetylene	$^3J(H, F)$	21
Fluorobenzene	$^3J(H, F)$ $^4J(H, F)$ $^5J(H, F)$	9.0 5.7 0.2

Tab. 3.11 ¹H, ³¹P coupling constants of selected phosphorus compounds

Compound	type of coupling	$|J|$ in Hz
H—PH—C_6H_5 Phenylphosphine	$^1J(\text{H.P})$	201
H—P$(CH_3)_2$ Dimethylphosphine	$^1J(\text{H.P})$	192
H—$\overset{\oplus}{P}(CH_3)_3$ Trimethylphosphonium	$^1J(\text{H.P})$	506
(H—CH_2—CH—$)_3$P Triethylphosphine	$^2J(\text{H.P})$ $^3J(\text{H.P})$	0.5 14
(H—CH_2—CH—$)_4$P$^+$Cl$^-$ Tetraethyl- phosphonium	$^2J(\text{H.P})$ $^3J(\text{H.P})$	13 18
(H—CH_2—CH—O—$)_3$P Triethyl phosphite	$^3J(\text{H.P})$ $^4J(\text{H.P})$	8 1
(H—CH_2—CH$)_3$P=O Triethylphosphine oxide	$^2J(\text{H.P})$ $^3J(\text{H.P})$	16 12
H—P$(OCH_3)_2$ ‖ O Dimethyl phosphite	$^1J(\text{H.P})$	710
H—P$(OC_2H_5)_2$ ‖ O Diethyl phosphite	$^1J(\text{H.P})$	688
C=C P(O)$(OC_2H_5)_2$ Diethyl ethenylphosphonate	$^3J(Z\text{—H.P})$ $^3J(E\text{—H.P})$	14 30
Phosphabenzene (phosphorine)	$^2J(\text{H.P})$ $^3J(\text{H.P})$ $^4J(\text{H.P})$	38 8 4

From Tab. 3.1 it can be seen that the only nuclei of importance in organic chemistry with appreciable abundances and with $I=1/2$ are ¹⁹F and ³¹P. These both have a natural abundance of 100%; therefore, proton spectra of fluorine or phosphorus compounds show splittings due to the ¹H,¹⁹F or ¹H,³¹P couplings. Summaries of the sizes of the coupling constants observed are given in Tab. 3.10 and Tab. 3.11. (For ¹H,D and ¹H,¹³C couplings see p. 127 and Sec. 4.3, p. 149 ff.)

3.5 Correlation of ¹H Shifts with Structural Features

Methyl protons. A summary of the chemical shifts of methyl groups in different chemical environments is given in Tab. 3.12. (Here and in the following tables typical values only are given. Extreme chemical shift values are ignored.)

Methylene protons. As explained in Sec. 2.2 (p. 88) the two protons of a methylene group are only chemically equivalent when they are related by a symmetry element in the molecule. This includes the effect of internal motion in the molecule. Even when the two protons are chemically equivalent, this does not imply that they are necessarily magnetically equivalent. In Tab. 3.13 the shift ranges of methylene groups are given as a function of the substitution.

Methine protons. The shift ranges for methine protons are considerably wider than those for methyl or methylene groups as a consequence of the larger number of substitution possibilities. The most common combinations of substituents are summarised in Tab. 3.14.

Protons on double and triple bonds. Tab. 3.15 gives information on shifts of **protons on (C=C) bonds**.

The shift range for **acetylenic protons** is at ca. $1.8 \leq \delta \leq 3.2$, shifted to higher field than for olefinic protons.

$$H_3C-C\equiv C-H \qquad \delta=1,8$$

70

$-C\equiv C-H \qquad \delta=3,1$

123

In Tab. 3.16 chemical shifts of aldehyde and aldimine protons are summarised.

Tab. 3.12   Chemical shifts of methyl groups (δ values measured in CCl_4 or $CDCl_3$)

Tab. 3.13 Chemical shifts of methylene groups (δ values measured in CCl_4 or $CDCl_3$)

δ	7	6	5	4	3	2	1	Group
							▭	$-\overset{\mid}{C}-CH_2-\overset{\mid}{C}-\overset{\mid}{C}-X$
						▭		$-\overset{\mid}{C}-CH_2-\overset{\mid}{C}-X$
								$-C=C\big\langle$
								$-C\equiv C-$
								$-$Aryl, Heteroaryl
								$-CO-$
								$-N\big\langle$
								$-O-$
								$-NO_2$
								$-$Hal
								$-\overset{\mid}{C}-CH_2-\overset{\mid}{C}X_{2(3)}$
								$X-\overset{\mid}{C}-CH_2-\overset{\mid}{C}-X$
								$-\overset{\mid}{C}-CH_2-X$
								$-\overset{\mid}{C}-CH_2-X$
								$-C=C\big\langle$
								$-C\equiv C-,\ -C\equiv N$
								$-$ Aryl, Heteroaryl
								$-CO-$Alkyl
								$-CO-$Aryl
								$-CO-O-,\ -CO-N\big\langle$
								$-CO-$Hal
								$-\overset{\mid}{C}=N-$
								$-S-$
								$-SO_2-$
								$-NH_{(2)}/$Alkyl$_{(2)}$
								$-NAryl_{(2)}$
								$-N-CO-$
								$-\overset{+}{\underset{\mid}{N}}-$
								$-N=$

Tab. 3.13 continued

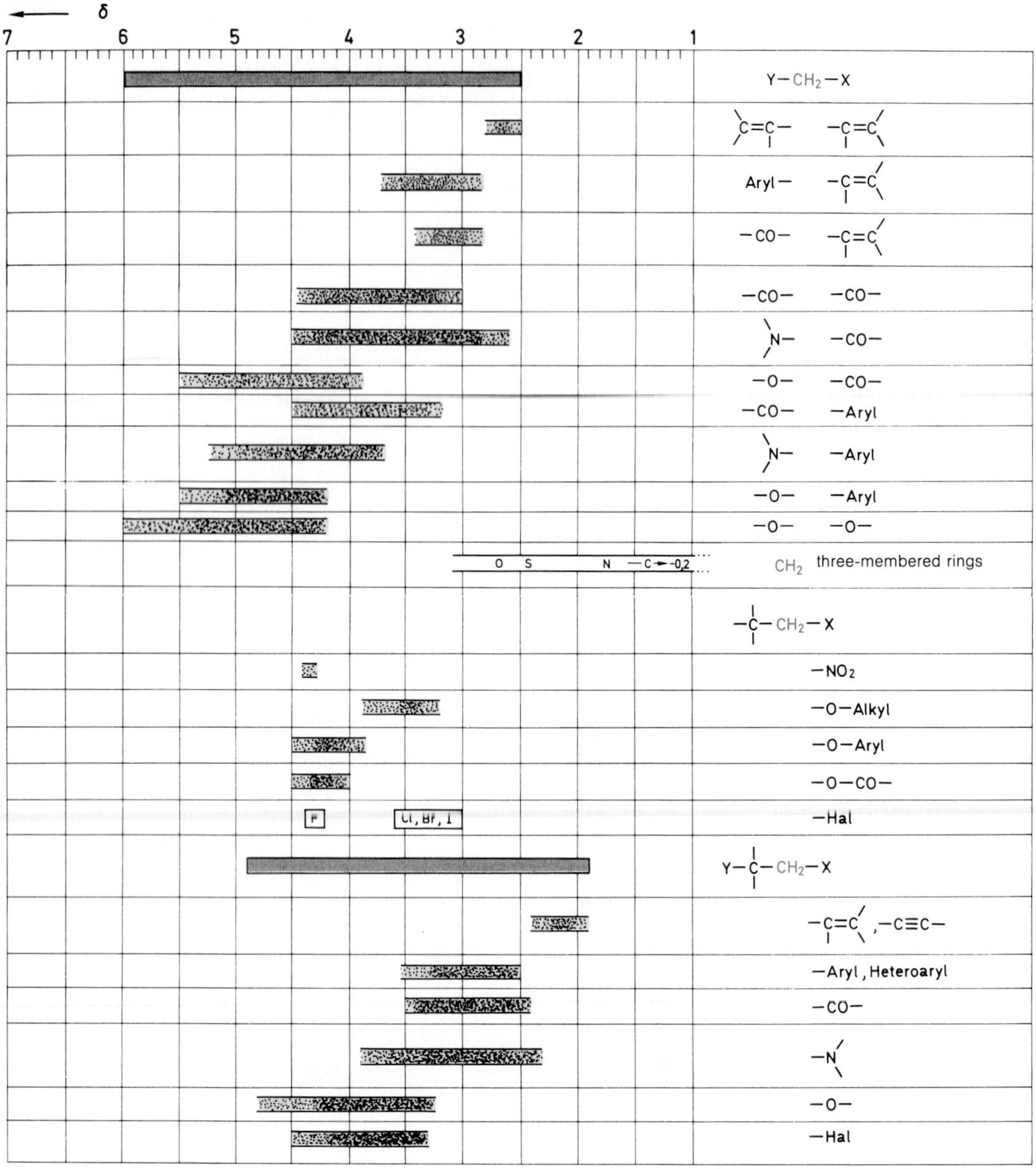

Tab. 3.14   Chemical shifts of methine groups (δ values measured in CCl₄ or CDCl₃)

Tab. 3.15 Chemical shifts of olefinic protons (δ values measured in CCl_4 or $CDCl_3$)

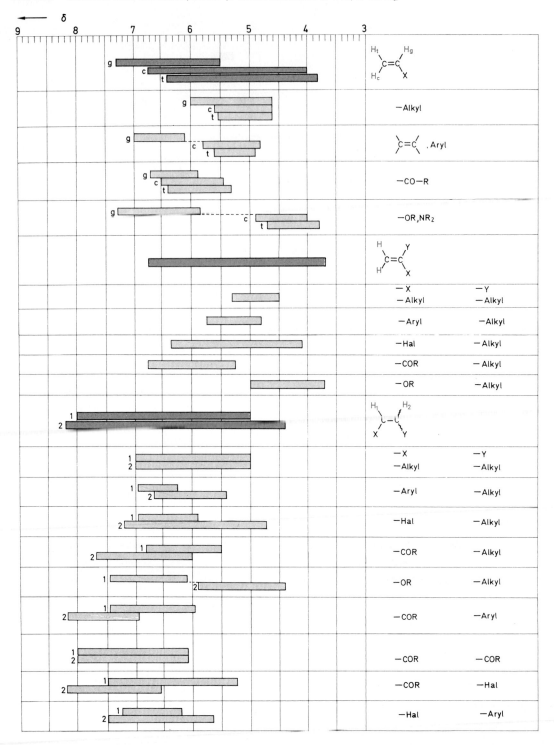

Tab. 3.15   continued

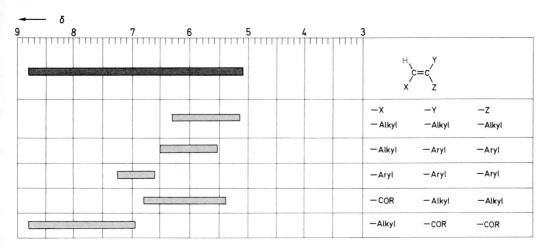

Tab. 3.16   Chemical shifts of aldehyde and aldimine protons (δ values measured in CCl₄ or CDCl₃)

Protons on aromatic and heteroaromatic rings. Protons bound to C atoms in aromatic or heteroaromatic rings have shifts between $\delta=6.0$ and 9.5, the majority being between 7 and 8. The variety of possibilities in this class of compounds is so great that producing tables for this introduction is inappropriate. Selected examples may be found in Sec. 5 (p. 185).

OH-, SH-, and NH-Protons. The chemical shifts of **protons** bound to heteroatoms such as O, S, or N depend strongly on the measurement conditions (concentration, temperature, solvent). The signals are often broad and/or show no couplings. The reason for this is the rapid intermolecular **proton exchange** (see also p. 109). If a rapid exchange occurs between two chemically non-equivalent groups within a molecule, only a single, averaged signal is observed.

Tab. 3.17 Chemical shifts of OH–, SH– and NH– protons (δ values measured in CCl_4 or $CDCl_3$)

To identify the signals of XH-protons **exchange with D_2O or trifluoroacetic acid** is often useful.

In Tab. 3.17 ranges for chemical shifts of XH-protons are summarised. The δ-values given are for carbon tetrachloride or deuteriochloroform as solvent. The OH-signals of alcohols generally shift to lower field with increasing concentration or decreasing temperature; both effects increase the association of the alcohol molecules and therefore the deshielding effect of the hydrogen bonds. The OH-signal of a very dilute solution of ethanol in CCl_4 lies at $\delta \approx 0.9$. With increasing concentration it shifts towards $\delta \approx 4.6$, the value of pure ethanol. **When assigning X−H signals the strong dependence of shift on concentration, temperature, and solvent must be taken into consideration.**

3.6 Increment Systems for Estimating ¹H Shifts

The large amount of known ¹H NMR data allows empirical rules for the dependence of the shifts of protons on their chemical environment to be established. These are based on the assumption of additivity of substituent effects, though this will not apply perfectly. Except in cases where particularly strong steric or electronic interactions occur reasonable estimates of δ-values can however be obtained.

A simple increment system for methylene and methyl protons is given in Tab. 3.18.

A few examples (Tab 3.19) serve to illustrate the comparison of calculated and observed values.

On a similar basis and with similar "quality" the shifts of olefinic protons can be estimated (Tab. 3.20).

The effects of substituents have been particularly carefully studied in the case of benzene derivatives. The limits of this approach are met where there are steric interactions of substituents (1,2-disubstituted, 1,2,3-trisubstituted benzenes, etc.) or unusual electronic interactions.

Tab. 3.18 Increment system for the estimation of chemical shifts of methylene and methine protons (modified Shoolery rules)

$R^1-CH_2-R^2$	$R^1-\underset{R^3}{\overset{}{C}H-R^2}$
$\delta = 1{,}25 + I_1 + I_2$	$\delta = 1.50 + I_1 + I_2 + I_3$
Substituent	Increment I
—Alkyl	0.0
—C=C—	0.8
—C≡C—	0.9
—C_6H_5	1.3
—CO—H, —CO-Alkyl	1.2
—CO—C_6H_5	1.6
—COOH	0.8
—CO—O-Alkyl	0.7
—C≡N	1.2
—NH_2, NH-Alkyl, N(Alkyl)$_2$	1.0
—NO_2	3.0
—SH, —S—Alkyl	1.3
—OH	1.7
—O-Alkyl	1.5
—O—C_6H_5	2.3
—O—CO-Alkyl	2.7
—O—CO—C_6H_5	2.9
—Cl	2.0
—Br	1.9
—I	1.4

Tab. 3.19 Calculated and measured δ-values in ¹H NMR of methylene and methine protons

Compound	$\delta_{calculated}$	$\delta_{measured}$
CH_2Br_2 Dibromomethane	$1.25 + 2 \cdot 1.9 = 5.05$	4.94
$C_6H_5-O-CH_2-C_2H_5$ Phenyl propyl ether	$1.25 + 2.3 + 0 = 3.55$	3.86
$Cl-\underset{a}{CH_2}-\overset{\overset{O}{\|\|}}{C}-O-\underset{b}{CH_2}-CH_3$ Ethyl chloroacetate	a: $1.25 + 2.0 + 0.7$ $= 3{,}95$ b: $1.25 + 2.7 + 0$ $= 3{,}95$	4.05 4.25
$H_3C-CHCl_2$ 1,1-Dichloroethane	$1.50 + 2 \cdot 2.0 + 0$ $= 5{,}50$	5.75
$C_2H_5-\underset{NO_2}{CH}-Cl$ 1-Choro-1-nitropropane	$1.50 + 2.0 + 3.0 + 0$ $= 6.50$	5.80
$(C_6H_5)_3CH$ Triphenylmethane	$1.50 + 3 \cdot 1.3 = 5.40$	5.56
$H_3C-\underset{b}{\overset{\overset{OH}{\|}}{C}H}-CH_2-\underset{a}{CH_2}-OH$ 1,3-Butanediol	a: $1.25 + 1.7 + 0$ $= 2.95$ b: $1.50 + 1.7 + 0 + 0$ $= 3.20$	3.80 4.03

Tab. 3.20 Increment system for the estimation of the chemical shifts of olefinic protons (Matter, U.E. et al.)

$$\delta = 5.25 + I_{gem} + I_{cis} + I_{trans}$$

Substituent	Increments		
	I_{gem}	I_{cis}	I_{trans}
–H	0	0	0
–Alkyl	0.45	−0.22	−0.28
–Alkyl ring*	0.69	−0.25	−0.28
–CH$_2$-Aryl	1.05	−0.29	−0.32
–CH$_2$OR	0.64	−0.01	−0.02
–CH$_2$NR$_2$	0.58	−0.10	−0.08
–CH$_2$–Hal	0.70	0.11	−0.04
–CH$_2$–CO–R	0.69	−0.08	−0.06
–C(R)=CR$_2$ (Diene)	1.00	−0.09	−0.23
extended conjugation	1.24	0.02	−0.05
–C≡C–	0.47	0.38	0.12
–Aryl	1.38	0.36	−0.07
–CHO	1.02	0.95	1.17
–CO–R (Enone)	1.10	1.12	0.87
extended conjugation	1.06	0.91	0.74
–CO–OH (α,β-unsaturated carboxylic acid)	0.97	1.41	0.71
extended conjugation	0.80	0.98	0.32
–CO–OR (α,β-unsaturated ester)	0.80	1.18	0.55
extended conjugation	0.78	1.01	0.46
–CO–NR$_2$	1.37	0.98	0.46
–CO–Cl	1.11	1.46	1.01
–C≡N	0.27	0.75	0.55
–OR (saturated)	1.22	−1.07	−1.21
–OR (other)	1.21	−0.60	−1.00
–O–CO–R	2.11	−0.35	−0.64
–S–R	1.11	−0.29	−0.13
–SO$_2$-R	1.55	1.16	0.93
–NR$_2$ (saturated)	0.80	−1.26	−1.21
–NR$_2$ (other)	1.17	−0.53	−0.99
–N–CO–R	2.08	−0.57	−0.72
–NO$_2$	1.87	1.32	0.62
–F	1.54	−0.40	−1.02
–Cl	1.08	0.18	0.13
–Br	1.07	0.45	0.55
–I	1.14	0.81	0.88

*applies to double bonds in five- or six-membered rings

Tab. 3.21 Calculated and measured δ values in ^{1}H NMR of olefinic protons

Compound	$\delta_{calculated}$		$\delta_{measured}$	
Methyl acrylate	5.80 6.43	6.05 –	5.82 6,38	6.20 –
4-Chlorostyrene	5.18 5.61	6.63 –	5.28 5.73	6.69 –
Methyl 2-methacrylate	5.58 6.15	– –	5.57 6.10	– –
3,4-Dihydro-2H-pyran	4.73 6.19	– –	4.65 6,37	– –
(E)-Cinnamic acid	7.61 –	– 6.41	7.82 –	– 6.47
Ethyl 2-cyano-(Z)-cinnamate	7.84 –	– –	8.22 –	– –

*The order of the shift values corresponds to the position of the hydrogen atoms in the structural formula

Tab. 3.22 Increment system for the estimation of chemical shifts of benzene protons

$\delta = 7{,}26 + \Sigma I$

Substituent	I_{ortho}	I_{meta}	I_{para}
—H	0	0	0
—CH$_3$	−0.18	−0.10	−0.20
—CH$_2$CH$_3$	−0.15	−0.06	−0.18
—CH(CH$_3$)$_2$	−0.13	−0.08	−0.18
—C(CH$_3$)$_3$	0.02	−0.09	−0.22
—CH$_2$Cl	0.00	0.01	0.00
—CH$_2$OH	−0.07	−0.07	−0.07
—CH$_2$NH$_2$	0.01	0.01	0.01
—CH=CH$_2$	0.06	−0.03	−0.10
—C≡CH	0.15	−0.02	−0.01
—C$_6$H$_5$	0.30	0.12	0.10
—CHO	0.56	0.22	0.29
—CO—CH$_3$	0.62	0.14	0.21
—CO—CH$_2$—CH$_3$	0.63	0.13	0.20
—CO—C$_6$H$_5$	0.47	0.13	0.22
—COOH	0.85	0.18	0.25
—COOCH$_3$	0.71	0.11	0.21
—CO—O—C$_6$H$_5$	0.90	0.17	0.27
—CO—NH$_2$	0.61	0.10	0.17
—COCl	0.84	0.20	0.36
—CN	0.36	0.18	0.28
—NH$_2$	−0.75	−0.25	−0.65
—NH—CH$_3$	−0.80	−0.22	−0.68
—N(CH$_3$)$_2$	−0.66	−0.18	−0.67
—N$^+$(CH$_3$)$_3$I$^-$	0.69	0.36	0.31
—NH—COCH$_3$	0.12	−0.07	−0.28
—NO	0.58	0.31	0.37
—NO$_2$	0.95	0.26	0.38
—SH	−0.08	−0.16	−0.22
—SCH$_3$	−0.08	−0.10	−0.24
—S—C$_6$H$_5$	0.06	−0.09	−0.15
—SO$_2$—OH	0.64	0.26	0.36
—SO$_2$—NH$_2$	0.66	0.26	0.36
—OH	−0.56	−0.12	−0.45
—OCH$_3$	−0.48	−0.09	−0.44
—OCH$_2$—CH$_3$	−0.46	−0.10	−0.43
—O—C$_6$H$_5$	−0.29	−0.05	−0.23
—O—CO—CH$_3$	−0.25	0.03	−0.13
—O—CO—C$_6$H$_5$	−0.09	0.09	−0.08
—F	−0.26	0.00	−0.20
—Cl	0.03	−0.02	−0.09
—Br	0.18	−0.08	−0.04
—I	0.39	−0.21	−0.03

Tab. 3.23 Calculated and measured δ-values in ¹H NMR of benzene protons

Compound	$\delta_{calculated}$			$\delta_{measured}$		
p-Xylene	6.98 6.98	— 6.98 —	6.98 6.97 6.97	7.05	— — —	7.05 7.05 7.05

p-Xylene: — calc: 6.98 / 6.98 / — / 6.98 / 6.98; meas: — / 6.98 / 6.97 / 6.97 ... 7.05

Compound	$\delta_{calculated}$			$\delta_{measured}$		
o-Xylene	6.96 / 6.96 / 6.98 (6.98)	—	7.05 / 7.05 (7.05)			
1-Chloro-4-nitrobenzene	8.19 / 7.55	—	8.19 / 7.55 / 8.17 / 7.52	—		8.17 / 7.52
4-Chloroaniline	6.49 / 7.04	—	6.49 / 7.04 / 6.57 / 7.05	—		6.57 / 7.05
1,3-Diaminobenzene (m-phenylenediamine)	5.76 / 5.86 / 6.76	—	5.86 / 6.11	6.03 / 6.11 / 6.93	—	
1,3,5-Trimethylbenzene (mesitylene)	6.78 / 6.78	—	6.78 / 6.78	6.78 / 6.78	—	
2,4-Dinitro-1-methoxybenzene	7.30 / 8.38	— / 9.07	7.28 / 8.47	— / 8.72		

Figures in Tab. 3.23 (structures with calculated and measured δ-values):

p-Xylene

o-Xylene

1-Chloro-4-nitrobenzene

4-Chloroaniline

1,3-Diaminobenzene (m-phenylenediamine)

1,3,5-Trimethylbenzene (mesitylene)

2,4-Dinitro-1-methoxybenzene

Tab. 3.23 continued

Compound	$\delta_{calculated}$		$\delta_{measured}$	
2,4,5-Trichlorotoluene	7.07 —	— 7.20 —	7.19 —	— 7.31 —
3,5-Diphenyl-4-hydroxybenzaldehyde	8.10 —	8.10 —	7.83 —	7.83 —

3.7 ^{1}H NMR Data of Representatives of the Commoner Classes of Compounds

A collection of **^{1}H NMR data** in tabular form is given together with ^{13}C NMR data of selected structural examples in Sec. 5 (p. 185).

3.8 Specialised Techniques

Increasing the Measurement Frequency/Field Strength

As described in Sec. 1.1 (p. 71), the chemical shift measured in Hz increases linearly with the magnetic field B_0. The coupling constants J however are independent of B_0. If for example a 60 MHz spectrum shows groups of signals which overlap, then for an exact analysis a higher field spectrum, for example at 400 MHz, is recommended. Such high-field instruments employ **superconducting magnets**, cooled with liquid helium, which produce a very strong, but nevertheless homogeneous magnetic field. The improved resolution capability allows a better separation of neighbouring signals; the increased sensitivity allows the observation of very weak lines. Of decisive importance is the change in the ratio $\Delta v/J$. In Sec. 1.3 (p. 74) the ratio $\Delta v/J$ was used as the criterion for the appearance of first or second order spectra. The increase of this ratio on increasing the measurement frequency (field strength) can change the order and thus the appearance of the spectrum drastically. For example an ABC spectrum can change into an AMX spectrum. In general this is a decisive advantage; but

information can be lost as a result: the signs of the coupling constants cannot be determined from an AMX spectrum.

In Fig. 3.38 the 60 MHz ^{1}H NMR spectrum of strychnine (**124**) is reproduced. For comparison the regions of the aromatic and saturated aliphatic protons from a 250 MHz measurement are reproduced below the 60 MHz spectrum.

With the aid of methods described later in this chapter, such as selective deuteration, double resonance experiments, etc., it is possible to assign the signals of the individual protons and their couplings from the 250 MHz spectrum of even such complicated molecules as strychnine.

Spectrometers are currently commercially available for fields up to 750 MHz (17.63 T).

^{1}H NMR data for strychnine (**124**)

Position	δ (± 0.004)	J (Hz) (± 0.10)
2	3.846	$J_{2-16} = 10.47$
3	3.924	
5a	2.861	$J_{5a-5b} = 9.88$
5b	3.185	
6a	1.869	$J_{6a-6b} = 0.02$
		$J_{6a-5a} = 10.06$
		$J_{6a-5b} = 4.85$
6b	1.870	$J_{6b-5a} = 8.54$
		$J_{6b-5b} = 3.33$
9	7.145	$J_{9-10} = 7.41$
		$J_{9-11} = 1.18$
		$J_{9-12} = 0.45$
10	7.076	$J_{10-11} = 7.46$
		$J_{10-12} = 1.12$
11	7.230	$J_{11-12} = 8.10$
12	8.085	
14a	1.430	$J_{14a-14b} = 14.37$
		$J_{14a-3} = 1.82$
14b	2.338	$J_{14b-3} = 4.11$
15	3.126	$J_{15-14a} = 1.98$
		$J_{15-14b} = 4.58$
		$J_{15-18a} = 2.5$
16	1.252	$J_{16-15} = 3.10$
17	4.266	$J_{17-16} = 3.12$
18a	4.047	$J_{18a-18b} = 14.19$
18b	4.127	
19	5.881	$J_{19-18a} = 5.74$
		$J_{19-18b} = 6.88$
21a	2.712	$J_{21a-21b} = 14.83$
21b	3.691	$J_{21b-19} = 1.2$
23a	3.105	$J_{23a-23b} = 17.40$
		$J_{23a-17} = 8.44$
23b	2.657	$J_{23b-17} = 3.28$

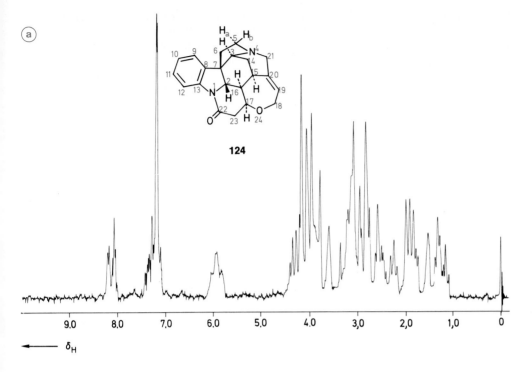

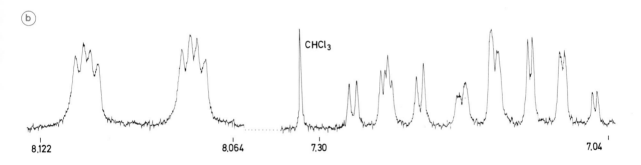

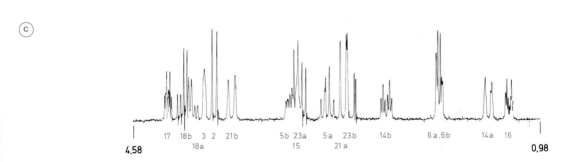

Fig. 3.38 ¹H NMR spectra of strychnine (**124**) in CDCl₃, **a** 60 MHz spectrum **b** and **c** sections of a 250 MHz spectrum (Carter, J. C., Luther, G. W., Long, T. C. (1974), J.Magn. Reson. **15**, 122)

Measurement of Limited Amounts of Sample

For the measurement of ¹H NMR spectra of samples which are only available in limited amounts microcells with a volume of 100 μl can be used instead of the conventional 5 mm diameter tubes. For samples which are less than 0.1 mg special probes can be employed, in conjunction with as high a field B_0 as possible, since the achievable sensitivity is proportional to $B_0^{3/2}$. Microgram quantities are then measurable on a routine basis, sub-microgram quantities are possible with overnight accumulations.

Variation of Solvent

Solvent effects can be deliberately used in NMR for solving specific problems. In the sections on p. 100 and p. 109 mention

has already been made of the recognition of acidic protons by exchange with D_2O or trifluoroacetic acid and the slowing down of proton exchange in dimethylsulphoxide as solvent. Here the use of aromatic solvents such as hexadeuteriobenzene or pentadeuteriopyridine will be discussed. The formation of preferred encounter complexes in the solvation of the compounds studied combines with strong intermolecular anisotropy effects to produce large changes in shifts compared with those measured in CCl_4 or $CDCl_3$ solutions. An illustrative example is *O*-methyl phenylthioacetate (**125**) (Fig. 3.39).

In $CDCl_3$ the signals of the methyl and methylene protons are coincident. Such coincidental isochronicities can easily lead to errors in structural interpretation. The spectrum measured in C_6D_6 however shows the expected two separate signals, with intensity ratio 3 : 2.

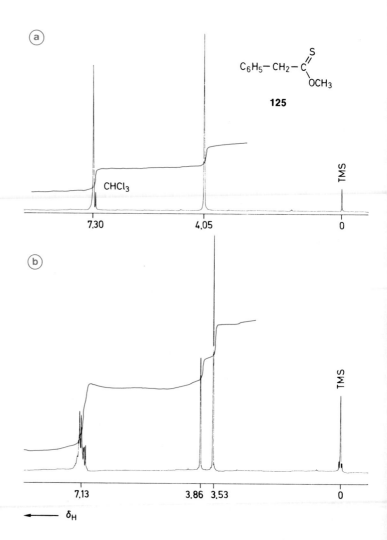

Fig. 3.39   ¹H NMR spectra of *O*-methyl phenyl-thioacetate (**125**) with integration **a** in $CDCl_3$; **b** in C_6D_6

To conclude this section brief mention will be made of the use of **chiral solvents**, such as (R)- or (S)-2,2,2-trifluoro-1-phenyl-ethanol. They can be used to determine optical purity and in suitable cases can give indications of absolute configuration. The NMR spectra of enantiomers (+)A and (−)A are identical in optically inactive solvents. In an optically active medium S **diastereomeric solvation complexes** can be formed, which will have different spectra

$$(+)A \cdots (+)S \neq (-)A \cdots (+)S.$$

In principle enantiotopic protons in an achiral molecule can become anisochronous when complexed to a chiral (solvent) partner. Alternatively, chemically non-equivalent protons may become coincidentally isochronous. Only homotopic groups remain chemically equivalent in chiral media (see Sec. 2.2, p. 88).

Selective Deuteration

The simplification of NMR spectra can occasionally be usefully achieved by the replacement of selected H atoms by deuterium. **Acidic protons** undergo such exchange simply on shaking with D_2O (see p. 121). In other cases a **specific synthesis of the relevant deuteriated compound** must be undertaken. In the ¹H NMR spectrum of the deuteriated compound the signal of the substituted proton will then be absent. The other shifts remain practically unchanged by the deuteriation. However the different coupling behaviour of ¹H and D must be taken into account. Deuterium has a nuclear spin $I=1$, which changes the number of lines and the intensities of the splitting pattern (see Tab. 3.2, p. 77).

Furthermore the ¹H,D couplings are considerably smaller than the corresponding ¹H,¹H couplings and are therefore often not observed in the spectrum.

$$J(H, H) \approx 6.5 \; J(H, D)$$

Apart from the simplification of the spectrum the deuteriation can also be utilised for the determination of couplings between magnetically equivalent nuclei. A proton is replaced by a deuteron, the J(H,D) coupling is measured, and the J(H,H) coupling calculated.

For a better understanding of the changes in ¹H NMR spectra caused by H/D exchange the spectra of cyclopropene (**10**) and its two deuteriated derivatives (**126**) and (**127**) will be discussed (Fig. 3.40).

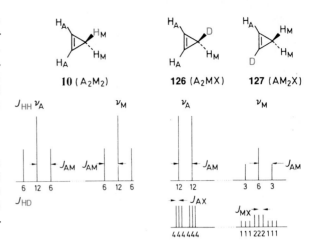

Fig. 3.40 ¹H NMR "stick" spectra of cyclopropene (**10**) and its deuterium derivative (**126**) and (**127**)
(The numbers below the lines indicate the relative intensities)

From $J_{MX} = {}^2J(H,D)$ in (**126**) the *geminal* coupling between the chemically and magnetically equivalent H_M protons in cyclopropene (**10**) can be calculated.

If an H_a proton is replaced by deuterium, an AM_2X spin system (**127**) is obtained, corresponding to the right hand part of Fig. 3.40 with A and M exchanged. ${}^3J_{AA}$ can then be determined in an analogous fashion to ${}^2J_{MM}$.

A further, very interesting application of deuterium incorporation is seen in the **isotopic perturbation method** developed by Saunders. Because of the identical lineshapes, even at low temperature, it is impossible to distinguish between a degenerate exchange process with sufficiently low activation barrier and a single symmetric structure. The two indistinguishable alternatives are the dynamic model with a double potential energy minimum and the static model with a single minimum. The dimethylisopropyl carbonium ion (**128**) can be considered as an example. The twelve methyl protons give just one doublet at $\delta = 2.93$ when measured in SbF_5/SO_2ClF at $-100°$. This can be explained either by a rapid hydride shift or by a bridged ionic structure. The first hypothesis was shown to be correct.

If one of the methyl protons is replaced by deuterium, the degeneracy is lifted, and two doublets, separated by v, are observed.

A   **129**   B

The lower field signal is due to the six protons of (**129A**) shown in blue, which are in fast exchange with the protons of (**129B**) also shown in blue. At slightly higher field is the signal of the five remaining protons, also a doublet, and also undergoing fast exchange.

The fact that the methyl signal of the six green protons lies at lower field proves that (**129A**) is preferred over (**129B**). The separation of the signals and the equilibrium constant K are related by

$$v = \frac{[\delta_1 c(\mathbf{129\,A}) + \delta_2 c(\mathbf{129\,B})] - [\delta_2 c(\mathbf{129\,A}) + \delta_1 c(\mathbf{129\,B})]}{c(\mathbf{129\,A}) + c(\mathbf{129\,B})}$$

$$K = \frac{c(\mathbf{129\,B})}{c(\mathbf{129\,A})} = \frac{\omega + v}{\omega - v}$$

v Shift difference between the two signals (from isotopic perturbation)

$\omega = \delta_2 - \delta_2$ hypothetical shift difference if equilibrium were frozen out (must be estimated!)

In this particular case at $-56°C$ $K = 1.132 = 53:47$. A splitting of the doublet signal would also be expected in the case of a single bridged carbonium ion; the separation would however be very small, since the effect of deuterium substitution on the ^1H shift is very small.

Use of Shift Reagents

The presence of paramagnetic ions in an NMR sample often causes drastic shifts of the signals of nucleophilic species. **Shift reagents** use this effect in a systematic fashion. The most useful ions have proved to be Eu(III) and Yb(III), which both generally cause low-field shifts, and Pr(III), which causes a high-field shift. Chelate complexes of these ions with β-diketones are relatively well soluble in organic solvents. Commonly used are:

Eu(dpm)₃

dpm: 2,2,6,6-tetramethyl-3,5-heptanedione (dipivaloylmethane)

      130

Eu(fod)₃ and Eu(fod)₃−d_{27}

fod: 6,6,7,7,8,8,8-heptafluoro-2,2-dimethyl-3,5-octanedione

      131

and the chiral shift reagents

Eu(facam)₃

facam: 3-trifluoroacetyl-d-camphor

      132

Eu(hfbc)₃

hfbc: 3-heptafluorobutyryl-d-camphor

      133

Compounds with nucleophilic groups complex reversibly with the central lanthanide atom, e.g.

The resultant shifts of the NMR signals increase with the stability of the complexes and with increased concentration of the shift reagent. In Fig. 3.41 the ^1H NMR spectrum of dibutylether (**134**) is reproduced. With increasing dosage of Eu(fod)₃ the signals become more and more separated.

A quantitative relationship between the change in chemical shift Δv_i for a nucleus i and the location of this nucleus in a so-called **pseudocontact complex** is given by the **McConnell-Robertson equation**

$$\frac{\Delta v_i}{v_i} = K \cdot \frac{3 \cos^2 \Theta_i - 1}{r_i^3}.$$

According to this equation Δv_i is inversely proportional to the third power of the distance r_i of the nucleus i and the central lanthanide atom. K is a proportionality constant and Θ the angle between the principal magnetic axis of the complex and the line joining the nucleus i to the lanthanide ion. It is usually assumed that the principal magnetic axis is coincident with the bond between the central atom and the nucleophilic centre.

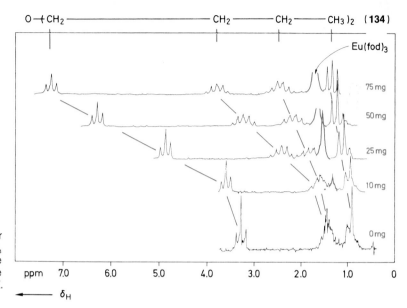

Fig. 3.41 ¹H NMR spectra of a 10^{-4} molar solution of dibutyl ether (**134**) in 0.5 ml CCl$_4$ with increasing addition of Eu(fod)$_3$. The blue signal arises from the *t*-butyl group in the shift reagent (Rondeau, R. E., Rievers, R. E. (1971), J. Am. Chem. Soc. **93**, 1522)

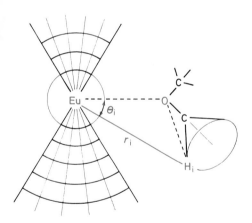

Fig. 3.42 Visualisation of the McConnell-Robertson relation in the pseudo-contact complex $(C_4H_9)_2O \cdots Eu(fod)_3$

The angular term in the McConnell-Robertson equation can in no way be ignored: in the region from 0 and 55° it is positive, from 55 and 125° it is negative.

The statement that europium causes a low-field shift should not be true for the sector shown dark in Fig. 3.42. (In the case of the pseudo-contact complex with di-*n*-butyl ether there is however no proton in this region.)

The potential usefulness of shift reagents in the analysis of NMR spectra is shown by the example of 2β-androstanol (**135**,

Fig. 3.43). In the normal spectrum only the two methyl groups and the $H_{2\alpha}$ proton can be distinguished. The remaining signals overlap. On addition of Eu(dpm)$_3$ all the protons in the vicinity of the complexation site can be identified.

On p. 127 it has already been noted that the determination of enantiomer ratios (optical purity) can be carried out by measuring NMR spectra in an optically active solvent. The differences in the chemical shifts of corresponding protons in the two enantiomers are however often very small. This disadvantage can often be overcome by measurement in CCl$_4$ or CDCl$_3$ with addition of a chiral shift reagent. Fig. 3.44 shows the ¹H NMR spectrum of racemic 2-phenyl-2-butanol (**136**) in the presence of an achiral and a chiral shift reagent. The 1:1 ratio of the two enantiomers is clearly shown by the splitting of the methyl signal in the lower spectrum.

Measurement of ¹³C Satellites

Coupling between magnetically equivalent nuclei has no visible effect on the spectrum. The ¹H spectrum of (*E*)-1,2-dichloroethylene (**137**) is an A$_2$ singlet.

137   **138**

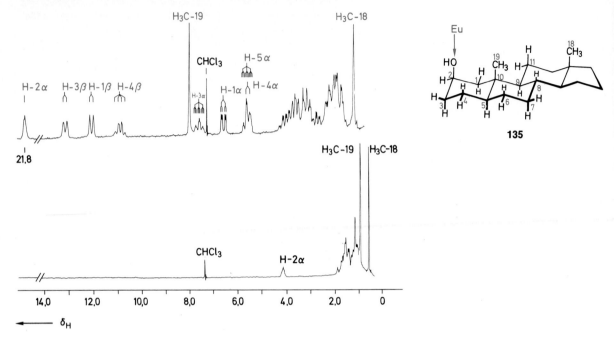

Fig. 3.43 ¹H NMR spectra of 2β-androstanol (**135**; 0.73 · 10⁻⁴ molar solution in 0.4 ml CDCl₃). The lower part shows the normal spectrum; for the upper spectrum 40 mg Eu(dpm)₃ were added (Demarco, P. V. et al. (1970), J.Am. Chem. Soc. **92** 5737)

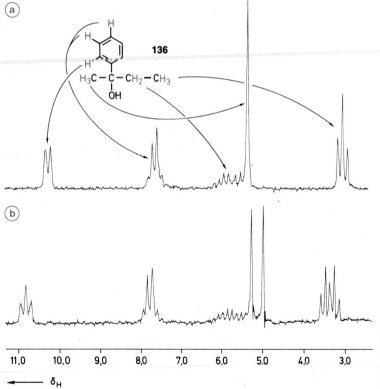

Fig. 3.44 ¹H NMR spectra of a 0.54 molar solution of 2-phenyl-2-butanol (**136**) in CCl₄ (Goering, H. L. et al. (1971), J.Am. Chem. Soc. **93**, 5913) **a** with 0.13 molar added Eu(dpm)₃ **b** with 0.42 molar added Eu(facam)₃

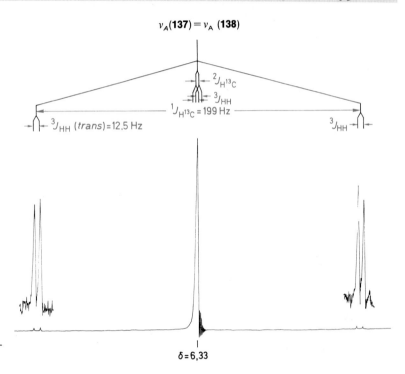

Fig. 3.45 ¹H NMR spectra of (E)-1,2-di-chloroethylene (**137**) with the ¹³C satellites

$\delta = 6,33$

If the natural ¹³C content of 1.1% is considered however, about 2% of (E)-1,2-dichloroethylene is (**137**), only the remaining 98% being (**138**). (**138**) is an AA'X system. (The isotope effect on the ¹H shift is negligibly small). Fig. 3.45 shows the A-part (¹H signals) of the **satellite spectrum**, which lies symmetrically about the singlet signal of (**137**) with the corresponding intensity. Since $^2J_{A'X}$ is very small, the inner lines lie under the strong main signal of the A_2 system.

From the satellite spectrum the coupling constant $^1J_{XA}$ for the direct ¹³C,¹H coupling can be extracted. The same value can naturally be obtained by measuring the X part, i.e. the coupled ¹³C spectrum.

Furthermore the ¹H,¹H coupling can be extracted from the satellite spectrum. In contrast to (**137**) the two protons in (**138**) are no longer magnetically equivalent so the coupling constant between them can be determined from the spectrum. For the molecule (**137**) one can assume again neglecting any isotope effect the same value. In this case it is found to be 12.5 Hz, corresponding to the (E)-configuration. In this way the configuration of symmetrically 1,2-disubstituted ethylenes can be determined.

The coupling between *geminal*, magnetically equivalent protons can neither be determined from the normal ¹H spectrum nor from the ¹³C satellites.

The ^2J(H,H) coupling in methyl or methylene groups can however be determined by deuteriation (see p. 127).

Spin Decoupling (Multiple Resonance)

The spin-spin coupling of magnetically non-equivalent nuclei causes, as described in Sec. 1.3 (p. 74) splitting of the signals. This often leads to complex multiplets. To simplify such spectra and to determine coupling partners **spin decoupling** can be performed by **double resonance**.

Let us consider the simple example of an AMX system ($I = 1/2$). The corresponding spectrum consists of four lines of equal intensity for each nucleus.

If the sample is subjected to an additional irradiation at the resonance frequency v_M, then the nuclei A and X do not "see" two separate spin states for M, because the spin orientation of M is changing too rapidly. The average value is zero, i.e. M no longer couples with A and X. At the position of the signal of M an oscillation occurs (in CW spectra); otherwise the spectrum is simplified to two doublets with the coupling constant J_{AX}. The AMX spectrum is thus reduced to an AX spectrum (Fig. 3.46). Similarly the irradiation can be set to v_X or v_A. In **triple resonance** the irradiation is carried out at **two** additional frequencies. The AMX spectrum then reduces to a singlet.

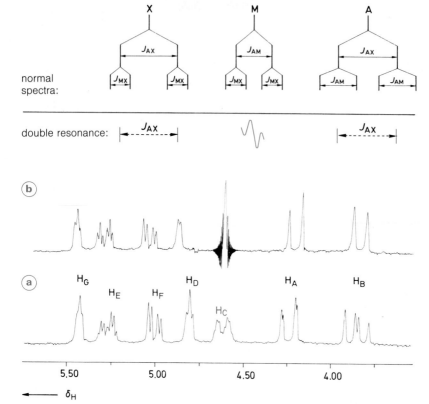

Fig. 3.46 Stick spectra of an AMX system and its simplification on additional irradiation at v_m

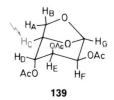

139

Fig. 3.47 **a** 100 MHz ^{1}H NMR spectrum of mannose triacetate (**139**)
b Double resonance spectrum in frequency sweep on irradiating at v_{HC} (L. F. Johnson, (1965), Varian Inform. Bull, 5)

If the signals lie close together, it is not possible to irradiate at the frequency of one nucleus without perturbing the transitions of other nuclei. Consider an ABX case. If the amplitude of the decoupling field is reduced so that not the whole X part, but only two of the X lines are affected, then only a part of the AB system is affected. This is called **selective decoupling**.

Finally, if the amplitude of the decoupling field is so much reduced that only a single line (e.g. in the X part) is affected, then all the other lines in the spectrum which have common energy levels with the irradiated line are split. This effect is known as **spin tickling**. Further information on these methods is available from the literature quoted in the bibliography. Only the normal double resonance experiment will be illustrated by a few specific examples.

Compound (**139**), mannose triacetate, has the ^{1}H NMR spectrum shown in Fig. 3.47 Apart from the acetate methyl groups there are seven non-equivalent types of protons, H_A to H_G. By irradiating at the frequency of H_C all couplings to H_C are eliminated.

Because of the dihedral angle the largest vicinal coupling is J_{CB}. The effect of the decoupling is therefore greatest for H_B. But the signals of H_A, H_D, and H_E are also recognisably simplified. The ABCDEFG system is transformed to an AB system and a DEFG system. At the position of $v(H_C)$ an oscillation is seen which results from the interference between the observation and irradiation frequencies.

In Fig. 3.47 it is also apparent that the irradiation causes a slight shift of the observed signals (**Bloch-Siegert effect**).

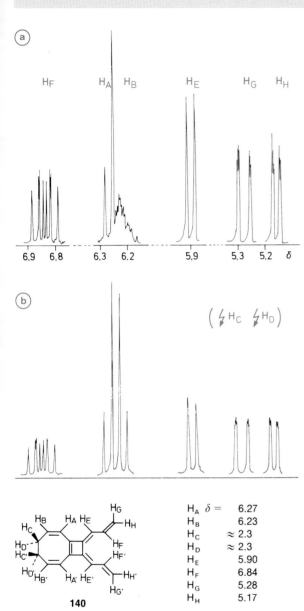

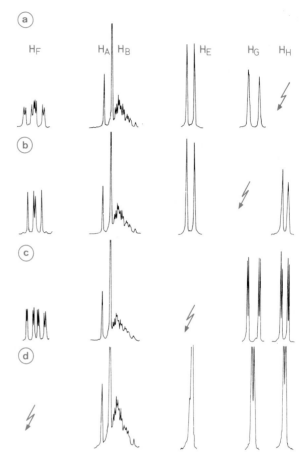

Fig. 3.49 Double resonance measurements on (**140**) (400 MHz, CDCl₃) Irradiations:
a at H$_H$ **b** at H$_G$ **c** at H$_F$

H$_A$ $\delta =$	6.27
H$_B$	6.23
H$_C$	≈ 2.3
H$_D$	≈ 2.3
H$_E$	5.90
H$_F$	6.84
H$_G$	5.28
H$_H$	5.17

140

Fig. 3.48 **a** 400 MHz ¹H NMR spectrum (olefinic part) of (**140**) in CDCl₃
b Triple resonance (irradiation at H$_C$ and H$_D$)

A further example is the polyolefin (**140**), which has a complicated ¹H NMR spectrum. In the low field region a 400 MHz spectrum shows six signals for the olefinic protons H$_A$, H$_B$, H$_E$, H$_F$, H$_G$, and H$_H$ (Fig. 3.48). On irradiation at the frequencies of the saturated protons H$_C$ and H$_D$ all the multiplets remain unchanged except for H$_A$ and H$_B$, the signals of which (AB part of an AA'BB'CC'DD' system) are simplified to an AB

system with $^3J_{AB} = 11.0$ Hz. The assignment of the signals of the protons in the side chains attached to the four-membered ring is possible from the double resonance experiments in Fig. 3.49. Irradiating H$_H$ leads to the disappearance of the couplings $^2J_{G,H} = 1.8$ Hz, $^3J_{F,H} = 10.1$ Hz, and $^4J_{E,H} = 0.8$ Hz; irradiating H$_G$ eliminates $^3J_{F,G} = 16.6$ Hz and $^4J_{E,G} = 0.9$ Hz as well as $^2J_{G,H}$; irradiating H$_E$ removes $^3J_{E,F} = 11.3$ Hz as well as $^4J_{E,G}$ and $^4J_{E,H}$. Irradiating H$_F$ achieves the greatest simplification of the spectrum and serves as a control experiment.

All the cases discussed here involve ¹H NMR. Apart from these homonuclear double resonance experiments heteronuclear double resonance is also possible. In ¹H NMR this is occasionally used for compounds with NH groups. Irradiation at the ¹⁴N frequency can remove the line broadening usually observed for such systems. An excellent example is the ¹H NMR spectrum of formamide (**141**) (Fig. 3.50).

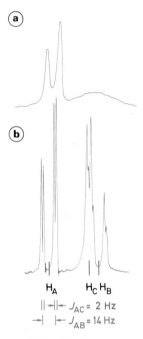

(a)

(b)

H_A H_C H_B

$\| {\dashv}\!\|{\vdash}\ J_{AC} = 2\ Hz$

${\dashv}\ \ \vdash\ J_{AB} = 14\ Hz$

Fig. 3.50 ^{1}H NMR spectrum of formamide (**141**)
a normal spectrum
b Double resonance experiment (irradiation at the ^{14}N Absorption frequency)

Heteronuclear double resonance is especially important in ^{13}C NMR spectroscopy (see Sec. 4.1, p. 142, and p. 163 ff.).

$$H_A\!\!-\!\!\overset{O}{\underset{}{C}}\qquad\longleftrightarrow\qquad H_A\!\!-\!\!\overset{\overset{..}{O}{}^-}{\underset{}{C}}$$

141

NOE Difference Spectroscopy

As mentioned in Sec. 1.5 (p. 84) the **nuclear Overhauser effect (NOE)** is apparent in double resonance ^{1}H NMR experiments. The intensity of an observed ^{1}H signal ν_A can be increased by additional irradiation at ν_B. The precondition for this effect is that the distance r between the nuclei A and B is small, since the dipole-dipole interaction which is responsible for longitudinal relaxation in such cases is proportional to $1/r^6$.

To decide, for example, whether compound (**142**) exists in the *exo*- or *endo*-configuration, the intensities of the ^{1}H signals on irradiating the signal of the bridgehead proton H_e can be compared with the normal spectrum where no irradiation is carried out. For the *exo*-compound (**142a**) the intensity of the signal of H_c (as well as H_d) should be increased on irradiation; for the *endo*-compound, on the other hand, the signal of H_h should increase in intensity. Since the effects are often small,

it is recommended to use **difference spectroscopy**; that is, the normal spectrum is subtracted from the double resonance spectrum. (Even better comparison is achieved by using as the "normal" spectrum one irradiated where there is no signal). Fig. 3.51 shows an increase of intensity for H_c and H_d i.e. **142** has the exo-configuration.

142 a
exo

142 b
endo

Apart from configurational determinations NOE measurements can be used with advantage in conformational analysis.

INDOR Technique

A further useful method is the **INDOR technique (INternuclear DOuble Resonance).** This involves monitoring the intensity of one individual line (monitor line) while an additional irradiating field is swept across the remaining spectral region. Whenever this irradiating field reaches the frequency of a line in the spectrum which shares an energy level with the monitor line, the intensity of the monitor line is altered as a consequence of the **general Overhauser effect.** If the lower level (starting level) of the monitor line is more strongly populated, an intensity increase is observed; if the upper level is more strongly populated, then the intensity decreases. The "**spin pumping**" leads in the case of **progressive transitions** to a positive line in the difference spectrum, in the case of **regressive transitions** to a negative line. If a transition has no energy level in common with the monitor line, then its line is not observed in the **INDOR difference spectrum.** To illustrate the effect consider the simple example of an AX spin system. The four possible spin orientations and the corresponding energy levels are shown in Fig. 3.3. If the case where $J < 0$ is considered, then the A_2 line as monitor line will be increased in intensity when the X_1 transition is irradiated, conversely it will be decreased in intensity when the X_2 transition is irradiated. If X_2 is chosen as the monitor line, then at A_1 a positive signal is obtained, at A_2 a negative signal (Fig. 3.52).

To determine the coupling partners of a nucleus A with a decoupling experiment the exact frequency of A must be found. When the signals are complex and/or unsymmetrical this is far more difficult than selecting an individual line for an INDOR experiment. In cases where signals overlap badly the INDOR technique is often preferable to a decoupling experiment. As described above INDOR spectroscopy is exclusively a CW experiment; there are however pseudo-INDOR experiments using the PFT method, so that INDOR difference spectroscopy is a routine experiment even with an FT spectrometer.

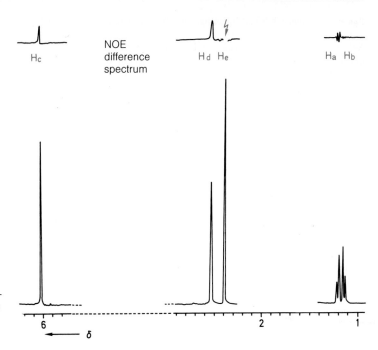

Fig. 3.51 400 MHz ¹H NMR spectrum of the nor-bornene derivative (**142a**) and NOE difference spectrum in a deoxygenated CDCl₃ solution

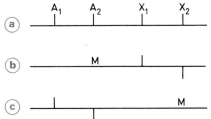

Fig. 3.52 Representation of INDOR difference spectroscopy on an AX spin system with $J < 0$
a Four line spectrum
b INDOR difference spectrum using A_2 as the monitor line
c INDOR difference spectrum using X_2 as the monitor line

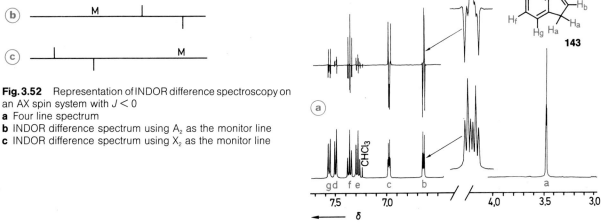

Fig. 3.53 400 MHz ¹H NMR spectrum of indene (**143**) in CDCl₃;
a normal spectrum, **b** INDOR difference spectrum

The INDOR spectrum of indene (**143**) will be discussed as a representative example. The methylene protons give a complex, symmetrical multiplet at $\delta = 3.48$. If its maximum is used for an INDOR experiment, the spectrum shown in Fig. 3.53 is obtained. The principal coupling partners of the CH₂ protons are the olefinic protons ($^3J = +2.20$ Hz, $^4J = 1.98$ Hz); couplings to the aromatic protons, particularly H_f, are however also apparent.

Two-dimensional ^{1}H NMR Spectroscopy (2D-^{1}H-NMR)

As explained in Sec. 3.1 (p. 102), the PFT method involves the measurement of the decay of the transverse magnetisation (FID). The detected signal is a function of the **detection time** t_2. If the **evolution time** t_1, i.e the time between the first pulse of the chosen pulse sequence and the start of data acquisition, is systematically varied, then the detected signal is a function of t_1 and t_2. The Fourier transformation now has two frequency variables F_1 and F_2, the requirement for a two-dimensional spectrum.

There is a basic distinction between **J-resolved** and **shift-correlated 2D spectra**.

For **homonuclear J-resolved 2D-experiments** the initial 90° pulse is followed by an evolution time t_1, in the middle of which a 180° pulse is inserted. In successive measurements t_1 is increased by a constant amount. During the evolution time t_1 only the scalar coupling develops, i.e. the F_1 information only contains the couplings. The F_2 information contains the whole spectrum as usual. After suitable mathematical processing the result is presented graphically as a **contour plot**, as shown in Fig. 3.54 for a triplet and a quartet. It can be considered as a view from above onto the pattern of peaks, which appear as contour lines.

Projection onto the F_2 axis produces a **fully decoupled ^{1}H NMR spectrum**. Each signal appears as a singlet. It is obvious that this method is interesting for large molecules with many overlapping multiplets. The projection of individual signals onto the F_1 axis (slices, cross-sections) allows the multiplets to be drawn out separately. Weakly coupled spectra with much overlap of multiplets can be resolved with this J,S-spectroscopy. Disadvantages of the method are a rather long measurement time and the appearance of artefacts for strongly coupled spin systems. Shift-correlated 2D spectra have become much more widely used.

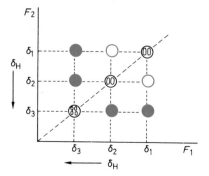

Fig. 3.55 Schematic contour plot for a ^{1}H shift correlated 2D NMR spectrum of a three spin system with two couplings

In **shift-correlated 2D NMR spectra** both frequency axes represent chemical shifts. Without going into the detail of the individual experiments, several different methods can be used (COSY, SECSY, FOCSY, NOESY, ROESY), each with their own pulse sequence, and with different graphical forms. The **homonuclear ^{1}H shift correlations** obtained from these experiments result from coupling between (COSY, SECSY, FOCSY) or spatial proximity of (NOESY, ROESY) the two correlated nuclei. A single such experiment can replace a whole series of double resonance experiments and can also be superior at recognising long-range or weak correlations.

Consider for example the COSY spectrum of a three spin system (Fig. 3.55). In the contour plot the normal spectrum lies on the diagonal; these contours are surrounded by black circles.

Further contours, such as those in the solid blue circles, result from coupling between nuclei; in this case between H_3 and H_1 and H_2. The two empty crossing points (empty blue circles) show that there is no detectable coupling between H_1 and H_2.

As an explicit example of two dimensional ^{1}H NMR the spectra of sucrose (**144**) will be considered. Fig 3.56a shows the normal (one dimensional) spectrum. Underneath it is shown the projection on the F_2 axis of the J-resolved 2D spectrum. For the α-D-glucose part G and the β-D-fructose part F a total of 11 signals is observed ($g_1 - g_6$ and $f_1, f_3 - f_6$). The diastereotopic protons of the methylene groups are effectively isochronous.

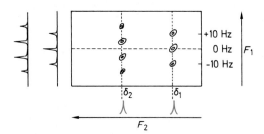

Fig. 3.54 Schematic contour plot for a J-resolved 2D ^{1}H NMR spectrum with the projections onto the F axes

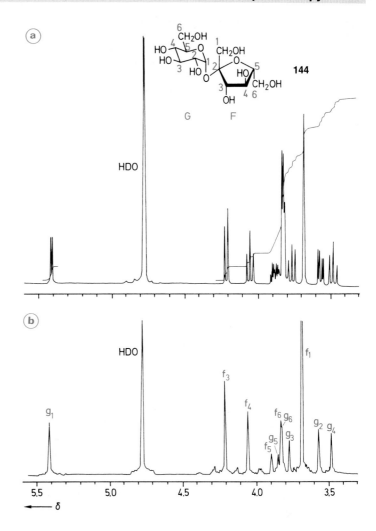

Fig. 3.56 400 MHz ¹H NMR spectrum of sucrose (**144**) in D₂O
a Normal (1D) spectrum
b 2D *J*-resolved spectrum
(from Bruker brochure: Two-Dimensional NMR)

The ¹H shift correlated 2D spectrum (**COSY**: *C*orrelated *S*pec-*troscopy*) in Fig. 3.57 provides the key to the total coupling network. For example, proton g₁ couples with g₂ (³*J*), g₃ (⁴*J*), g₅ (⁴*J*), and f₁ (⁵*J*). In the normal 400 MHz spectrum g₁ only appears as a doublet, i.e. only ³*J* is resolved. The doublet for f₃ however is "genuine"; the 2D spectrum shows no other correlations than that to f₄, the ³*J* coupling.

¹H,¹H COSY spectra are nowdays a routine procedure for the structural determination of organic compounds. The measurement often takes no longer than a few minutes and provides valuable information about ¹H,¹H coupling, the larger (in magnitude) geminal and vicinal couplings being easiest to observe. An experimental modification, **long-range COSY**, allows observation of correlations due to the smaller long-range couplings, often even when these cannot be directly observed in the 1D spectrum. Fig. 3.58 shows the long-range

COSY spectrum of 6-hexyloxy-10-methylphenanthrene-2-carboxyaldehyde (**145**).

Apart from the cross peaks for the two vicinal couplings it contains cross peaks for the ⁴*J* couplings of 1-H with 3-H, 5-H with 7-H, 8-H with 9-H and CH₃, and for the ⁵*J* couplings of 1-H with 4-H, 4-H with 5-H, 4-H and CHO, and 5-H with 9-H; weak indications of the couplings ⁵*J*(1-H,9-H) and ⁵*J*(5-H,8-H). With this information an unambiguous assignment of the signals is relatively easy.

145

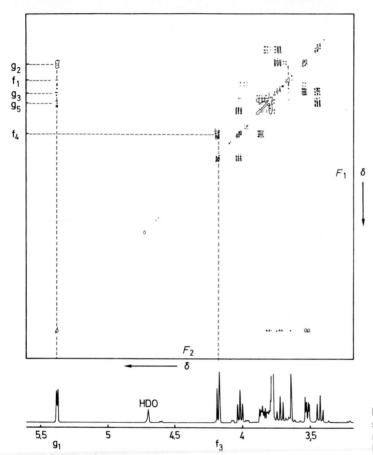

Fig. 3.57 500 MHz ^{1}H shift correlated 2D NMR spectrum of sucrose (**144**) in D$_2$O (COSY-45 experiment, from Bruker brochure: Two-Dimensional NMR)

It is not always possible to dispense with the normal COSY spectrum and only measure a long-range COSY spectrum. Under the measurement conditions required to observe long-range couplings the cross peaks for larger couplings can disappear.

A further example, L-serine (**146**) is a simple molecule, but its behaviour in 2D experiments is more complicated. This is always the case when strong couplings are present, as is the case for all the three relevant spin-spin interactions in L-serine.

$$HOOC - \underset{\underset{NH_2}{|}}{\overset{\overset{H_{C/X}}{|}}{C}} - \underset{\underset{H_B}{|}}{\overset{\overset{H_A}{|}}{C}} - OH$$

146

At 60 MHz the signals of H$_A$, H$_B$, and H$_C$ lie practically together (ABC system). At 360 MHz the spectrum is that of an ABX system (Fig. 3.59a). The 2D *J*-resolved spectrum is shown in Fig. 3.59c as a stack plot (panoramic view). As a result of the strong coupling more than three multiplets can be observed, and correspondingly there are more than three singlets in the projection onto the δ axis (Fig. 3.59b). The slices through the individual multiplets (cross sections) in the *J* direction can be seen in Fig. 3.59d. Despite the extra signals caused by the strong coupling the chemical shifts and couplings can be unambiguously determined.

In the **relayed technique** the magnetisation is not directly transferred from one nucleus to the coupling nuclei, as in the normal H,H COSY; instead an intervening nucleus serves as a "relay". The polarisation can be transferred stepwise across several nuclei. As an example a ^{1}H$\curvearrowright$^{1}H$\curvearrowright$^{1}H case will be discussed. Fig 3.60a shows a section of the ^{1}H shift-correlated NMR spectrum of sucrose (**144**). Underneath is reproduced the same section from a measurement using the relayed COSY method. Fig. 3.60b clearly shows correlations which are not present in a. They correspond to the ^4J couplings f$_3$/f$_5$, f$_4$/f$_6$, g$_2$/g$_4$, and g$_3$/g$_5$; The correlation g$_1$/g$_3$ is also observed, but lies outside the area shown. In contrast no correlation is observed for f$_1$/f$_3$, since C-2 of fructose has no "relay proton".

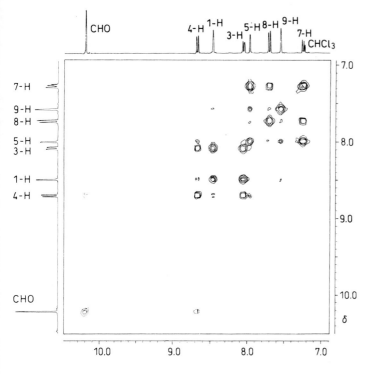

	1-H	CHO	3-H	4-H	5-H	7-H	8-H	9-H	CH₃
1-H	-----			^4J	^5J			^5J	
CHO		-----			^5J				
3-H	^4J		-----	^3J					
4-H	^5J	^5J	^3J	-----	^5J				
5-H				^5J	-----	^4J	^5J	^5J	
7-H					^4J	-----	^3J		
8-H					^5J	^3J	-----	^4J	
9-H	^5J				^5J		^4J	-----	^4J
CH₃								^4J	-----

Fig. 3.58 400 MHz ¹H, ¹H long range COSY spectrum (aromatic part) of 6-hexyloxy-10-methylphenanthrene-2-carboxaldehyde (**145**). The coupling matrix shows the detectable long range couplings ^4J and ^5J in blue

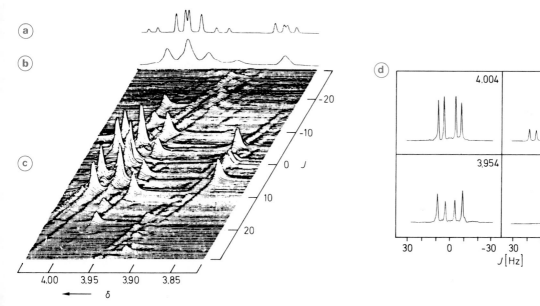

Fig. 3.59 360 MHz ¹H NMR spectrum of L-serine (**146**) in D₂O (Wider, G., Baumann, R., Nagayama, K., Ernst, R. R., Wüthrich, K. (1981), J. Magn. Res. **42**, 73)
a Normal (1D) spectrum
b 2D J-resolved projection spectrum
c Stack plot ot the 2D J-resolved spectrum
d Cross sections (slices down the J axis) at the given δ values (e.g. $δ_x = 3.857$)

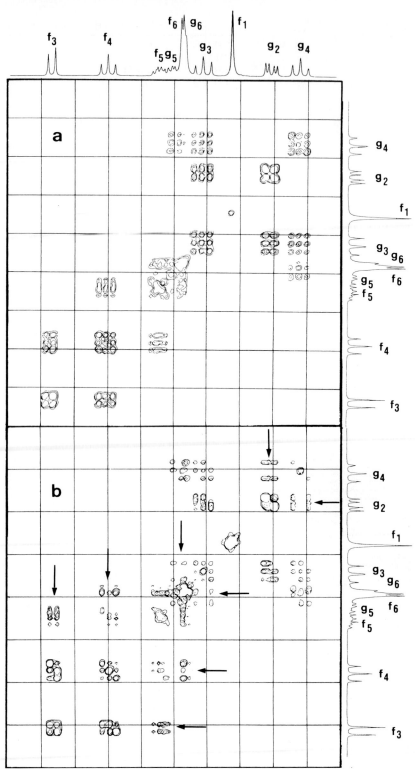

Fig. 3.60 a Expansion of part of Fig. 3.57: ^{1}H shift-correlated 2D NMR spectrum COSY-45 of sucrose (**144**)
a Same expansion of the H-relayed (H,H) COSY spectrum

Whereas H,H COSY and related methods rely on correlation of signals by through-bond couplings, **NOESY** measurements make use of the nuclear Overhauser effect. The cross peaks in the 2D spectrum then show the spatial proximity of nuclei. The through-space connectivities are not only useful in difficult structural problems, but also particularly useful for the determination of conformations present in solution. Since the NOE enhancements can be either positive or negative, particular attention must be paid to the possibility of zero values occurring. This difficulty can be avoided by the **ROESY** experiment (rotating frame NOESY) where the enhancements are always positive. The condensed [18]annulene (**147**) with three phenanthrene systems serves as an example. As in unsubstituted [18]annulenes the problem is the identification of the inner and outer protons of the eighteen-membered ring. The cross peaks in Fig. 3.61 show on the one hand the spatial proximity of 8-H and 28-H and on the other hand the proximity of 9-H to 7-H and 1-H. There is no exchange between internal and external protons.

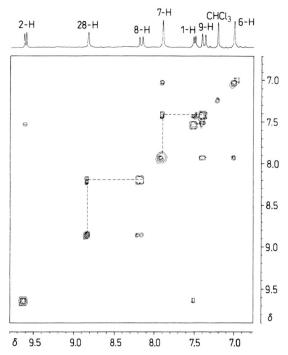

Fig. 3.61 Expansion of part of the 2D ROESY spectrum of **147**, measured at 400 MHz in CDCl₃ (Kretzschmann, H., Müller, K., Kolshorn, H., Schollmeyer, D., Meier, H. (1994) Chem Ber. **127**, 1735)

147

2D spectroscopy provides such a wealth of structural and conformational information that it has now established itself as one of the foremost experimental methods. Several experiments are now routine NMR techniques, and 3D spectroscopy is also starting to prove its usefulness as an analytical method (see also p. 179).

An example is the ¹H NMR spectrum of the hydrocarbon (**44**).

44   C₂   Cₛ

Spectral Simulation

The interpretation of the spectra of complicated spin systems often requires the use of **simulated spectra** calculated by computer. Starting with estimated values for the chemical shifts and coupling constants of the nuclei involved iteration is then used to improve the comparison of the experimental and simulated spectra. Many programs have been developed for such purposes (LAOCOON etc.). When the observed and simulated spectra agree in both position and intensity the values of the parameters used for the final simulation can be taken as valid.

At room temperature an AA'BB' spectrum is observed for the aliphatic protons. The non-equivalence of the four protons shows that the system is rigid and there is no rapid inversion of the ring. The spin system can be caused by either a chiral C₂-conformation or an achiral Cₛ-conformation. The Karplus equation or a modified form thereof (see p. 108) yields very different values of the *vicinal* couplings. From the parameters obtained from a spectral simulation it is clear that the compound must have the C₂-conformation (Fig. 3.62).

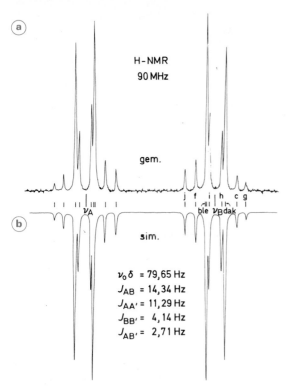

Fig. 3.62 ¹H NMR spectrum of **44** (aliphatic portion); **a** measured, **b** simulated AA'BB' system (Meier, H., Gugel, H., Kolshorn, H. (1976), Z. Naturforsch., Part B, **31**, 1270)

In the figure:

H–NMR
90 MHz

gem.

sim.

$$\nu_0 \delta = 79{,}65 \text{ Hz}$$
$$J_{AB} = 14{,}34 \text{ Hz}$$
$$J_{AA'} = 11{,}29 \text{ Hz}$$
$$J_{BB'} = 4{,}14 \text{ Hz}$$
$$J_{AB'} = 2{,}71 \text{ Hz}$$

ferred orientation. Apart from the **thermotropic liquid crystalline phases** there are **lyotropic liquid crystalline phases**, which can be formed by amphiphilic compounds, e.g. tensides, and water or other solvents.

Nematic phases are often composed of linear molecules, which are orientated in a specific direction. Guest molecules introduced into the phase then have their Brownian motion restricted and are forced into a similar preferred orientation. In the NMR spectra this preferred orientation is apparent from the direct dipolar couplings between the nuclear spins of the guest molecule; the result is additional splittings of the lines. Partially oriented molecules consequently give multi-line spectra, several kHz broad, since the dipolar couplings are much larger than the coupling constants which result from scalar interactions. The analysis of such spectra can yield valuable information about structural parameters such as bond angles and bond lengths, which are related to the geometry of the molecules in liquid phase.

On going from partially oriented phases to the solid phase, the number of spin-spin interactions is drastically increased, because now the **intermolecular** interactions also come into play; for guest molecules in oriented liquid crystalline phases these are eliminated by the translational and rotational motions. The signals therefore become very broad in solids. In Sec. 4.8 (see p. 180 ff.) the use of **magic-angle spinning** to obtain high resolution spectra from solids is described.

Finally it should be mentioned that NMR can be used not only for observations on individual molecules or aggregates of molecules, but that methods such as NMR imaging and nuclear spin tomography have been developed which can provide pictures of macroscopic objects. For biology and above all for medicinal purposes these methods have become immensely important, allowing images (cross sections) of internal organs to be acquired without physical intervention or the use of dangerous radiation.

NMR Spectra of Oriented Phases and Solids

Certain compounds form liquid crystalline states (nematic, smectic, or cholesteric) between the melting point and the clarification point; in these phases the molecules have a pre-

4. ¹³C NMR Spectroscopy

4.1 Sample Preparation and Measurement of Spectra

For the measurement of a ¹³C spectrum the sample needs to be fairly concentrated, but not viscous. As a rule of thumb, a fast routine measurement requires 3 mg of sample (dissolved in 0.6 ml of solution) for each carbon atom in the molecules. The PFT method is used (see Sec. 3.1, p. 100).

The usual **solvent** is deuteriochloroform. A summary of alternative solvents is given in Tab. 3.24. Deuterated solvents are

used to facilitate the measurement by employment of the deuterium resonance of the solvent as a **lock signal** to stabilise the field-frequency relationship in the spectrometer. A further advantage of deuterated solvents is that their ¹³C signals are weaker than those of protonated solvents, as a combined result of the splitting of the ¹³C signal by deuterium (see Tab. 3.24), the lack of the NOE enhancement (see p. 85), and the longer relaxation times T_1. The relatively strong ¹³C signals of protonated solvents or of deuterated solvents in dilute solutions can lead to problems of dynamic range (ability to

Tab. 3.24   Solvents for ¹³C NMR spectroscopy

Solvent	¹³C NMR shift δ	Multi- plicity	$J(^{13}C,D)$ (Hz)	
[D₁] Chloroform	77.0	triplet	32	
[D₄] Methanol	49.3	septet	21	T
[D₆] Acetone	29.3	septet	20	T
	206.3	multiplet	<1	
[D₆] Benzene	128.0	triplet	24	
[D₂] Dichloromethane	53.5	quintet	27	T
[D₃] Acetonitrile	1.3	septet	21	
	117.7	multiplet	<1	
[D₁] Bromoform	10.2	triplet	31.5	H
[D₂] 1,1,2,3-Tetra- chloroethane	74.0	triplet		
[D₈] Tetra- hydrofuran	25.5	quintet	21	
	67.7	quintet	22	
[D₈] Dioxane	66.5	quintet	22	
[D₆] Dimethyl sulfoxide	39.7	septet	21	
[D₅] Pyridine	123.5	triplet	25	
	135.5	triplet	24	
	149.5	triplet	27	
[D₂] Water	–	–	–	
[D₄] Acetic acid	20.0	septet	20	
	178.4	multiplet	<1	
[D₁₈] Hexamethyl- phosphoric- triamide (HMPT)	35.8	septet	21	H
Carbon tetrachloride	96.0	singlet	–	
Carbon disulfide	192.8	singlet	–	T
Trichlorofluoromethane	117.6	doublet	($^1J(C,F) = 337$	T
[D₁] Trifluoroacetic acid	116.5	quartet	($^1J(C,F) = 283$)	
	164.4	quartet	($^2J(C,F) = 44$)	

T suitable for low temperature measurements (see Tab. 3.6)
H suitable for high temperature measurements (see Tab. 3.6)

As in ¹H NMR tetramethylsilane (TMS) may be used as a **reference** to establish the zero point of the δ scale (internal or external standard). Frequently the ¹³C signals of the solvent with their known shifts suffice for the determination of the δ values of the sample.

The low natural ¹³C content of 1.1% and the low magnetic moment mean that the ¹³C nucleus has a low NMR sensitivity (see Sec. 1.1, especially Tab. 3.1, p. 72).

In measurements of ¹³C spectra of organic compounds there will normally be hydrogen nuclei present in the molecule, which give rise to spin-spin couplings; that means that for a set of isochronous ¹³C nuclei a multiplet will be expected, caused by direct, *geminal*, and *vicinal* couplings. The signal intensity will thereby be distributed over several lines. In ¹H NMR these ¹³C,¹H couplings are normally not apparent because of the low natural abundance of ¹³C (see however Sec. 3.9, p. 129 on ¹³C satellites). This problem is generally avoided in ¹³C NMR by using **¹H broadband decoupling**. In contrast to **homonuclear spin decoupling** (Sec. 3.8, p. 131) this is a **heteronuclear spin decoupling** technique.

In order to simultaneously remove all the ¹³C,¹H couplings present, a powerful irradiation is used which covers the whole proton chemical shift range. The decoupler frequency is modulated with low frequency noise. This has led to the use of the term **¹H noise decoupling** as an alternative to **¹H broadband decoupling**. Even more effective methods of decoupling using phase modulation are now replacing noise modulation (Waugh, GARP, etc.).

The decoupling effect, as described on p. 131 ff, depends on the coupling ¹H nuclei changing their precession direction (spin orientation) so rapidly as a result of the irradiation at their resonance frequency that all their coupling partners (here the coupling ¹³C nuclei) only experience an average value of zero. All the lines of the multiplet of a ¹³C signal form a singlet as a result. Its intensity can amount to 300% of the intensity of the individual lines of the multiplet. Part of this increased intensity is a consequence of the **heteronuclear Overhauser effect** (see Sec. 1.5, p. 84).

observe weak signals in the presence of strong ones) in the computer or can obscure weak signals of the sample. The first problem is largely avoided in modern spectrometers by the use of 32 bit technology (older spectrometers often use 24 or 16 bits) and the second can be avoided by the use of commercially available solvents which are depleted in ¹³C. In these the natural ¹³C content of 1.1% is reduced to ca. 0.1%. (Additionally there are instrumental techniques available to reduce the intensity of solvent signals: pre-saturation, Redfield technique, etc.)

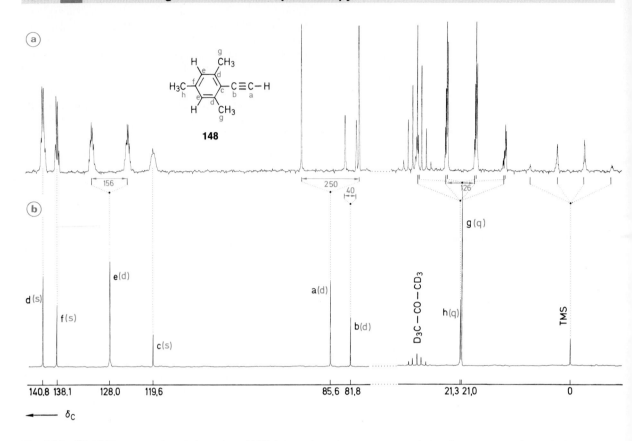

Fig. 3.63 ^{13}C NMR spectra of mesitylacetylene (**148**) in hexadeuterioacetone **a** coupled **b** ^{1}H broadband decoupled

The ^{1}H broadband decoupling simplifies the ^{13}C NMR spectra considerably, and the signals gain in intensity. These advantages outweigh the loss of information from ^{13}C,^{1}H coupling so much that routine ^{13}C spectra are always acquired with broadband decoupling. The ^{13}C,D couplings remain unaffected. Fig. 3.63 shows as an example a comparison of coupled and ^{1}H broadband decoupled ^{13}C spectra for mesitylacetylene (**141**).

In the decoupled spectrum 3.63b the signals d, f, and c remain in the same positions as in 3.63a. They are singlets (s). The corresponding C atoms are not bonded to hydrogen. In the coupled spectrum 3.63a their signals are only broadened by couplings to more distant protons. The signals e, a, and b each appear in the coupled spectrum as a doublet (d), i.e. the corresponding C nuclei couple, apart from long range couplings, to only one proton. The three coupling constants, 156, 250, and 40 Hz, differ widely. For e and a the coupling is to a directly bonded proton (^1J(C,H)) for an aromatic carbon and an acetylenic carbon respectively. For b an unusually large ^2J(C,H) is observed (see Sec. 4.3, p. 149). The methyl C nuclei g and h each show a quartet (q) from coupling to the three methyl protons. Only in the decoupled spectrum is it easily apparent that

there are two signals, with such different intensities that the higher peak can be assigned to the two carbon atoms g. (A CH$_2$ group would give a triplet (t), but there is none in this molecule.)

For the coupled spectrum some ten times as many scans were accumulated as for the broadband decoupled spectrum, although the concentration of **148** remained unchanged. This is apparent from the increased intensity of the solvent signal (septet for hexadeuterioacetone at δ=29.3. The CO signal at δ=206.3 is not reproduced). The measurement of coupled spectra therefore requires a longer measurement time or a higher sample concentration. For a ^{1}H broadband decoupled spectrum of a reasonably concentrated solution a few hundred or thousand scans are needed. To determine the time required it must be remembered that the signal:noise ratio $S:N$ is proportional to $\sqrt{n}$, where n is the number of scans. The practical effect of this is that if the concentration is halved, then not twice as many, but four times as many scans are necessary to attain the same **signal-noise ratio**.

From the application of these considerations to the acquisition of fully coupled spectra, it can be appreciated that at least 10 times as many scans will be required as for a ^{1}H broadband decoupled spectrum.

4.2 ¹³C Chemical Shifts

In the introductory Sec. 1.2 (p. 73) the relationship between the **chemical shift** of a nucleus and the **shielding constant** σ was described.

$$10^{-6} \cdot \delta\,(X) = \sigma\,(\text{TMS}) - \sigma\,(X)$$

$$\sigma = \sigma_{\mathrm{d}} + \sigma_{\mathrm{p}} + \sigma'$$

In contrast to ¹H NMR the σ_{para} term is particularly important in ¹³C NMR. Since this term involves **electronic excitation**, the necessary energy ΔE is involved in σ_{para}. With decreasing ΔE a low field shift is observed.

The **hybridisation** of a ¹³C atom is of decisive importance for the chemical shift. sp^3 C atoms absorb at highest field, followed by sp C atoms and finally by sp^2 C atoms at lowest field. This order is the same as for ¹H shifts of saturated, acetylenic, and olefinic protons. The shifts of ¹³C nuclei and the protons attached to them often show parallel behaviour. This comparison should not however be taken too far. The ¹³C and ¹H shifts of cyclobutane (**149**) and cyclopropane (**93**) for example do show parallel behaviour.

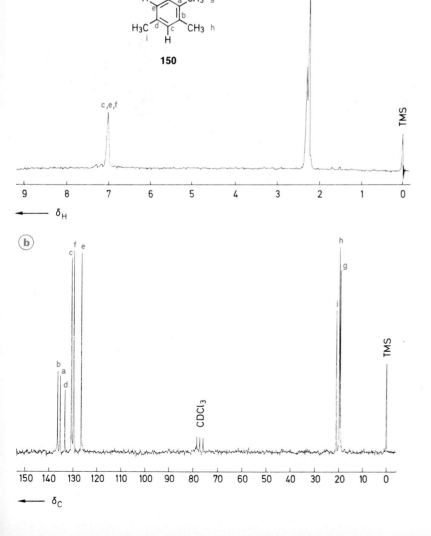

Fig. 3.64   NMR spectra of 1,2,4-tri-methylbenzene (**150**) in CDCl$_3$
a ¹H spectrum
b ¹³C spectrum (broadband decoupled)

δ			
a	135.0	f	129.5
b	136.2	g	19.2
c	130.5	h	19.5
d	133.1	i	20.8
e	126.5		

δ_H: 1.96
δ_C: 23.1

149

δ_H: 0.22
δ_C: – 2,8

93

Benzene (**75**) and cyclooctatetraene (**90**) show opposite effects.

δ_H: 7.26
δ_C: 128,5

75

δ_H: 5,80
δ_C: 131,5

90

The **ring current**, a special case of an anisotropy effect, causes a marked low field shift in the ^1H NMR of aromatics as compared to olefinic protons. In ^{13}C NMR this effect is apparently negligible for the ring carbon atoms. Olefinic and aromatic carbons absorb in the same region. In the example above the ^{13}C shift of cyclooctatetraene is even 3 ppm to low field of that of benzene. Since a comparison of ^1H and ^{13}C shifts shows that the **shift range** of ^{13}C, ignoring extreme cases, is some 200 ppm whereas the ^1H range is only some 10 ppm, it is apparent that changes in the chemical environment will in general be more noticeable in ^{13}C than in ^1H NMR. As an example the spectra of 1,2,4-trimethylbenzene (**150**) are reproduced (Fig. 3.64). Whereas in the ^1H spectrum the three chemically non-equivalent ring protons overlap, and the three different methyl groups are very close together, well separated signals are observed in the ^{13}C spectrum.

The value of ^{13}C spectroscopy is particularly apparent for more complex molecules, for example steroids and alkaloids. In Fig. 3.65 quinine (**151**) is shown as an example. The twenty ^{13}C signals for the twenty different C atoms are easily distinguished.

Functional groups generally deshield the ^{13}C nucleus they are directly bound to. (A high field shift is however observed for heavy atoms such as iodine (**155**).) In the neighbouring β-position the deshielding is usually less. There are however exceptions, as shown by a comparison of 1-octanol (**152**) and 1-octanethiol (**153**):

Iodine causes a low field shift in the β-position. (Compare C-1 with C-8, C-2 with C-7, etc.) The effect of substituents is not restricted, as in ^1H NMR, to the immediate vicinity of the measured nucleus, but also includes more distant groups.

All X substituents cause an increase of the shielding of the γ-C atom of the carbon chain. The resulting high field shift is known as the γ-**effect**. The substituent effect on the δ or further positions is small in open chain compounds, but not necessarily in bi- and polycyclic compounds. Steric factors also play an important role, as shown by the examples of *cis*-decalin (**156**) and *trans*-decalin (**157**) and the stereoisomers of 2,3-dibromosuccinic acid (**158–160**). The enantiomers *RR* and *SS* with C_2 symmetry have identical ^{13}C spectra in an achiral medium; the achiral *meso*-form with a centre of symmetry has different shifts.

36,8 29,7 24,5

156

44,0 34,6 27,1

157

H COOH
167,9
HOOC
42,8 Br
Br H

158
RR (C_2)

H Br
169,6
HOOC
47,3 COOH
Br H

160
meso (C_i)

HOOC H
167,9
HOOC
42,8 Br
H Br

160
SS (C_2)

In the case of *geminal* multiple substitution the deshielding effects are not necessarily additive. A comparison of the shifts of the halogenomethanes shows that an increasing number of fluoro- or chloro-substituents causes an increasing low field shift. With increasing substitution with iodine the reverse effect, a high field shift, is observed. The trend for brominated methanes is inconsistent (Tab. 3.25).

Increasing alkyl substitution generally leads to a low field shift.

$$\delta\,(CH_4) < \delta\,(C_{prim}) < \delta\,(C_{sec}) < \delta\,(C_{tert}) < \delta\,(C_{quart})$$

		α	β	γ	δ	ε			
X—		—CH_2	—CH_2	—CH_2	—CH_2	—CH_2	—CH_2	—CH_2	—CH_3
152	(X = OH):	63.1	32.9	25.9	29.5	29.4	31.9	22.8	14.1
153	(X = SH):	24.7	34.2	28.5	29.2	29.1	31,9	22.8	14.1
154	(X = Br):	33.8	33.0	28.3	28.8	29.2	31.8	22.7	14.1
155	(X = I):	6.9	33.7	30.6	28.6	29.1	31.8	22.7	14.1

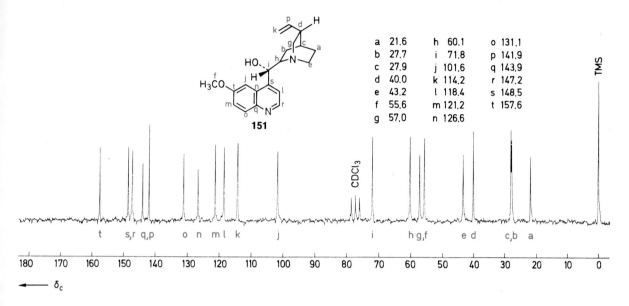

a	21.6	h	60.1	o	131.1	
b	27.7	i	71.8	p	141.9	
c	27.9	j	101.6	q	143.9	
d	40.0	k	114.2	r	147.2	
e	43.2	l	118.4	s	148.5	
f	55.6	m	121.2	t	157.6	
g	57.0	n	126.6			

Fig. 3.65   ¹³C NMR spectra of quinine (**151**) in CDCl₃ (¹H broadband decoupled)

Tab. 3.25   Relationship between ¹J(C,H) and s-character of the hybrid orbitals of carbon

Compound	X=F	Cl	Br	I
CH₃X	75.0	24.9	9.8	− 20.8
CH₂X₂	109.0	54.0	21.4	− 54.0
CHX₃	116.4	77.0	12.1	− 139.9
CX₄	118.6	96.5	− 29.0	− 292.5

The magnitude of the substituent effect is, as in ¹H NMR, dependent on the electronegativity of the substituent X. With increasing negativity of X the α-C atom is shifted further to low field. Tab. 3.25 shows this in the halogenated methane series. This rule no longer holds for the β-position and more distant carbons. The γ-effect in haloalkanes is exactly reversed. Fluorine causes the strongest high field shift, iodine the weakest.

Independent of the various factors which determine ¹³C shifts there is a direct correlation between the δ values and the **charge density** at the relevant C atom. This relationship is very useful in spectral interpretation and will therefore be illustrated by some examples. If for example ethylene is compared with its formyl or methoxy derivatives, effects are observed which are typical for unsaturated carbonyl compounds (e.g. acrolein, **161**) and enol ethers (e.g. methyl vinyl ether, **162**):

$$O=CH-CH=CH_2 \qquad\qquad H_3CO-CH=CH_2$$

$$H_2C=CH_2$$
$$123.3 \quad 123.3$$
$$\mathbf{15}$$

$$\bar{I}\bar{O}-CH=CH-CH_2^+ \qquad H_3\overset{+}{CO}=CH-\bar{C}H_2$$
$$136.4 \quad 136.0 \qquad\qquad 152.7 \quad 84.4$$
$$\mathbf{161} \qquad\qquad\qquad \mathbf{162}$$

While the inductive effect is largely effective in the α-position, the mesomeric effect in the β-position causes a reduction of the electron density in acrolein (**161**) and an increase in methyl vinyl ether (**162**). The stronger deshielding is apparent as a low field shift, the stronger shielding as a high field shift. Extreme cases are observed for keteneacetals and related compounds:

163   **145.1   39.3   32.1**

Further examples are nitrobenzene (**164**) and aniline (**165**). As a result of the mesomeric effect the electron density in the o- and p-positions is changed. The same considerations can be applied to pyridine (**83**) and its N-oxide (**166**). The influence of charge density on ¹³C shifts is clearly apparent. The shift of the o-C atoms in nitrobenzene (**164**) however also shows clearly that other factors need to be considered.

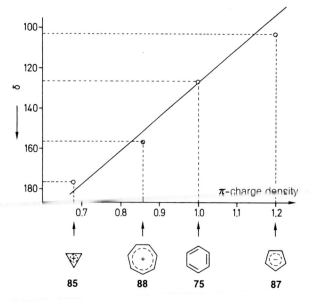

164

165

83

166

169

170

171

172

173

The relationship between ^{13}C shift and electron density is particularly noticeable for ions. In Fig. 3.66 this is demonstrated using a comparison between uncharged benzene and aromatic ions as an example.

Fig. 3.66   Relationship between ^{13}C shifts and π-electron density in ions

In studies of carbocations the measurement of their ^{13}C spectra in "magic acid" played a decisive role. As shown by the examples (**167–173**), the extent of delocalisation of the charge is easily recognised from the ^{13}C shift of the central C atom.

330.0

167

255.7

168

After these comments about the effect of charges on chemical shift it is understandable that the signals of acidic or basic compounds will be strongly pH dependent. Simplified assumptions of electron densities can however easily lead to errors. The protonation of linear alkyl amines generally causes a high-field shift of C_α, C_β, and C_γ; conversely a low field shift is observed on deprotonation of carboxylic acids!

$$-\overset{|}{C}_\gamma-\overset{|}{C}_\beta-\overset{|}{C}_\alpha-NH_2/\overset{+}{N}H_3 \qquad \textbf{174}$$

$\Delta\delta \approx -0.5 \quad -3.8 \quad -2.3$

$$-\overset{|}{C}_\gamma-\overset{|}{C}_\beta-\overset{|}{C}_\alpha-COOH/COO^- \qquad \textbf{175}$$

$\Delta\delta \approx +0.6 \quad +1.6 \quad +3.5 \qquad +4.7$

In the amphoteric amino acids the δ values generally increase as the pH increases, as shown by the example of alanine:

$$H_3\overset{3}{C}-\overset{2}{\underset{NH_3^+}{\overset{H}{C}}}-\overset{1}{C}OOH \underset{+H^+}{\overset{-H^+}{\rightleftharpoons}} H_3C-\underset{NH_3^+}{\overset{H}{C}}-COO^-$$

23

$$\underset{+H^+}{\overset{-H^+}{\rightleftharpoons}} H_3C-\underset{NH_2}{\overset{H}{C}}-COO^-$$

pH	0.43	4.96	12.52
C–1	174.0	177.0	185.7
C–2	50.1	51.9	52.7
C–3	16.5	17.5	21.7

To conclude this section a few comments about **medium effects** on ^{13}C shifts are appropriate. These can be described by an additional shielding constant $\delta_{Med.}$. Where there are no special interactions, such as acidbase activity, solvent and concentration shifts are generally less than 3 ppm (see however the section on p. 171 on shift reagents).

The influence of temperature on ^{13}C shifts is also small, unless temperature-dependent processes (internal molecular motion, chemical rearrangements) alter the molecular structure (see Sec. 2, p. 86 ff.).

Tab. 3.26 shows measured coupling constants and values calculated from this formula for ethane, ethylene, and acetylene.

4.3 ^{13}C,^{1}H Couplings

The coupling $^1J(C,H)$ is a measure of the s-character of the hybrid orbitals of the relevant C–H bonds. The following empirical relation holds approximately

$$^1J(C,H) = 500\,p \quad \text{with} \quad p = \begin{cases} 0.25 \text{ for } C - sp^3 \\ 0.33 \text{ for } C - sp^2 \\ 0.50 \text{ for } C - sp \end{cases}$$

Tab. 3.26 Relationship between $^1J(C,H)$ and s-character of the hybrid orbitals of carbon

	Hybridi-sation sp^{λ^2}	λ^2	$p=\dfrac{1}{1+\lambda^2}$	$^1J(C,H)$ calculated	measured
Ethane	sp^3	3	0.25	125	124.9
Ethylene	sp^2	2	0.33	167	156.4
Acetylene	sp	1	0.50	250	248

Tab. 3.27 $^1J(C,H)$ coupling constants in substituted methanes

Compound	$^1J(C,H)$ (Hz)
CH_4	125
CH_3F	149
CH_3Cl	150
CH_3Br	152
CH_3I	151
CH_3NH_2	133
$CH_3N^+H_3$	145
CH_3NO_2	147
CH_3OH	141
CH_3O^-	131
CH_3OCH_3	140
CH_3SCH_3	138
$CH_3Si(CH_3)_3$	118
CH_3Li	98
CH_2Cl_2	177
$CHCl_3$	209
CHF_3	239

Tab. 3.28 $^1J(C,H)$ couplings of selected compounds

Compound	$^1J(C,H)$ (Hz)
$H-CH_2-CH_3$	125
$H-CH(CH_3)_2$	119
$H-C(CH_3)_3$	114
$H-CH_2-CH=CH_2$	122
$H-CH_2-C_6H_5$	129
$H-CH_2-C\equiv CH$	132
$H-CH_2-C\equiv N$	136
$H-CH_2-COOH$	130
$H-CH(OH)-C_6H_5$	140

Tab. 3.28 continued

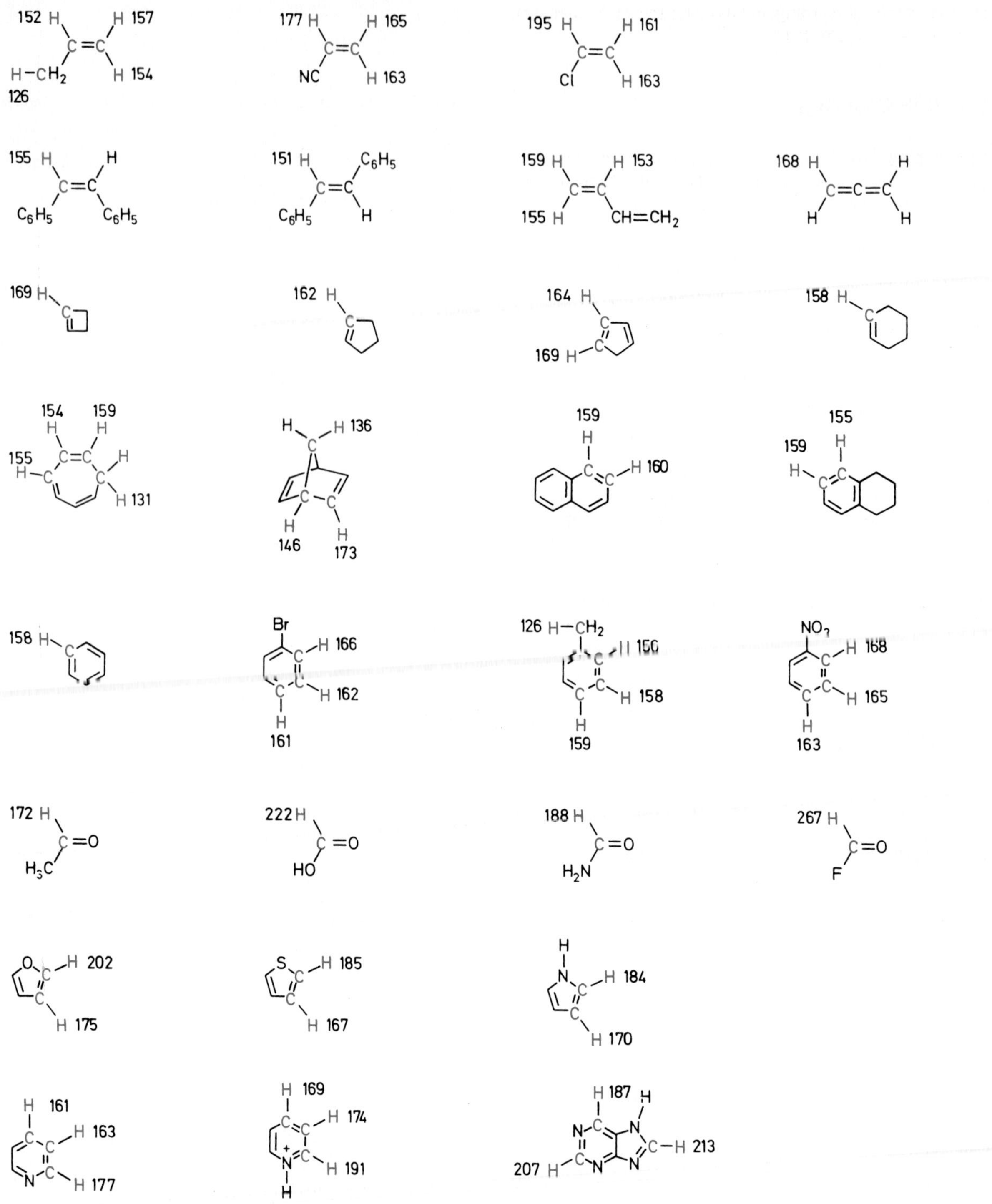

Row 1:
- 152 H, H 157, H−CH₂, H 154, 126
- 177 H, H 165, NC, H 163
- 195 H, H 161, Cl, H 163

Row 2:
- 155 H, H, C₆H₅, C₆H₅
- 151 H, C₆H₅, C₆H₅, H
- 159 H, H 153, 155 H, CH=CH₂
- 168 H, H, H, H

Row 3:
- 169 H, C
- 162 H, C
- 164 H, C, 169 H
- 158 H, C

Row 4:
- 154, 159, 155 H, H, H, H 131
- H, H 136, C, C, H, H, 146, 173
- 159, H, H 160
- 155, 159, H, H, C

Row 5:
- 158 H, C
- Br, H 166, C, H 162, H, 161
- 126 H−CH₂, H 156, C, H 158, H, 159
- NO₂, H 168, C, H 165, H, 163

Row 6:
- 172 H, C=O, H₃C
- 222 H, C=O, HO
- 188 H, C=O, H₂N
- 267 H, C=O, F

Row 7:
- O, C, H 202, C, H 175
- S, C, H 185, C, H 167
- H, N, C, H 184, C, H 170

Row 8:
- H 161, C, C, H 163, N, C, H 177
- H 169, H 174, C, N⁺, H 191, H, H
- H 187 H, N, C, N, C, H 213, 207 H, N

If the $^1J(C,H)$ coupling constants of cyclohexane (125 Hz) and cyclopropane (160 Hz) are compared, the similarity of cyclopropane to olefinic systems, as suggested by the Walsh orbital model, for example, is apparent. These estimates of s-character of hybrid orbitals should however be restricted to hydrocarbons.

The effect of substituents on the $^1J(C,H)$ couplings constants is shown in Tab. 3.27 for a series of methane derivatives.

Tab. 3.29 $^2J(C,H)$-, $^3J(C,H,)$-, and higher couplings of selected compounds

Compound	$^2J(C,H)$ (Hz)	$^3J(C,H)$ (Hz)	$^nJ(C,H)$ (Hz)
H_3C-CH_2 (with H)	-4.5		
$H_2C=CH$ (with H)	-2.4		
$HC\equiv C-H$	$+49.6$		
$H_3C-CH_2-CH_2-H$	-4.4	$+5.8$	
cyclopropane H_2C, C, H, H	-2.5		
$H_b\backslash C=C / H_c$, $H_a / \ CH_2-H_d$ (C1=C2–C3)	C_1H_c: $+0.4$ C_2H_a: -2.6 C_2H_b: -1.2 C_2H_d: -6.8 C_3H_c: $+5.0$	C_1H_d: $+6.7$ C_3H_a: $+7.6$ C_3H_b: $+12.7$	
$H_b\backslash C=C / H_c$, $H_a / \ C\equiv C-H_d$ (C1=C2–C3≡C4)	C_1H_c: $+8.8$ C_2H_a: -3.7 C_2H_b: -0.3 C_3H_c: $+2.0$	C_2H_d: $+4.8$ C_3H_a: $+9.5$ C_3H_b: $+16.3$	C_1H_d: $+2.8$ $(n = 4)$ C_4H_a: <1 $(n = 4)$ C_4H_b: <1 $(n = 4)$
$Cl\backslash C=C / H_a$, $H_c / \ H_b$	C_1H_a: -8.3 C_1H_b: $+7.1$ C_2H_c: $+6.8$		
$Cl\backslash C=C / H$, $H / \ Cl$	$+0.8$		
$Cl\backslash C=C / Cl$, $H / \ H$	$+16.0$		
$H_2C-C(=O)$, H_b, H_a (C2–C1)	C_1H_b: -6.6 C_2H_a: $+26.7$		
$Cl_3C-C(=O)(H)$	$+46.3$		

Tab. 3.29   continued

Compound	$^2J(C,H)$ (Hz)	$^3J(C,H)$ (Hz)	$^nJ(C,H)$ (Hz)
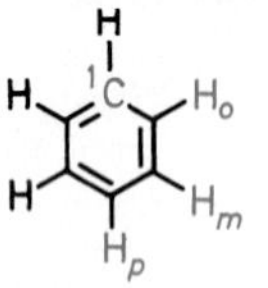	CH_o: $+1.1$	CH_m: $+7.6$	CH_p: -1.2 ($n = 4$)
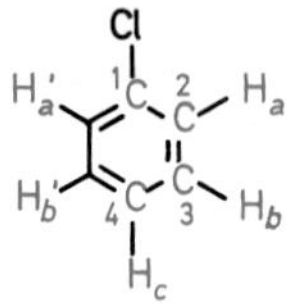	C_1H_a: -3.4 C_2H_b: $+1.4$ C_3H_a: $+0.3$ C_3H_c: $+1.6$ C_4H_b: $+0.9$	C_1H_b: $+11.1$ C_2H_c: $+7.8$ C_2H_a': $+5.1$ C_3H_b': $+8.2$ C_4H_a: $+7.4$	C_1H_c: -2.0 ($n = 4$) C_2H_b': -1.2 ($n = 4$) C_3H_a': -0.9 ($n = 4$)
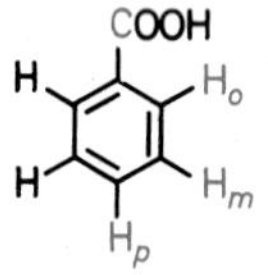		CH_o: $+4.1$	CH_m: $+1.1$ ($n = 4$) CH_p: $+0.5$ ($n = 5$)
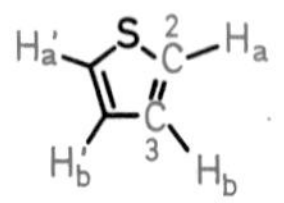	C_2H_b: $+7.4$ C_3H_a: $+4.7$ C_3H_b: $+5.9$	C_2H_b': $+10.0$ C_2H_a': $+5.0$ C_3H_a': $+9.8$	
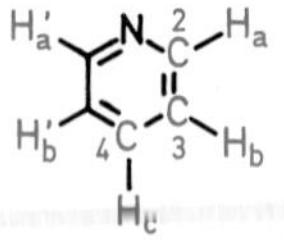	C_2H_b: $+3.1$ C_3H_a: $+8.5$ C_3H_c: $+0.9$ C_4H_b: $+0.7$	C_2H_a': $+11.1$ C_2H_c: $+6.8$ C_3H_b': $+6.6$ C_4H_a: $+0.4$	C_2H_b': -0.9 ($n = 4$) C_3H_a': -1.7 ($n = 4$)

Since $^1J(C,H)$ is of some importance for spectral interpretation, some further examples are given in Tab. 3.28.

Whereas the $^1J(C,H)$ coupling constants lie between ca. $+320$ and $+100$ Hz, values of $^2J(C,H)$ between ca. $+70$ and -20 are known. The *vicinal* couplings $^3J(C,H)$ are always positive and less than 15 Hz. Their values depend on the dihedral angle according to the Karplus curve (see p. 108). Some characteristic data for $^2J(C,H)$, $^3J(C,H)$, and $^nJ(C,H)$ ($n \geq 4$) are given in Tab. 3.29.

4.4 Coupling of ^{13}C to Other Nuclei (D,F,N,P)

In Tab. 3.24 a series of $^1J(C,D)$ **coupling constants** have already been given for deuterated solvents. Approximately,

$$J(C,H) : J(C,D) \approx \gamma_H : \gamma_D \approx 6.5 : 1$$

The ^{13}C,D couplings are therefore considerably smaller than the corresponding ^{13}C,^{1}H couplings.

Fig. 3.67 (see p. 153) shows the ^{13}C spectrum of trifluoroacetic acid (**176**) in deuteriochloroform.

The direct ^{13}C,^{19}F **couplings** have a larger magnitude than the comparable ^{13}C,^{1}H couplings, but have a negative sign. The $^1J(C,F)$ values lie between -150 and -400 Hz. A selection of $^nJ(C,F)$ coupling constants is given in Tab. 3.30.

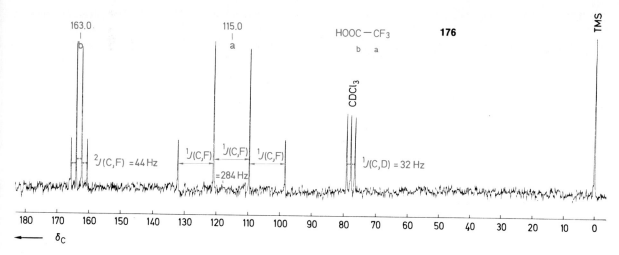

Fig. 3.67   ¹³C NMR spectra of trifluoroacetic acid (**176**) in deuteriochloroform

Tab. 3.30 $^nJ(C,F)$ coupling constants of selected compounds (in Hz)

Compound	$\lvert^1J(C,F)\rvert$	$\lvert^2J(C,F)\rvert$	$\lvert^nJ(C,F)\rvert$
F—CH₃	162		
F—CFH₂	235		
F—CF₂H	274		
F—CF₃	259		
F—CF₂—C(OH)₂—CF₃	286	34	
F—CH₂—CH₂—CH₂—C₃H₇	167	20	^3J (C,F): 5 ^4J (C,F): <2 ^5J (C,F): ≈0
F₂C=CH₂	287		
FHC=O	369		
F(H)C cyclohexane (CH₂, CH₂, CH₂)	171	19	^3J (C,F): 5 ^4J (C,F): ≈0
F cyclohexene (CH, CH)	245	21	^3J (C,F): 8 ^4J (C,F): 3
F—CF₂ cyclohexene (CH, CH)	272	32	^3J (C,F): 4 ^4J (C,F): 1 ^5J (C,F): 0

Tab. 3.31 ^{13}C, ^{15}N coupling constants of selected compounds (in Hz)

Compound	$|^1J(C,N)|$	$|^2J(C,N)|$	$|^nJ(C,N)|$
H_3C-NH_2	5		
$H_3C-\overset{+}{N}(CH_3)_3$	6		
$H_3C-\underset{O}{\overset{NH_2}{C}}$	14	10	
$H_3C-C\equiv N$	18	3	
(aniline ring)	11.4	2.7	$^3J(C,N)$: 1,3 $^4J(C,N)$: <1
(pyridine ring)	0.5	2.4	$^3J(C,N)$: 3,9
(pyridinium ring)	12.0	2.1	$^3J(C,N)$: 5,3

A few comments on ^{13}C, ^{15}N and ^{13}C, ^{31}P couplings will complete this section. Whereas ^{13}C, ^{19}F and ^{13}C, ^{31}P couplings can be directly measured from ^1H broadband decoupled ^{13}C spectra, ^{13}C, ^{15}N couplings are only accessible *via* ^{15}N enrichment. The natural abundance of both ^{19}F and ^{31}P is 100%, but only 0.37% for ^{15}N (see Tab. 3.1, p. 72). Some idea of typical values of ^{13}C, ^{15}N and ^{13}C, ^{31}P coupling constants can be gained from the examples given in Tabs. 3.31 and 3.32.

4.5 ^{13}C, ^{13}C Couplings

The coupling of a ^{13}C nucleus with a neighbouring ^{13}C nucleus is generally not observable in routine ^{13}C NMR spectra. The low natural abundance of ^{13}C of 1.1% means that the nuclear combination $^{13}C...^{13}C$ is only ca. 1/100 times as likely as $^{13}C...^{12}C$ and only 10^{-4} times as likely as $^{12}C...^{12}C$. Apart from weak satellites in long accumulations of intense signals of pure liquids the measurement of ^{13}C, ^{13}C couplings requires the use of ^{13}C enriched compounds or the use of the INADEQUATE method (see p. 179).

Tab. 3.32 ^{13}C, ^{31}P coupling constants of selected compounds (in Hz)

Compound	$|^1J(C,P)|$	$|^2J(C,P)|$	$|^nJ(C,P)|$
$P(CH_2-CH_2-CH_2-CH_3)_3$	11	12	$^3J(C,P)$: 13 $^4J(C,P)$: ≈ 0
$^+P(CH_2-CH_2-CH_2-CH_3)_4\,Br^-$	48	4	$^3J(C,P)$: 15 $^4J(C,P)$: ≈ 0
$Cl_2P-CH_2-CH_2-CH_2-CH_3$	44	14	$^3J(C,P)$: 11 $^4J(C,P)$: ≈ 0
$O{=}P(CH_2-CH_2-CH_2-CH_3)_3$	66	5	$^3J(C,P)$: 13 $^4J(C,P)$: ≈ 0
$O{=}\underset{OC_2H_5}{\overset{OC_2H_5}{P}}{-}CH_2-CH_2-CH_2-CH_2-CH_2-CH_3$	141	5	$^3J(C,P)$: 16 $^4J(C,P)$: 1 $^{5,6}J(C,P)$: ≈ 0
$O{=}\underset{OC_2H_5}{\overset{OC_2H_5}{P}}{-}C\equiv C-CH_3$	300	53	$^3J(C,P)$: 5
$(H_5C_2O)_2P-O-CH_2-CH_3$		11	$^3J(C,P)$: 5

Tab. 3.32 continued

Compound	$|^1J(C,P)|$	$|^2J(C,P)|$	$|^nJ(C,P)|$
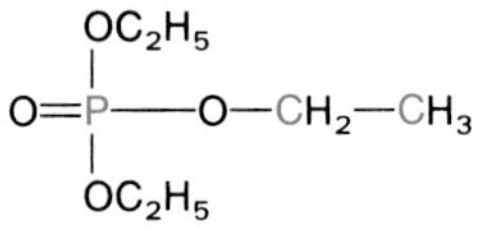 $O{=}\overset{\displaystyle OC_2H_5}{\underset{\displaystyle OC_2H_5}{P}}{-}O{-}CH_2{-}CH_3$		6	$^3J(C,P)$: 7
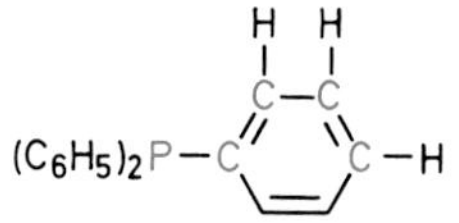 $(C_6H_5)_2P{-}C_6H_4{-}H$	13	20	$^3J(C,P)$: 7 $^4J(C,P)$: 0.3
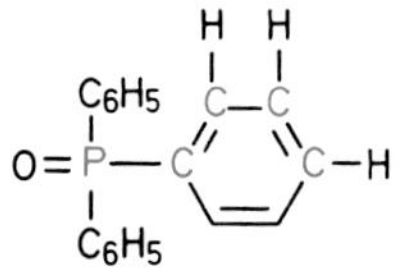 $O{=}\overset{\displaystyle C_6H_5}{\underset{\displaystyle C_6H_5}{P}}{-}C_6H_4{-}H$	104	10	$^3J(C,P)$: 12 $^4J(C,P)$: 2
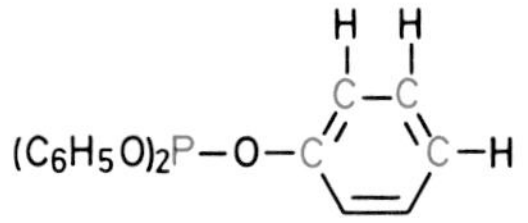 $(C_6H_5O)_2P{-}O{-}C_6H_4{-}H$		3	$^3J(C,P)$: 7 $^4J(C,P)$: ≈ 0 $^5J(C,P)$: 1
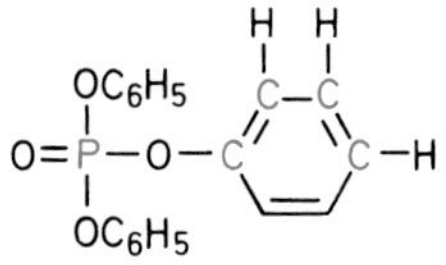 $O{=}\overset{\displaystyle OC_6H_5}{\underset{\displaystyle OC_6H_5}{P}}{-}O{-}C_6H_4{-}H$		7	$^3J(C,P)$: 5 $^4J(C,P)$: 1 $^5J(C,P)$: 2

^{13}C,^{13}C couplings show a similar dependence on hybridisation and electronic effects as ^{1}H,^{13}C couplings.

The following series of compounds shows that $^1J(C,C)$ increases strongly with the s-character of the orbitals involved:

sp-sp
- $HC{\equiv}C{-}C{\equiv}CH$ + 190.3 Hz   **177**
- $HC{\equiv}CH$ + 171.5 Hz   **178**
- $HC{\equiv}C{-}C{\equiv}CH$ + 153.4 Hz   **177**

sp²-sp
- $H_2C{=}C{=}CH_2$ + 98.7 Hz   **179**
- $HC{\equiv}C{-}CH{=}CH_2$ + 86.7 Hz   **180**

sp²-sp²
- $H_2C{=}CH{-}CH{=}CH_2$ + 68.6 Hz   **118**
- $H_2C{=}CH_2$ + 67.6 Hz   **15**

sp³-sp
- $H_3C{-}C{\equiv}CH$ + 67.4 Hz   **70**

sp²-sp²
- (C₆H₅ ring with C=C)   + 56.0 Hz   **75**
- $H_2C{=}CH{-}CH{=}CH_2$ + 53.7 Hz   **118**

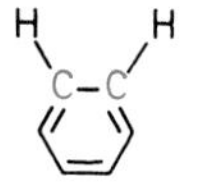

sp³-sp²
- $H_3C{-}C_6H_5$ + 44.2 Hz   **77**
- $H_3C{-}CH{=}CH_2$ + 41.9 Hz   **68**

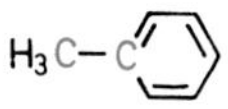

sp³-sp³
- $H_3C{-}CH_3$ + 34.6 Hz   **92**

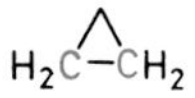

- $H_2C{-}CH_2$ (cyclopropane)   + 12.4 Hz   **93**

The effect of substituents on $^1J(C,C)$ is generally small for saturated C atoms; larger effects are observed for olefinic, aromatic, or carbonyl carbons.

	X = H	CH₃	Cl	OC₂H₅
H₃C—CH₂—X	34.6	34.6	36.1	38.9
H₂C=CH—X	67.6	70.0	77.6	78.1
HC≡C—X	171.5	175.0		216.5
C₆H₅—X	56.0	57.1	65.2	67.0
H₃C—CO—X	39.4	40.1	56.1	58.8

Geminal couplings $^2J(C,C)$ can be positive or negative; their magnitude is generally less than 5 Hz. Exceptions occur particularly for carbonyl compounds, alkynes, and organometallic compounds.

$$H_3C-CO-CH_3 \quad H_3C-C\equiv CH \quad H_3C-CH_2-CN$$

16 Hz	12 Hz	33 Hz
181	**70**	**182**

20 Hz
183

$$H_3C-Hg-CH_3$$

22 Hz
184

$^3J(C,C)$ coupling constants are positive and generally smaller than 5 Hz. Exceptions occur for conjugated systems; thus the ^3J coupling between C-1 and C-4 in butadiene for example is 9.1 Hz.

A few further $^{13}C,^{13}C$ couplings (values in Hz) are tabulated below.

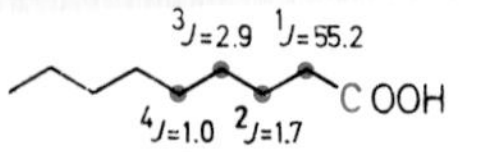

185

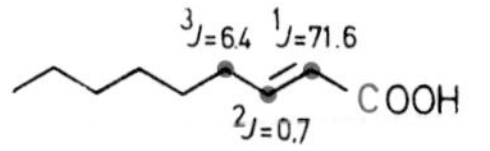

186

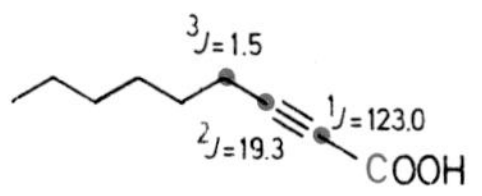

187

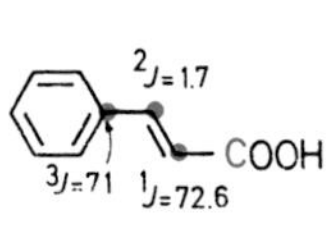

8

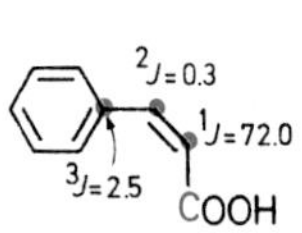

109

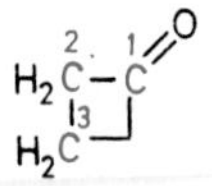

188

$^1J(C-1, C-2) = 29,5$ $^2J(C-1, C-3) = 9,7$

$^1J(C-2, C-3) = 28,5$

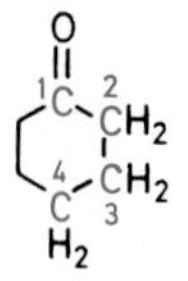

189

$^1J(C-1, C-2) = 37.7$ $^2J(C-2. C-4) = 2.2$

$^1J(C-2, C-3) = 30.4$

$^1J(C-3, C-4) = 33.1$

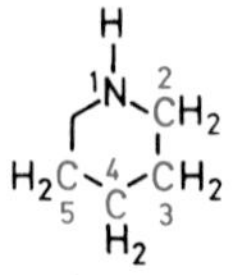

190

$^1J(C-2, C-3) = 35.2$ $^2J(C-2. C-4) = 2.6$

$^1J(C-3, C-4) = 33.0$ $^3J(C-2. C-5) = 1.7$

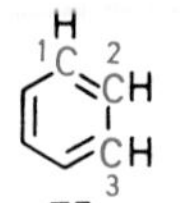

75

$^1J(C-1, C-2) = 56.0$ $^2J(C-1. C-3) = 2.5$

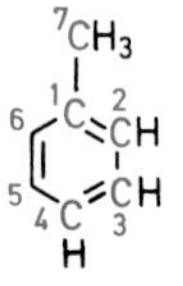

77

$^1J(C-1, C-2) = 57.3$ $^2J(C-1. C-3) = 0.8$

$^1J(C-1, C-7) = 44.2$ $^2J(C-2. C-4) = 2.6$

$^2J(C-2. C-7) = 3.1$

$^3J(C-1. C-4) = 9.5$

$^3J(C-3. C-7) = 3.8$

$^4J(C-4. C-7) = 0.9$

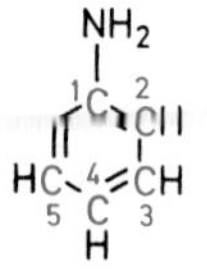

165

$^1J(C-1, C-2) = 61.3$ $^3J(C-2. C-5) = 7.9$

$^1J(C-2, C-3) = 58.1$

$^1J(C-3, C-4) = 56.2$

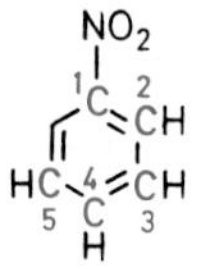

164

$^1J(C-1, C-2) = 55.4$ $^3J(C-2. C-5) = 7.6$

$^1J(C-2, C-3) = 56.3$

$^1J(C-3, C-4) = 55.8$

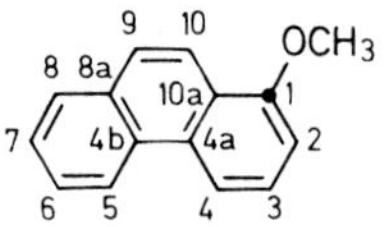

191

$^1J(C-1, C-2) = 70.0$ $^3J(C-1, C-9) = 4.4$

$^1J(C-1, C-10a) = 67.2$ $^4J(C-1, C-5) = 0.4$

$^2J(C-1, C-3) = 0.6$ $^4J(C-1, C-8a) = 0.9$

$^2J(C-1, C-4a) = 2.1$ $^5J(C-1, C-6) = 0.7$

$^2J(C-1, C-10) = 0.7$ $^5J(C-1, C-8) = 0.3$

$^3J(C-1, C-4) = 7.7$ $^6J(C-1, C-7) = 0.4$

$^3J(C-1, C-4b) = 3.8$

4.6 Correlation of ¹³C Shifts With Structural Features

The ranges of the ¹³C chemical shifts for the most important structural types of organic compounds are collected in Tab. 3.33. Extreme values are ignored. For the interpretation of ¹³C

spectra it is recommended that this table is used in conjunction with the data in Sec. 5 (see p. 185) arranged by type of compound and the increment systems in Sec. 4.7 (see p. 159).

If more than one functional group is attached to a single saturated C atom, then to a first approximation the additivity of the shift effects can be assumed (see also Sec. 4.7).

Tab. 3.33   ¹³C shifts of important structural elements; δ values (ppm)

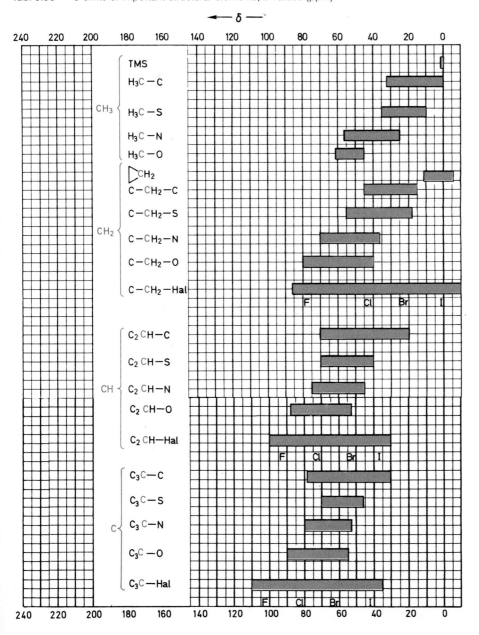

Tab. 3.33 continued

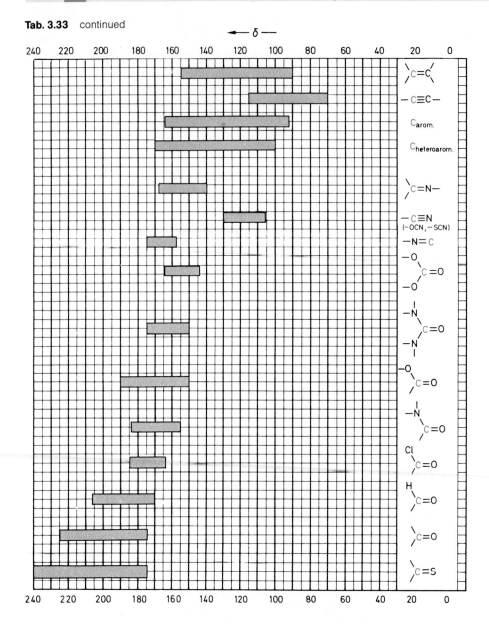

For the effect of functional groups on (C=C) bonds or aromatic rings see Sec. 4.2 (p. 145) and Sec. 4.7 (p. 159).

It is already apparent from the ranges given in Tab. 3.33 that the chemical shifts of carbonyl carbons vary greatly with the compound type. An explicit comparison is possible from the data in Tab. 3.34.

Conjugation of double bonds has relatively little effect, unless charge shifts are effective. Two examples underline this: cyclohexene (**107**)/1,3-cyclohexadiene (**74**) and cyclohexanone

(**189**)/2-cyclohexen-1-one (**192**).

$$23,0 \quad \overset{25,4}{\diagdown} 127,4$$

107

$$\overset{126,2}{\diagdown} \underset{22,3}{\diagdown} 124,6$$

74

$$\overset{O}{\overset{\|}{\diagdown}} \overset{208,5}{\diagdown} 40,4 \quad \underset{23,8}{\diagdown} 26,5$$

189

$$\overset{O}{\overset{\|}{\diagdown}} 199,5 \quad 37,7 \diagdown 129,5 \quad 22,3 \diagdown 150,6 \quad 25,3$$

192

Cumulated double bonds on the other hand lead to very characteristic δ-values of the chemical shifts (Tab. 3.35).

Tab. 3.34 ^{13}C shifts of carbonyl C-atoms in various types of compounds

Type of compound	R—CO—X	δ values R = CH$_3$	R = C$_6$H$_5$
Ketones	R—CO—CH$_3$	206.0	195.7
Aldehydes	R—CO—H	199.7	197.6
Thiocarboxylic acids S-esters	R—CO—SC$_2$H$_5$	195.0	191.2
Carboxylate salts	R—CO—O$^-$	181.7	175.5
Carboxylic acids	R—CO—OH	178.1	172.6
Carboxylic acid amides	R—CO—NH$_2$	172.7	169.7
Carboxylic acid esters	R—CO—OCH$_3$	170.7	166.8
Carboxylic acid chlorides	R—CO—Cl	170.5	168.0
Carboxylic acid anhydrides	R—CO—OCOR	166.9	162.9

Type of compound	X—CO—X	δ values
Ureas	R$_2$N—CO—NR$_2$	161.2 (R = H).
		165.4 (R = CH$_3$)
Urethanes	RNH—CO—OR	157.8 (OR = OC$_2$H$_5$,
		NR = NCH$_3$)
Carbonate esters	RO—CO—OR	156.5 (R = CH$_3$)
Chloroformate esters	Cl—CO—OR	149.9 (R = C$_2$H$_5$)
Phosgene	Cl—CO—Cl	142.1

Tab. 3.35 ^{13}C shifts of compounds with cumulated double bonds

Allene	H$_2$C=C=CH$_2$
	73,5 212.6
Ketene	H$_2$C=C=O
	2,5 194.0
Diazomethane	H$_2$C=N=$\bar{\text{N}}$I
	23.1
Methyl isocyanate	H$_3$C—N=C=O
	26.3 121.5
Methyl isothiocyanate	H$_3$C—N=C=S
	29.3 128.7
Dicyclohexylcarbodiimide	25.5 ⟨⟩—N=C=N—⟨⟩
	55.8 139.9
	24.8 35.0
Carbon dioxide	O=C=O
	123.9
Carbon disulfide	S=C=S
	192.3

Extreme shift values are observed for thio-, seleno-, and telluroketones.

x	δ
O	226
S	282
Se	294
Te	301

193

4.7 Increment Systems for the Estimation of ^{13}C Shifts

The ^{13}C shifts of aliphatic compounds and benzene derivatives can be estimated with **empirical increment systems**. If there is more than one substituent an additive behaviour is assumed. The following increment systems are a useful aid for the assignment of ^{13}C signals. The tabulated examples however show the possible deviations between calculated and observed shifts.

For the estimation of the ^{13}C shifts of saturated C atoms the simplest procedure is to base the calculation on a corresponding hydrocarbon and add the increments for the various functional groups.

If the ^{13}C shifts δ_i of the hydrocarbon itself are not known, they can be calculated from the **Grant-Paul rules** as follows

$$\delta_i = -2.3 + \sum_k A_k n_k + S_{i\alpha}$$

The increments $A_k n_k$ are added to the shift value for methane $\delta = 2.3$. Increments are added for all the positions k = $\alpha, \beta, \gamma, \delta, \varepsilon$ relative to the relevant carbon. n_k is the number of C atoms at position k. The increments A_k have the following values:

$$A_\alpha = +9.1 \qquad A_\gamma = 2.5 \qquad A_\varepsilon = +0.2$$
$$A_\beta = +9.4 \qquad A_\delta = +0.3$$

For tertiary and quaternary C atoms and their immediate neighbours a steric correction factor $S_{i\alpha}$ must be added. This is derived from the most substituted carbon C_α next to the carbon C_i for which the shift is being calculated. The correction values $S_{i\alpha}$ are then as follows.

C_i (Carbon atom under consideration)		C_α (highest substituted neighbouring C atom)			
		—CH$_3$	—CH$_2$—	—CH—	—C—
primary	—CH$_3$	0	0	− 1.1	− 3.4
secondary	—CH$_2$—	0	0	− 2.5	− 7.5
tertiary	—CH—	0	− 3.7	− 9.5	(− 15.0)
quaternary	—C—	− 1.5	− 8.4	(− 15.0)	(− 25.0)

The system will become clearer from a consideration of 2 methylbutane (**194**).

194

C_i	-2.3	$+\sum_k A_k n_k$	$+S_{i\alpha}$	$\equiv \delta_{calculated}$	$\delta_{observed}$
C–1	-2.3	$+9.1$ $+9.4 \cdot 2$ -2.5	-1.1	$= 22.0$	21.9
C–2	-2.3	$+9.1 \cdot 3$ $+9.4$	-3.7	$= 30.7$	29.7
C–3	-2.3	$+9.1 \cdot 2$ $+9.4 \cdot 2$	-2.5	$= 32.2$	31.7
C–4	-2.3	$+9.1$ $+9.4$ $-2.5 \cdot 2$		$= 11.2$	11.4

In sterically hindered hydrocarbons the agreement between $\delta_{calculated}$ and $\delta_{observed}$ is poorer. If rotation around a C–C bond is restricted or prevented, then additional **conformational corrections** must be made.

Tab. 3.36 Increment system for the estimation of ^{13}C shifts of aliphatic compounds
$\delta(\text{RX}) = \delta(\text{RH}) + I_{xk} + S_{i\alpha}$ $(k = \alpha, \beta, \gamma, \delta)$ for all C_i

Substituent X	$k = \alpha$	β	γ	δ
—C=C— (| |)	20.0	6.9	-2.1	0.4
—C≡C—	4.4	5.6	-3.4	-0.6
—C$_6$H$_5$	22.1	9.3	-2.6	0,3
—CH=O	29.9	-0.6	-2.7	0
—C=O (| R)	22.5	3.0	-3.0	0
—COOH	20.1	2.0	-2.8	0
—COOR	22.6	2.0	-2.8	0
—CO—NR$_2$	22.0	2.6	-3.2	-0.4
—COCl	33.1	2.3	-3.6	0
—C≡N	3.1	2.4	-3.3	-0.5
—OH	49.0	10.1	-6.2	0
—OR	58.0	7.2	-5.8	0
—O—CO—R	54.0	6.5	-6.0	0
—NR$_2$	28.3	11.3	-5.1	0
—$\overset{+}{N}$R$_3$	30.7	5.4	-7.2	-1.4
—NO$_2$	61.6	3.1	-4.6	-1.0
—SH	10.6	11.4	-3.6	-0.4
—SCH$_3$	20.4	6.2	-2.7	0
—F	70.1	7.8	-6.8	0
—Cl	31.0	10.0	-5.1	-0.5
—Br	18.9	11.0	-3.8	-0.7
—I	-7.2	10.9	-1.5	-0.9

Tab. 3.37 Calculated and observed δ values for selected aliphatic compounds

Compound		$\delta_{calculated}$ [a]	$\delta_{observed}$
1-Hexyne			
	C–1	–	67.4
	C–2	–	82.8
HC≡C–CH$_2$–CH$_2$–CH$_2$–CH$_3$ (1 2 3 4 5 6)	C–3	17.4 (18.1)	17.4
	C–4	30.4 (30.9)	29.9
	C–5	21.4 (21.9)	21.2
	C–6	12.4 (13.1)	12.9
Basis			
H$_3$C–CH$_2$–CH$_2$–CH$_3$ (1 2)	C–1	13.7	13.0
	C–2	25.3	24.8
2-Butanol[h]			
H$_3$C–CH–CH$_2$–CH$_3$ (1 2 3 4), OH	C–1	22.0 (22.7)	22.6
	C–2	70.1 (70.6)	68.7
	C–3	32.4 (32.9)	32.0
	C–4	6.8 (7.5)	9.9
Basis			
H$_3$C–CH$_2$–CH$_2$–CH$_3$ (1 2)	C–1	13.7	13.0
	C–2	25.3	24.8
2-Chloro-2-methylbutane			
H$_3$C–$\overset{Cl}{\underset{CH_3}{C}}$–CH$_2$–CH$_3$ (1 2 3 4, 1)	C–1	31.9 (32.0)	32.0
	C–2	60.7 (61.7)	71.1
	C–3	41.7 (42.2)	38.8
	C–4	6.3 (6.1)	9.4
Basis			
H$_3$C–CH–CH$_2$–CH$_3$ (1 2 3 4), CH$_3$ (1)	C–1	22.0	21.9
	C–2	30.7	29.7
	C–3	32.2	31.7
	C–4	11.2	11.4
Leucine[b]			
HOOC–CH–CH$_2$–CH–CH$_3$ (1 2 3 4 5), NH$_2$, CH$_3$ (6)	C–1	–	176.6
	C–2	56,1 (55.9)	54.8
	C–3	42.5 (43.0)	41.0
	C–4	21.8 (22.8)	25.4
	C–5		
	C–6	21.9 (22.0)	23.2/22.1
Basis			
H$_3$C–CH–CH$_2$–CH$_3$ (1 2 3 4), CH$_3$ (1)	C–1	22.0	21.9
	C–2	30.7	29.7
	C–3	32.2	31.7
	C–4	11.2	11.4

[a] The basis values used are the **observed** ^{13}C shifts of the hydrocarbon; the δ values in brackets are calculated from the **calculated** ^{13}C shifts of the hydrocarbon

[b] For C-1, C-2 and C-3 in 2-butanol and for C-2 and C-3 in leucine steric correction factors $S_{i\alpha}$ have been included

Using the measured or calculated δ_i values of an alkane C_nH_{2n+2} as a basis the ^{13}C shifts of the substituted compounds $C_nH_{2n+1}X$, $C_nH_{2n}XY$, etc. can be calculated. Tab. 3.36 gives the increments I for a selection of substituents X depending on the position of substitution relative to the carbon C_i for which the shift is being calculated.

In Tab. 3.37 the observed shifts of some representative examples are given, together with the values calculated with the increment system. (For the substituents $-OR$, $-NR_2$ and $-SR$ it is recommended to apply the steric corrections $S_{i\alpha}$ as for hydrocarbons.)

The ^{13}C shifts of olefinic carbons can be calculated using the values given in Tab. 3.38.

Tab. 3.38   Increment system for the estimation of ^{13}C shifts of olefinic carbons

$$X-\overset{1}{C}H=\overset{2}{C}H_2 \qquad \delta_1 = 123.3 + I_1. \; \delta_2 = 123.3 + I_2$$
$$X-\overset{1}{C}H=\overset{2}{C}H-Y \qquad \delta_1 = 123.3 + I_{X1} + I_{Y2}.$$
$$\delta_2 = 123.3 + I_{Y1} + I_{X2}$$

Substituent	Increments	
	I_1	I_2
—H	0	0
—CH$_3$	10.6	− 8.0
—C$_2$H$_5$	15.5	− 9.7
—CH$_2$—CH$_2$—CH$_3$	14.0	− 8.2
—CH(CH$_3$)$_2$	20.3	− 11.5
—(CH$_2$)$_3$—CH$_3$	14.7	− 9.0
—C(CH$_3$)$_3$	25.3	− 13.3
—CH=CH$_2$	13.6	− 7.0
—C≡C—R	7.5	8.9
—C$_6$H$_5$	12.5	− 11.0
—CH$_2$Cl	10.2	− 6.0
—CH$_2$Br	10.9	− 4.5
—CH$_2$OR	13.0	− 8.6
—CH=O	13.1	12.7
—CO—CH$_3$	15.0	5.9
—COOH	4.2	8.9
—COOR	6.0	7.0
—CN	− 15.1	14.2
—OR	28.8	− 39.5
—O—CO—R	18.0	− 27.0
—NR$_2$	16.0	− 29.0
—$\overset{+}{N}$(CH$_3$)$_3$	19.8	− 10.6
—NO$_2$	22.3	− 0.9
—SR	19.0	− 16.0
—F	24.9	− 34.3
—Cl	2.6	− 6.1
—Br	− 7.9	− 1.4
—I	− 38.1	7.0

Tab. 3.39 gives an impression of the "quality" of the estimations possible with this increment system. (The application of the system to trisubstituted olefins, not to mention tetrasubstituted olefins, is not recommended.)

Tab. 3.39   Calculated and observed δ values for the ^{13}C shifts of olefinic C atoms

Compound	δ_i calculated		δ_i observed	
	C–1	C–2	C–1	C–2
2-Butene				
	125.9	125.9	124.8	124.8
	125.9	125.9	123.4	123.4
Methyl methacrylate				
	122.3	139.9	124.7	136.9
(E)-Crotonaldehyde				
	146.6	128.4	153.7	134.9
Fumaric acid				
	136.4	136.4	134.5	134.5
Maleic acid				
	136.4	136.4	130.8	130.8
2-Phenyl-1-butene				
	102.6	151.3	109.7	148.9

The increment system for benzene derivatives in Tab. 3.40 works on similar principles. The increments I_1, I_2 etc. for each substituent X_1, X_2 etc. are added to the basis value for benzene

($\delta = 128.5$). The value of I depends on the nature of the substituent and its position relative to the C atom for which the shift is being calculated.

Several illustrative examples are given in Tab. 3.41. Larger deviations between calculated and observed values can be expected when the additivity of the substituent effects is disturbed by steric or electronic interactions.

Tab. 3.40 Increment system for the estimation of ^{13}C shifts of substituted benzenes

$$\delta_i = 128{,}5 + I_{1i} + I_{2j} + \ldots$$

Substituent	ipso	ortho	meta	para
—H	0.0	0.0	0.0	0.0
—CH$_3$	9.3	0.6	0.0	−3.1
—C$_2$H$_5$	15.7	−0.6	−0.1	−2.8
—CH(CH$_3$)$_2$	20.1	−2.0	0.0	−2.5
—C(CH$_3$)$_3$	22.1	−3.4	−0.4	−3.1
—CH=CH$_2$	7.6	−1.8	−1.8	−3.5
—C≡CH	−6.1	3.8	0.4	−0.2
—C$_6$H$_5$	13.0	−1.1	0.5	−1.0
—CF$_3$	2.6	−2.6	−0.3	−3.2
—CH$_2$Cl	9.1	0.0	0.2	−0.2
—CH$_2$Br	9.2	0.1	0.4	−0.3
—CH$_2$OR	13.0	−1.5	0.0	−1.0
—CH$_2$—NR$_2$	15.0	−1.5	−0.2	−2.0
—CH=O	7.5	0.7	−0.5	5.4
—CO—CH$_3$	9.3	0.2	0.2	4.2
—COOH	2.4	1.6	−0.1	4.8
—COOR	2.0	1.0	0.0	4.6
—CO—NR$_2$	5.5	−0.5	−1.0	5.0
—COCl	4.6	2.9	0.6	7.0
—C≡N	−16.0	3.5	0.7	4.3
—OH	26.9	−12.6	1.6	−7.6
—OCH$_3$	31.3	−15.0	0.9	−8.1
—OC$_6$H$_5$	29.1	−9.5	0.3	−5.3
—O—CO—R	23.0	−6.0	1.0	−2.0
—NH$_2$	19.2	−12.4	1.3	−9.5
—NR$_2$	21.0	−16.0	0.7	−12.0
—NH—CO—CH$_3$	11.1	−9.9	0.2	−5.6
—N=N—C$_6$H$_5$	24.0	−5.8	0.3	2.2
—N=C=O	5.7	−3.6	1.2	−2.8
—NO$_2$	19.6	−5.3	0.8	6.0
—SH	2.2	0.7	0.4	−3.1
—SCH$_3$	10.1	−1.6	0.2	−3.5
—SC$_6$H$_5$	6.8	0.5	2.2	−1.6
—SO$_3$H	15.0	−2.2	1.3	3.8
—F	35.1	−14.3	0.9	−4.4
—Cl	6.4	0.2	1.0	−2.0
—Br	−5.4	3.3	2.2	−1.0
—I	−32.3	9.9	2.6	−0.4

Tab. 3.41 Calculated and observed δ values for the ^{13}C shifts of substituted benzenes

Compound	C atom	$\delta_{calculated}$	$\delta_{observed}$
p-Xylene	C–1	134.7	134.5
	C–2	129.1	129.1
Mesitylene	C–1	137.8	137.6
	C–2	126.6	127.4
Hexamethylbenzene	C	135.9	132.3
p-Cresol	C–1	152.3	152.6
	C–2	115.9	115.3
	C–3	130.7	130.2
	C–4	130.2	130.5
3,5-Dimethoxybenzalde-hyde	C–1	137.8	138.4
	C–2	106.1	107.0
	C–3	160.2	161.2
	C–4	103.9	107.0
1-Chloro-2,4-dimethoxy-5-nitrobenzene	C–1	112.5	113.8
	C–2	166.9	159.9
	C–3	100.3	97.2
	C–4	153.4	154.5
	C–5	126.0	131.9
	C–6	125.2	127.7

4.8 Special Methods

Many of the methods already described for ^{1}H NMR are also applicable to ^{13}C NMR. These include the variation of the magnetic field strength and the solvent (see p. 124 and 126) and the measurement of temperature dependent spectra (see Sec. 2.2 and 2.3). In the following sections spin decoupling, the spin echo, spectral integration, the use of shift reagents, specific isotopic labelling, NOE experiments, polarisation transfer, double quantum transitions, two-dimensional ^{13}C NMR spectroscopy, solid state spectroscopy (magic angle spinning), and the use of ^{13}C databases will be described.

Spin Decoupling: Heteronuclear Double Resonance

In ^{1}H NMR coupled spectra are normally observed. Decoupling is only employed as a special experiment to aid signal assignment in difficult structural problems (see p. 131 ff.). Routine ^{13}C NMR spectra are in contrast **proton broadband decoupled** (see Sec. 4.1, p. 142). This method of measurement is frequently referred to as ^{13}C–{^{1}H} NMR (**proton noise decoupling**).

Because of the low natural abundance of ^{13}C of 1.1% ^{13}C,^{13}C couplings are normally not observed. Couplings to ^{1}H, D, ^{19}F, ^{31}P, etc. however are observed. The **proton broadband decoupling** causes all couplings to protons to be removed, so that all the multiplets caused by ^{13}C,^{1}H coupling collapse to singlets. This simplifies the ^{13}C spectra enormously. There is also a considerable gain in signal intensity, caused by the removal of the splittings and additionally by the **nuclear Overhauser effect** (see Sec. 4.1, p. 142 ff.).

To achieve proton broadband decoupling a high intensity of radiation must be applied to the sample to cover the whole of the proton frequency range. This is in fact a case of heteronuclear multiple resonance. If the decoupling power is too low, only the quaternary C atoms give sharp, intense signals, all other C atoms give relatively broad signals. This phenomenon can be used systematically to identify the quaternary C atoms (**low power noise decoupling**).

A serious disadvantage of proton broadband decoupling is the total loss of information about the multiplicity of the signals derived from direct ^{13}C,^{1}H couplings, i.e. the differentiation between primary (CH$_3$), secondary (CH$_2$), tertiary (CH), and quaternary (C) carbon atoms. This disadvantage can be overcome by **proton off-resonance decoupling**, the J-modulated spin-echo experiment described on p. 168, or the DEPT experiment described on p. 174. For off-resonance decoupling a decoupling frequency outside the shift range of the protons is chosen and no noise modulation is applied. This leads to a reduction of the coupling constants, so that in general only the direct $^1J^R$(C,H) couplings are visible. The magnitude of the reduced coupling constants $^1J^R$ increases with the magnitude

of ^1J, with the difference between the relevant ^{1}H resonance frequency and the irradiation frequency, and with decreasing irradiation power; typical values of $^1J^R$ are ca. 30–50 Hz. It is useful to vary the conditions as appropriate to produce the minimal overlap of the **quartet, triplet, doublet,** and **singlet signals** of the CH$_3$*, CH$_2$*, CH groups, and quaternary C atoms.

Since the nuclear Overhauser effect is still effective in the case of off-resonance decoupling, the measurement time required is less than for a fully coupled ^{13}C NMR spectrum (but still longer than for a normal, broadband decoupled spectrum).

In Fig. 3.68 the proton broadband decoupled, off-resonance, and coupled spectra of ethylbenzene (**195**) are shown for comparison.

In the off-resonance spectrum the methyl signal a is split into a quartet, the methylene signal b into a triplet, and the signals of the protonated benzene C atoms c, d, and e into doublets. The quaternary C atom remains as a singlet. All the coupling constants are reduced. (Since the decoupler frequency was placed on the low field side, the reduction of the coupling constants is stronger on the low-field side of the spectrum than on the high-field side.)

In the coupled spectrum 3.68c the multiplets overlap. As a result, the aromatic part is very confusing. Figs. 3.68d and e show expansions of the aliphatic and aromatic parts of the spectrum respectively. More splittings than those due to ^1J(C,H) couplings are apparent. The exact interpretation of coupled ^{13}C spectra is often made difficult by two factors: firstly by the already mentioned overlap of multiplets (a particular problem in the case illustrated for the signals d and e) and secondly by the appearance of coupling patterns which do not obey the **rules for first order spectra**. This is clearest for C atom c. The left and right hand parts of this doublet are different. The asymmetry of such spin multiplets is a result of the mixing of nuclear spin states of similar energies. This phenomenon can even occur when the ^{1}H NMR spectrum of the **^{12}C isotopomer** is first order.

---

* This assumes the equivalence of geminal protons. An

$$H_A-\overset{|}{\underset{|}{C}}-H_B$$ group has the spectrum of the

X part of an ABX system, which is not always a 1 : 2 : 1 triplet.

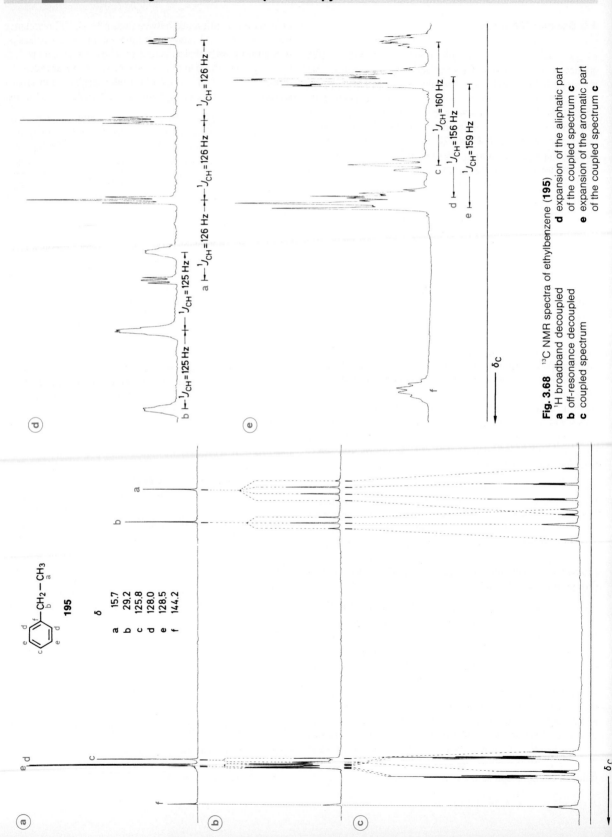

195

δ

a 15.7
b 29.2
c 125.8
d 128.0
e 128.5
f 144.2

b |—$^1J_{CH}=125$ Hz—|—$^1J_{CH}=125$ Hz—|

a |—$^1J_{CH}=126$ Hz—|—$^1J_{CH}=126$ Hz—|—$^1J_{CH}=126$ Hz—|

c |—$^1J_{CH}=160$ Hz—|
d |—$^1J_{CH}=156$ Hz—|
e |————$^1J_{CH}=159$ Hz————|

$\longrightarrow \delta_C$

Fig. 3.68 ^{13}C NMR spectra of ethylbenzene (**195**)
a ^{1}H broadband decoupled
b off-resonance decoupled
c coupled spectrum
d expansion of the aliphatic part
of the coupled spectrum **c**
e expansion of the aromatic part
of the coupled spectrum **c**

A heteronuclear double resonance experiment can naturally also be carried out in such a way that the frequency of a single proton signal is irradiated. In the ¹³C spectrum only the ¹³C,¹H couplings arising from this proton signal are removed. For the remaining ¹³C signals the conditions are those of off-resonance decoupling, i.e. multiplets with reduced coupling constants are observed. This **selective decoupling (single frequency decoupling, SFD)** relies on the proton signals being reasonably well separated. (If necessary this may be achieved by employing a higher magnetic field strength or by the use of lanthanide shift reagents.)

As an aid to understanding the various possibilities a schematic representation of the different decoupling methods is given in Fig. 3.69.

A further example of the practical application of selective heteronuclear double resonance is shown in Fig. 3.70 for the heterocyclic compound **196**.

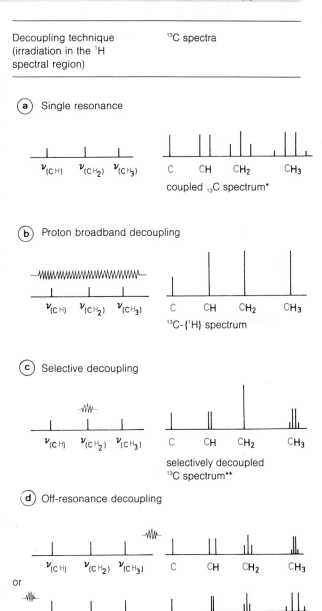

The ¹H NMR spectrum is shown in Fig. 3.70a. Whereas clearly separated signals are observed for the *t*-butyl groups, the methoxy groups, and the protons H_A and H_B of the oxepine ring, the signals of the benzene protons H_C are coincidentally isochronous.

In the proton broadband decoupled ¹³C spectrum (Fig. 3.70 b) 18 singlets are observed, as expected for the 18 types of chemically non-equivalent carbon atoms in the molecule (a-r). The interesting question is which ¹³C signals correlate with the signals of H_A, H_B, and H_C. This was determined by successively irradiating at the frequency v of the protons; the ¹³C spectra obtained from these heteronuclear double resonance experiments are shown in Figs. 3.70 c, d, and e. On irradiating at the frequency of H_A (Fig. 3.70c) C_j appears as a singlet and C_g, C_i, and C_k as doublets. This allows the first unambiguous assignment: H_A is bonded to C_j. From the magnitudes of the reduced couplings for C_k, C_i, and C_g it can additionally be concluded, that H_B belongs with C_g. The final proof is provided by irradiating at v_B (Fig. 3.70d). For the protons H_C the C atoms C_i and C_k are the only remaining possibilities. As a control experiment an irradiation was carried out at v_C (Fig. 3.70e). As expected, both the doublets of C_i and C_k collapse to singlets. The quaternary C-atoms can be assigned with the aid of increment systems and from comparisons with related systems. Overall this gives the following assignments:

Decoupling technique (irradiation in the ¹H spectral region)   ¹³C spectra

(a) Single resonance

$v_{(CH)}$ $v_{(CH_2)}$ $v_{(CH_3)}$ C   CH   CH₂   CH₃

coupled ₁₃C spectrum*

(b) Proton broadband decoupling

$v_{(CH)}$ $v_{(CH_2)}$ $v_{(CH_3)}$ C   CH   CH₂   CH₃

¹³C-{¹H} spectrum

(c) Selective decoupling

$v_{(CH)}$ $v_{(CH_2)}$ $v_{(CH_3)}$ C   CH   CH₂   CH₃

selectively decoupled ¹³C spectrum**

(d) Off-resonance decoupling

$v_{(CH)}$ $v_{(CH_2)}$ $v_{(CH_3)}$ C   CH   CH₂   CH₃

or

$v_{(CH)}$ $v_{(CH_2)}$ $v_{(CH_3)}$ C   CH   CH₂   CH₃

¹³C-{¹H} off-resonance spectrum

Fig. 3.69   Schematic diagram showing the various decoupling techniques in the measurement of ¹³C NMR spectra

*   Only direct ¹³C, ¹H couplings are represented in the diagrams
** Similar decoupling at the frequency v(CH) collapses the doublet, at v(CH₃) the quartet

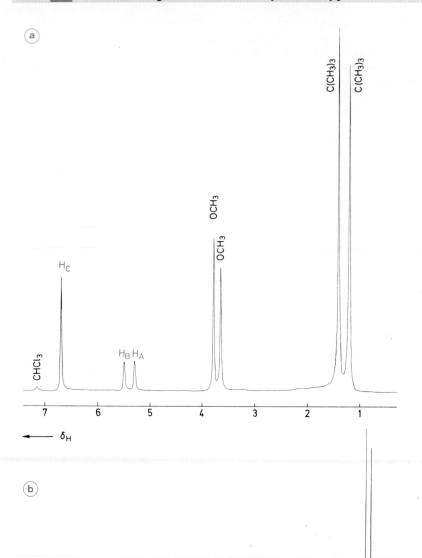

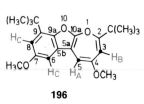

196

Fig. 3.70 NMR spectra of **196** in CDCl₃ (Meier, H., Schneider, H.-P., Rieker, A., Hitchcock, P. B. (1978), Angew. Chem. **90**, 128)
a ¹H spectrum
b ¹³C-{¹H} spectrum (proton broadband decoupling)

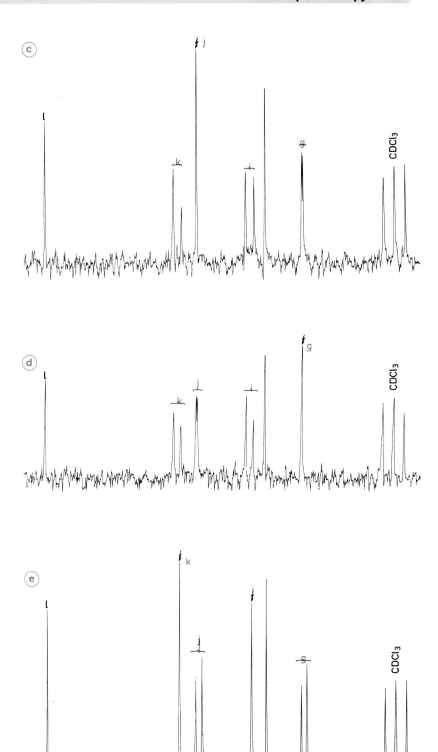

Fig. 3.70 continued
c Expansion of ¹³C spectrum from double resonance experiment ν_A
d Expansion of ¹³C spectrum from double resonance experiment ν_B
e Expansion of ¹³C spectrum from double resonance experiment ν_C

C–2:   r: $\delta = 166.7$
C–3:   j: $\delta = 106.0$ OCH$_3$ = e, f: $\delta = 54.8/55.6$
C–4/7: p, q: $\delta = 154.5/$
                        156.0

C–5:   g: $\delta = 90.6$
C–5a: h: $\delta = 96.1$ C(CH$_3$)$_3$ = a, b: $\delta = 27.5/29.6$
C–5b: 1: $\delta = 128.0$ = c, d: $\delta = 34.1/36.9$
C–6:   i: $\delta = 98.3$
C–7/4: p, q: $\delta = 154.5/$
                        156.0

C–8:   k: $\delta = 108.9$
C–9:   m: $\delta = 135.6$
C–9a: n: $\delta = 142.4$
C–10a: o: $\delta = 152.9$

exo-2-Chloro-1,7,7-trimethylbicyclo[2.2.1]-heptane
(isobornyl chloride, **197**)

endo-2-Chloro-1,7,7-trimethylbicyclo[2.2.1]-heptane
(bornyl chloride, **198**)

Position	δ	δ
C–1	49.7	50.7
C–2	68.2	67.1
C–3	42.4	40.3
C–4	46.0	45.3
C–5	26.9	28.3
C–6	36.2	28.3
C–7	47.3	47.8
1-CH$_3$	13.4	13.3
7-CH$_3$	20.4/20.1	20.5/18.4

The quaternary C atoms C–1 and C–7 and the methylene groups H$_2$C–3, H$_2$C–5, and H$_2$C–6 have positive signals when $\tau = 1/J$, the methine groups HC–2 and HC–4 and the methyl groups at C–1 and C–7 in contrast negative signals.

Fig. 3.71 shows the normal broadband decoupled ^{13}C spectrum of **197** and the *J*-modulated spin-echo spectrum.

J-modulated Spin-echo

The *J*-modulated spin-echo experiment is an alternative to off-resonance decoupling (see p. 163) for the differentiation of the ^{13}C signals of CH$_3$-, CH$_2$-, CH-groups and quaternary carbons. This method has particular advantages for compounds which have closely spaced ^{13}C signals, where the off-resonance technique can lead to a confusing overlap of the partially decoupled multiplets. The method is based on the spin-echo pulse sequence: A 90° excitation pulse is followed after a time τ by a 180° pulse, which inverts the transverse magnetisation. After a further time interval τ the FID is observed. The ^{1}H decoupler is switched off during the second τ-period, which causes a modulation of the signal intensity by the ^{13}C,^{1}H coupling. The decoupler is switched back on immediately before the FID is accumulated, so the observed spectrum is decoupled, but the intensities of the signals of the CH$_n$-groups depend on τ, or more exactly on expressions containing the functions cos $(n\pi \tau J)$. If $\tau = 1/J$ is chosen ($\tau \approx 8$ ms for $^1J(^{13}$C, ^{1}H) = 125 Hz), then positive signals are obtained for C- and CH$_2$-groups (cos 0 and cos $2\pi > 0$) and negative signals for CH- and CH$_3$-groups (cos π and cos $3\pi < 0$). (An analogous experiment with $\tau = 1/2J$ leads to a spectrum only showing the signals of quaternary carbons).

As an example the ^{13}C spectra of the bicyclic compounds **197** and **198** will be discussed.

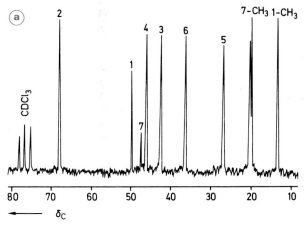

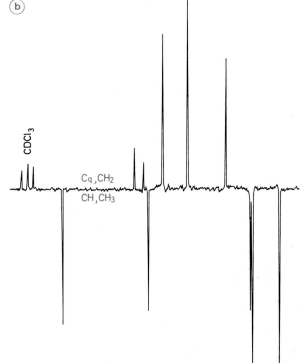

Fig. 3.71 ¹³C spectrum of isobornyl chloride (**197**) in CDCl₃
a ¹H broadband decoupling
b J-modulated spin-echo spectrum ($\tau = 1/J = 8$ ms)

Spectrum Integration

In contrast to ¹H spectra ¹³C NMR spectra are not normally integrated (see Sec. 1.5, p. 84 f.).

This is because the signal **intensities** depend on the **relaxation times** of the various ¹³C nuclei and in decoupled spectra on the different **nuclear Overhauser effects**.

Fig. 3.72 shows the marked **influence of the measurement conditions**. All four spectra were measured on the same solution of 4-ethyl-5-methyl-1,2,3-thiadiazole (**199**) using different values of the pulse length (PW). The other instrumental conditions remained unchanged (800 scans, 8K etc.). It can be seen that the relative intensity of the quaternary C atoms increases as the pulse length decreases. Decreasing the pulse length also worsens the signal/noise ratio considerably.

One possibility of obtaining spectra which yield useful integrations is the addition of paramagnetic **relaxation reagents**. Chromium and iron acetylacetonate have been found to be particularly useful (ca. 0.05 molar solution). Non-degassed samples contain dissolved oxygen, which also increases relaxation rates, particularly for the most slowly relaxing ¹³C nuclei.

The magnetic moment of the unpaired electrons makes a new relaxation process effective, which dominates over other mechanisms. The different relaxation times of the different nuclei all become the same, and the nuclear Overhauser effect also becomes ineffective. The relaxation reagent must not react with th. sample or even form weak complexes with it; otherwise the signals would be shifted as when paramagnetic shift reagents are used. Fig. 3.73 shows the integrated spectrum of 4-ethyl-5-methyl-1,2,3-thiadiazole (**199**) with added Cr(acac)₃.

PW	Relative peak heights h and integrals I				
(μs)	C	a :	b :	c :	d :	e
12.0	h	44	100	67	18	9
	I	44	100	78	11	8
3.5	h	53	100	92	28	18
	I	53	100	97	21	18
1.0	h	66	100	83	69	39
	I	62	100	101	43	43
0.5	h	65	100	86	81	44
	I	57	100	120	52	48

The second method of obtaining useful integrations from ¹³C spectra is the **inverse gated decoupling method**. (Normal **gated decoupling** is a method for measuring coupled spectra. The decoupler is only switched on during the period between the end of accumulation of one FID and the next pulse [pulse delay or relaxation delay]. This allows the nuclear Overhauser

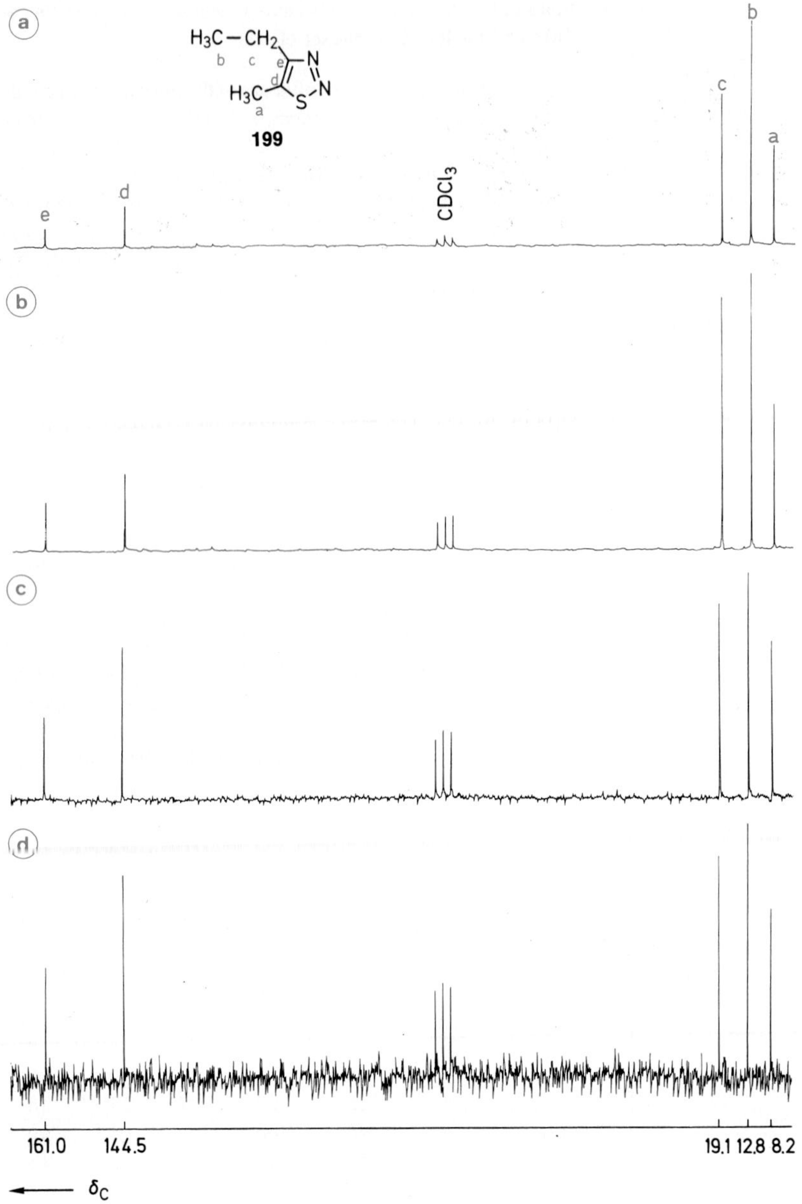

Fig. 3.72 ^{1}H broadband decoupled ^{13}C spectrum of 4-ethyl-5-methyl-1,2,3-thiadiazole (**199**) in CDCl$_3$ under identical measurement conditions apart from the pulse length; PW: 12 μ s (**a**), 3.5 μ) **b**), 1.0 μ s (**c**), 0.5 μ s (**d**)

effect to enhance the intensity of signals of ^{13}C nuclei with attached protons without decoupling the signals. Fully coupled ^{13}C spectra are therefore best measured using this method of **gated proton decoupling**.) For inverse gated decoupling the decoupler is switched on during the accumulation of the FID (so that a decoupled spectrum is obtained) but switched off during the pulse delay. The latter needs to be long enough to allow complete relaxation of the nuclei and decay

of the nuclear Overhauser effect, which means that it must be longer than the longest ^{13}C T_1 in the sample (occasionally $T_1 \geqq$ 100 s!). Spin lattice relaxation and NOE then have no effect on the populations of the nuclear energy levels and therefore no effect on the signal intensities. The disadvantage of this method compared to the use of relaxation reagents is the long measurement times caused by the long pulse delays needed to ensure full relaxation.

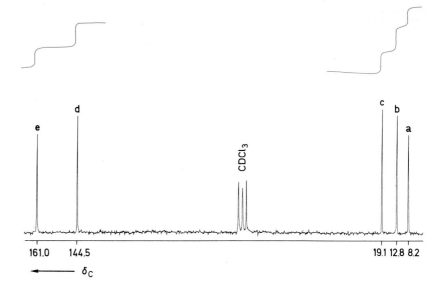

Fig. 3.73 ^{13}C NMR spectrum (proton broadband decoupled) of 4-ethyl-5-methyl-1,2,3-thiadiazole (**199**) in CDCl$_3$ containing Cr(acac)$_3$ with integration curve

Use of Lanthanide Shift Reagents

An introduction to the use of **lanthanide shifts reagents** in ^{1}H NMR was given on p. 128 ff. Essentially the same considerations apply to their use in ^{13}C spectroscopy. The total shift is made up of a **pseudo-contact** and a **contact term**. For chelate complex- es of the lanthanides the (Fermi) contact term is generally small, and the **McConnell-Robertson equation** applies approximately (see p. 128). It therefore follows that ^{1}H

and ^{13}C nuclei in comparable positions (relative to the lanthanide central atom) should have the same shift. The literature values for isoborneol (**200**) have been chosen as an example. The upper shift value refers to Eu(fod)$_3$, the middle value to Eu(dpm)$_3$, and the lower value to Pr(fod)$_3$. The blue figures give the lanthanide-induced shifts for ^{13}C, the black figures the ^{1}H values. (Eu(fod)$_3$ and Eu(dpm)$_3$ shift to low field, Pr(fod)$_3$ to high field.)

Specific Isotopic Labelling

If one or more H atoms in a compound are substituted by **deuterium**, the ^{13}C spectrum changes in a very characteristic fashion.

Whereas CH, CH$_2$, and CH$_3$ groups only show singlets in the ^{1}H broadband decoupled ^{13}C spectrum, a 1:1:1 triplet is observed for CD-, CHD-, and CH$_2$D-groups, a 1:2:3:2:1 quintet for CD$_2$- and CHD$_2$-groups, and a 1:3:6:7:6:3:1 septet for CD$_3$-groups (see p. 77).

The ^{13}C,D couplings are smaller than the corresponding ^{13}C,^{1}H couplings by a factor $\gamma_H/\gamma_D \approx 6.5$. ^{13}C,D long range couplings are therefore usually not observed. The H/D exchange is also noticeable by a small isotope effect on the ^{13}C shift: for ^{13}C-D a high-field shift of ca. 0.2 to 0.7 ppm is observed, for ^{13}C-C-D the isotope shift is even less (0.11 to 0.15 ppm).

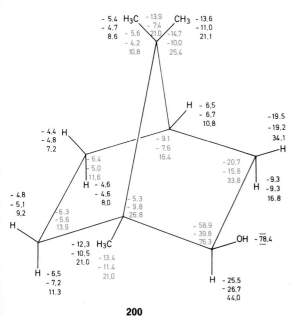

200

More important is the fact that the nuclear Overhauser effect of fully deuteriated ^{13}C nuclei (i.e. with no C-H bond) is reduced and the relaxation time is increased. These two factors combine to cause a considerable loss of signal intensity. Together with the splitting caused by $^{13}C,D$ coupling described above this often leads to the signals of fully deuteriated ^{13}C nuclei disappearing partially or wholly into the noise. The signals of undeuteriated ^{13}C nuclei are scarcely affected; at most they are split or broadened by the small $^{13}C,D$ coupling over two or three bonds.

H/D exchange is therefore a very valuable method for making unambiguous **signal assignments**. This goal can be approached even more directly with ^{13}C **labelling**. The incorporation of ^{13}C

enriched carbon into specific positions in a molecule leads to intense ^{13}C signals. With a small number of accumulations these may often be the only signals observed. ^{13}C labelling also affords a convenient method of measuring $^{13}C,^{13}C$ **coupling constants**.

In Fig. 3.74 this is demonstrated using cyclopentanecarboxylic acid (**201**) as an example. Fig. 3.74a shows the ^{13}C NMR spectrum of the normal, unlabelled material. In Fig. 3.74b, where C atom a has been enriched with ^{13}C to a level of 50%, the signal of this carboxy C atom has greatly increased in intensity. The tertiary C atom b shows three peaks in the spectrum. The central line is the same as in Fig. 3.74a, being due to a ^{13}C nucleus at b with only ^{12}C neighbours. The two remaining lines are a

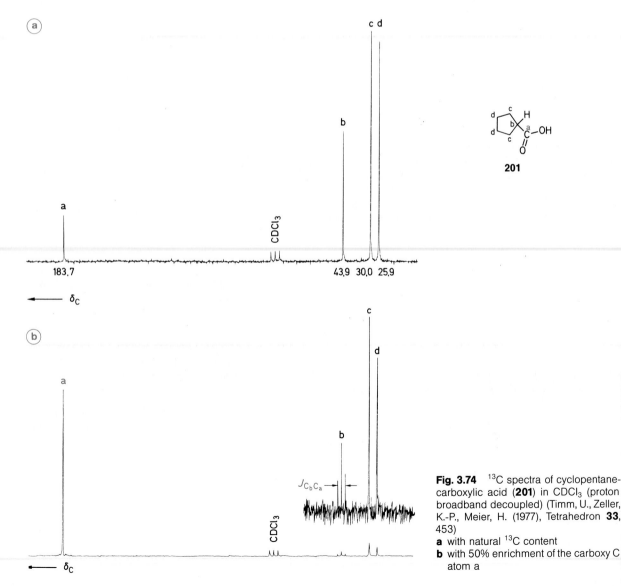

Fig. 3.74 ^{13}C spectra of cyclopentane-carboxylic acid (**201**) in CDCl₃ (proton broadband decoupled) (Timm, U., Zeller, K.-P., Meier, H. (1977), Tetrahedron **33**, 453)
a with natural ^{13}C content
b with 50% enrichment of the carboxy C atom a

doublet, caused by ^{13}C nuclei at b with neighbouring ^{13}C nuclei at a. From the separation of the lines of this doublets the coupling constant $^1J(C_b,C_a) = 56.8$ Hz can be obtained. (Because of the natural abundance of ^{13}C of 1.1 % at nucleus b and the enrichment of nucleus b to a level of 50 % every other ^{13}C nucleus at b has a ^{13}C neighbour at a, but only every ninety ninth nucleus a has a ^{13}C neighbour b).

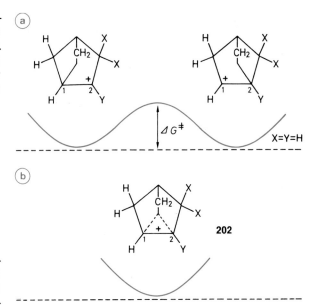

Molecules (201)	a: ^{12}C; b ^{12}C	a: ^{13}C; b: ^{12}C
^{13}C signals	–;–	s;–
natural abundance	97.796%	1.096%
enrichment	49.45 %	49.45%
Molecules (201)	a: ^{12}C; b: ^{13}C	a: ^{13}C; b: ^{13}C
^{13}C signals	–; s	d; d
natural abundance	1.096 %	0.012%
enrichment	0.55%	0.55%

The converse procedure, ^{13}C **depletion** below the natural abundance of 1.1 %, is also of interest. The incorporation of ^{13}C depleted carbon leads to a disappearance of the relevant signals in the ^{13}C spectrum (For the use of ^{12}C enriched solvents see also p. 143).

Since a wide variety of D-, ^{13}C-, and ^{12}C -labelled compounds are commercially available, the synthesis of suitable labelled compounds is becoming an increasingly popular method for the unambiguous assignment of ^{13}C spectra. Such labelling techniques are of especial importance for following reaction pathways of organic and biochemical processes by NMR.

The **isotopic perturbation** method already described on p. 127 f. for ^{1}H NMR spectroscopy can naturally also be applied to ^{13}C NMR. The central problem of **non-classical ions**, the norbornyl cation (**202**) will be discussed as an example. There are two alternatives

a) the double energy minimum model, with two "classical" ions in equilibrium

b) the single energy minimum model of a symmetrical, non-classical ion.

After the incorporation of deuterium (X=D, Y=H/ X=H, Y = D) only a very small splitting of the ^{13}C signals of C-1 and C-2 is observed.

This is only compatible with the static model b. If there were a "perturbed equilibrium" of a fast Wagner-Meerwein rearrangement (Model a) the difference between the chemical shifts would be an order of magnitude larger.

Here borderline cases are naturally also feasible, involving a fast equilibrium between two species which only deviate very slightly from the symmetrical form.

NOE Measurements

In order to measure the ^{13}C,^{1}H nuclear Overhauser effect (NOE) the signal intensities of ^{1}H decoupled and ^{1}H coupled spectra would normally need to be measured. The intensity determination is particularly inaccurate for coupled spectra. Therefore it is preferable in practice to compare the intensities of two decoupled spectra, one with, one without nuclear Overhauser effect. To produce the decoupled spectrum without Overhauser effect gated decoupling is used, with the decoupler only switched on during the pulse and the measurement of the FID. In the (long) pulse delay following accumulation of the FID the decoupler is switched off. This method relies on the fact that the nuclear Overhauser effect builds up relatively slowly, whereas the decoupling occurs almost spontaneously. If the ^{13}C relaxation only takes place through the dipole–dipole mechanism (i.e. as a result of the effect of the protons), the nuclear Overhauser effect is given by

$$\eta_C = \frac{\gamma_H}{2\gamma_C} = 1.988.$$

Much smaller values than this maximum are observed if other relaxation mechanisms are important.

The importance of NOE measurements in structure determination has already been indicated (see p. 134 ff.). An explicit example of the use of a heteronuclear NOE experiment is dimethylformamide (**45**) (Fig. 3.24, p. 94). At 25°C the following NOE factors are observed.

CO   CH$_3$(b)   CH$_3$(a)

η 1.4   1.8   1.4

(a) is the CH$_3$-group *anti* to the formyl hydrogen. Its signal lies at higher field and has a lower intensity than that of (b) in the broadband decoupled spectrum. With increasing temperature the exchange rate of the two methyl groups increases, and the difference between the two NOE factors decreases.

Polarisation Transfer

From Tab. 3.1 (see p. 72) it can be seen that the natural abundance and the sensitivity of certain NMR measurable nuclei like ^{13}C and ^{15}N are very low. One reason for the low sensitivity is the low population difference between the two spin states (see Sec. 1.1, p. 71). Various methods are available (**SPI: selective population inversion, INEPT, DEPT pulse sequences**) which allow the transfer of the larger population difference for protons to a less sensitive nucleus (^{13}C, ^{15}N, etc.) present in the same molecule. This **polarisation transfer** causes the transitions (absorption and emission!) to become stronger. The effect exceeds that due to the **nuclear Overhauser effect**.

One use of polarisation transfer is therefore to increase the signal intensity of ^{1}H decoupled or ^{1}H coupled ^{13}C spectra. A second application of **INEPT (insensitive nuclei enhanced by polarisation transfer)** or **DEPT (distortionless enhancement by polarisation transfer)** is the measurement of spectra in which peaks can be selected by their multiplicity (spectral editing). For example a ^{1}H decoupled or ^{1}H coupled ^{13}C spectrum of cholesteryl acetate (**203**) can be obtained in which only the CH-groups are present (Fig. 3.75), or alternatively only the CH$_2$- or CH$_3$-groups. **203** has a total of 29 chemically non-equivalent C atoms. The congestion of signals in the region between $\delta = 20$ and $\delta = 40$ is particularly high. In the normal decoupled or off-resonance decoupled spectra the signals are hopelessly overlapped.

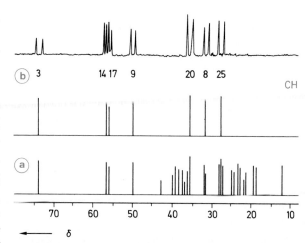

203

CH$_3$	δ	CH$_2$	δ	CH	δ	C$_q$	δ
C–18	12.0	C–1	37.3	C–3	73.7	C–5	139.9
C–19	19.3	C–2	28.2	C–6	122.6	C–10	36.7
C–21	18.9	C–4	38.4	C–8	32.2	C–13	42.5
C–26	22.7	C–7	32.2	C–9	50.4	C–28	169.6
C–27	22.9	C–11	21.3	C–14	57.0		
C–29	20.9	C–12	32.5	C–17	56.6		
		C–15	24.6	C–20	36.1		
		C–16	24.6	C–25	28.2		
		C–22	36.7				
		C–23	24.3				
		C–24	39.8				

Fig. 3.75 ^{13}C spectra of cholesteryl acetate (**203**) in CDCl$_3$

a Normal broadband decoupled spectrum of the saturated carbon atoms

b Sub-spectrum of the methine groups (CH) coupled and decoupled (DEPT technique)

Multidimensional ^{13}C NMR Spectra

As a complement to the section on p. 136 the applications of two-dimensional spectra to ^{13}C NMR will be briefly described.

J-resolved 2D ¹³C spectra give a separation of the parameters δ(¹³C) and J(¹³C,¹H). Chemical shifts and ^1J(C,H) coupling constants (multiplicities) can be instantly read off from the F_2 and F_1 axes respectively. Signal overlap, common in coupled ¹³C spectra, which often causes difficulties with interpretation, is thereby avoided.

Using several examples more detail will be given about **¹³C,¹H shift correlation (H,C COSY, HETCOR)**.

In Fig. 3.76a the normal broadband decoupled ¹³C spectrum of α-tetralone (**204**) is shown. The signals of the quaternary C atoms are immediately recognisable by their low intensity. There remain the signals of the CH$_2$ groups H$_2$C-2,3,4 and the CH groups HC-5,6,7,8.

The contour plots reproduced in Fig. 3.76 b and c allow an unambiguous assignment of the correspondence of the ¹³C and ¹H signals. Thus it can be directly deduced that the ¹³C nucleus with $\delta = 126.7$ bears a proton that absorbs at $\delta = 8.00$, or that the carbon of the methylene group which absorbs at $\delta = 2.85$ in the ¹H spectrum has $\delta = 29.3$ etc. Finally, the individual proton signals are shown in Fig. 3.76 d. The doublet, triplet, or quintet structure, resulting from *vicinal* ¹H, ¹H couplings, is clearly visible without overlap. If the assignments of the ¹H spectrum are known, the shift correlation affords the ¹³C assignments, and *vice versa*. If neither assignment is completely known, then the shift correlation, together with splitting patterns and perhaps increment tables, is an especially valuable aid to the complete assignment of the ¹³C and ¹H signals of a compound.

	δ(¹³C)	δ(¹H)
C–1	197.7	–
C–2	38.8	2.55
C–3	22.9	2.04
C–4	29.3	2.85
C–4a	144.1	–
C–5	128.4	7.24
C–6	133.0	7.45
C–7	126.2	7.30
C–8	126.7	8.00
C–8a	132.2	–

204

To extend the technique to quaternary C atoms variants of the experiment (**long-range HETCOR, CH-COLOC**) are avail-

able, which depend on the existence of smaller CH couplings ($^2J_{C,H}$ and $^3J_{C,H}$). 4,7-Dimethoxy-2,3-dimethylindole (**205**) will be discussed as an example.

Fig. 3.77 a shows the "normal" HETCOR spectrum (contour plot) with the correlations between directly bonded C and H atoms. This relies on the $^1J_{C,H}$ couplings. The assignment of the six quaternary C atoms of **205** is possible by a long-range HETCOR measurement. The plot in Fig. 3.77 b shows the correlations arising from $^3J_{C,H}$ and $^2J_{C,H}$ couplings.

^2J (---➤)	
5-H	C-4
6-H	C-5
6-H	C-7
3-CH$_3$	C-3
2-CH$_3$	C-2

^3J (→)	
5-H	C-3a
5-H	C-7
6-H	C-4
6-H	C-7a
3-CH$_3$	C-3a
3-CH$_3$	C-2
2-CH$_3$	C-3
4-OCH$_3$	C-4
7-OCH$_3$	C-7

205

No correlation is visible for 5-H to C-6 ($^2J_{C,H}$). The assignment of the individual quaternary C atoms is best done starting from a reliable reference point. C-7a for example can only show a single correlation, ^3J, to 6-H.

In an analogy to the H-relayed (H,H) COSY experiment the magnetisation can be carried over from one proton, via another proton, to a ¹³C nucleus. In Fig. 3.78 is shown part of an H-relayed (H,C) COSY spectrum of sucrose (**144**). By comparison with a normal ¹³C,¹H shift correlated 2D NMR spectrum the contour plot contains additional peaks, arising from the **relayed technique** employed, which correspond to ¹³C nuclei and protons on neighbouring carbon atoms ($^2J_{C,H}$):

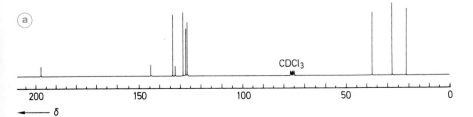

(a)

CDCl$_3$

200   150   100   50   0

δ

Fig. 3.76 2D NMR spectroscopy of α-tetralone (**204**); (in CDCl$_3$
a ¹³C spectrum, broadband decoupled (1D spectrum)

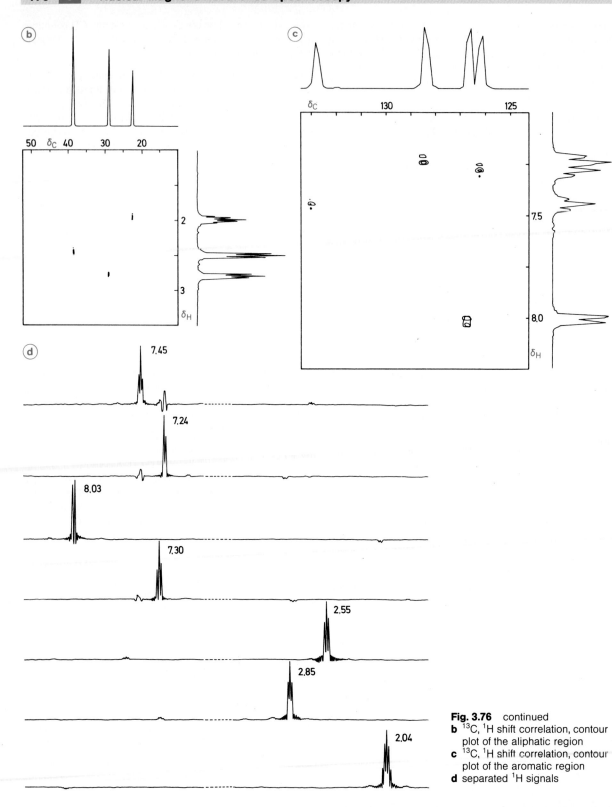

Fig. 3.76 continued
b ^{13}C, ^{1}H shift correlation, contour plot of the aliphatic region
c ^{13}C, ^{1}H shift correlation, contour plot of the aromatic region
d separated ^{1}H signals

a

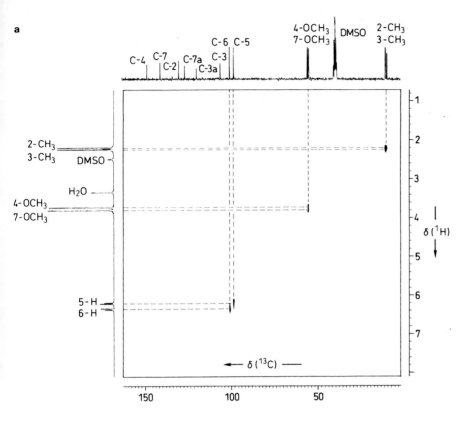

b

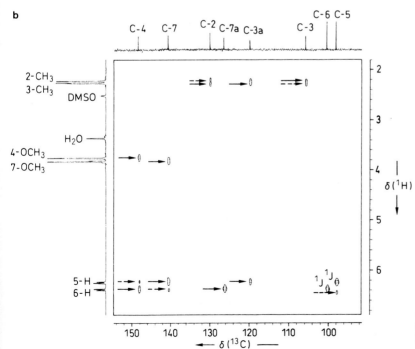

Fig. 3.77 ¹H ¹³C Heteronuclear shift correlation of 4,7-dimethoxy-2,3-dimethylindole (**205**) in DMSO-d₆
a HETCOR
b Long-range HETCOR

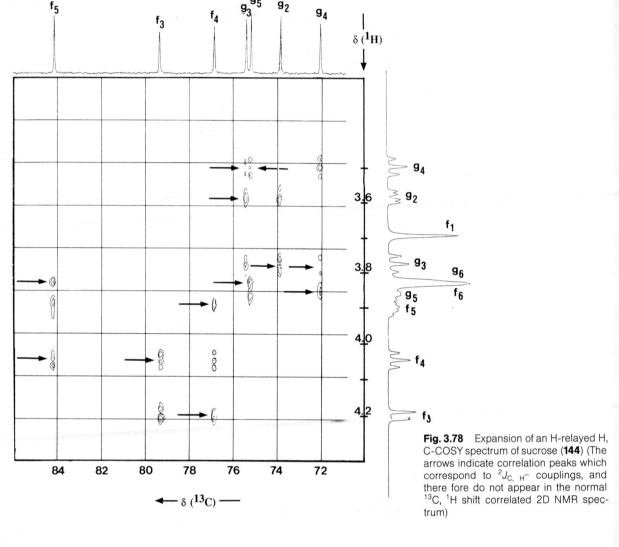

Fig. 3.78 Expansion of an H-relayed H, C-COSY spectrum of sucrose (**144**) (The arrows indicate correlation peaks which correspond to $^2J_{C,\,H}-$ couplings, and there fore do not appear in the normal ^{13}C, ^1H shift correlated 2D NMR spectrum)

Fructose ring   C-3/4-H ($f_3 \rightarrow f_4$)
                    C-4/3-H and 5-H
                    C-5/4-H and 6-H
                    (C-1,2,6 not shown)

Glucose ring   C-2/3-H
                    C-3/2-H and 4-H
                    C-4/3-H and 5-H
                    C-5/4-H and 6-H
                    (C-1,6 and 1-H not shown)

For very complex structures a variety of three-dimensional techniques have been developed. For example it is of interest to combine the connectivity through bonds (2D) and the connectivity through space (2D) in a 3D measurement. A further application is the correlation of three different nuclear types. Fig. 3.79 shows a **three-dimensional ^1H-^{13}C-^{15}N correlation spectrum** of ^{13}C and ^{15}N labelled ribonuclease T1, measured at 750 MHz. Each crosspeak indicatess the ^1H and ^{15}N shift (cf. Sec. 6.3) of an N–H group in this higher protein. In the third dimension the connectivity to the α-C atom of the next amino acid is indicated. The method allows a signal assignment related to the amino acid sequence.

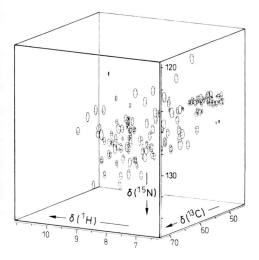

206

about 0.01% of the molecules contain **two** anisochronous ¹³C nuclei and therefore fulfil the condition for the appearance of ¹³C,¹³C coupling in the spectrum. The main signal arising from molecules with **one** ¹³C nucleus is accompanied by low intensity satellites, which are however difficult to measure. This is particularly so when the coupling constants are small and the satellites lie under the wings of the main signal. The **INADEQUATE technique (incredible natural abundance double quantum transfer experiment)** suppresses the main signal. A special pulse sequence excites the **double quantum transitions**. A practical example, piperidine (**190**), is shown in Fig. 3.80. Four couplings are observed, arising from the isotopomers (**190a-d**) which are present in low concentration in normal piperidine.

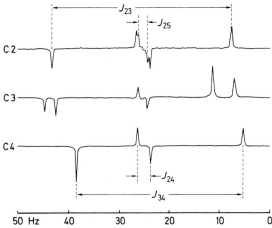

$^1J_{2,3} = 35.2\,\text{Hz}$

190a

$^2J_{2,4} = 2.6\,\text{Hz}$

190c

$^1J_{3,4} = 33.0\,\text{Hz}$

190b

$^3J_{2,5} = 1.7\,\text{Hz}$

190d

Each AX (AB) spectrum is made up of four lines. In Fig. 3.80 the two components of each doublet appear in antiphase. The experimental conditions can be adjusted so that they appear as in phase doublets.

Fig. 3.79 750 MHz 3D ¹H-¹³C-¹⁵N-NH(CO)CA correlation spectrum of a 2 mM solution of ¹³C and ¹⁵N enriched ribonuclease T1 in H₂O/D₂O (Bruker, Analytische Messtechnik)

Double Quantum Coherence for the Measurement of ¹³C,¹³C Couplings

As outlined in Sec. 4.5 (see p. 154), ¹³C,¹³C couplings give valuable information about the C–C bonds present in a molecule. Because of the low natural abundance of ¹³C (1.1%), only

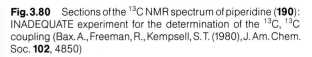

Fig. 3.80 Sections of the ¹³C NMR spectrum of piperidine (**190**): INADEQUATE experiment for the determination of the ¹³C, ¹³C coupling (Bax. A., Freeman, R., Kempsell, S. T. (1980), J. Am. Chem. Soc. **102**, 4850)

If the one-dimensional INADEQUATE spectrum becomes too complicated, a two-dimensional INADEQUATE spectrum is recommended. This separates the AX spin systems of neighbouring ^{13}C nuclei on the F_1 axis. (See Fig. 3.81). The observed d values of the ^{13}C nuclei lie on the F_2 axis, the double quantum frequencies on the F_1 axis. The **connectivities** over the C–C bonds follow the blue arrow and prove in this case that the carbon atoms are arranged in the chain in the sequence C_a–C_d–C_c–C_b.

While the INADEQUATE method gives information about **connectivity over bonds** NOE experiments answer questions of **connectivity through space**. The two methods therefore complement each other and provide the basis for the solution of demanding structural problems.

Solid State Spectra

For various reasons, high resolution spectra cannot be obtained from solids using the normal PFT technique. Dipole-dipole and quadrupole-field gradient interactions, which average to zero in solution, lead to extremely broad lines: the anisotropy of the chemical shift leads to broad, complex line shapes: and the spin-lattice relaxation times of nuclei such as ^{13}C are very long in solids.

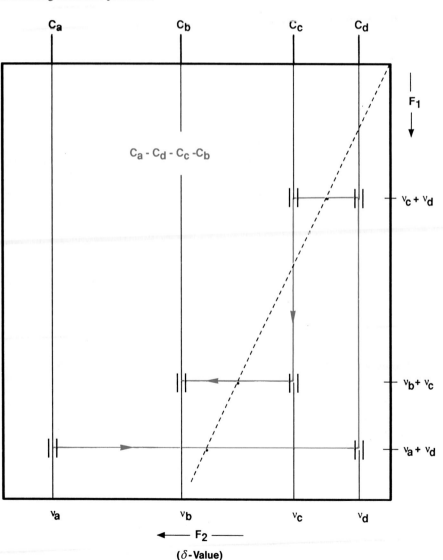

Fig. 3.81 Schematic diagram of a 2D INADEQUATE spectrum of a C_4 chain with unknown connectivity. Each pair of ^{13}C nuclei connected by a ^1J coupling gives an AX spin pattern at the F_1 position which corresponds to the sum of the chemical shifts. (The centre points of the connected contours lie on the straight line $F_1 = 2F_2$).

The chemical shift is described by a second order **shift tensor**, the matrix of which can be diagonalised. In solution the average value σ_{iso} is measured, which is a third of the trace of the matrix.

$$\begin{vmatrix} \sigma_{11} & 0 & 0 \\ 0 & \sigma_{22} & 0 \\ 0 & 0 & \sigma_{33} \end{vmatrix} \qquad \sigma_{iso} = 1/3\,(\sigma_{11} + \sigma_{22} + \sigma_{33})$$

In a single crystal the individual components σ_{ii} can be determined. The completely symmetrical case $\sigma_{11} = \sigma_{22} = \sigma_{33}$ occurs in methane; cylindrical symmetry, e.g. in acetylene, leads to $\sigma_{\perp}$ and $\sigma_{\parallel}$; otherwise the tensor is asymmetric with the three main components σ_{ii} ($i = 1,2,3$). For polycrystalline or amorphous solids broad signals are observed, which cover the whole of the range of the main components σ_{ii} (Fig. 3.82). Particularly large **chemical shift anisotropies** are observed for sp^2 and sp carbon atoms. In the diagram below the main components σ_{11}, σ_{22}, and σ_{33} of the ^{13}C shift tensor of naphthalene (**104**) are given for the individual carbon atoms, one below the other. The ppm values are referred to TMS. Their average in each case gives values close to the σ values of the solution spectrum (given in blue). It is apparent that in the crystalline state of naphthalene the symmetry elements, except the centre of symmetry, are absent. C-1 and C-8 for example have quite different values of the component σ_{22}.

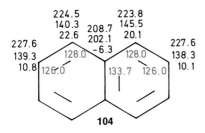

104

The **dipolar coupling**, a through-space effect, can also be described by a tensor. Since its trace is zero, this coupling has no effect in solution. The magnitude of the dipolar coupling depends on the magnetogyric ratios γ of the nuclei involved, the internuclear distance r, and the angle Θ between the line joining the nuclei and the magnetic field $\boldsymbol{B}_0$. For powder measurements with a statistical distribution of the angle line widths are typically up to 30 KHz. Normal ^{13}C,^{1}H couplings in CH-, CH$_2$-, or CH$_3$-groups are around 10 kHz. Because of the low natural abundance ^{13}C,^{13}C coupling can be ignored. Double ^{13}C labelling can afford an accurate method of determining the C–C distance from this coupling.

The angular dependence of the chemical shift σ and the dipolar coupling D is expressed by the term $(3\cos^2\Theta - 1)$:

$$\sigma = \sigma_{iso} + 1/3 \sum_{i=1}^{3} (3\cos^2\Theta - 1)\sigma_{ii}$$

$$D = \pm h/2 \cdot \gamma_1 \cdot \gamma_2 \cdot \frac{1}{r^3}(3\cos^2\Theta - 1)$$

High resolution solid state spectra can therefore be obtained, if $3\cos^2\Theta - 1 = 0$. This can be achieved by spinning the sample at high speed about an axis which is at the "magic" angle Θ_m to the field $\boldsymbol{B}_0$. From $\cos^2\Theta_m = 1/3$ it follows that $\Theta_m = 54.73°$. This **magic angle spinning (MAS)** makes $\sigma = \sigma_{iso}$ and $D = 0$; this means that the powder spectrum is essentially the same as the solution spectrum. (The symmetry properties in the crystal, which may be different from the free molecule, are naturally unaffected).

Scalar couplings and the remnants of the dipolar coupling between ^{13}C and ^{1}H nuclei can be eliminated by high power proton broadband decoupling.

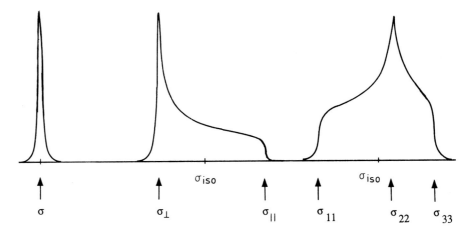

Fig. 3.82 ^{13}C NMR signals in powder measurements for the cases
$\sigma_{11} = \sigma_{22} = \sigma_{33} = \sigma$
$\sigma_{11} = \sigma_{22} = \sigma_{\perp}, = \sigma_{33} = \sigma_{\parallel}$
$\sigma_{11} = \sigma_{22} = \sigma_{33}$ different.
(Ignoring dipolar couplings)

σ $\sigma_{\perp}$ σ_{iso} $\sigma_{\parallel}$ σ_{11} σ_{iso} σ_{22} σ_{33}

The final problem, of long relaxation times, can be solved by **cross polarisation (CP)**. This utilises polarisation transfer from surrounding protons to the ^{13}C nuclei. The effective relaxation times are thus drastically reduced, and the intensity of the ^{13}C signals is increased. The signal/noise ratio increases by a maximum factor of $\gamma_H/\gamma_C = 3.98$.

Fig. 3.83 shows a comparison of the ^{13}C CP-MAS and solution spectra of the enol (**207**), which can exist in the chelated (Z)-form as well as the (E)-configuration.

The spectrum observed for the solid corresponds to the pure (Z)-form with an intramolecular hydrogen bond. Carbons 1 and 3 are coincidentally isochronous within the limits of resolution. The restricted rotation of the mesityl groups in the solid state causes the two o- and m-carbons and the two o-methyl groups of each mesityl group to become chemically non-equivalent. This increases the number of aromatic signals from 8 to 12 and the number of methyl signals from 4 to 6. 10 and 4 signals respectively are resolved, some showing double intensity because of overlap.

Temperature dependent dynamic processes can also take place in crystals. The valence tautomerism of bullvalene in the solid state has already been discussed on p. 96. There are additional jumps around the C_3 axis, which have an activation energy of ca. 88 kJ · mol^{-1}. Many benzene derivatives show processes involving flipping by 180°.

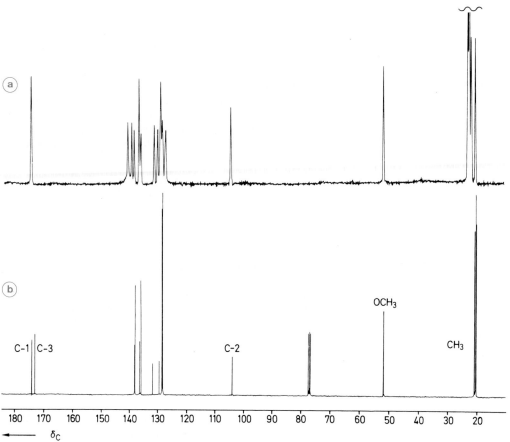

Fig. 3.83 ^{13}C NMR spectra of methyl (Z)-3-hydroxy-2,3-dimesityl-2-propenoate (**207**)
a CP-MAS solid state spectrum (without spinning sidebands)
b normal PFT spectrum of a solution in CDCl$_3$

High resolution solid state NMR spectroscopy is useful for structural determination particularly for insoluble or almost insoluble compounds, e.g. for polymers or biopolymers, and also for molecular structure problems where solvation has an important effect on the configuration or conformation. Low temperature measurements on solid state matrices can also provide useful information about short-lived intermediates. An example of an accurate structural assignment for an insoluble material is shown for the polyesters **208** and **209** in Fig. 3.84. The introduction of a second methoxy group into the repeat unit leads to $\Delta\delta$ values which are accurately reproduced by the increment system for benzene derivatives. Special measurement conditions even allow the distinction between amorphous and polycrystalline states of polymers.

Use of ¹³C Databases

The use of information technology (electronic data processing) is becoming increasingly important in the area of analytical instrumentation. Many spectrometer suppliers offer comprehensive libraries of spectra with IR, MS, and NMR spectrometers, which can be consulted using the built-in computer. More interesting is the use of large, internationally accessible, databases.

Here the use of ¹³C NMR databases like SPECINFO will be described.

It contains the ¹³C shifts, and, where known, also coupling constants and relaxation times of very many organic compounds. A fundamental problem is that the spectral data are very dependent on sample-specific (solvent, concentration, purity, temperature) and instrument-specific measurement conditions. This fact can be allowed for by defining a tolerance range ("window method") for the chemical shifts.

SPECINFO permits a variety of searches:
a) search for reference spectra on the basis of measured chemical shifts,

b) search for reference spectra on the basis of names of compounds or compound types,

c) search for reference spectra on the basis of molecular formula,

d) search for similar spectra,

e) search for compounds and spectra for defined structures or partial structures (sub-structures),

f) the estimation of chemical shifts (and also coupling constants where appropriate) for assumed structures

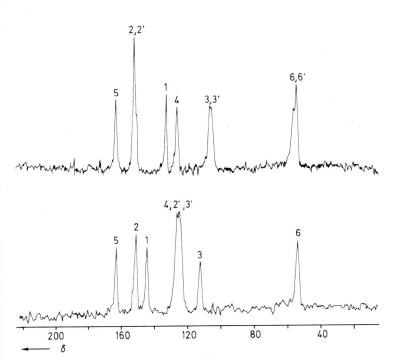

Fig. 3.84 ¹³C CP-MAS NMR spectrum of methoxysubstituted homopolymers of 4-hydroxybenzoic acid (Fyfe, C. A., Lyeria, J. J., Volksen, W., Yanoni, C. S. 1979, Macromol. **12**, 757)

As an illustration a simple example will be chosen. The chemical shifts of a compound are known, with the signal multiplicities derived from ^1J(C,H) couplings:

observed δ values 154.7 (singlet)
139.5 (singlet)
129.1 (doublet)
121.5 (doublet)
115.9 (doublet)
112.3 (doublet)
20.8 (quartet)

Which compound can it be? To answer this question the choice a), search for reference spectra, is chosen. The specific question is: Are there any of the ^{13}C NMR spectra (currently some 100,000 in SPECINFO) which agree with the observed shifts of the seven signals within a chosen error limit (here 0.5 ppm for example)? As each signal is input the number of possible compounds must reduce. Finally in this case only one compound remains, namely m-cresol (**22**) with the following stored data (δ values in CDCl$_3$):

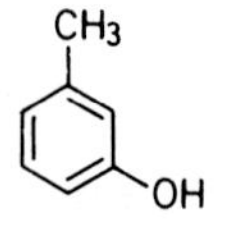

C	δ
1	139.8
2	116.2
3	155.0
4	112.5
5	129.4
6	121.8
7	21.1

A more important application is offered by choice f). For a newly synthesised compound a specific structure is assumed. What ^{13}C chemical shifts are predicted? The example of 3,8-dithiabicyclo[8.3.1]tetradeca-1(14),10,12-trien-5-yne **210** and its fluoro derivative **211** show the possibilities but also the limits of such a question. The blue δ values give the predictions with their standard deviations, the δ values printed in black the ^{13}C shifts observed in CDCl$_3$.

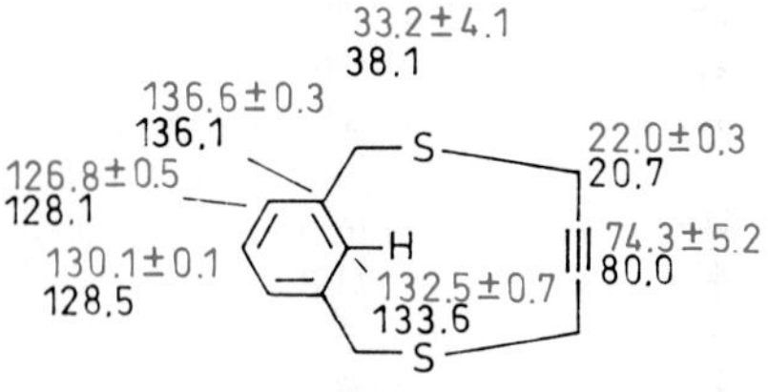

For the aromatic C atoms the agreement is good. The calculated δ values in the bridge of the m-cyclophane are interpolated; the criteria for similarity, on which the incorporation of reference data is based, are apparently dubious. The program gives unsatisfactory values for the bridge carbon atoms when fluorine is included in the structure. Nonetheless, the large differences between observed and calculated δ values are usually associated with larger standard deviations. The same considerations apply here as in the application of increment systems: a critical use can still be of great assistance in structure determination.

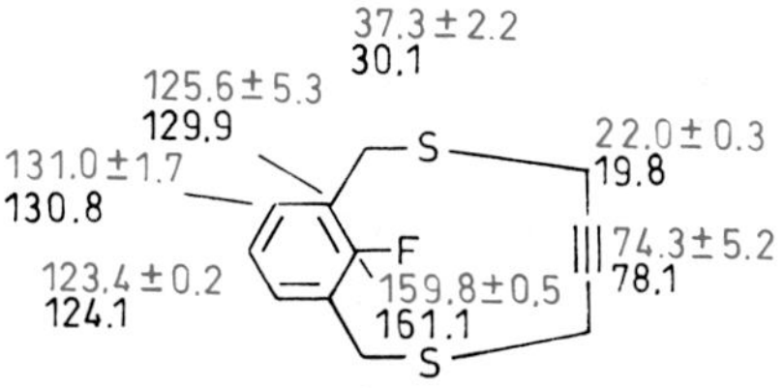

Every structure determination using analytical instrumentation, including computer-aided methods, should be based on as many methods as possible. The idealised objective is to use computers to derive a structure directly from the observed UV, IR, NMR, and MS data. This requires programs which not only rely on databases, but which can adopt the logical and empirical thought processes of the analytical chemist. The automatic operation of such programs can at best be expected only for relatively simple structural problems; for more complex problems a dialogue with the analytical chemist, a semiautomatic solution, is desirable. This offers new, promising perspectives for instrumental analysis.

5. ¹H and ¹³C NMR Data: Representative Examples of the Most Important Classes of Compounds[a]

Alkanes

CH_4	C_2H_6	$H_3C-CH_2-CH_3$	$H_3C-CH_2-CH_2-CH_3$	$H_3C-CH_2-CH_2-CH_2-CH_3$
- 2.3	6.5	16.0 16.3	13.0 24.8	14.1 22.4 34.2
0.23	0.86	0.91 1.33	0.90 1.23	0.89 ~1.28 ~1.28
Methane	Ethane	Propane	Butane	Pentane

$H_3C-CH_2-CH_2-CH_2-CH_2-CH_3$
14.1   22.7   31.7
0.89 ~1.27 ~1.27
Hexane

$H_3C-CH_2-CH_2-CH_2-CH_2-CH_2-CH_3$
14.1   22.8   32.0   29.1
0.89 ~1.27 ~1.27 ~1.27
Heptane

$H_3C-CH_2-CH_2-CH_2-CH_2-CH_2-CH_2-CH_3$
14.1   22.8   32.1   29.5
0.88 ~1.27 ~1.27 ~1.27
Octane

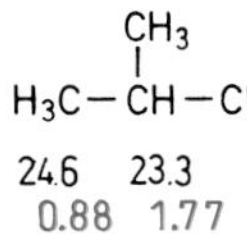

2-Methylpropane (Isobutane)
$H_3C-CH-CH_3$, CH_3
24.6   23.3
0.88   1.77

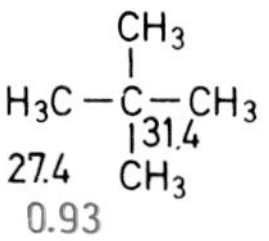

2,2-Dimethylpropane (Neopentane)
$H_3C-C-CH_3$, CH_3 , CH_3
31.4
27.4   CH₃
0.93

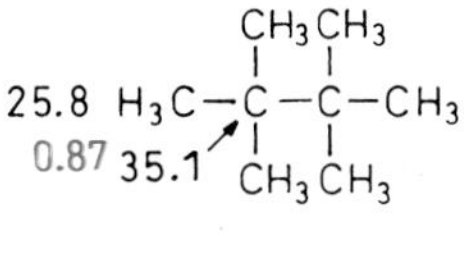

2,2,3,3-Tetramethylbutane (Hexamethylethane)
CH_3 CH_3
25.8 $H_3C-C-C-CH_3$
0.87 35.1 CH_3 CH_3

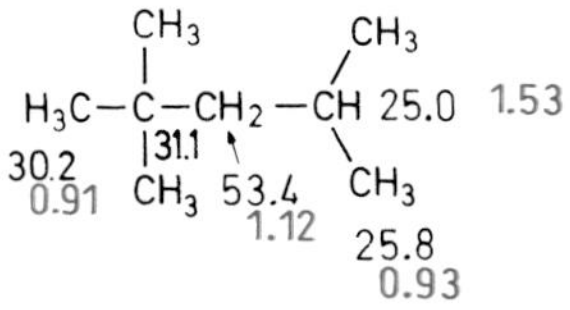

2,2,4-Trimethylpentane (Isooctane)
CH_3 , CH_3
$H_3C-C-CH_2-CH$ 25.0   1.53
30.2   31.1
0.91 CH_3 53.4 CH_3
1.12   25.8
0.93

Alkenes

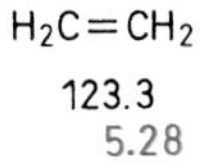

$H_2C=CH_2$
123.3
5.28
Ethylene

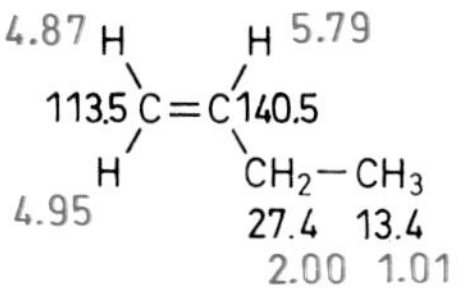

4.87 H   H 5.79
113.5 $C=C$ 140.5
H CH_2-CH_3
4.95   27.4   13.4
2.00   1.01
1-Butene

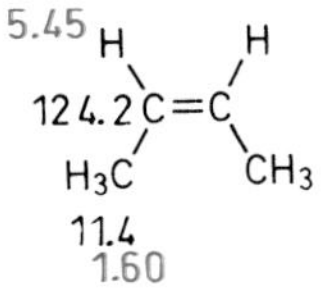

5.45 H   H
124.2 $C=C$
H_3C CH_3
11.4
1.60
(Z)-2-Butene

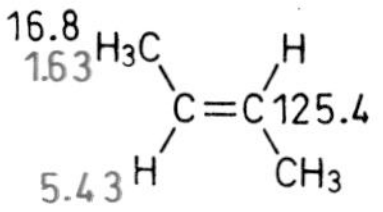

16.8 H_3C H
1.63 $C=C$ 125.4
5.43 H CH_3
(E)-2-Butene

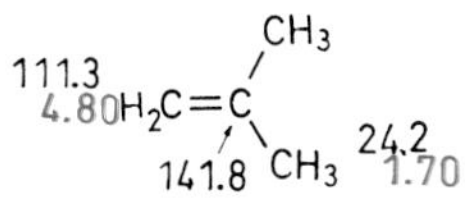

2-Methylpropene
111.3 CH_3
4.80 $H_2C=C$
141.8 CH_3 24.2
1.70

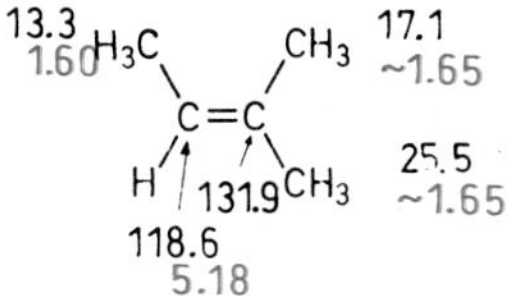

2-Methyl-2-butene
13.3 H_3C CH_3 17.1
1.60   ~1.65
$C=C$
H   131.9 CH_3 25.5
118.6   ~1.65
5.18

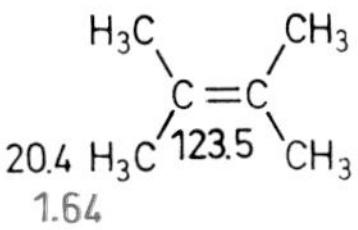

2,3-Dimethyl-2-butene
H_3C CH_3
$C=C$
20.4 H_3C 123.5 CH_3
1.64

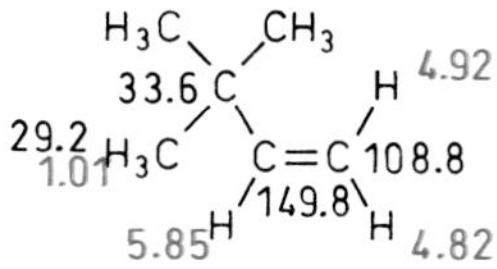

3,3-Dimethyl-1-butene
H_3C CH_3
33.6 C   H 4.92
29.2 H_3C $C=C$ 108.8
1.01   149.8
5.85 H   H 4.82

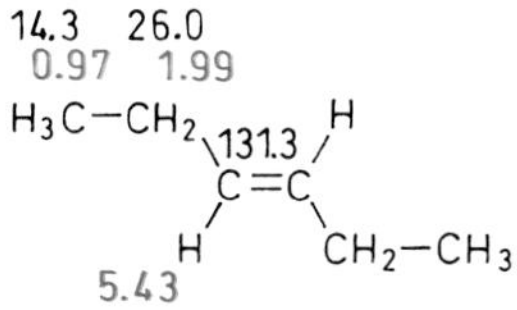

(E)-3-Hexene
14.3   26.0
0.97   1.99
H_3C-CH_2 131.3   H
$C=C$
H CH_2-CH_3
5.43

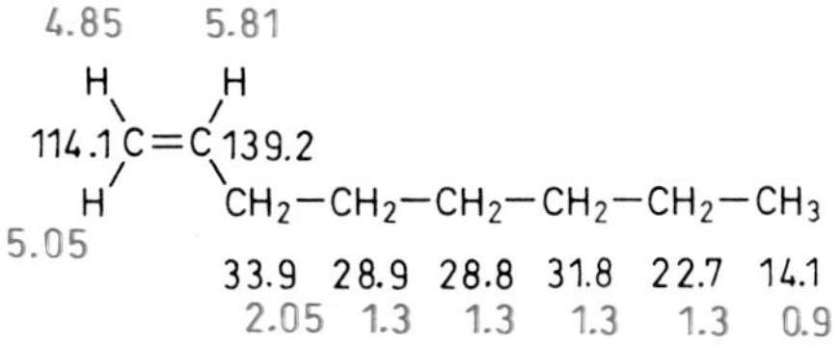

1-Octene
4.85   5.81
H   H
114.1 $C=C$ 139.2
H $CH_2-CH_2-CH_2-CH_2-CH_2-CH_3$
5.05   33.9 28.9 28.8 31.8 22.7 14.1
2.05 1.3   1.3   1.3   1.3   0.9

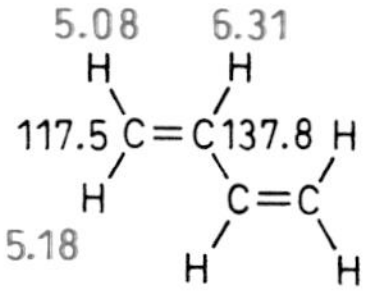

1,3-Butadiene
5.08   6.31
H   H
117.5 $C=C$ 137.8 H
H $C=C$
5.18   H   H

[a] The figures printed in black give the δ values of the ¹³C signals, the blue figures the δ values of the ¹H signals. Most values are for measurements in CDCl₃ or CCl₄ with TMS as internal standard.

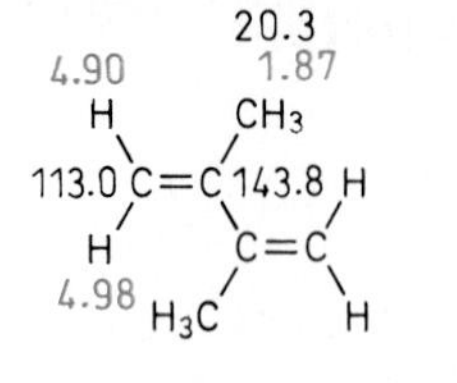

2,3-Dimethyl-1,3-butadiene

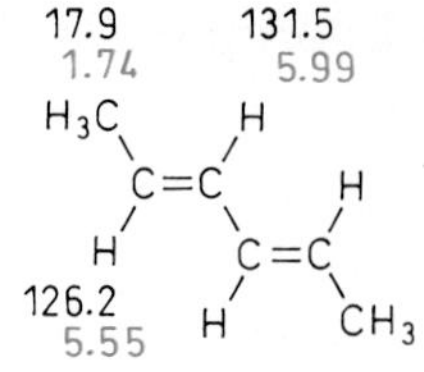

(E,E)-2,4-Hexadiene

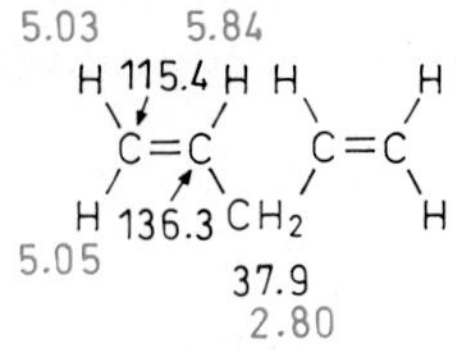

1,4-Pentadiene

Alkynes

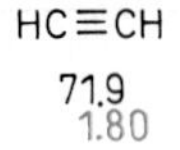

$HC\equiv CH$

71.9
1.80

Acetylene

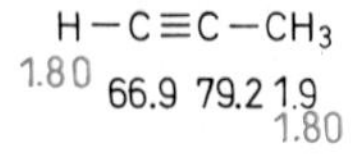

$H-C\equiv C-CH_3$

1.80   66.9 79.2 1.9
          1.80

Propyne

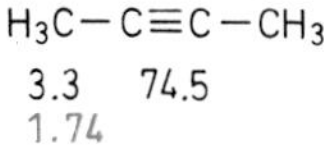

$H_3C-C\equiv C-CH_3$

3.3   74.5
1.74

2-Butyne

$HC\equiv C-CH_2-CH_2-CH_2-CH_3$

68.0 84.7 18.1 30.6 21.9 13.6
1.92   2.18 1.51 1.44 0.91

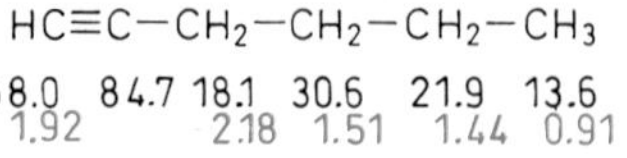

1-Hexyne

$H_3C-CH_2-C\equiv C-CH_2-CH_3$

14.4 12.4   80.9
1.11 2.16

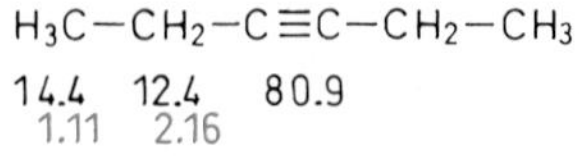

3-Hexyne

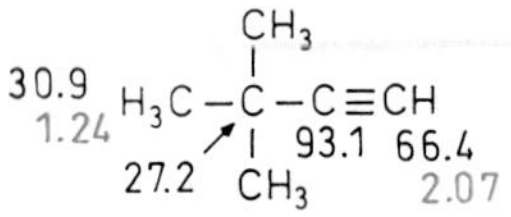

3,3-Dimethyl-1-butyne

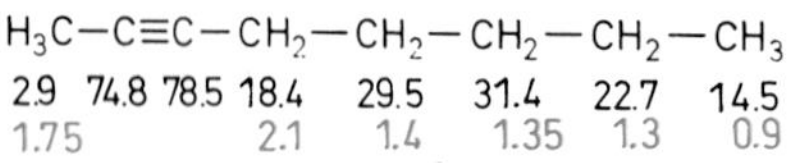

$H_3C-C\equiv C-CH_2-CH_2-CH_2-CH_2-CH_3$

2.9 74.8 78.5 18.4   29.5   31.4   22.7   14.5
1.75   2.1   1.4   1.35   1.3   0.9

2-Octyne

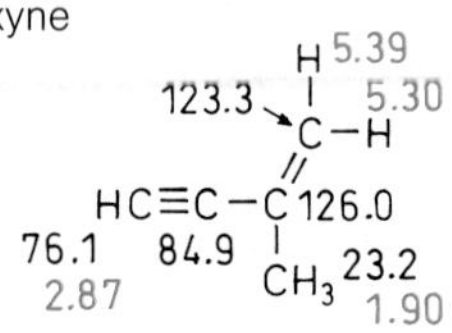

2-Methyl-1-buten-3-yne

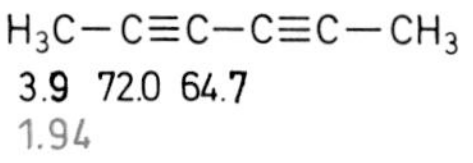

$H_3C-C\equiv C-C\equiv C-CH_3$

3.9 72.0 64.7
1.94

2,4-Hexadiyne

Allenes

$H_2C=C=CH_2$

73.5 212.6
4.67

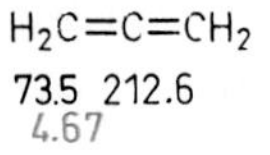

Allene

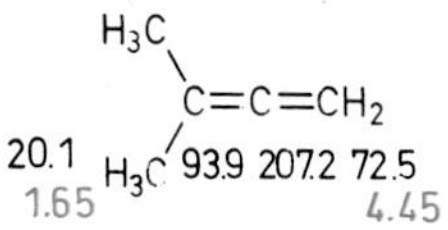

3-Methyl-1,2-butadiene

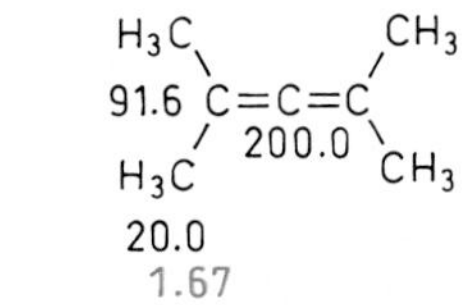

2,4-Dimethyl-2,3-pentadiene

Cyclic Aliphatic
Hydrocarbons (monocyclic)

Cyclopropane	C_3H_6	−2.8	0.22		Cyclohexane	C_6H_{12}	26.9	1.44
Cyclobutane	C_4H_8	22.3	1.96		Cyclooctane	C_8H_{16}	27.5	1.54
Cyclopentane	C_5H_{10}	25.8	1.51		Cyclododecane	$C_{12}H_{24}$	23.8	1.34

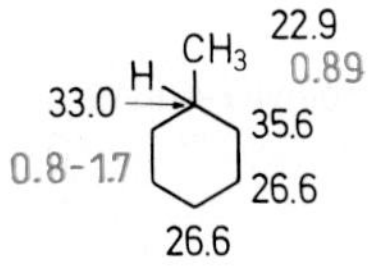

Methylcyclohexane

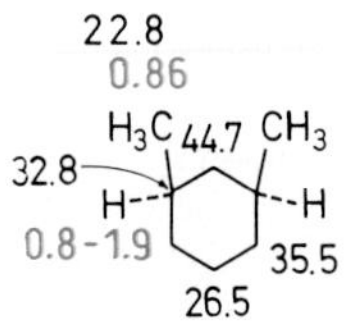

cis-1,3-Dimethyl-
cyclohexane

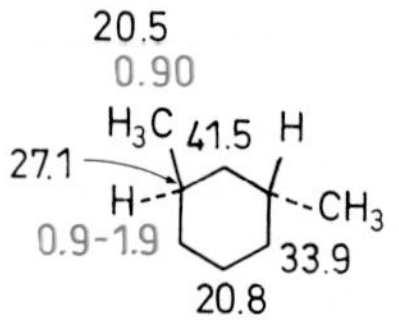

trans-1,3-Dimethyl-
cyclohexane

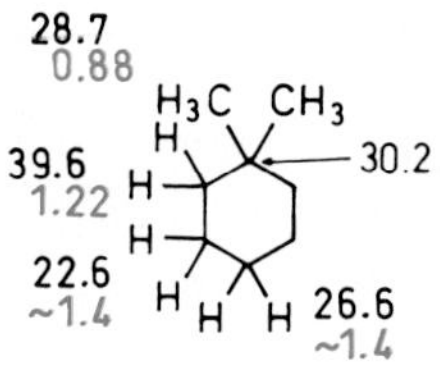

1,1-Dimethyl-
cyclohexane

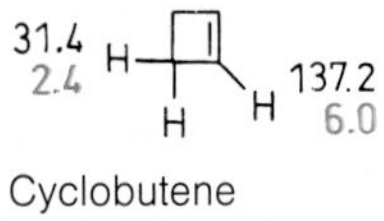

Cyclobutene

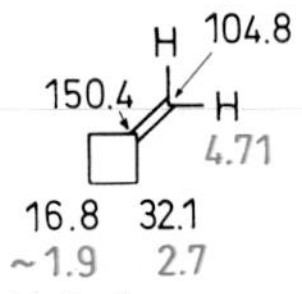

Methylene-
cyclobutane

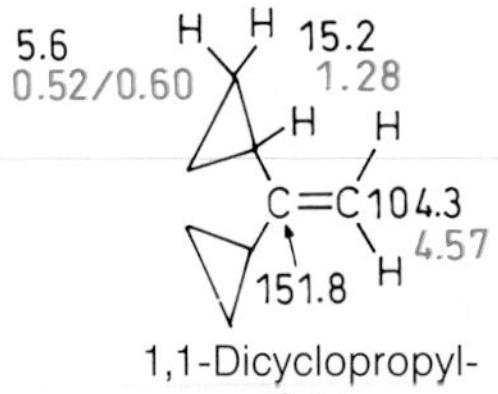

1,1-Dicyclopropyl-
ethene

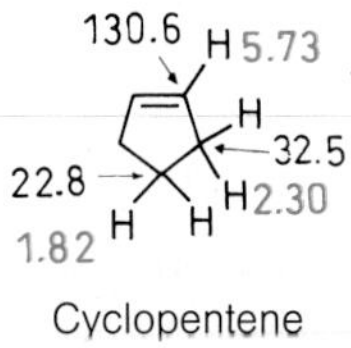

Cyclopentene

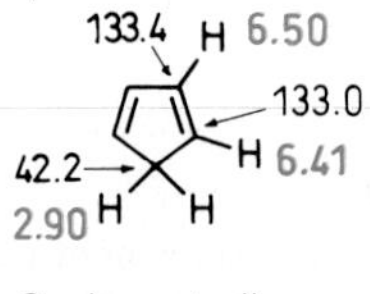

Cyclopentadiene

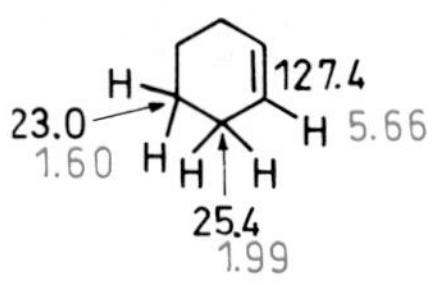

Cyclohexene

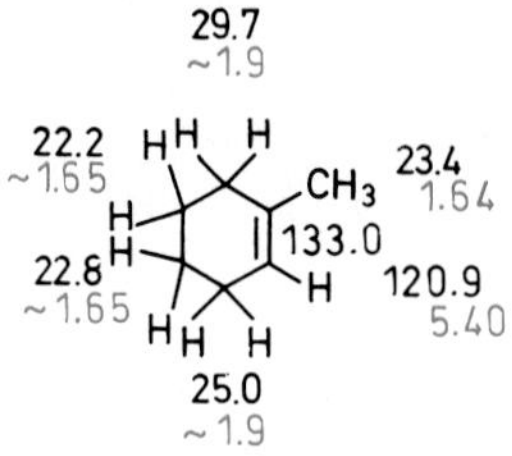

1-Methylcyclohexene

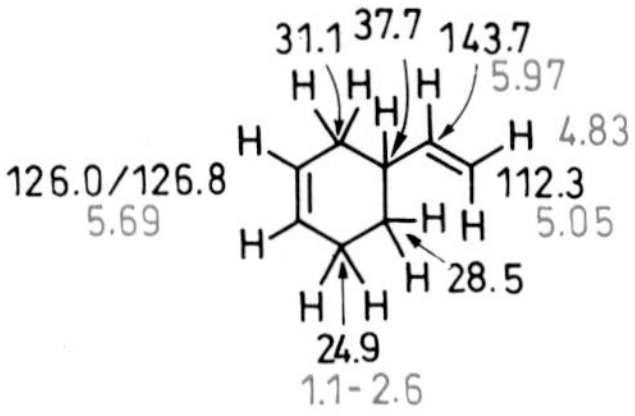

4-Vinyl-1-cyclohexene

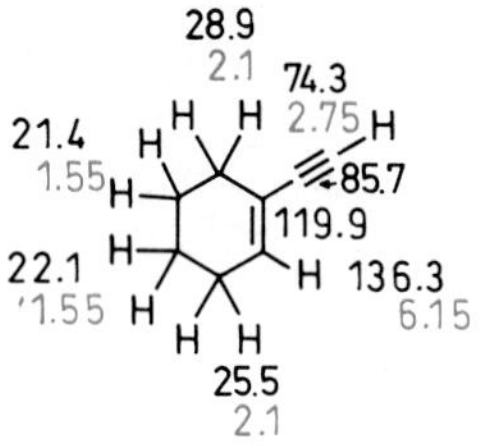

1-Ethynylcyclohexene

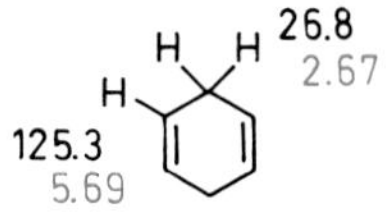

1,4-Cyclohexadiene

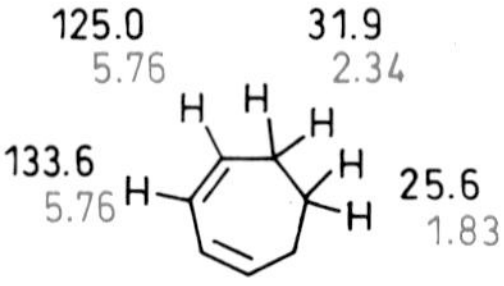

1,3-Cycloheptadiene

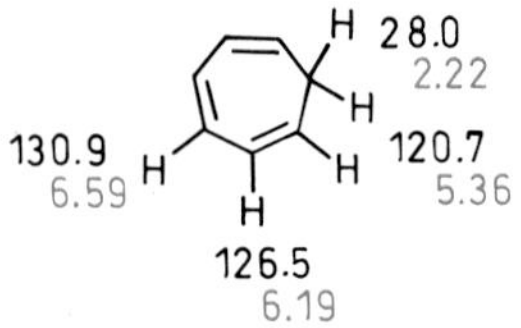

1,3,5-Cycloheptatriene

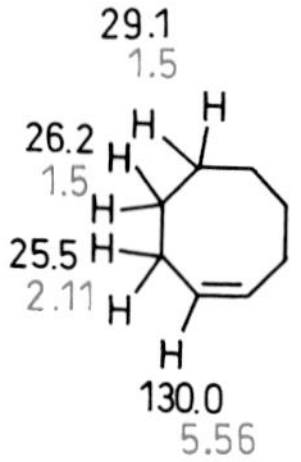

Cycloocthepene

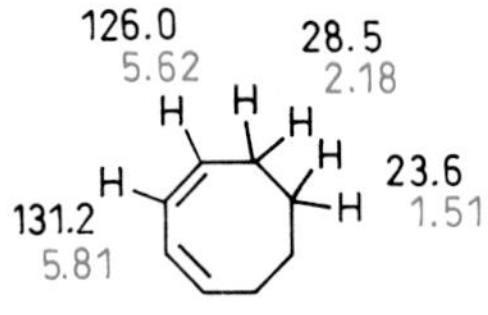

1,3-Cyclooctadiene

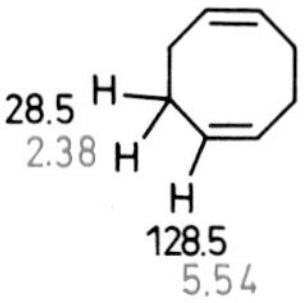

1,5-Cyclooctadiene

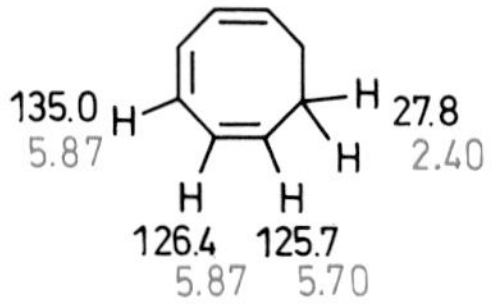

1,3,5-Cyclooctatriene

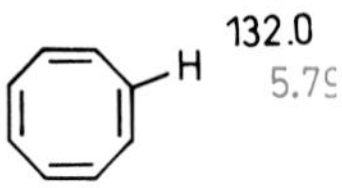

Cyclooctatetraene

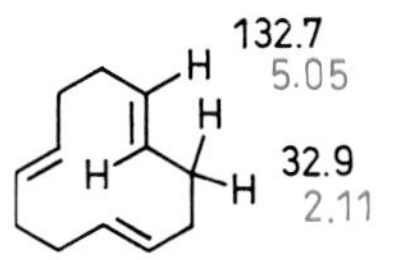

(E,E,E)-1,5,9-Cyclododeca-
triene

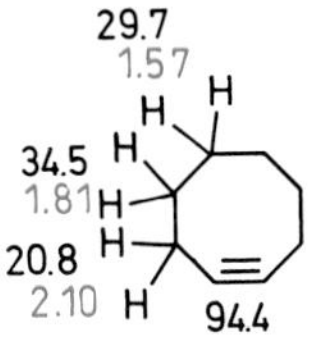

Cyclooctyne

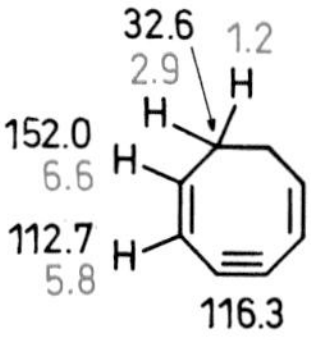

1,5-Cyclooctadien-3-yne

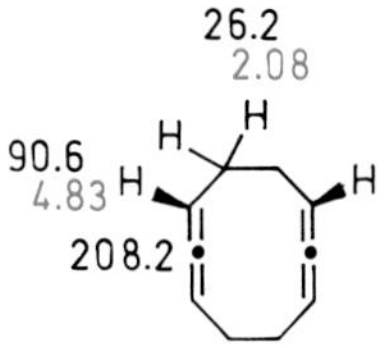

meso-1,2,6,7-Cyclodecatetraene
[(1R*,6S*)-1,2,6,7-Cyclodecatetraene]

Bi- and Polycyclic Hydrocarbons

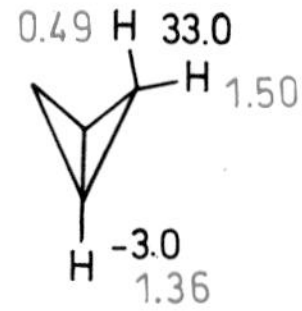

Bicyclo[1.1.0]butane

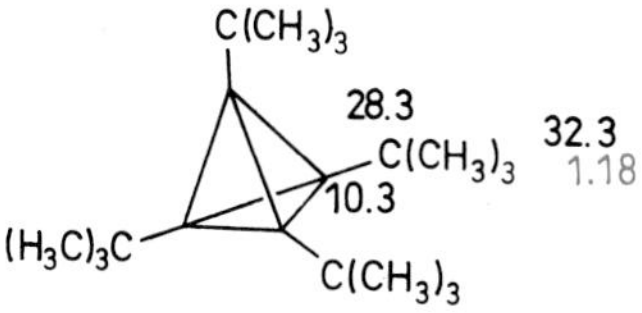

Tetra-t-butyltetrahedrane

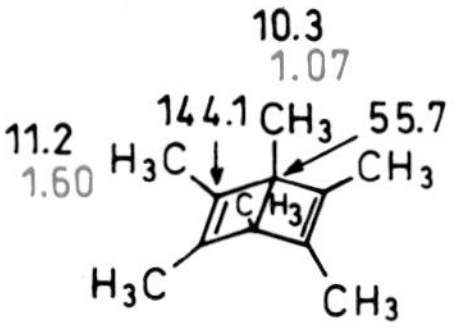

Hexamethyl Dewar benzene

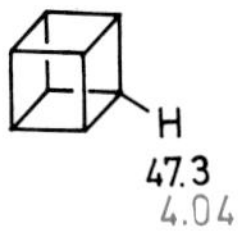

Cubane
(cyclo[4.2.0.0²·⁵.-
0³·⁸.0⁴·⁷]octane)

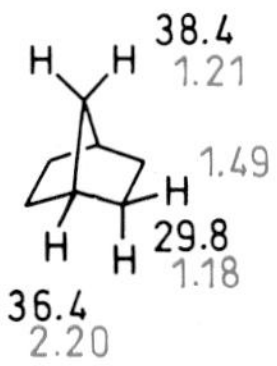

Norbornane
(Bicyclo[2.2.1]-heptane)

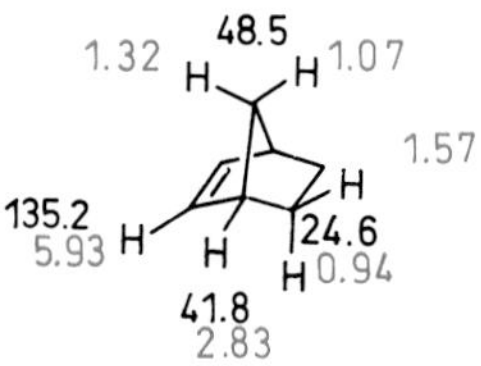

Norbornene
(Bicyclo[2.2.1]-hept-2-ene)

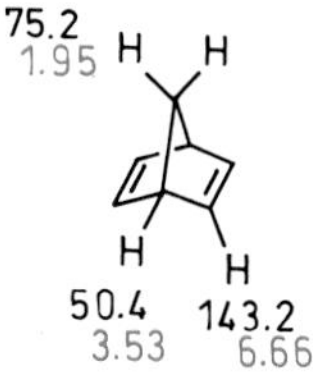

Norbornadiene
(Bicyclo[2.2.1]-hepta-2,5-diene)

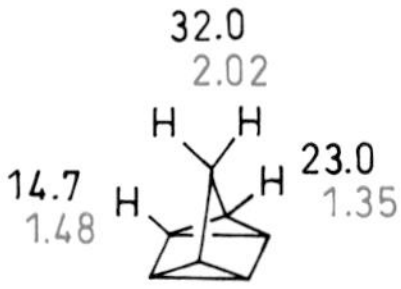

Quadricyclane

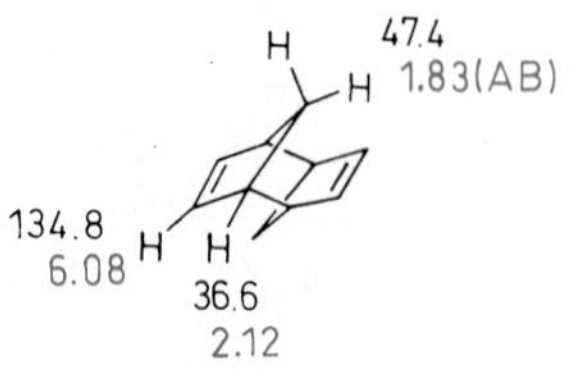

47.4
1.83(AB)

134.8
6.08

36.6
2.12

anti-Tricyclo[4.2.1.1²·⁵]
deca-3,7-diene

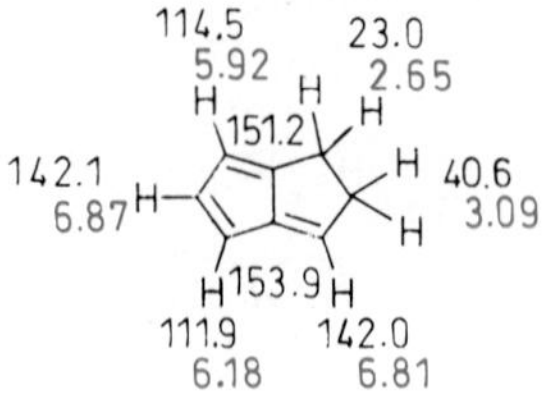

114.5
5.92
151.2
142.1
6.87
153.9
111.9
6.18
142.0
6.81
23.0
2.65
40.6
3.09

1,2-Dihydropentalene

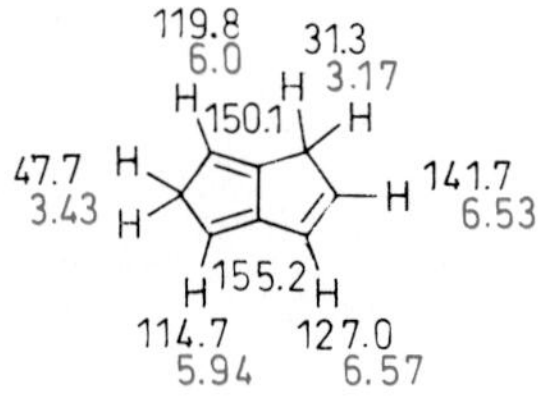

119.8
6.0
150.1
47.7
3.43
155.2
114.7
5.94
127.0
6.57
31.3
3.17
141.7
6.53

1,5-Dihydropentalene

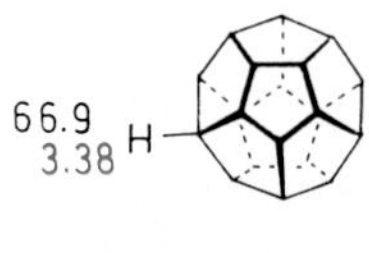

66.9
3.38

Dodecahedrane

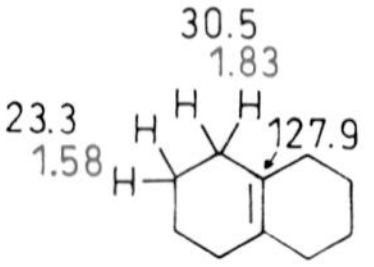

30.5
1.83
23.3
1.58
127.9

1,2,3,4,5,6,7,8-
Octahydronaphthalene

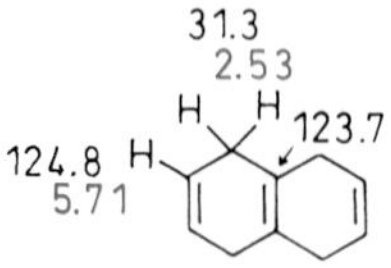

31.3
2.53
124.8
5.71
123.7

1,4,5,8 Tetrahydronaphthalene

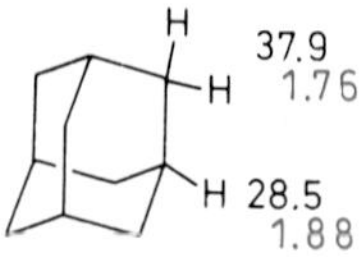

37.9
1.76
28.5
1.88

Adamantane

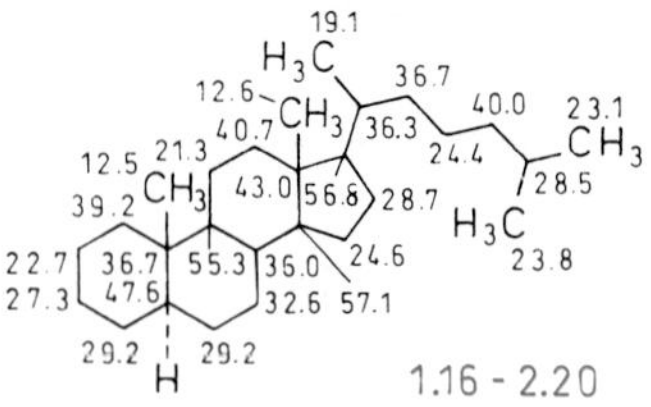

19.1
H_3C
12.6
CH_3
36.7
40.0
23.1
12.5
21.3
40.7
36.3
24.4
28.5
CH_3
39.2
43.0
56.8
28.7
H_3C
22.7
36.7
55.3
36.0
24.6
23.8
27.3
47.6
32.6
57.1
29.2
29.2
H
1.16 – 2.20

5α-Cholestane

Aromatic Hydrocarbons

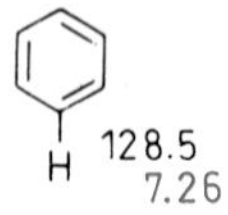

128.5
7.26

Benzene

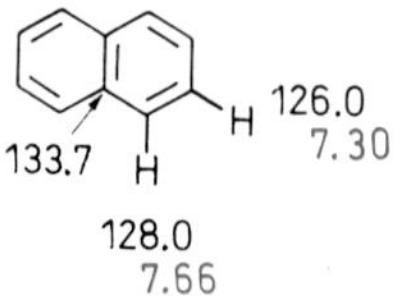

133.7
128.0
7.66
126.0
7.30

Naphthalene

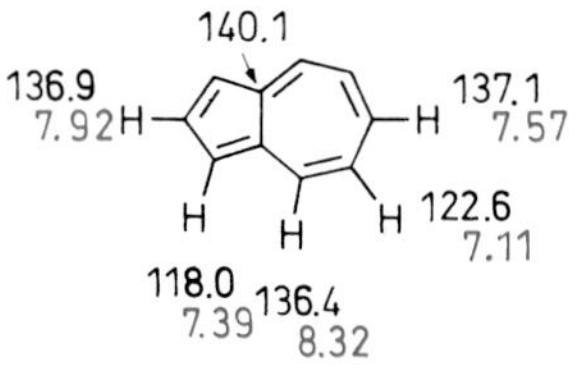

140.1
136.9
7.92
137.1
7.57
118.0
7.39
136.4
8.32
122.6
7.11

Azulene

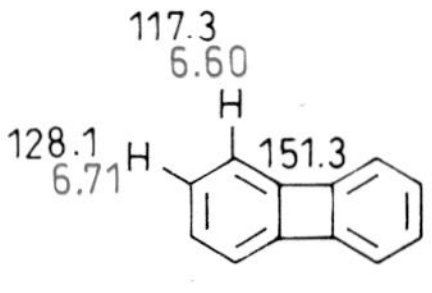

117.3
6.60
128.1
6.71
151.3

Biphenylene

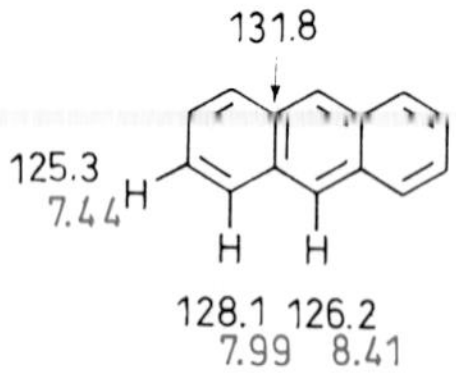

131.8
125.3
7.44
128.1
7.99
126.2
8.41

Anthracene

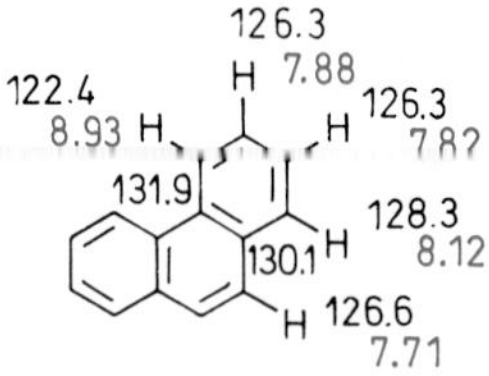

126.3
7.88
122.4
8.93
126.3
7.82
131.9
128.3
8.12
130.1
126.6
7.71

Phenanthrene

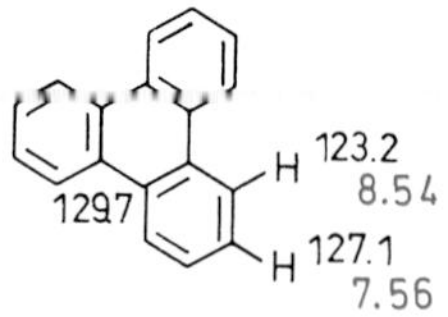

123.2
8.54
129.7
127.1
7.56

Triphenylene

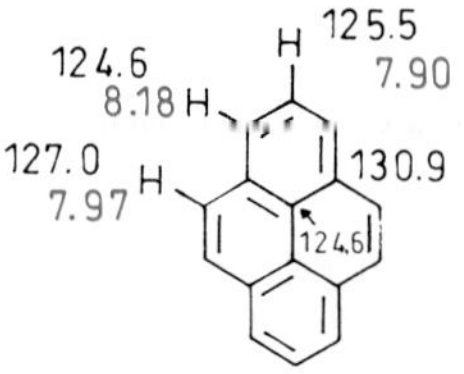

124.6
8.18
127.0
7.97
125.5
7.90
130.9
124.6

Pyrene

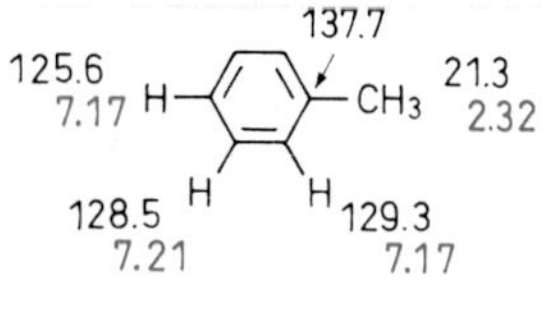

137.7
125.6
7.17
CH_3
21.3
2.32
128.5
7.21
129.3
7.17

Toluene

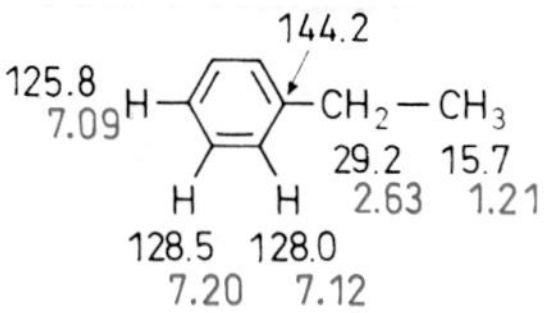

144.2
125.8
7.09
CH_2—CH_3
29.2
15.7
2.63
1.21
128.5
7.20
128.0
7.12

Ethylbenzene

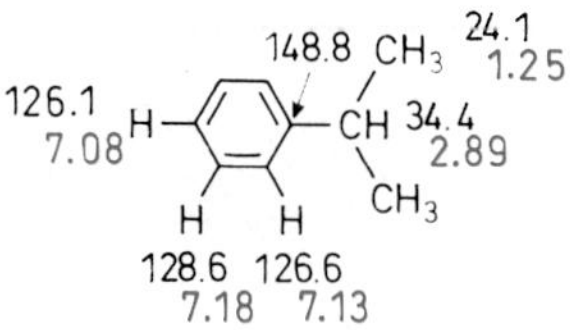

148.8
CH_3
24.1
1.25
126.1
7.08
CH
34.4
2.89
CH_3
128.6
7.18
126.6
7.13

Isopropylbenzene

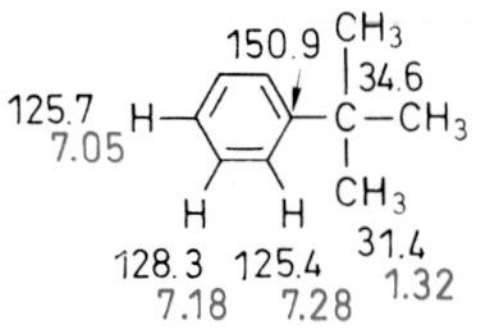

150.9
CH_3
34.6
125.7
7.05
C—CH_3
CH_3
128.3
7.18
125.4
7.28
31.4
1.32

t-Butylbenzene

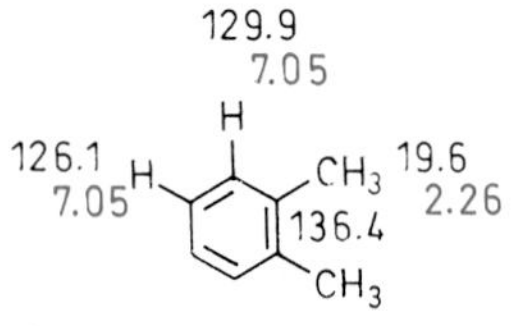

129.9
7.05
126.1
7.05
CH_3
19.6
2.26
136.4
CH_3

1,2-Dimethylbenzene
[*o*-Xylene]

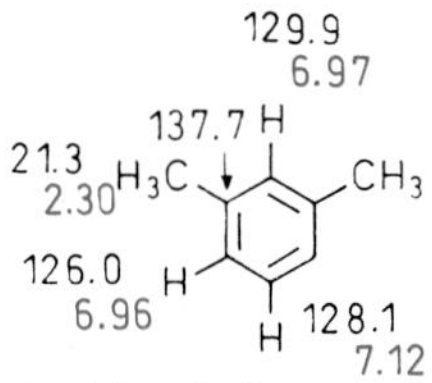

129.9
6.97
137.7
21.3
2.30
H_3C
CH_3
126.0
6.96
128.1
7.12

1,3-Dimethylbenzene
[*m*-Xylene]

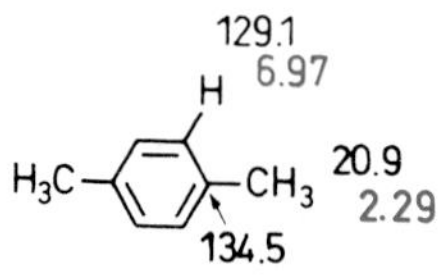

129.1
6.97
H_3C
CH_3
20.9
2.29
134.5

1,4-Dimethylbenzene
[*p*-Xylene]

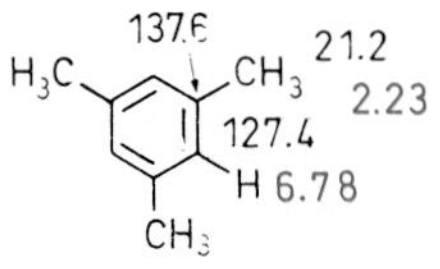

137.6
H_3C
CH_3
21.2
2.23
127.4
6.78
CH_3

1,3,5-Trimethylbenzene
[Mesitylene]

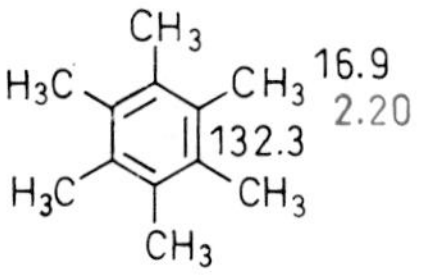

Hexamethyl-
benzene

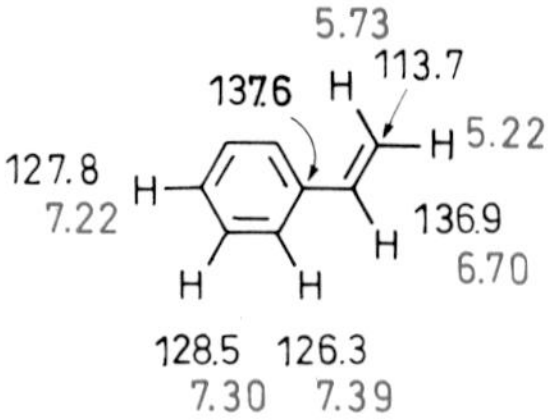

H₃C CH₃ 16.9
H₃C CH₃ 2.20
      132.3
H₃C CH₃
      CH₃

Styrene

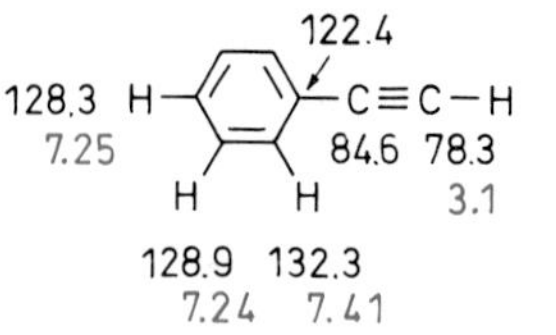

5.73
137.6 113.7
H
      H 5.22
127.8 H
7.22   136.9
           6.70
H   H
128.5 126.3
7.30   7.39

Phenylacetylene

122.4
128.3 H   C≡C−H
7.25   84.6 78.3
H   H   3.1
128.9 132.3
7.24   7.41

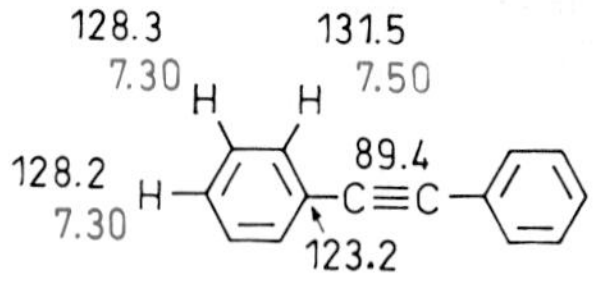

Diphenylacetylene
(Tolane)

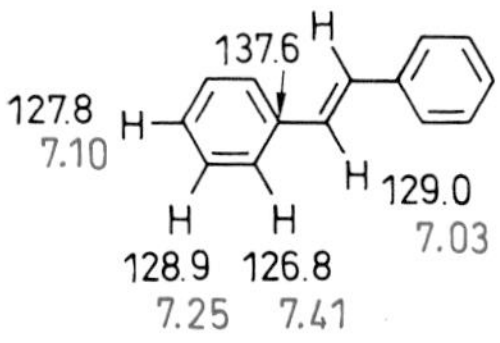

128.3   131.5
7.30 H   H 7.50
128.2   89.4
7.30 H   C≡C
           123.2

(E)-1,2-Diphenylethylene
(E-Stilbene)

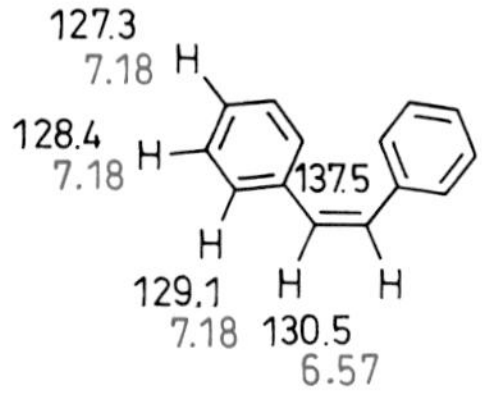

137.6 H
127.8 H
7.10   H 129.0
           7.03
H   H
128.9 126.8
7.25   7.41

(Z)-1,2-Diphenylethylene
(Z-Stilbene)

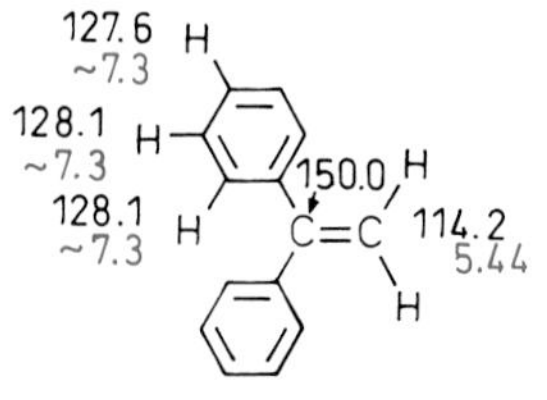

127.3
7.18 H
128.4   137.5
7.18 H
H   H
129.1 130.5
7.18   6.57

1,1-Diphenylethylene

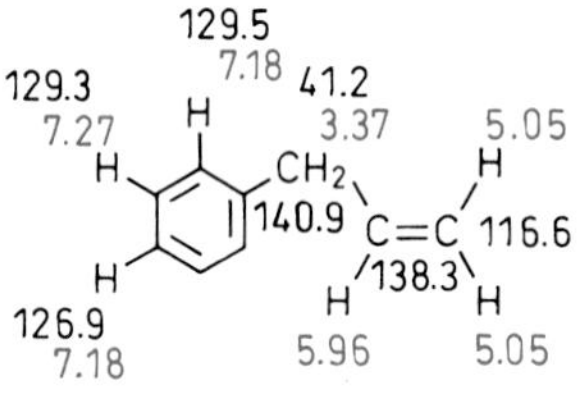

127.6
~7.3 H
128.1
~7.3 H
128.1   150.0 H
~7.3 H C=C 114.2
           H 5.44

Allylbenzene

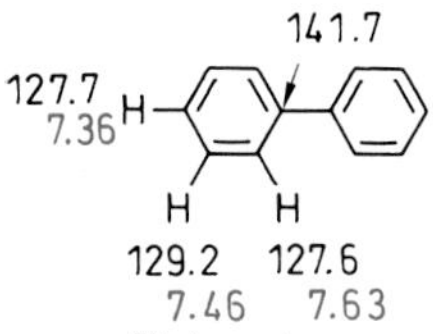

129.5
129.3 7.18 41.2
7.27 H   3.37   5.05
H   CH₂
     140.9 C=C 116.6
           138.3
H   H
126.9 5.96 5.05
7.18

Biphenyl

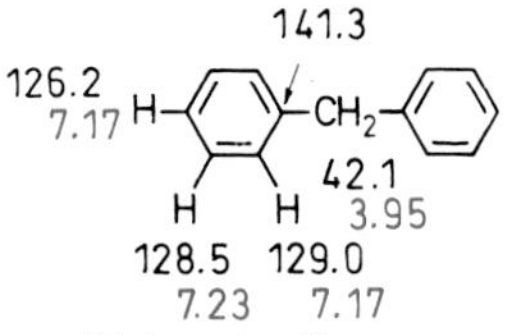

141.7
127.7 H
7.36
H   H
129.2 127.6
7.46   7.63

Diphenylmethane

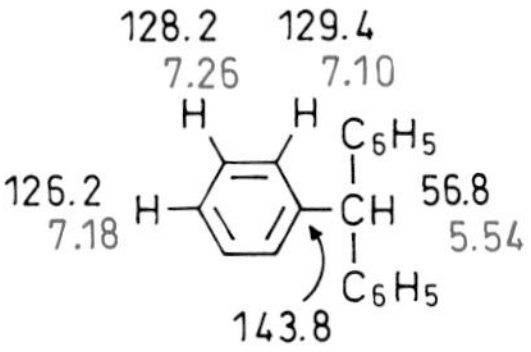

141.3
126.2
7.17 H   CH₂
           42.1
H   H   3.95
128.5 129.0
7.23   7.17

Triphenylmethane

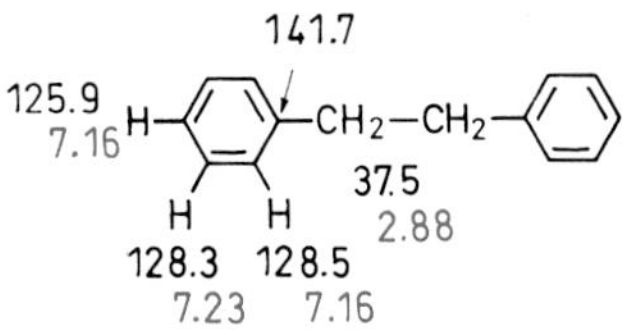

128.2   129.4
7.26   7.10
H   H C₆H₅
126.2   CH 56.8
7.18 H   5.54
H   C₆H₅
143.8

1,2-Diphenylethane

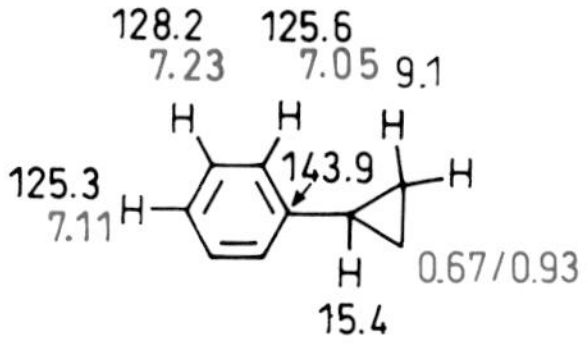

141.7
125.9 H   CH₂−CH₂
7.16
H   H   37.5
128.3 128.5 2.88
7.23   7.16

Cyclopropylbenzene

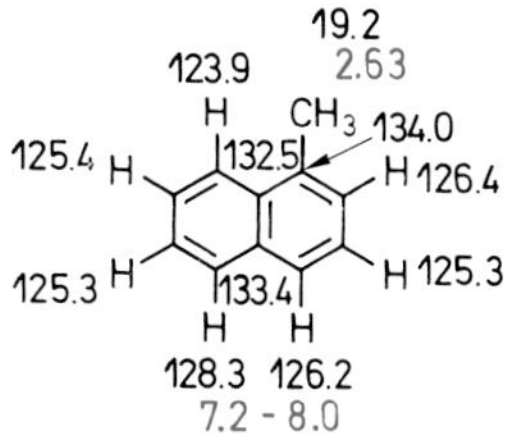

128.2 125.6
7.23   7.05 9.1
H   H   H
125.3 143.9 H
7.11 H
H   H 0.67/0.93
15.4

1-Methylnaphthalene

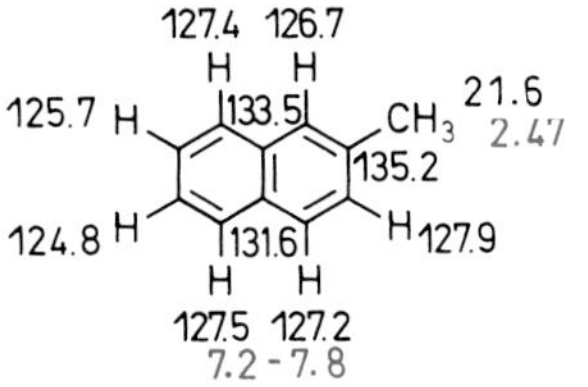

19.2
123.9 2.63
H CH₃ 134.0
125.4 H 132.5
       H 126.4
125.3 H 133.4 H 125.3
H   H
128.3 126.2
7.2 - 8.0

2-Methylnaphthalene

127.4 126.7
H   H
125.7 H 133.5
           CH₃ 21.6
           2.47
           135.2
124.8 H 131.6 H 127.9
H   H
127.5 127.2
7.2 - 7.8

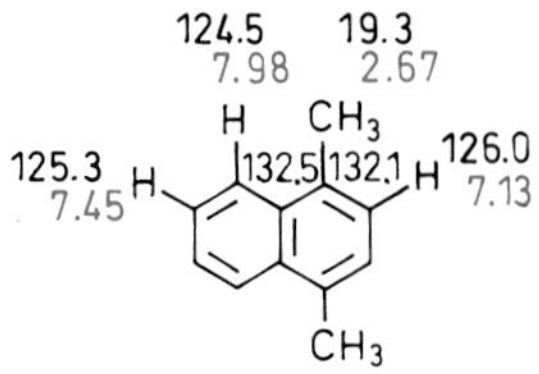

1,4-Dimethylnapthalene

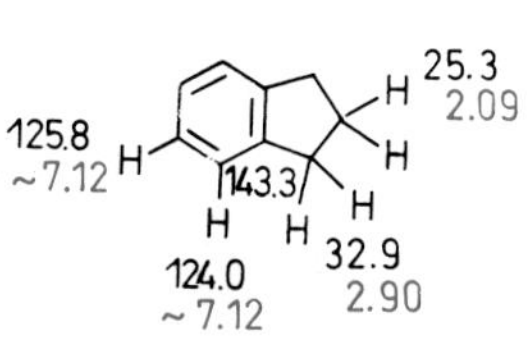

124.5   19.3
7.98   2.67
H CH₃
125.3 H 132.5 132.1 H 126.0
7.45   7.13
CH₃

Indane

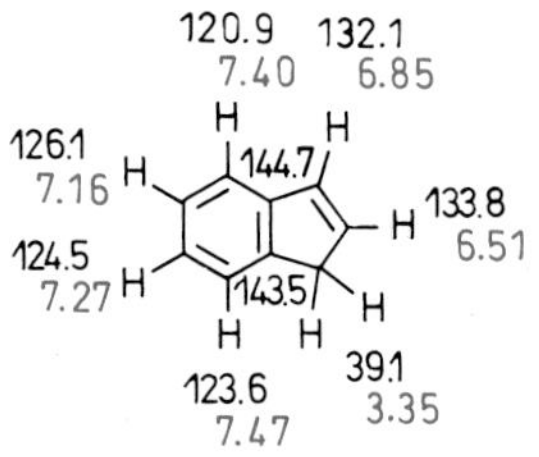

25.3
2.09
125.8 H
~7.12 143.3
H   H 32.9
124.0   2.90
~7.12

Indene

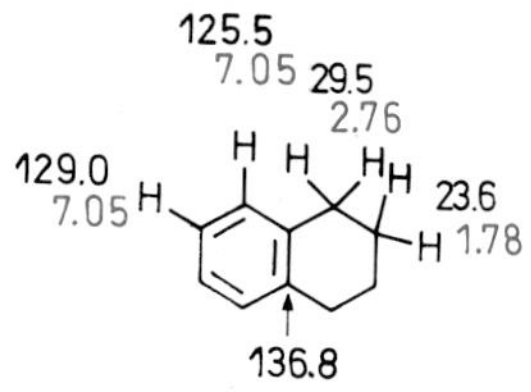

120.9 132.1
7.40   6.85
126.1 H 144.7
7.16 H
           H 133.8
           6.51
124.5 H 143.5
7.27 H
H   H 39.1
123.6   3.35
7.47

Tetralin

125.5
7.05 29.5
      2.76
129.0 H H H
7.05 H   23.6
           H 1.78
136.8

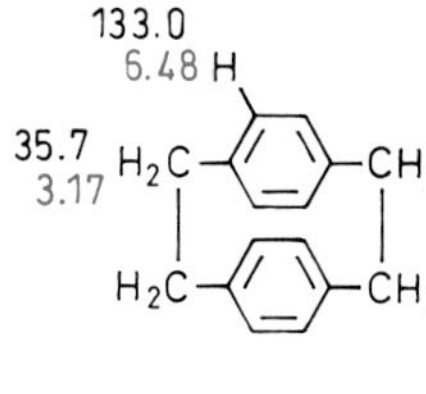

[2.2](1,4)-Benzocyclophane
(Paracyclophane)

133.0
6.48 H
35.7 H₂C   CH₂
3.17
      H₂C   CH₂

Heterocycles

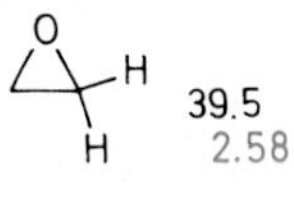

Oxirane
(Ethylene oxide)

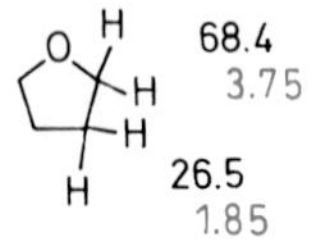

Tetrahydrofuran

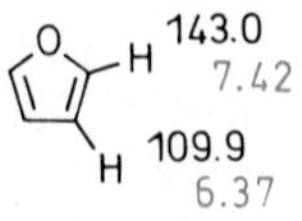

Furan

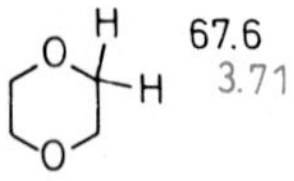

1,4-Dioxane

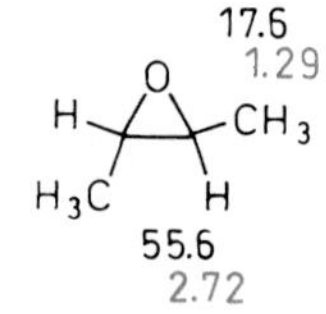

trans-2,3-
Dimethyloxirane

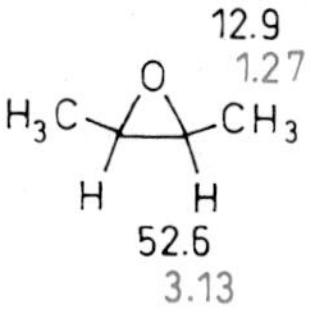

cis-2,3-
Dimethyloxirane

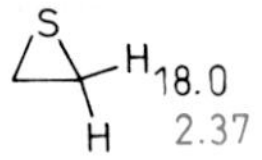

Thiirane
(Ethylene sulfide)

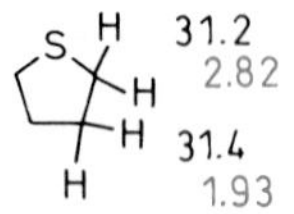

Tetrahydrothiophene

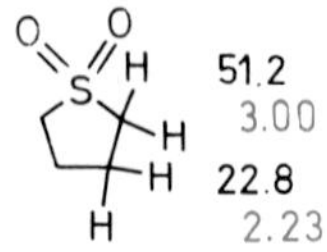

Sulfolane

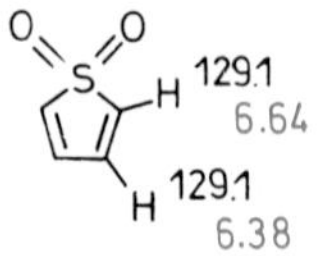

Thiophene 1,1-dioxide
(not isolable)

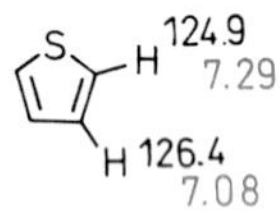

Thiophene

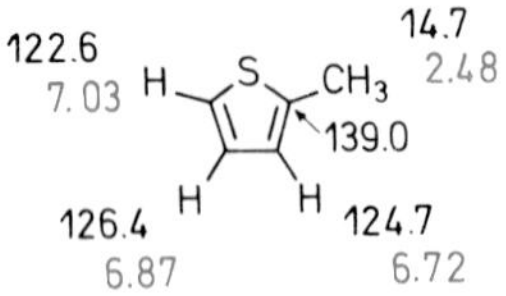

2-Methylthiophene

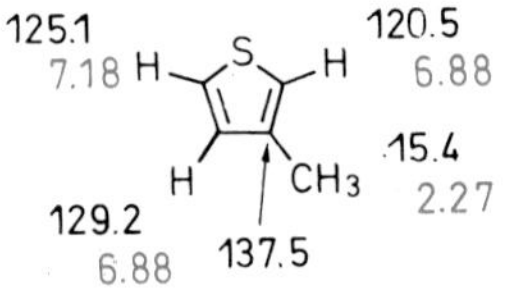

3-Methylthiophene

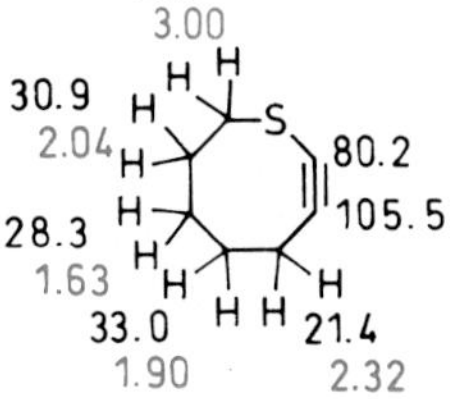

1-Thia-2-cyclooctyne

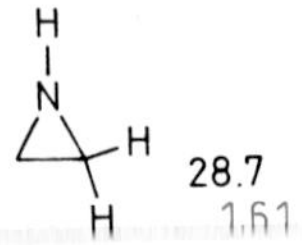

Aziridine
(Ethylene imine)

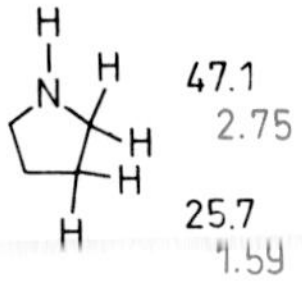

Pyrrolidine

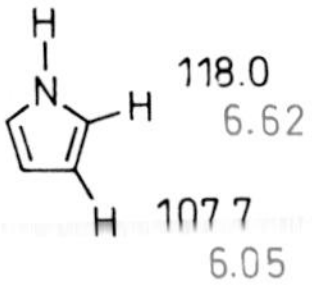

Pyrrole

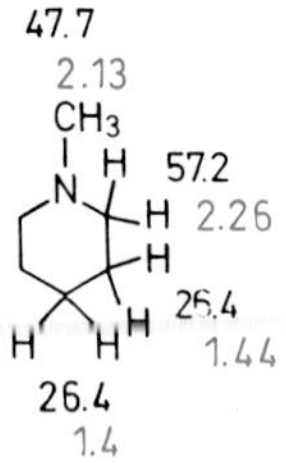

N-Methylpiperidine

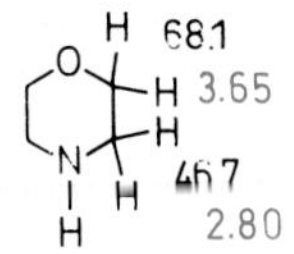

Morpholine

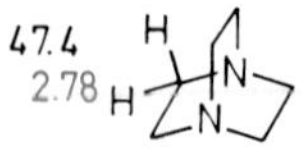

1,4-Diazabicyclo-
[2.2.2]octane

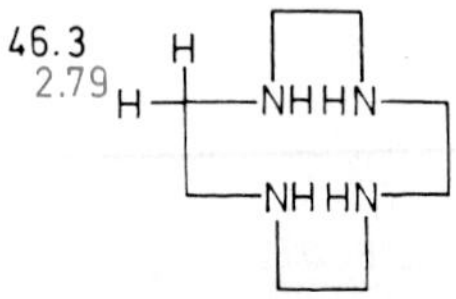

1,4,7,10-Tetraazacyclo-
dodecane (Cyclene)

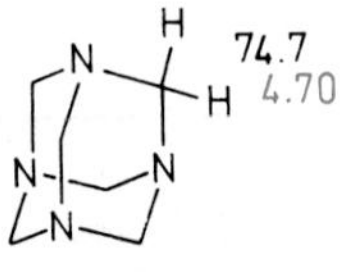

Hexamethylenetetramine
(Urotropine)

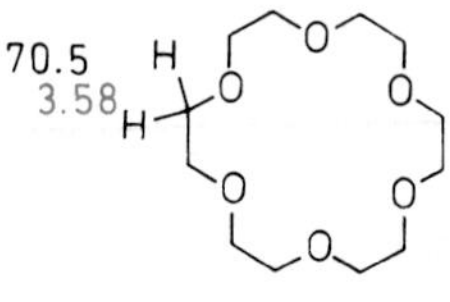

18-Crown-6

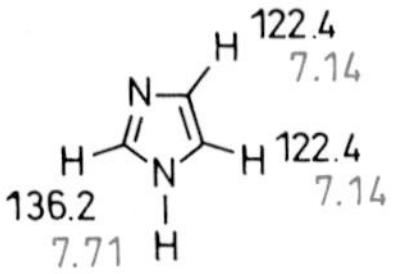

Imidazole[a]

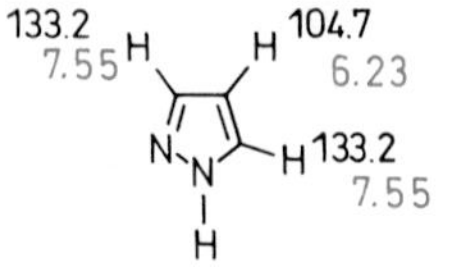

Pyrazole[a]

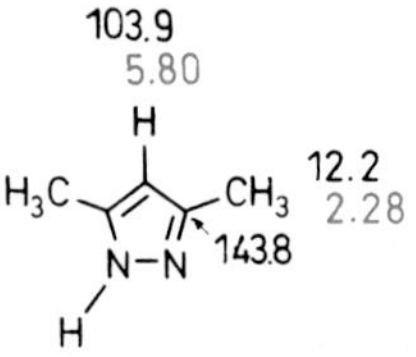

3,5-Dimethylpyrazole[a]

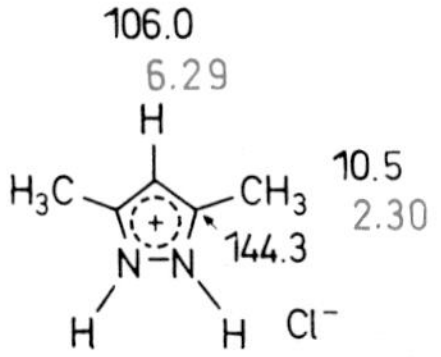

3,5-Dimethylpyrazole
hydrochloride (DMSO)

---

[a] In pyrazole and imidazole there is a fast (on the NMR time-scale) transfer of protons between the two N atoms,
  which increases the symmetry of the system.

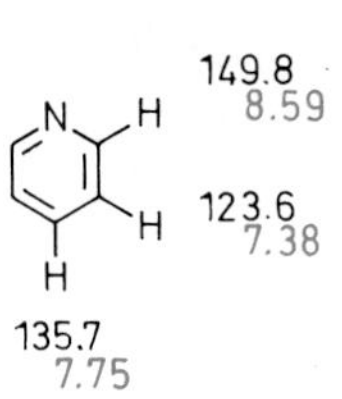

Pyridine
149.8 / 8.59
123.6 / 7.38
135.7 / 7.75

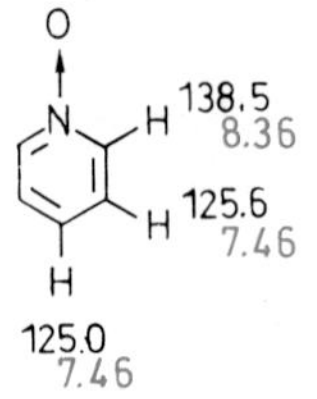

Pyridine N-oxide
138.5 / 8.36
125.6 / 7.46
125.0 / 7.46

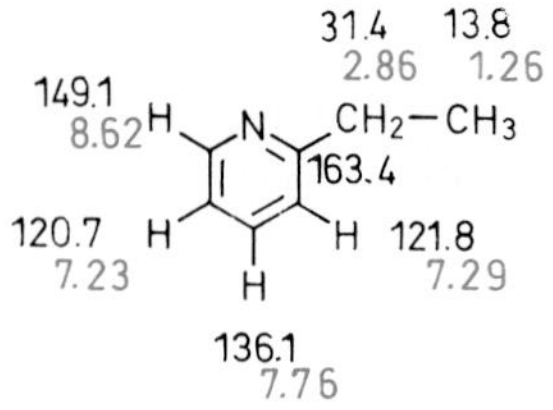

2-Ethylpyridine
31.4 / 2.86   13.8 / 1.26
149.1 / 8.62
163.4
120.7 / 7.23
136.1 / 7.76
121.8 / 7.29

Pyrimidine
159.0 / 9.26
121.4 / 7.36
156.4 / 8.78

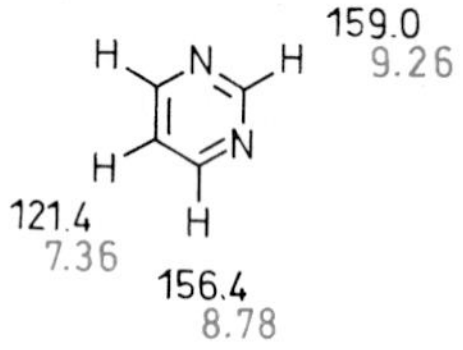

2H-Benzo[b]thiete
122.7 / 6.86   139.5   36.5 / 4.29
123.9 / 7.12
128.4 / 7.26   142.5
120.8 / 7.06

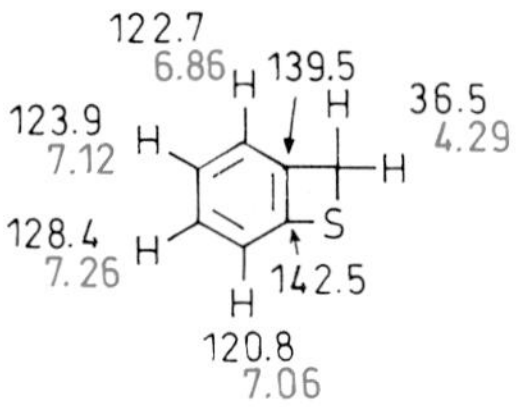

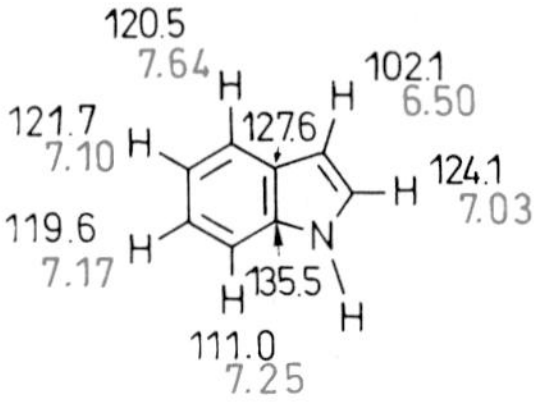

Indole
120.5 / 7.64   102.1 / 6.50
121.7 / 7.10   127.6
124.1 / 7.03
119.6 / 7.17   135.5
111.0 / 7.25

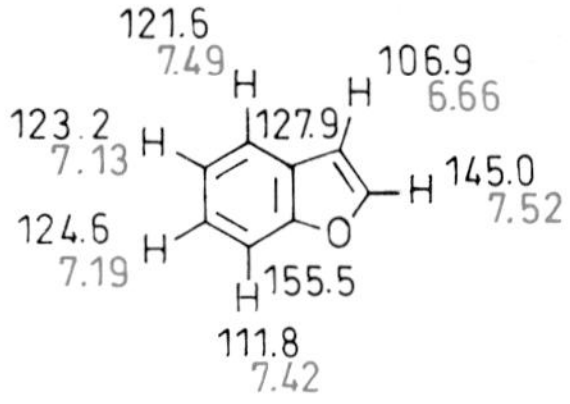

Benzo[b]furan
121.6 / 7.49   106.9 / 6.66
123.2 / 7.13   127.9
145.0 / 7.52
124.6 / 7.19   155.5
111.8 / 7.42

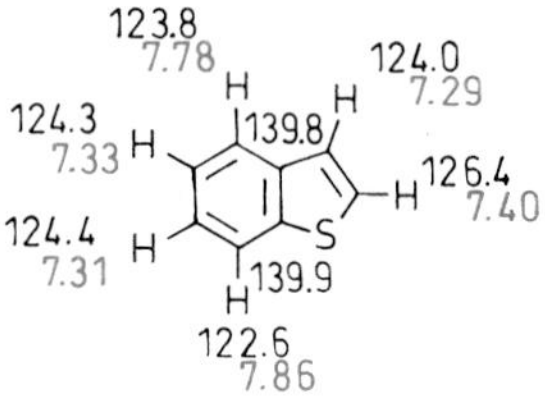

Benzo[b]thiophene
123.8 / 7.78   124.0 / 7.29
124.3 / 7.33   139.8
126.4 / 7.40
124.4 / 7.31   139.9
122.6 / 7.86

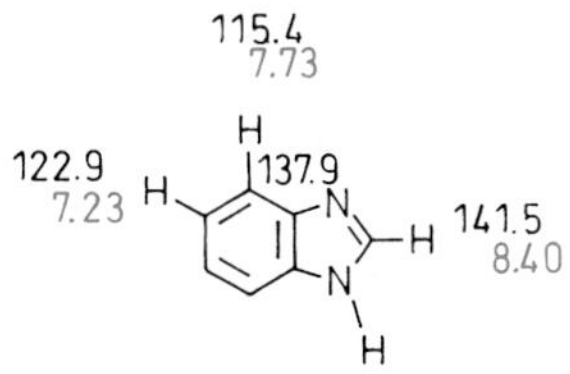

Benzimidazole[a]
115.4 / 7.73
122.9 / 7.23   137.9
141.5 / 8.40

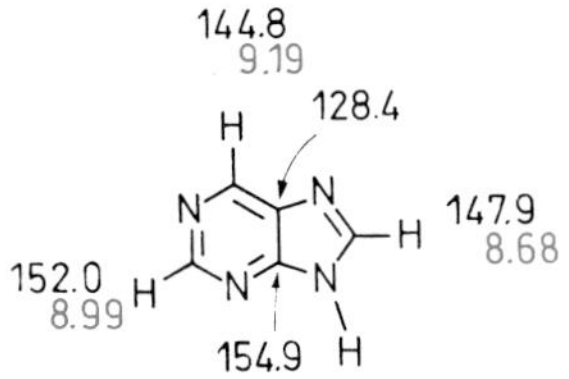

Purine
144.8 / 9.19
128.4
147.9 / 8.68
152.0 / 8.99
154.9

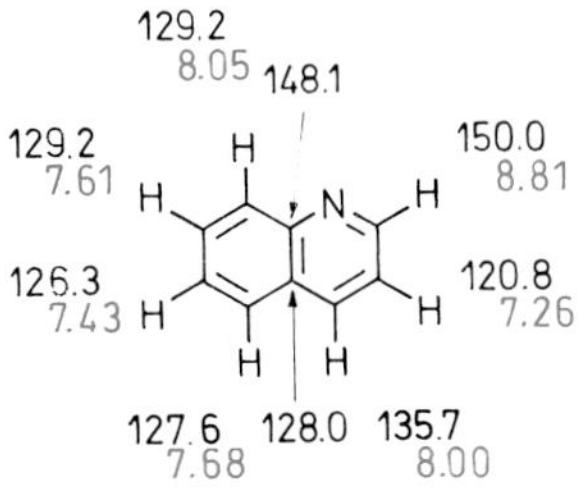

Quinoline
129.2 / 8.05   148.1
129.2 / 7.61
150.0 / 8.81
126.3 / 7.43
120.8 / 7.26
127.6 / 7.68   128.0   135.7 / 8.00

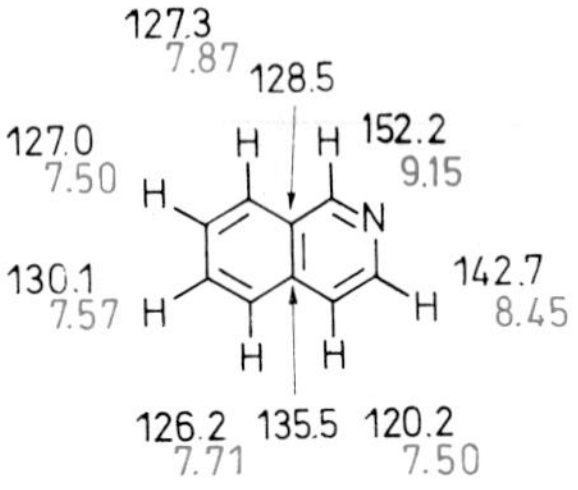

Isoquinoline
127.3 / 7.87   128.5
127.0 / 7.50   152.2 / 9.15
130.1 / 7.57
142.7 / 8.45
126.2 / 7.71   135.5   120.2 / 7.50

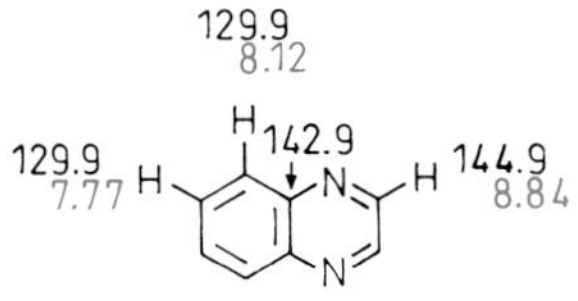

Quinoxaline
129.9 / 8.12
129.9 / 7.77   142.9
144.9 / 8.84

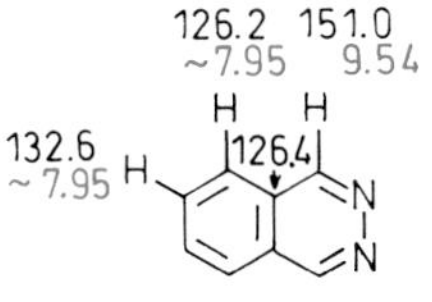

Phthalazine
126.2 / ~7.95   151.0 / 9.54
132.6 / ~7.95   126.4

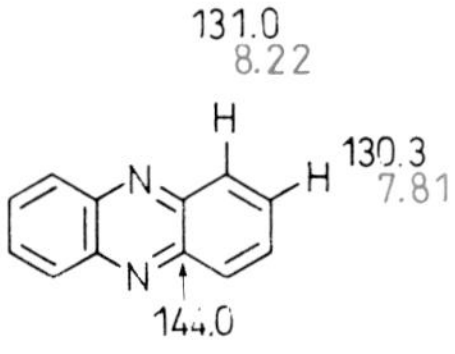

Phenazine
131.0 / 8.22
130.3 / 7.81
144.0

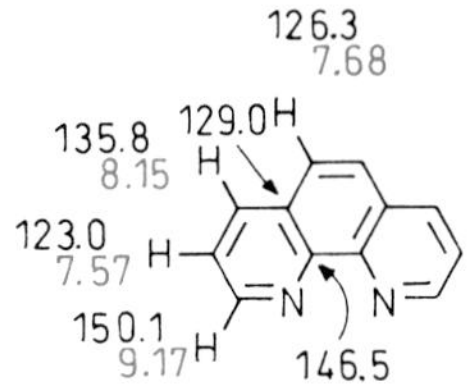

1,10-Phenanthroline
126.3 / 7.68
135.8 / 8.15   129.0
123.0 / 7.57
150.1 / 9.17   146.5

Halogen Compounds[b]

Fluoroethane
H_3C-CH_2F
13.3 / 1.24   78.0 / 4.36

Chloroethane
H_3C-CH_2Cl
17.5 / 1.33   38.7 / 3.47

Bromoethane
H_3C-CH_2Br
19.3 / 1.66   27.9 / 3.37

Iodoethane
H_3C-CH_2I
20.5 / 1.88   -1.0 / 3.18

2-Chlorobutane
$H_3C-CH-CH_2-CH_3$ (Cl)
25.0 / 1.50   60.4 / 3.94   33.4 / 1.74   11.1 / 1.03

2-Bromo-2-methylpropane (tert-Butyl bromide)
36.3 / 1.82
62.2

1-Bromo-2,2-dimethylpropane (Neopentyl bromide)
28.7 / 1.05   49.0 / 3.28
33.0

1,1-Dibromoethane
$H_3C-CHBr_2$
34.2 / 2.47   39.1 / 5.86

1,4-Dibromobutane
$Br-CH_2-CH_2-C_2H_4Br$
33.6 / 3.47   31.8 / 2.05

[b] For the halogenated methanes see Tab. 3.7 (p. 103) and Tab. 3.25 (p. 147).

Cl—CH₂—CH₂—CH₂—C₃H₆Cl
44.9 32.4 26.2
3.53 1.79 1.48

1,6-Dichlorohexane

CHCl₂—CH₂Cl
70.4   50.1
5.77   3.98

1,1,2-Trichloroethane

F₃C—CF₂—CF₂—C₂F₅
118.5 109.8 111.1

Dodecafluoropentane
(Perfluoropentane)

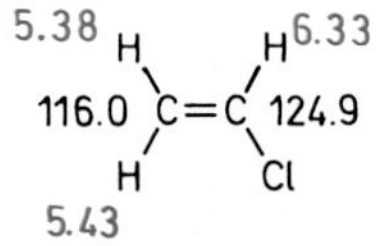

5.38 H   H 6.33
116.0 C=C 124.9
     H   Cl
     5.43

Vinyl chloride

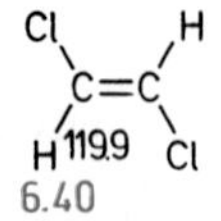

Cl   H
  C=C
H 119.9 Cl
6.40

(E)-1,2-Dichloroethylene

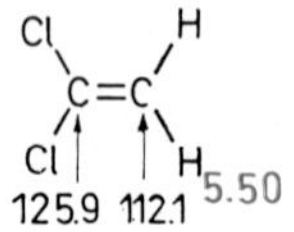

Cl   H
  C=C
Cl   H 5.50
125.9 112.1

1,1-Dichloroethylene

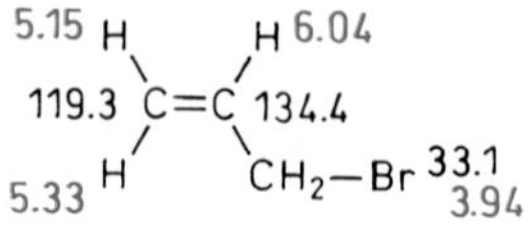

5.15 H   H 6.04
119.3 C=C 134.4
   H   CH₂—Br 33.1
   5.33   3.94

Allyl bromide

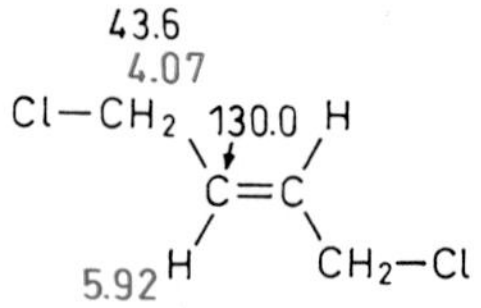

43.6
4.07
Cl—CH₂ 130.0 H
        C=C
  5.92 H   CH₂—Cl

(E)-1,4-Dichloro-2-butene

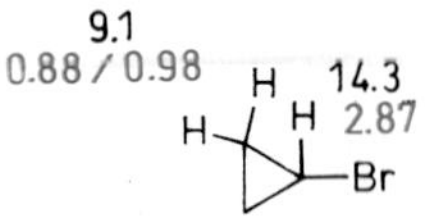

9.1
0.88 / 0.98 H   14.3
          H H 2.87
       H —Br

Bromocyclopropane

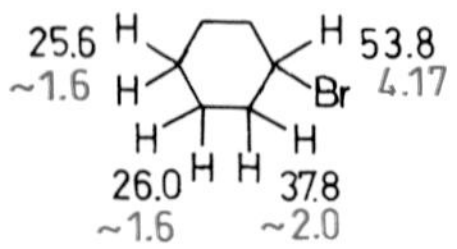

25.6 H   H 53.8
~1.6 H   Br 4.17
     H   H
     26.0 H H 37.8
     ~1.6   ~2.0

Bromocyclohexane

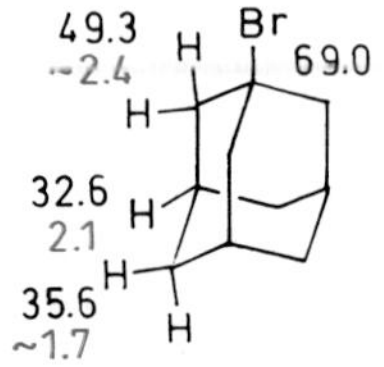

49.3   Br
~2.4 H 69.0
   H
32.6
2.1 H
35.6   H
~1.7 H

1-Bromoadamantane

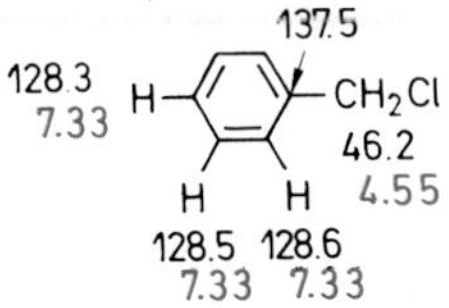

137.5
128.3 H   CH₂Cl
7.33   46.2
   H   H   4.55
128.5 128.6
7.33 7.33

Benzyl chloride

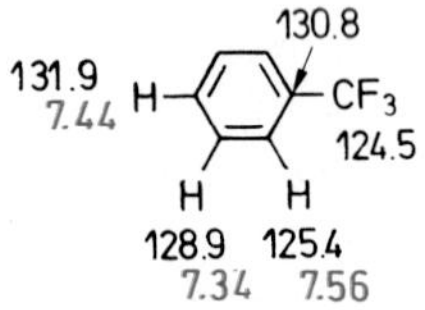

131.9 H   CF₃
7.44   130.8
          124.5
   H   H
128.9 125.4
7.34 7.56

Trifluoromethylbenzene

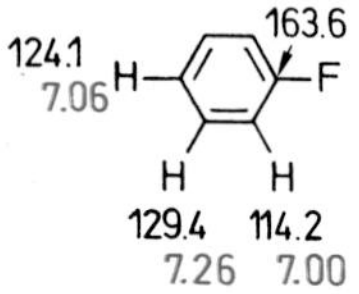

124.1   163.6
H   F
7.06
   H   H
129.4 114.2
7.26 7.00

Fluorobenzene

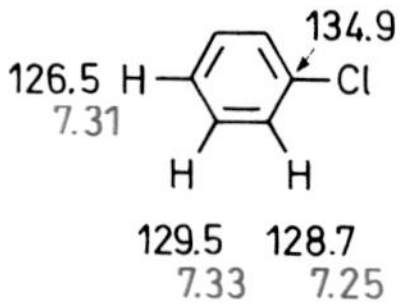

134.9
126.5 H   Cl
7.31
   H   H
129.5 128.7
7.33 7.25

Chlorobenzene

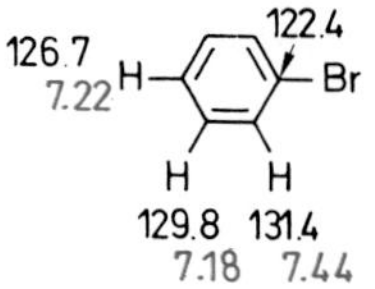

122.4
126.7 H   Br
7.22
   H   H
129.8 131.4
7.18 7.44

Bromobenzene

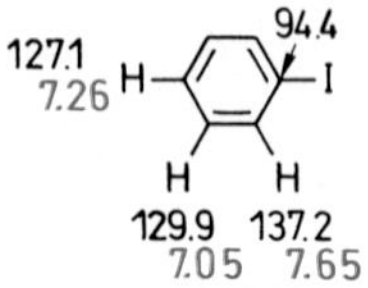

127.1   94.4
7.26 H   I
   H   H
129.9 137.2
7.05 7.65

Iodobenzene

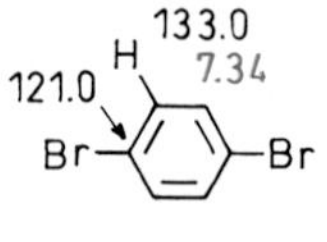

133.0
121.0 H 7.34
Br
        Br

1,4-Dibromobenzene

127.1   Cl
7.25 H   135.5
   Cl   Cl

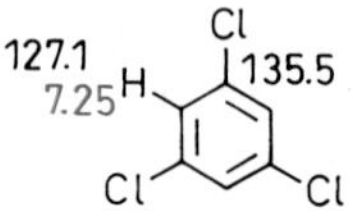

1,3,5-Trichlorobenzene

133.6
28.6   7.37
4.60 137.6 H
Br—CH₂   CH₂—Br

Br—CH₂   CH₂—Br

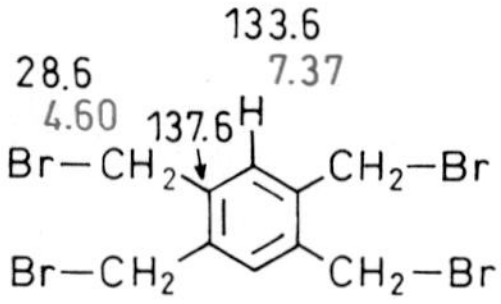

1,2,4,5-Tetrakis(bromomethyl)-
benzene

Alcohols, Phenols

CH₃OH
50.2
3.39

Methanol

H₃C—CH₂OH
17.6 57.0
1.18 3.59

Ethanol

H₃C—CH₂—CH₂OH
11.8 26.9 64.9
0.92 1.59 3.59

Propanol

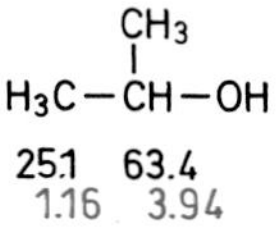

CH₃
|
H₃C—CH—OH
25.1 63.4
1.16 3.94

Isopropanol

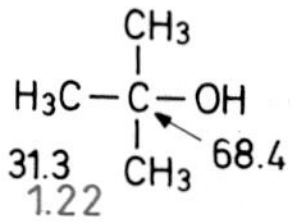

CH₃
|
H₃C—C—OH
31.3 CH₃ 68.4
1.22

tert-Butanol

HO—CH₂—CH₂—OH
67.3
3.58

Ethylene glycol

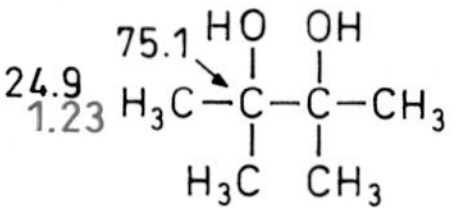

75.1 HO OH
24.9
1.23 H₃C—C—C—CH₃
     H₃C CH₃

Pinacol

HO—CH₂—CH₂—O—C₂H₄OH
61.5 72.6
~3.7 ~3.7

Diglycol

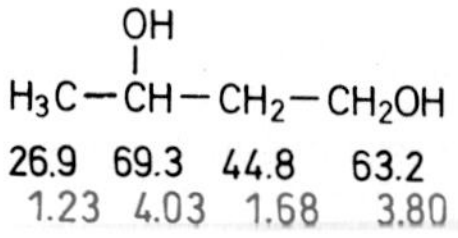

OH
|
H₃C—CH—CH₂—CH₂OH
26.9 69.3 44.8 63.2
1.23 4.03 1.68 3.80

1,3-Butandiol

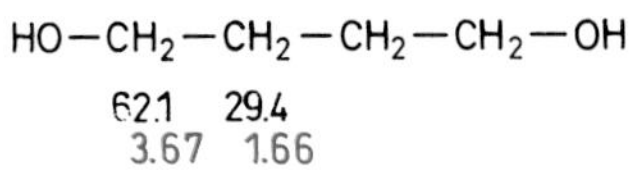

HO—CH₂—CH₂—CH₂—CH₂—OH
62.1   29.4
3.67   1.66

1,4-Butandiol

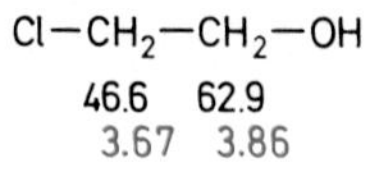

Cl—CH₂—CH₂—OH
46.6   62.9
3.67   3.86

2-Chloroethanol

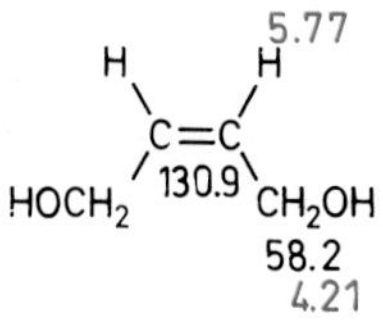

5.77
H   H
$\backslash$C=C$/$
130.9
HOCH₂   CH₂OH
58.2
4.21

(Z)-2-Butene-1,4-diol

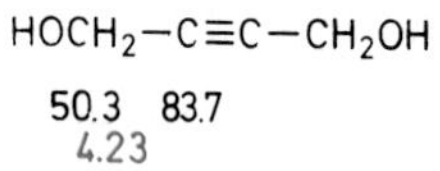

HOCH₂—C≡C—CH₂OH
50.3   83.7
4.23

2-Butyne-1,4-diol

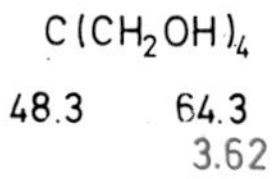

C(CH₂OH)₄
48.3   64.3
3.62

Pentaerythritol (D₂O)

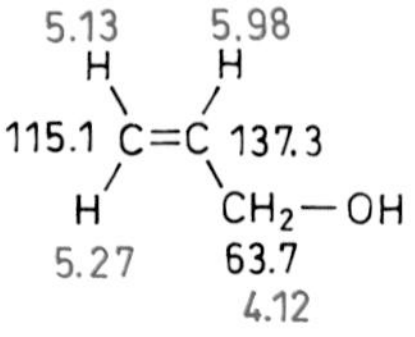

5.13   5.98
H   H
115.1 C=C 137.3
H   CH₂—OH
5.27   63.7
4.12

Allyl alcohol

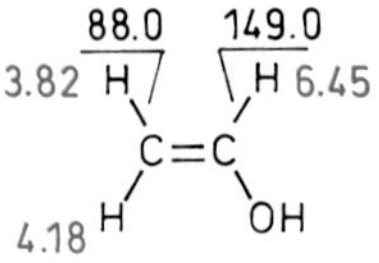

88.0   149.0
3.82 H   H 6.45
C=C
4.18 H   OH

Vinyl alcohol
(unstable)

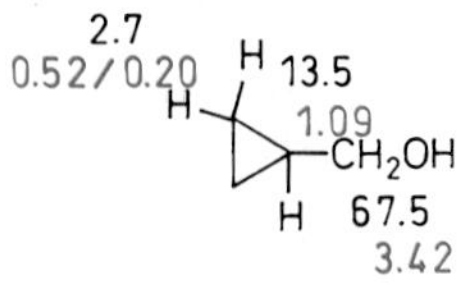

2.7
0.52/0.20   H 13.5
H
1.09
CH₂OH
H   67.5
3.42

Cyclopropane-
methanol

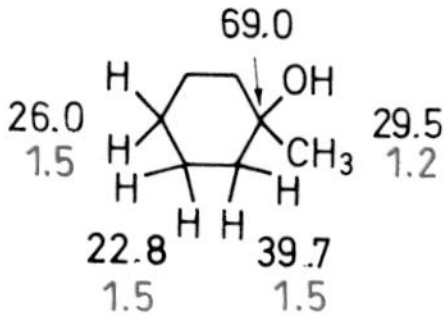

69.0
H   OH
26.0   H
1.5   H   CH₃ 29.5
H   H   1.2
22.8   H   39.7
1.5   1.5

1-Methylcyclohexanol

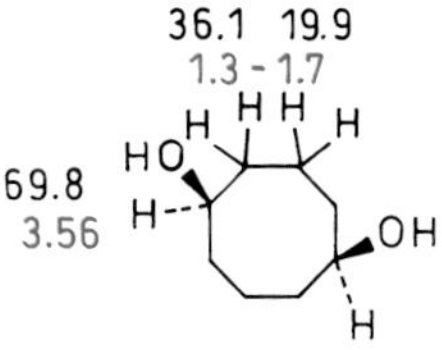

36.1   19.9
1.3 - 1.7
69.8   H H H
3.56   HO   H H H
H
OH
H

cis-1,5-Cyclooctandiol

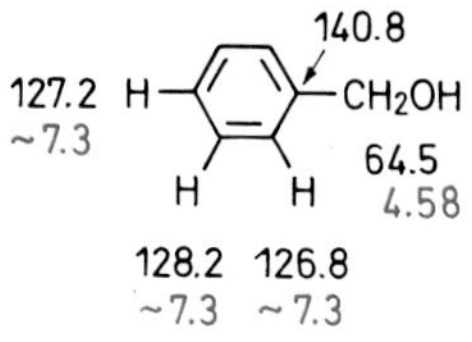

140.8
127.2 H   CH₂OH
~7.3
H   H   64.5
4.58
128.2   126.8
~7.3   ~7.3

Benzyl alcohol

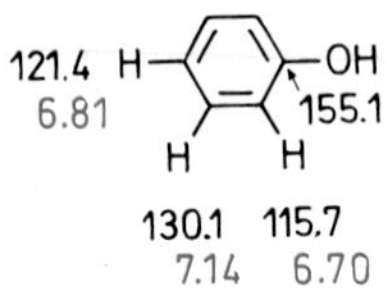

121.4 H   OH
6.81   155.1
H   H
130.1   115.7
7.14   6.70

Phenol

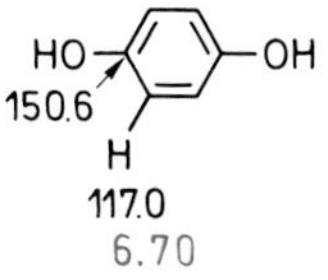

HO   OH
150.6
H
117.0
6.70

Hydroquinone

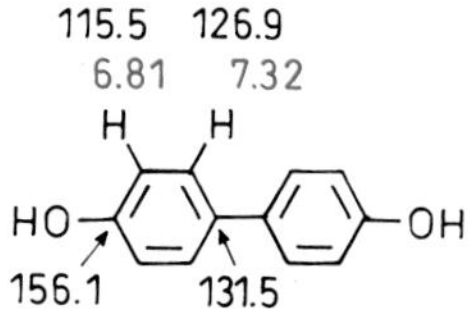

115.5   126.9
6.81   7.32
H   H
HO
H   H
OH
156.1   131.5

4,4'-Biphenol

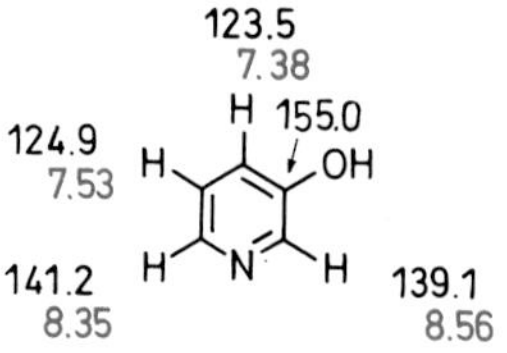

123.5
7.38
H 155.0
124.9   H   OH
7.53
141.2   H   N   H   139.1
8.35   8.56

3-Hydroxypyridine

Ethers (Acetals, Orthoesters)

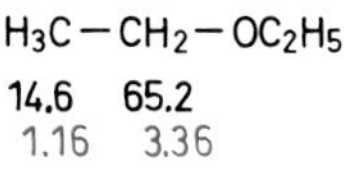

H₃C—CH₂—OC₂H₅
14.6   65.2
1.16   3.36

Diethyl ether

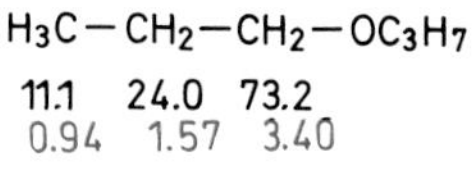

H₃C—CH₂—CH₂—OC₃H₇
11.1   24.0   73.2
0.94   1.57   3.40

Dipropyl ether

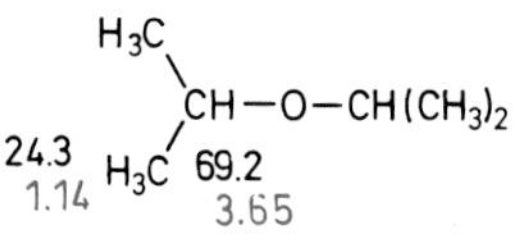

H₃C
$\backslash$
CH—O—CH(CH₃)₂
24.3   H₃C   69.2
1.14   3.65

Diisopropyl ether

CH₃
H₃C—O—C—CH₃
50.1   73.6 CH₃ 28.2
3.20   1.19

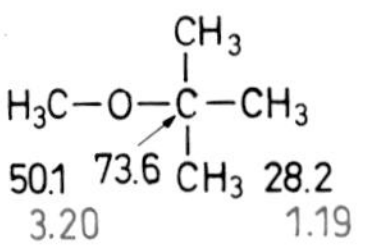

tert-Butyl methyl ether

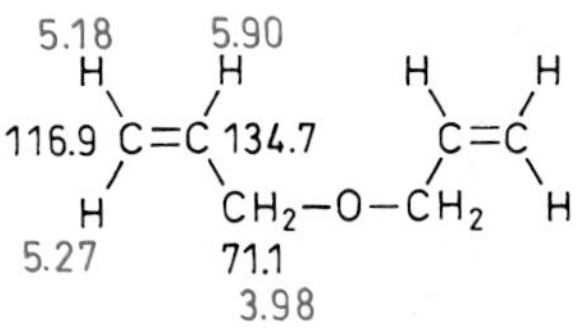

5.18   5.90
H   H
116.9 C=C 134.7
H   CH₂—O—CH₂
5.27   71.1
3.98

H   H
C=C
H   H

Diallyl ether

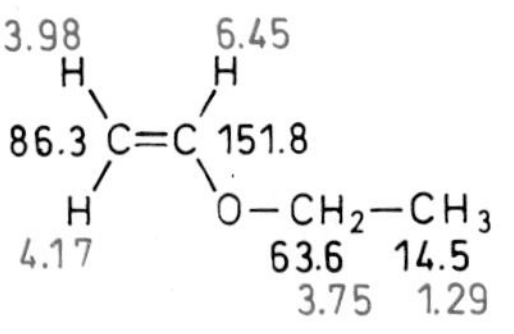

3.98   6.45
H   H
86.3 C=C 151.8
H   O—CH₂—CH₃
4.17   63.6   14.5
3.75   1.29

Ethyl vinyl ether

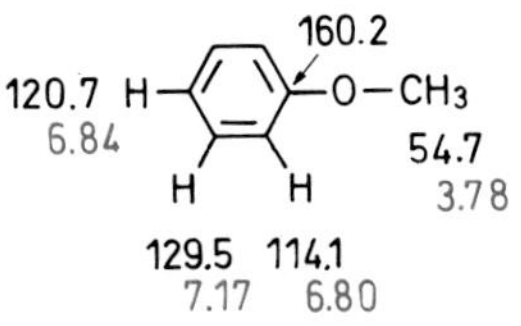

160.2
120.7 H   O—CH₃
6.84   54.7
H   H   3.78
129.5   114.1
7.17   6.80

Anisole

129.4 H   H 114.5
7.25   6.88
120.5 H
6.91   H   O—CH$_2$—CH$_3$
              63.2 14.8
    158.9   4.00 1.39

Phenetole

129.3 H   H 114.5
7.26   6.90
120.4 H
6.92   H   O—CH$_2$—CH$_2$—CH$_2$—CH$_3$
              67.5 31.4 19.3 13.8
    159.1   3.93 1.76 1.49 0.97

Butyl phenyl ether

157.6
123.2 H
7.03   O
    H   H
    128.8 119.0
    7.20 7.17

Diphenyl ether

OC$_2$H$_5$
H$_3$C—CH—O—CH$_2$—CH$_3$
19.9 99.5   60.6   15.3
1.3   4.7 3.54/3.65 1.2

Acetaldehyde
diethyl acetal

OC$_2$H$_5$
H—C—O—CH$_2$—CH$_3$
112.9 OC$_2$H$_5$ 59.5 15.2
5.14   3.61 1.22

Triethyl
orthoformate

C(OCH$_3$)$_4$
121.0   50.4
             3.31

Tetramethoxymethane
(Methyl orthocarbonate)

Nitro-compounds, Nitrite esters

H$_3$C—NO$_2$
61.2
4.28

Nitromethane

H$_3$C—CH$_2$—NO$_2$
12.3   70.8
1.58   4.43

Nitroethane

H$_3$C—CH$_2$—CH$_2$—NO$_2$
11.2   22.0   78.3
1.02   2.05   4.38

1-Nitropropane

NO$_2$
H$_3$C—CH—CH$_3$
20.8   78.8
1.56   4.65

2-Nitropropane

148.2
134.5 H   NO$_2$
7.69
    H   H
    129.3 123.4
    7.55   8.21

Nitrobenzene

132.6 H   H 124.9
7.68   8.11
Br   NO$_2$
    129.9   146.9

1-Bromo-4-nitrobenzene

115.6 H   H 125.8
6.90   8.08
HO   NO$_2$
    163.9   139.7

4-Nitrophenol

H$_3$C
    CH—CH$_2$—CH$_2$—O—N=O
H$_3$C 25.5 38.2 66.8
       ~1.71 ~1.63 4.71
22.5
0.95

Isoamyl nitrite

Amines

H$_3$C—NH$_2$
28.3
2.47

Methylamine

H$_3$C—CH$_2$—NH$_2$
19.0   36.8
1.10   2.74

Ethylamine

H$_3$C—CH$_2$—NH—C$_2$H$_5$
15.4   44.1
1.10   2.64

Diethylamine

H$_3$C—CH$_2$—N(C$_2$H$_5$)$_2$
12.4   47.0
1.00   2.54

Triethylamine

[H$_3$C—$\overset{+}{N}$H$_3$] Cl$^-$
24.4   (D$_2$O)
2.63

Methylamine
hydrochloride

[H$_3$C—$\overset{+}{N}$(CH$_3$)$_3$] Cl$^-$
54.3   (D$_2$O)
3.19

Tetramethyl-
ammonium chloride

H$_3$C—CH$_2$—CH$_2$—CH$_2$—NH$_2$
13.9   20.0   36.1   42.0
0.91   1.36   1.40   2.69

Butylamine

OCH$_3$ CH$_3$
113.2 H—C—N
4.36   CH$_3$ 37.5
    OCH$_3$ 2.29
    53.2
    3.33

N,N-Dimethylformamide
dimethyl acetal

H$_3$C
    N—CH$_2$—N(CH$_3$)$_2$
43.5 H$_3$C   83.7
2.23   2.71

N,N,N',N'-Tetramethyl-
diaminomethane

36.4 H$_3$C   CH$_3$
2.46
    N—CH$_2$—CH$_2$—N
H   52.0   H
             2.70

N,N'-Dimethylethylenediamine

HC—CH$_2$—CH$_2$—NH$_2$
64.2   44.6
3.68   2.88

Ethanolamine

[HO—CH$_2$—CH$_2$—$\overset{+}{N}$(CH$_3$)$_3$] Cl$^-$
56.6   68.3   54.8
4.01   3.51   3.21

Choline chloride (D$_2$O)

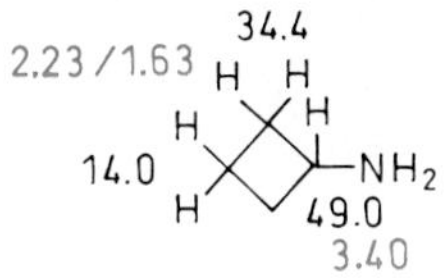

34.4
2.23 / 1.63 H H
H   H
14.0   —NH₂
H   49.0
3.40

Cyclobutylamine

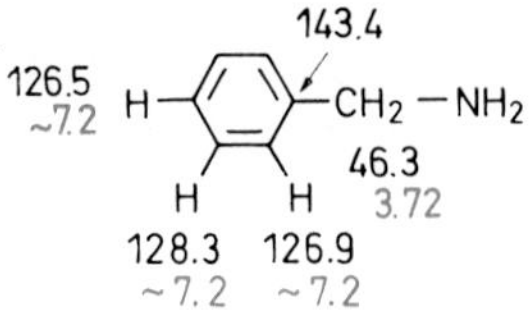

143.4
126.5 H—◯—CH₂—NH₂
~7.2
46.3
H   H   3.72
128.3   126.9
~7.2   ~7.2

Benzylamine

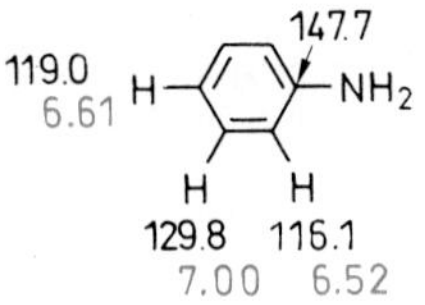

147.7
119.0 H—◯—NH₂
6.61
H   H
129.8   116.1
7.00   6.52

Aniline

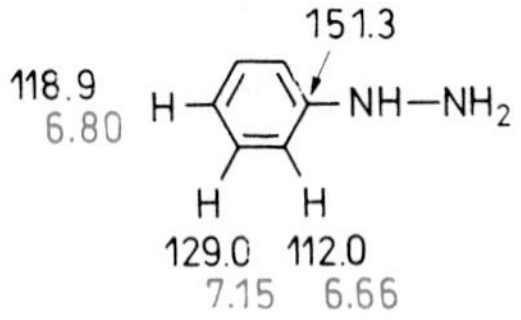

151.3
118.9 H—◯—NH—NH₂
6.80
H   H
129.0   112.0
7.15   6.66

Phenylhydrazine

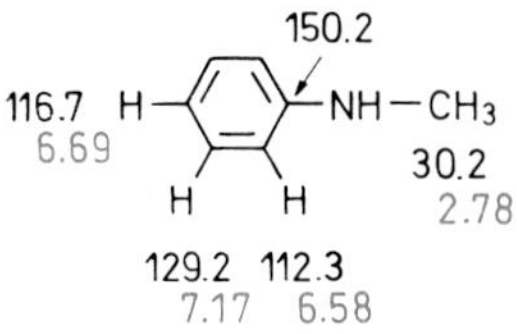

150.2
116.7 H—◯—NH—CH₃
6.69
30.2
H   H   2.78
129.2   112.3
7.17   6.58

N-Methylaniline

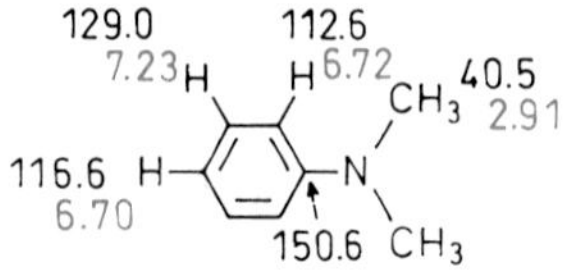

129.0   112.6
7.23 H   H 6.72
CH₃ 40.5
2.91
116.6 H—◯—N
6.70
150.6   CH₃

N,N-Dimethylaniline

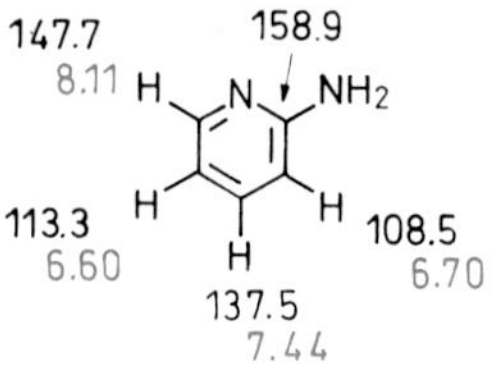

147.7   158.9
8.11 H   N   NH₂
113.3 H   H 108.5
6.60   6.70
137.5
7.44

2-Aminopyridine

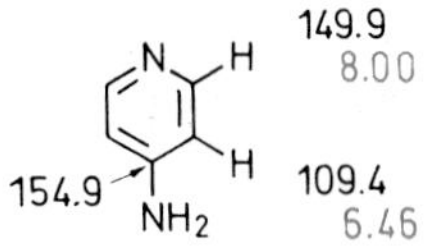

N   H   149.9
8.00
154.9   H   109.4
NH₂   6.46

4-Aminopyridine

Sulfur Compounds

H₃C—CH₂—SH
19.7   19.1
1.32   2.56
Ethanethiol

H₃C—CH₂—SC₂H₅
14.8   25.5
1.25   2.55
Diethyl sulfide

H₃C—CH₂—O—SO—OC₂H₅
15.4   58.3
1.33   4.06
Diethyl sulfite

H₃C—CH₂—SO₂Cl
9.1   60.2
1.59   3.63
Ethanesulfonyl chloride

H₃C—SCH₃
18.2
2.14
Dimethyl sulfide

H₃C—S—SCH₃
22.0
2.43
Dimethyl disulfide

H₃C—SO—CH₃
40.8
2.62
Dimethyl sulfoxide

H₃C—SO₂—CH₃
44.4
3.14
Dimethyl sulfone

H₃C—O—SO₂—OCH₃
59.0
4.00
Dimethyl sulfate

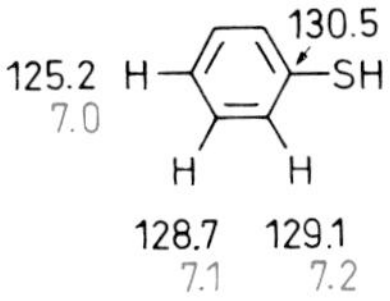

130.5
125.2 H—◯—SH
7.0
H   H
128.7   129.1
7.1   7.2
Thiophenol

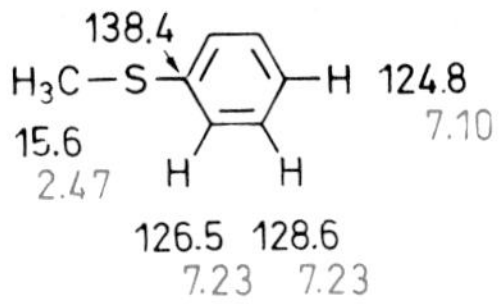

138.4
H₃C—S—◯—H 124.8
15.6
2.47   H   H
126.5   128.6
7.23   7.23
Methyl phenyl sulfide

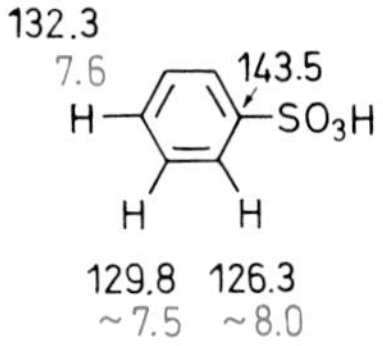

132.3
7.6   143.5
H—◯—SO₃H
H   H
129.8   126.3
~7.5   ~8.0
Benzenesulfonic acid

Phosphorus Compounds

H₃C—CH₂—P(C₂H₅)₂
10.3   19.5
0.99   1.27
Triethylphosphine

H₃C—CH₂—P(C₂H₅)Cl
8.9   27.6
1.18   1.85
Diethylchlorophosphine

H₃C—CH₂—PCl₂
7.2   36.2
1.30   2.31
Ethyldichlorophosphine

H₃C—CH₂—O—P(OC₂H₅)₂
17.4   58.0
1.28   3.91
Triethyl phosphite

O
‖
H₃C—CH₂—O—P(OC₂H₅)₂
15.9   63.4
1.35   4.12
Triethyl phosphate

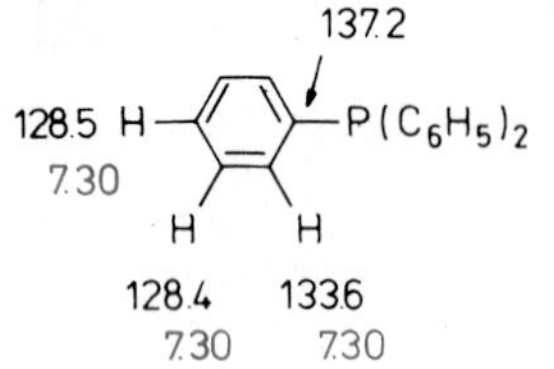

137.2

128.5 H
7.30
—P(C₆H₅)₂

128.4   133.6
7.30   7.30

Triphenylphosphine

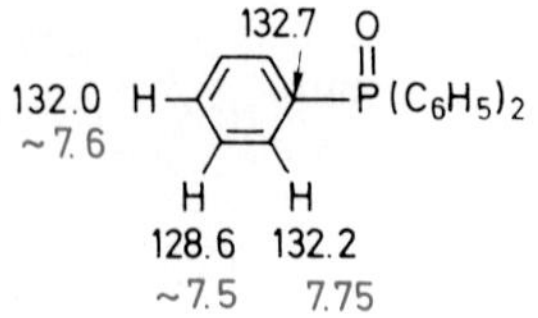

132.7 O
132.0 H   ‖
~7.6   —P(C₆H₅)₂

128.6 132.2
~7.5   7.75

Triphenylphosphine oxide

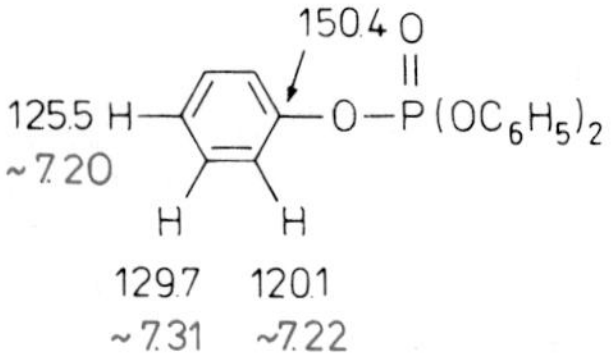

150.4 O
‖
125.5 H   —O—P(OC₆H₅)₂
~7.20

129.7 120.1
~7.31 ~7.22

Triphenyl phosphate

Aldehydes

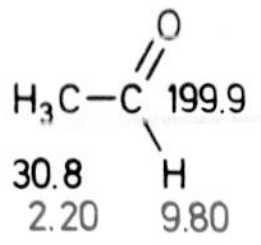

O
‖
H₃C—C 199.9
30.8   H
2.20   9.80

Acetaldehyde

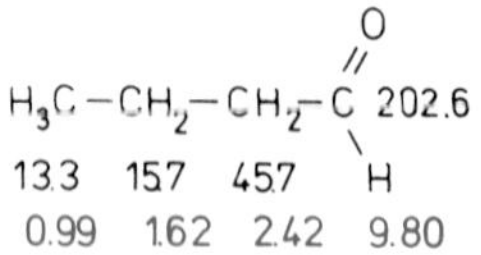

O
‖
H₃C—CH₂—CH₂—C 202.6
13.3 15.7 45.7 H
0.99 1.62 2.42 9.80

Butanal

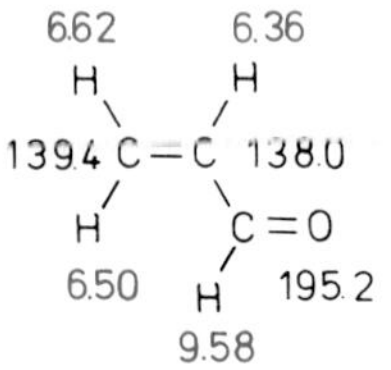

6.62   6.36
H   H
139.4 C=C 138.0
H
C=O
6.50 H 195.2
9.58

Acrolein

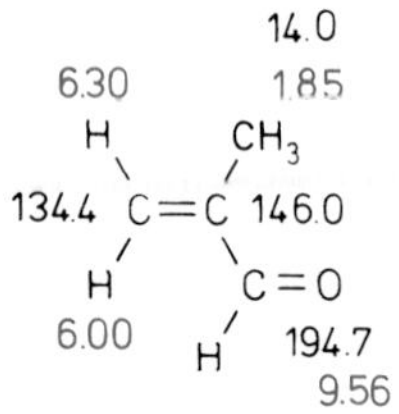

14.0
6.30   1.85
H   CH₃
134.4 C=C 146.0
H
C=O
6.00   H 194.7
9.56

Methacrolein

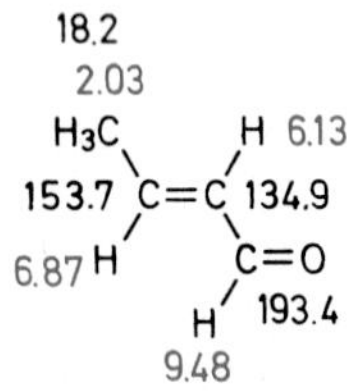

18.2
2.03
H₃C   H 6.13
153.7 C=C 134.9
6.87 H   C=O
H   193.4
9.48

(E)-Crotonaldehyde

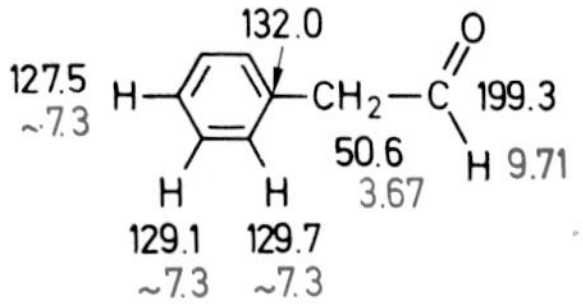

132.0
127.5 H   —CH₂—C 199.3
~7.3   50.6 H 9.71
H   H   3.67
129.1 129.7
~7.3 ~7.3

Phenylacetaldehyde

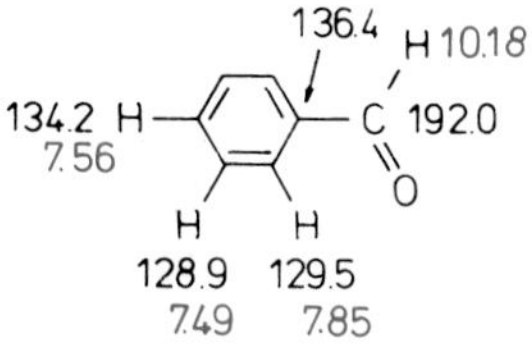

136.4 H 10.18
134.2 H   C 192.0
7.56
H   H   O
128.9 129.5
7.49   7.85

Benzaldehyde

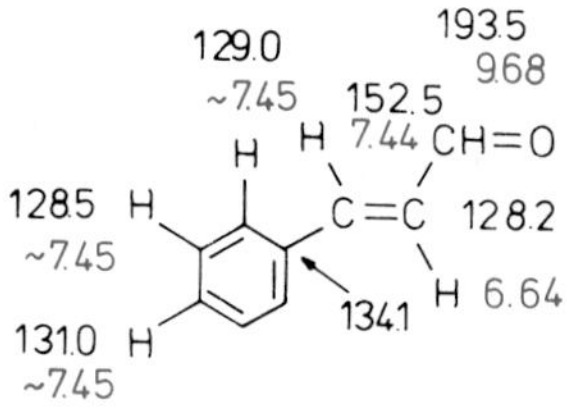

193.5
129.0   9.68
~7.45 152.5
H   H 7.44 CH=O
C=C 128.2
128.5 H
~7.45
131.0 H   H 6.64
~7.45   134.1

(Z) Cinnamaldehyde

133.8
7.53 120.9
119.9 H H   196.7
7.00   C=O 9.87
136.9 H
7.50   H 161.6
117.6
6.98

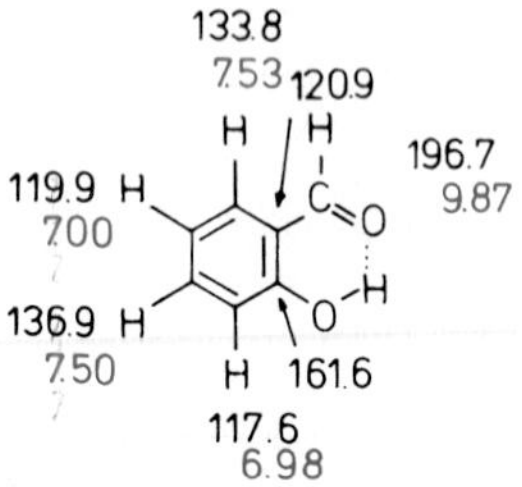

Salicylaldehyde

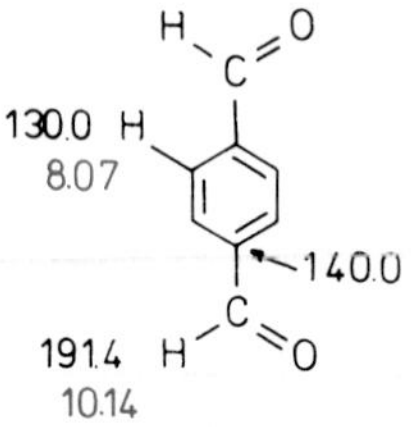

H   O
130.0 H   C
8.07
—140.0
191.4 H   C
10.14   O

Terephthaldialdehyde

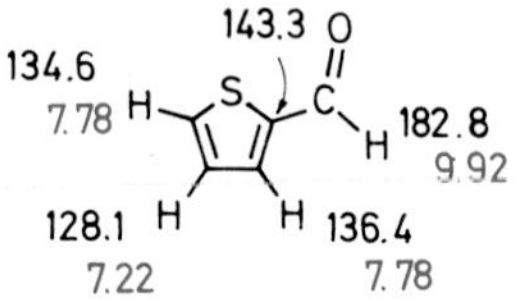

143.3 O
134.6 S   C
7.78 H   182.8
9.92
128.1 H   H 136.4
7.22   7.78

Thiophene-2-
carboxaldehyde

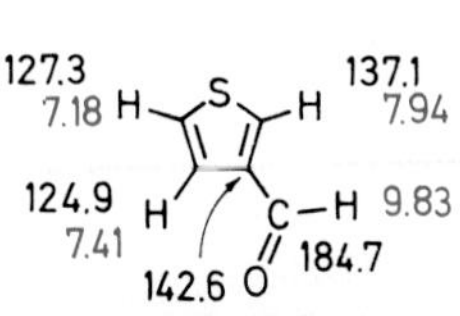

127.3   137.1
7.18 H S H 7.94
124.9 H   C—H 9.83
7.41   184.7
142.6 O

Thiophene-3-
carboxaldehyde

Ketones

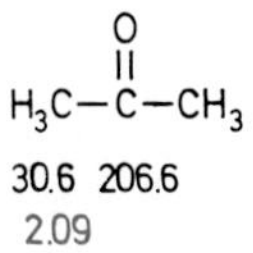

O
‖
H₃C—C—CH₃
30.6 206.6
2.09

Acetone

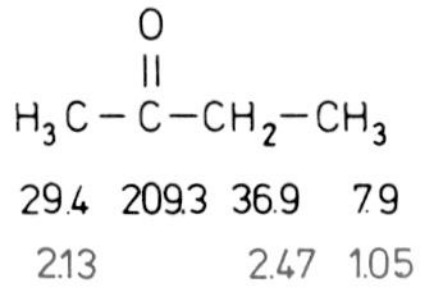

O
‖
H₃C—C—CH₂—CH₃
29.4 209.3 36.9 7.9
2.13   2.47 1.05

2-Butanone
(Methyl ethyl ketone)

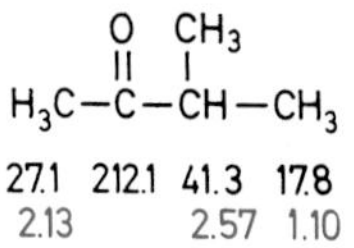

O CH₃
‖ |
H₃C—C—CH—CH₃
27.1 212.1 41.3 17.8
2.13   2.57 1.10

3 Methyl-2-
butanone

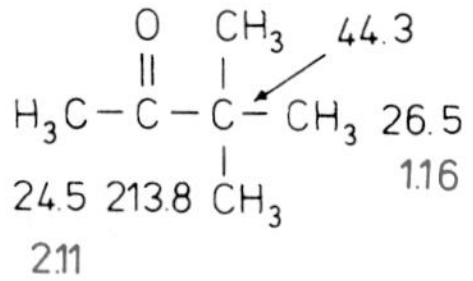

O CH₃ 44.3
‖ |
H₃C—C—C—CH₃ 26.5
24.5 213.8 CH₃   1.16
2.11

3,3-Dimethyl-2-butanone
(Pinacolone)

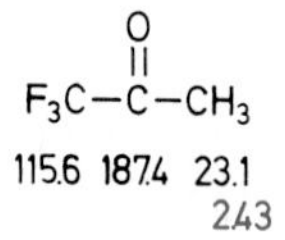

O
‖
F₃C—C—CH₃
115.6 187.4 23.1
2.43

1,1,1-trifluoroacetone

3-Hexanone

$$H_3C-CH_2-CH_2-\overset{\overset{O}{\|}}{C}-CH_2-CH_3$$

13.8   17.4   44.3   211.6   35.9   7.9
0.91   1.60   2.38   2.43   1.06

3-Hydroxy-2-butanone

$$H_3C-\overset{\overset{O}{\|}}{C}-\overset{\overset{OH}{|}}{CH}-CH_3$$

24.9   211.2   73.1   19.4
2.20   4.22   1.36

(E)-3-Penten-2-one

18.0
2.08 H₃C   H
197.0
H   CO–CH₃   26.6
     2.27
142.7 133.2
6.88 6.18

Cyclopentanone

213.6
O
H   H
22.0
2.02 H H   H
    36.7
    2.06

Cyclohexanone

208.5
23.8 H   O
1.74
H
H   H
26.5   H H   40.4
1.88   2.22

2-Cyclohexenone

150.6   129.5
7.01 H   H   6.01
H   199.5
25.3 →   O
~2.4 H
H   H
22.3   H H   37.7
2.02   ~2.4

Bicyclo[2.2.1]-heptan-2-one (Norcamphor)

37.6
49.7   1.51
2.41 H
1.76   H   H   1.69
24.2   H   217.4
1.44 H   O
1.76 H   H 1.95
27.1   45.1
1.41 H   H   H 1.73
    35.3
    2.61

2-3-Butandione (Diacetyl)

$$H_3C-\overset{\overset{O}{\|}}{C}-\overset{\overset{O}{\|}}{C}-CH_3$$

23.2   197.7
2.33

Acetylacetone (2,4-Pentandione)

$$\overset{O}{\overset{\|}{C}}-CH_2-\overset{O}{\overset{\|}{C}} \quad 201.9$$
H₃C   58.2   CH₃
   3.62
     30.2
     2.17

⇌

H···O
O   O   191.4
H₃C   C   C   CH₃
H   24.3
    2.00
100.3
5.57

2,5-Hexandione

$$H_3C-\overset{\overset{O}{\|}}{C}-CH_2-CH_2-\overset{\overset{O}{\|}}{C}-CH_3$$

29.6 206.9 37.0
2.21   2.72

1-Phenyl-2-propanone (Benzyl methyl ketone)

134.3
$$H_3C-\overset{\overset{O}{\|}}{C}-CH_2-\text{〈} \quad H \quad 126.9 \\ ~7.2$$
29.0 206.0 50.7
2.15   3.67 H   H
129.3   128.6
~7.2   ~7.2

Acetophenone

136.3
131.3 H   C   195.7
7.45
H   H   CH₃ 24.6
       2.55
128.1   128.1
7.40   7.91

Benzophenone

137.8 O
132.2 H   C 〈 〉
7.48
     195.2
H   H
128.2   130.1
7.39   7.73

Benzil

133.0 O   O
134.7 H   C–C 〈 〉
7.64   194.3
H   H
128.9   129.7
7.50   7.97

Squaric acid

HO   O
HO   O
189.5
(all C)

Quinones

p-Benzoquinone

187.0   O   H   136.4
       6.78
O

o-Benzoquinone

O
139.7 H   180.2
7.09   O
H
130.8
6.34

1,4-Naphthaquinone

131.8 O
     H
133.7 H   H 138.5
7.71   7.03
H   O 184.7
126.2
8.01

9,10-Anthraquinone

127.4   133.6
8.21 H   O 183.2
134.0 H
7.80
O

2,6-Di-tert-butyl-p-benzoquinone

O 35.5
(H₃C)₃C   187.7 C(CH₃)₃   28.3
    157.7   1.30
188.6   O   H 130.1
       6.47

Tetramethyl-p-benzoquinone (Duroquinone)

O 187.4
H₃C   CH₃ 12.4
      2.06
    140.4
H₃C   CH₃
O

Tetrachloro-p-benzoquinone (Chloranil)

O 169.4
Cl   Cl
   139.4
Cl   Cl
O

Tetrachloro-o-benzoquinone

143.8 Cl 131.9
Cl   O
     168.8
Cl   O
Cl

Carboxylic acids

$H_3C-COOH$
20.8   177.9
2.08

Acetic acid

H_3C-CH_2-COOH
8.7   27.5   180.1
1.16   2.36

Propionic acid

$H_3C-CH_2-CH_2-COOH$
13.1   18.2   36.0   179.3
1.00   1.58   2.31

Butyric acid

18.8 H_3C
1.19
     $CH-COOH$
H_3C 34.1 184.1
     2.57

2-Methylpropionic
acid (Isobutyric acid)

$HOOC-CH_2-CH_2-CH_2-CH_2-COOH$
174.2   33.9   24.0
2.20   1.50

Adipic acid
(Hexanedianoic acid)

5.72 H   CH$_3$ 17.5
126.2-C=C-136.3   1.97
6.30 H   COOH
172.3

Methacrylic acid

127.3 H—⟨⟩—CH$_2$—COOH 133.4
7.27   41.0 178.4
     H   H   3.61
   128.6 129.4
   7.27   7.27

Phenylacetic acid

133.7 H—⟨⟩—COOH 129.4
7.53   172.6
     H   H
   128.4 130.2
   7.44   8.19

Benzoic acid

$Cl-CH_2-COOH$
40.8   173.7
4.15

Chloroacetic
acid

$Cl_3C-COOH$
88.9   167.0

Trichloroacetic
acid

OH
|
$H_3C-CH-COOH$
19.9   66.0   176.8
1.56   4.40

2-Hydroxypropionic
acid (Lactic acid)

OH OH
| |
$HOOC-CH-CH-COOH$
172.7   73.8
     4.40

meso-Tartaric
acid

O
||
$H_3C-C-CH_2-CH_2-COOH$
29.8 206.9 37.7 27.8 178.5
2.20   2.76 2.62

Levulinic acid
(4-Oxohexanoic acid)

NH$_2$
|
$H_3C-CH-COOH$
18.9   52.9   178.5
1.49   3.79

Alanine (D$_2$O)

15.9
1.2
H_3C NH$_2$
| |
$H_3C-CH_2-CH-CH-COOH$
12.5   25.7   39.7   60.9   175.2
1.1   1.6/1.7   2.3   3.8

Isoleucine (D$_2$O)

OH   NH$_2$
| |
$H_3C-CH-CH-COOH$
22.3   68.7   63.2   175.6
1.33   4.27   3.58

Threonine (D$_2$O)

129.5 137.4 3.07 3.59
7.35   H   H
H—⟨⟩—C—C—COOH
      137.5 157.3
   H   H   NH$_2$
130.7 131.1 2.86
7.35   7.35

Phenylalanine (D$_2$O)

118.5
135.8 H 7.63
8.89 N   NH$_2$
H—⟨ ⟩
  N   CH$_2$-CH-COOH
1 130.0 27.6 55.1 174.0
H   3.61 4.56

Histidine (D$_2$O)

$HOOC-CH_2-COOH$
170.4   41.4
     3.23

Malonic acid

CH$_2$-COOH
|
HO-C-COOH
74.2 | 177.5
   CH$_2$-COOH
44.1   174.2
2.73

Citric acid (D$_2$O)

6.75 H   134.2 COOH
   C=C
HOOC   H
166.6

Fumaric acid

H   130.5   H 6.30
   C=C
HOOC   COOH
166.1

Maleic acid

$H-C≡C-COOH$
77.6   73.9   157.3
     3.10

Propiolic acid

Carboxylic acid Esters, Amides, Anhydrides, Chlorides, Lactones, Lactams

O
||
$H-C-OCH_2-CH_3$
161.0   60.0   14.2
8.04   4.22   1.30

Ethyl formate

O
||
$H_3C-C-OCH_3$
20.6 171.4 51.5
2.01   3.67

Methyl acetate

O
||
$H_3C-C-OCH_2-CH_3$
21.0 170.6 60.3 14.2
2.03   4.12 1.25

Ethyl acetate

4.55 H   H 7.25
   C=C   141.8
96.8
4.85 H   O-C-CH$_3$
       167.6 || 20.2
          O   2.12

Vinyl acetate

127.2 H—⟨⟩—O-C-CH$_3$ 151.9 168.5
7.18   211.1
   H   H   O   2.28
   130.9 123.3
   7.30   7.05

Phenyl acetate

O   CH$_3$
|| |
$H_3C-C-O-C-CH_3$
22.4 170.3 /| 28.1
1.96   80.0 CH$_3$ 1.44

tert-Butyl acetate

38.7 H$_3$C O
     | ||
$H_3C-C-C-O-CH_3$
27.2 | 179.0   51.7
1.20   CH$_3$ 3.67

Methyl trimethyl-
acetate

O
||
$H_3C-C-S-CH_2-CH_3$
30.6 195.8   23.5 14.7
2.32   2.88 1.25

S-Ethyl
thioacetate

S
||
$H_3C-C-S-CH_2-CH_3$
39.2 233.0   31.3 12.2
2.81   3.20 1.31

Ethyl dithioacetate

$$Cl-\overset{\overset{\displaystyle O}{\|}}{C}-OCH_2-CH_3$$
149.9   68.1   13.3
4.36   1.39

Ethyl chloroformate

$$Cl-CH_2-\overset{\overset{\displaystyle O}{\|}}{C}-OCH_3$$
40.7 167.8 53.0
4.10   3.83

Methyl
chloroacetate

$$Br-CH_2-\overset{\overset{\displaystyle O}{\|}}{C}-OCH_3$$
25.5 167.6 53.1
3.87   3.80

Methyl
bromoacetate

$$N_2CH-\overset{\overset{\displaystyle O}{\|}}{C}-OCH_2-CH_3$$
46.2 167.1 61.0   14.6
4.74   4.23   1.28

Ethyl
diazoacetate

$$H_3C-CH_2-CH_2-\overset{\overset{\displaystyle O}{\|}}{C}-O-CH_2-CH_2-CH_3$$
13.7 18.5 36.3 173.8   65.8   22.1 10.4
0.95 1.66 2.29   4.03   1.66 0.95

Propyl butyrate

$$H_3C-CH_2-O-\overset{\overset{\displaystyle O}{\|}}{C}-CH_2-COOC_2H_5$$
14.1 61.4   166.5 41.7
1.38 4.20   3.36

Diethyl malonate

$$H_3C-CH_2-O-\overset{\overset{\displaystyle O}{\|}}{C}-CH_2-CH_2-CH_2-CH_2-CO_2C_2H_5$$
14.3 60.3   173.2 34.0 24.4
1.26 4.12   2.32   1.67

Diethyl adipate

2-Hydroxyethyl acrylate
6.44
$$\underset{\underset{H}{5.87}}{\overset{131.3}{}}C=C\underset{\underset{H}{6.16}}{\overset{128.0}{H}}-\overset{\overset{\displaystyle O}{\| 166.5}}{C}-O-CH_2-CH_2-OH$$
66.1 60.8
4.29 3.86

Methyl (E)-crotonate
17.1
1.88 H₃C   H   122.3
                    5.82
C=C
H 144.1 C-OCH₃ 50.3
6.99 O   3.73
166.0

Diethyl fumarate
H
H₃C-CH₂O 164.9 C-CO₂C₂H₅
14.2 61.3   C=C 133.8
1.33 4.27 O   H 6.87

Dimethyl acetylenedicarboxylate
$$H_3CO-\overset{\overset{\displaystyle O}{\|}}{C}-C\equiv C-COOCH_3$$
53.6 152.3 74.6
3.88

Methyl benzoate
130.3
166.8
132.8 H———C-OCH₃
7.47   O   51.8
H   H   3.88
128.3 129.5
7.37   7.97

Methyl acetoacetate (Keto form, (Z)- and (E)-Enol)
$$H_3C-\overset{\overset{\displaystyle O}{\|}}{C}-CH_2-\overset{\overset{\displaystyle O}{\|}}{C}-OCH_3$$
29.7 200.3 49.7 167.9 51.9
1.93   3.16   3.46

⇌

O·H·O
H₃C   C   C   C   OCH₃
176.3 173.2
20.9   H   55.3
1.69 89.7 3.49
4.91

18.8 CH₃ O
2.28   C   C
HO 173.5 168.1 OCH₃
H   50.5
91.0   3.26
4.99

Methyl furan-2-carboxylate
111.9   117.9
6.46 H   H   7.14
146.4 H   159.0
7.53   O   C-OCH₃
144.8 O   51.7
              3.90

Acetamide
$$H_3C-\overset{\overset{\displaystyle O}{\|}}{C}-NH_2$$
22.3 172.7
1.92

N-Methylacetamide (syn)
$$H_3C-NH-\overset{\overset{\displaystyle O}{\|}}{C}-CH_3$$
25.9   170.9 22.4
2.77   2.0

N,N-Dimethylacetamide
35.0
CH₃ 2.93
$$H_3C-\overset{\overset{\displaystyle O}{\|}}{C}-N$$
21.5 170.5 CH₃ 38.0
2.09   3.03

Acetanilide
$$H_3C-\overset{\overset{\displaystyle O}{\|}}{C}-NH-$$
24.1 169.5   124.1
2.1   138.2 H   H   7.0
120.4 128.7
7.4   7.2

Isobutyramide
H₃C   O
CH-C-NH₂
19.5 H₃C 34.0 179.1
1.06   2.39

151.5 120.8
H₂N   C—NH₂
              168.0
112.3 H   H 129.0
6.54   7.62

4-Aminobenzamide

30.3
2.73
183.6

Succinimide

42.6
3.23
30.6/29.7 H
~1.7
O 179.4
36.8
2.46
23.2
~1.7

ε-Caprolactam

134.2
7.85
132.6
169.1
122.8
7.85

Phthalimide

H₃C—C—O—C—CH₃
20.7 166.0
2.2

Acetic anhydride

172.9
28.7
2.92

Succinic anhydride

22.2   27.8
2.28   2.50
68.6 H
4.36 H
177.8

γ-Butyrolactone

136.6
7.1 H
164.2

Maleic anhydride

131.1
163.1
136.0 H
8.04
125.2
8.10

Phthalic anhydride

H₃C—C—Cl
33.6 170.5
2.66

Acetyl chloride

H₃C
CH—C—Cl
18.9 46.3 178.1
1.30 H₃C 2.98

Isobutyryl chloride

Cl—C—C—Cl
159.4

Oxalyl chloride

Cl—C—CH₂—CH₂—CH₂—CH₂—C—Cl
173.2 46.5 23.8
      2.95 1.80

1,6-Hexandianoic acid dichloride

133.4
~7.8
134.3
167.3
C—Cl
130.1
~7.9

Phthalyl dichloride

Imines, Oximes, Hydrazones, Carbodiimides, Azo Compounds

29.1
1.98 H₃C
16.0
1.80 H₃C
168.0
C=N
CH₃
38.6
~3.0

N-Isopropylidene
methylamine

8.42
H 160.2
136.1
C=N
152.0 H
131.3
7.46 128.7
7.46
125.8
7.21
128.7
7.37
120.9
7.19
129.0
7.89

N-Benzylidene aniline

N—OH
H₃C 155.4 CH₃
21.7   15.0
1.90   1.90

Acetone oxime

32.3   N—OH 159.4
2.52 H
26.3
~1.65
24.6
~1.65
27.5
2.23
26.1
~1.65

Cyclohexanone oxime

NOH
H₃C—C—C—CH₃
9.2 153.1 NOH
2.02

2,3-Butanone dioxime

15.2 150.7
1.78
H₃C—C—CH₂—CH₂—CH₃
40.7 19.5 13.6
2.15 1.49 0.87
N
HN
CONH₂   E
158.4

151.2 31.9
2.13
H₃C—C—CH₂—CH₂—CH₃
23.2   18.6 14.0
1.87   1.50 0.87
N
NH
CONH₂   Z
158.2

2-Pentanone semicarbazone

NH₂
N
128.5 H
~5.75
143.9
N
H   H NH₂
127.3 35.6
~5.75 ~3.0

Cycloocta-4,6-diene-1,2-dione-(*E,E*)-dihydrazone

24.8 H H 35.0
25.5 H
H
139.9
N=C=N
55.8
1.0–2.1   3.25

Dicyclohexylcarbodiimide

152.5
130.7
7.46 H
N
N
128.8 122.7
7.46 7.93

Azobenzene

143.0   154.3   156.1   146.6
H₂N   N=N   NO₂
113.4 125.2 122.0 124.7
6.72   7.77   7.90   8.34

4-Amino-4'-nitroazobenzene

Nitriles, Isocyanides, (Isonitriles)

H₃C—CN
1.8 116.5
1.98

Acetonitrile

H₃C—CH₂—CN
10.6 10.8 120.8
1.31 2.35

Propionitrile

NC—CH₂—CH₂CN
118.0 14.6
2.90

Succinodinitrile

                C₂H₅
H₃C—CH₂—N—CN
12.9 45.9   117.3
1.28 3.05

Diethylcyanamide

H₃C—NC
26.8 158.2
2.85

Methyl isocyanide
(Methyl isonitrile)

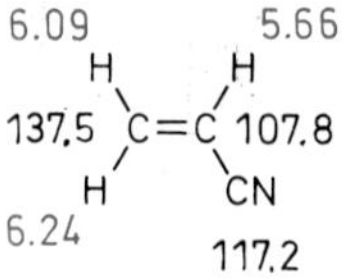

6.09   5.66
    H   H
137.5 C=C 107.8
    H   CN
6.24
        117.2

Acrylonitrile

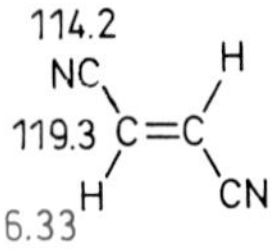

114.2
NC   H
119.3 C=C
        H   CN
6.33

Fumaric
dinitrile

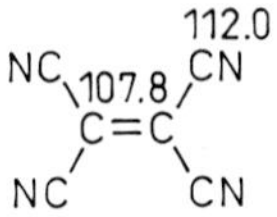

                112.0
NC   107.8   CN
    \C=C/
NC   CN

Tetracyano-
ethylene

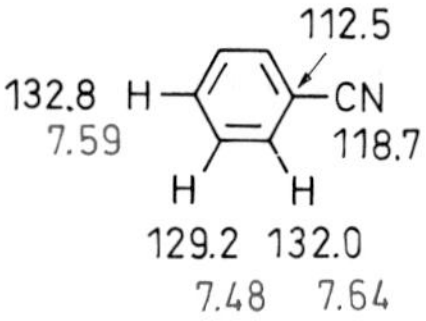

                112.5
132.8 H—⬡—CN
7.59   118.7
    H   H
129.2 132.0
7.48   7.64

Benzonitrile

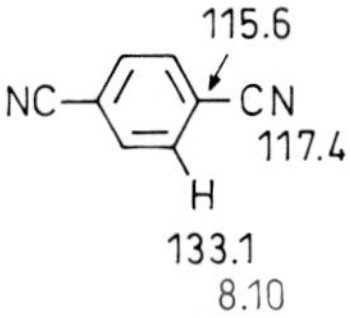

                115.6
NC—⬡—CN
            117.4
        H
        133.1
        8.10

1,4-Benzodinitrile

Organometallic Compounds, Complexes

H₃C—CH₂—CH₂—CH₂—Li
13.9 31.4 31.9 11.8
1.1   1.5   1.5   -0.8

Butyllithium

H₃C—Sn(CH₃)₃
-9.3
0.07

Tetra-
methyltin

H₃C—CH₂—CH₂—CH₂—Pb(C₄H₉)₃
13.7 27.7 31.5 18.3
0.9   1.0–2.0

Tetrabutyllead

H₃C—Hg—CH₃
21.4
0.29

Dimethyl-
mercury

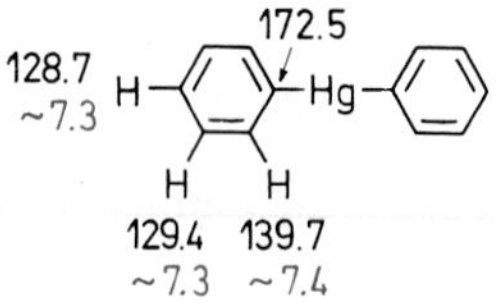

                172.5
128.7 H—⬡—Hg—⬡
~7.3
    H   H
129.4 139.7
~7.3   ~7.4

Diphenyl-
mercury

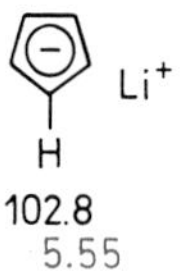

⬠ Li⁺
H
102.8
5.55

Lithium
cyclopentadienide

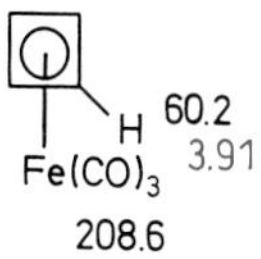

⬜   60.2
Fe(CO)₃   3.91
208.6

Cyclobutadiene
tricarbonyl iron

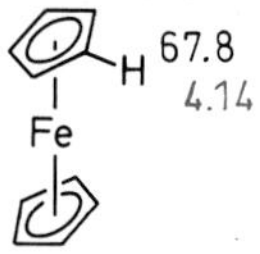

⬠   67.8
Fe   4.14
⬠

Ferrocene

            Cr(CO)₃ 232.7
            H 1.08
            24.2
            H 2.17
100.2 H   H 58.2
5.30   H   2.67
    102.3
    4.18

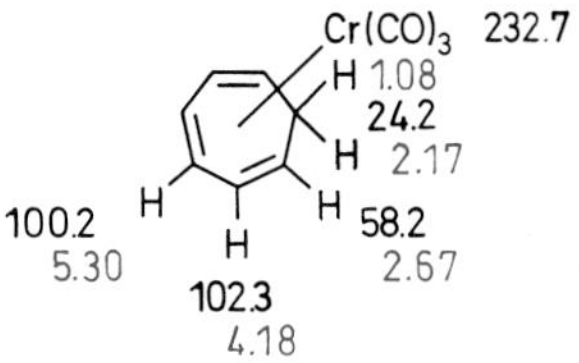

Cycloheptatriene
tricarbonyl chromium

Cl   Cl
  Pd   H 128.2
        5.50
    H 27.4
    H 2.30

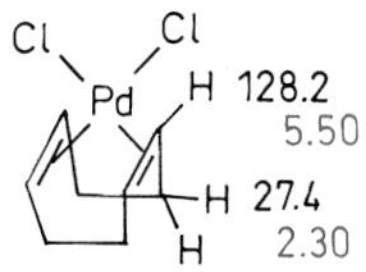

1,5-Cyclooctadiene
palladium(II)chloride

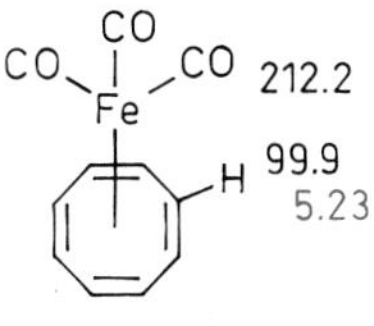

        CO
CO   CO 212.2
    Fe
    H 99.9
        5.23

Cyclooctatetraene
tricarbonyl iron

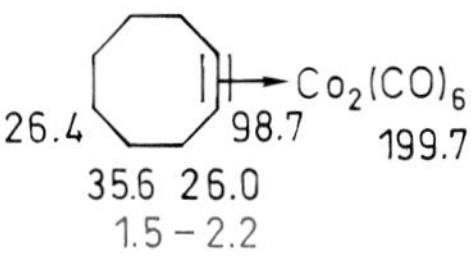

            Co₂(CO)₆
26.4   98.7   199.7
35.6 26.0
1.5 – 2.2

µ-Cyclooctyne
hexacarbonyl dicobalt

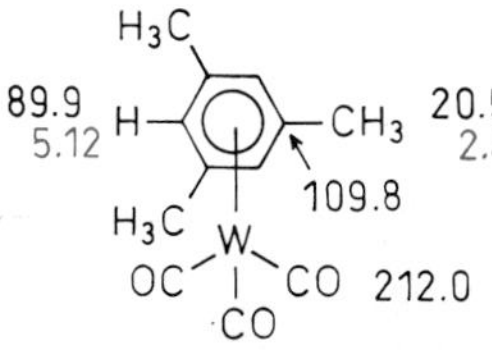

H₃C
89.9 H—⬡—CH₃ 20.5
5.12   2.44
        109.8
H₃C W
    OC CO 212.0
        CO

Mesitylene tricarbonyl tungsten

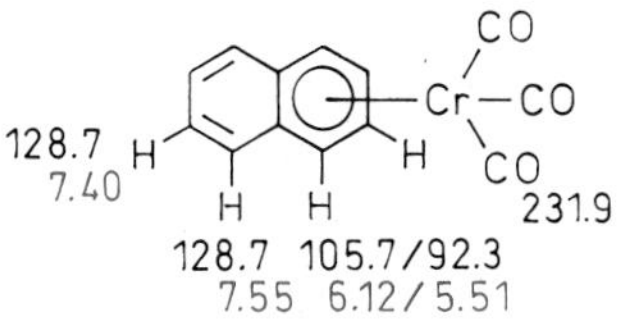

            CO
        ⬡—Cr—CO
128.7 H   CO
7.40   231.9
    H   H
128.7 105.7/92.3
7.55   6.12/5.51

Naphthalene tricarbonyl chromium

6. NMR Spectroscopy of Other Nuclei

6.1 ^{19}F NMR Spectroscopy

^{19}F, the only naturally occurring isotope of fluorine, has a **nuclear spin quantum number** of 1/2. The **relative sensitivity** is somewhat less than that of ^{1}H, (see Tab. 3.1, p. 72), the **shift range** however considerably wider. For carbon–fluorine compounds it extends over 300 to 400 ppm; if inorganic fluorine compounds are included, it is another factor of ten greater. The **shielding** is dominated by the σ_{para} term. The most common **reference compound** for establishing the zero-point of the δ-scale is trichlorofluoromethane (CFCl$_3$). Signals at higher field have a negative, signals at lower field a positive δ-value.

As a result of ^{19}F,^{1}H and/or ^{19}F,^{19}F coupling the ^{19}F NMR spectra of organofluorine compounds often show many lines. Since the quotients $\Delta v/J$ however are usually large, the spectra are mostly first order. Typical ^{19}F **chemical shifts** and ^{19}F,^{19}F **couplings** are summarised in Tables 3.42 and 3.43. (The ^{19}F,^{1}H and ^{19}F,^{13}C couplings have already been considered in Sections 3.4 and 4.4, see p. 112 and 152.)

The spectrum of 1,1,2-trichloro-1,2-difluoro-2-iodoethane (**212**) will be discussed as an example of a ^{19}F spectrum (Fig. 3.85).

At room temperature an AB spectrum is obtained, with the lines of the B part being particularly broad. These signals arise from averaging of the signals of the three rotamers **212 a, b, c**.

At -90°C the rotation around the C–C bond is frozen out, and three separate AB systems are seen. The intensities correspond to the **population** of the **rotamers**.

		Temperature	Chemical shifts (δ-values referred to CFCl$_3$)		^{3}J Coupling constants
		°C	F$_A$	F$_B$	(Hz)
212	CFCl$_2$CFClI	$+28$	-65.21	-63.20	-22.3
a		-90	-64.4	-59.6	-19.5
b		-90	-68.8	-63.5	-27.1
c		-90	-75.4	-67.6	-22.0

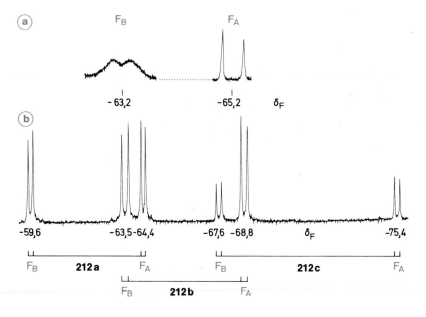

212

Fig. 3.85 ^{19}F NMR spectrum of CFCl$_2$–CFCl I (**212**) (94.1 MHz)
a at room temperature
b at -90°C (Cavalli, L. (1972), J. Magn. Res **6**, 298)

Tab. 3.42 ^{19}F shifts of selected organofluorine compounds (δ-values referred to $CFCl_3$)*

Group	Compound	δ	Group	Compound	δ
$-\overset{\mid}{\underset{\mid}{C}}-F$	$H-CH_2-F$	− 268	$=C-F$	$H_2C=CHF$	− 114
	CH_3-CH_2-F	− 212		(cis/geminal H_3C, H, H, F)	− 132
	$C_3H_7-CH_2-F$	− 219		(H_3C, H / H, F)	− 130
	$(CH_3)_2CH-F$	− 165		(H, CH_3 / H, F)	− 89
	$(CH_3)_3C-F$	− 131		(H_5C_6, C_6H_5 / F, F)	− 133
	CH_2Cl-CH_2-F	− 220		(H_5C_6, F / F, C_6H_5)	− 158
	CCl_3-CH_2-F	− 198		(H, H / F, F)	− 165
	$CH_3-CHCl-F$	− 123		(F, H / H, F)	− 186
	CH_3-CCl_2-F	− 46		(H, F / H, F)	81
	$CFCl_2-CCl_2-F$	− 68		(F_3C, F / F_3C, F)	− 66
	(cyclopropyl)–F	− 218		(F, F / F, F)	− 135
	(cyclohexyl)–F	− 174	$\equiv C-F$	$HC\equiv C-F$	− 273
$-CF_2-$	$H-CF_2-H$	− 144	$-\underset{\mathrm{O}}{\overset{\parallel}{C}}-F$	$H-CO-F$	+ 41
	CH_3-CF_2-H	− 110		CH_3-CO-F	+ 49
	$CH_3-CF_2-CH_3$	− 85		C_6H_5-CO-F	
	$C_6H_5-CF_2-C_6H_5$	− 89	**Aryl-F**		
	$H-CF_2-CN$	− 120		C_6H_5-F	− 113
	$H-CF_2-OCH_3$	− 88		$H_2N-C_6H_4-F$	− 129
	CH_3-CF_2-Cl	− 47			
	$Br-CF_2-Br$	+ 7			
	(cyclohexane CF_2)	− 96			
	(norbornane F)	− 87 / − 110			
	(perfluorocyclohexane)	− 133			
CF_3	$H-CF_3$	− 79			
	CH_3-CF_3	− 64			
	$C_6H_5-CF_3$	− 64			
	CF_3CO-CF_3	− 85			
	$HOOC-CF_3$	− 79			
	$HO-CF_3$	− 55			
	H_2N-CF_3	− 49			
	$Cl-CF_3$	− 29			
	$F-CF_3$	− 67			
	CF_3-CF_3	− 89			
	$F_3C-C\equiv C-CF_3$	− 57			

And for the right side under $=C-F$ there is also $=C\begin{smallmatrix}F\\F\end{smallmatrix}$ group heading.

* Solvent effects can cause large deviations of the δ-values; therefore only integer values of the δ-values are given.

Tab. 3.42 continued

Group	Compound	δ
	O_2N–⟨ ⟩–F	– 102
	F–⟨ ⟩–F	– 119
	(F, F naphthalene-type ring)	– 110
		– 139
	(perfluoro ring)	– 162
	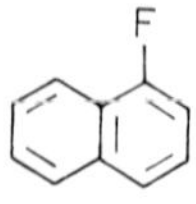	– 124
	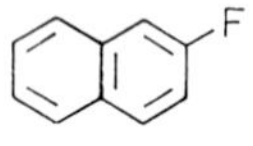	– 115
Heteroaryl-F		– 196
		– 137
		– 156
		– 165
	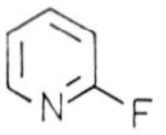	– 61
	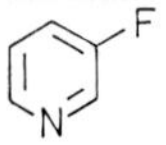	– 132
	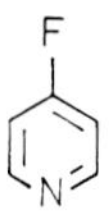	– 106
		– 134
		– 162
		– 88

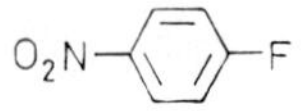

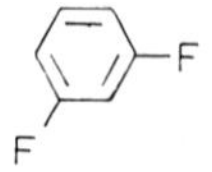

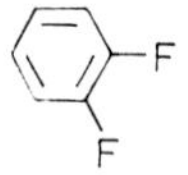

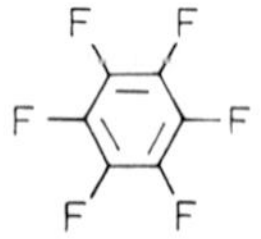

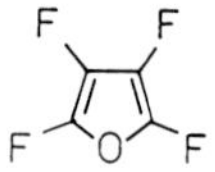

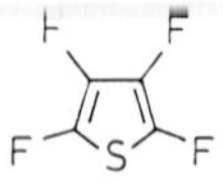

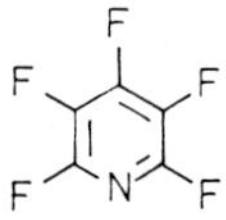

* Solvent effects can cause large deviations of the δ-values;
   therefore only integer values of the δ-values are given.

6.2 ^{31}P NMR Spectroscopy

^{31}P with the **nuclear spin quantum number** 1/2 is the only naturally occurring isotope of phosphorus. Its **relative sensitivity** is only 6.6% that of ^1H. The **shift range** is ca. 1000 ppm wide. This does however include extreme shift values, for example for P_4 and diphosphenes, which have shifts of −488 and up to 600 ppm respectively.

85% phosphoric acid is the most usual (external) standard for referencing the δ-scale. Signals at higher field have a negative, signals at lower field a positive δ-value.

Despite the wide total range of ^{31}P shifts many classes of compounds have relatively narrow shift ranges:

Primary phosphines	PH_2R	$-170 < \delta < -110$
Secondary phosphines	PHR_2	$-100 < \delta < -10$
Tertiary phosphines	PR_3	$-70 < \delta < +70$
Phosphonium salts	$PR_4^\oplus$	$-20 < \delta < +40$
Phosphates	$OP(OR)_3$	$-20 < \delta < 0$
Phosphonates	$RP(OR)_2$ $\overset{\|}{O}$	$-30 < \delta < +60$
Phosphinates	$R_2P(OR)$ $\overset{\|}{O}$	$0 < \delta < +70$
Phosphine oxides	OPR_3	$+10 < \delta < +70$

Some typical ^{31}P shifts are summarised in Tab. 3.44.

In the spectra of organic phosphorus compounds spin-spin coupling is most commonly seen to ^1H nuclei. (For sizes of $J(P,H)$ and $J(P,C)$ couplings see Sections 3.4 and 4.4, p. 112 and 152.) Some $^{31}P,^{31}P$ coupling constants are given in Tab. 3.45.

As an exercise a coupled spectrum will now be discussed. **213** forms an ABX system.

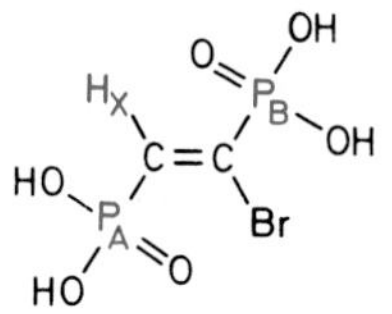

213

Tab. 3.43 ^{19}F, ^{19}F couplings of selected compounds (in Hz):

a) through bonds

Compound	$|^2J(F,F)|$	$|^3J(F,F)|$	Compound	$|^3J(F,F)|$	$|^4J(F,F)|$	$|^5J(F,F)|$
CBr—CHClBr	154	–		21	–	–
(cyclohexane difluoride)	244	–		–	7	–
(cycloheptane difluoride)	297	–	F—⟨⟩—F	–	–	18
CF_3—CFH_2	–	15				
CHF_2—CHF—CHF_2	–	13		21	0	13
CHF_2—CF_2—CHF_2	–	4				
C=C (F,F / H,H)	–	19		20	3	4
C=C (F,H / H,F)	–	133				
C=C (F,F / H,H)	33	–	CF_3 (perfluoro)		23	
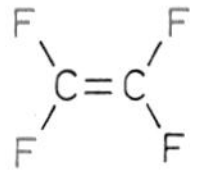 C=C (F,F / F,F)	124	73 (cis) 111 (trans)				

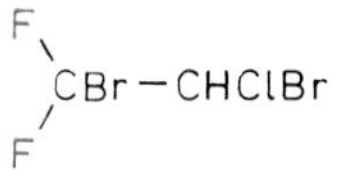

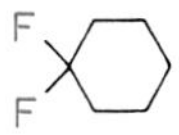

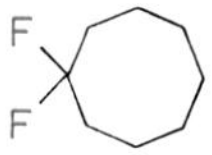

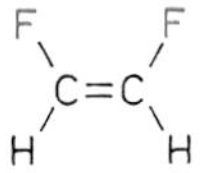

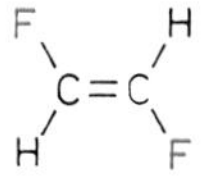

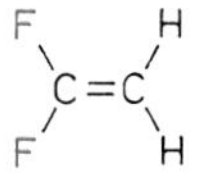

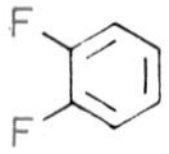

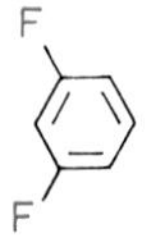

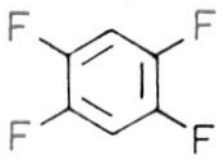

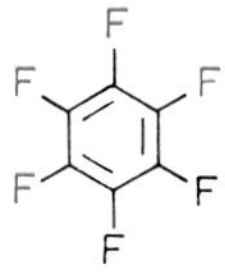

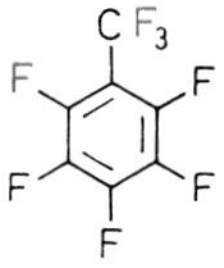

b) through space

Compound	$|J(F,F)|$	Compound	$|J(F,F)|$
	42		99
	59		170

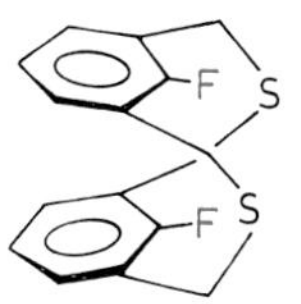

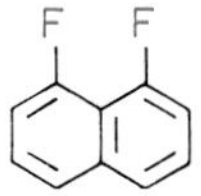

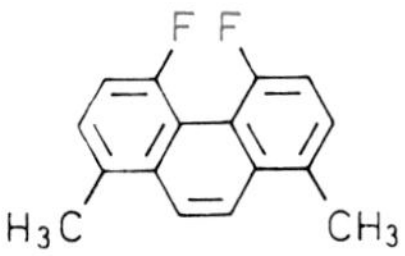

Tab. 3.44 ^{31}P shifts of selected of selected organophosphorus compounds (δ-values referred to 85% H_3PO_4 as external standard)*

Group	Compound	δ	Group	Compound	δ
 $\mid P-$	$PH_2(CH_3)$	− 164	$\mid P\equiv$	$P\equiv C-C(CH_3)_3$	− 69
	$PH_2(C_2H_5)$	− 127			
	$PH_2(C_6H_5)$	− 122	$-\overset{\mid}{\underset{\mid}{P}}{}^{\oplus}$	$PH_3(CH_3)^+ Cl^-$	− 62
	$PH(CH_3)_2$	− 99			
	$PH(C_2H_5)_2$	− 55		$PH(CH_3)_3^+ Cl^-$	− 3
	$PH(C_6H_5)_2$	− 41		$P(CH_3)_4^+ I^-$	+ 25
				$P(C_6H_5)_4^+ I^-$	+ 23
	$P(CH_3)_3$	− 62			
	$P(C_2H_5)_3$	− 20	$\overset{\mid}{\underset{\diagdown}{>}}P\diagdown$	$(C_6H_5)_3P(OC_2H_5)_2$	− 55
	$P(C_6H_5)_3$	− 6		$P(OCH_3)_5$	− 67
	$P(CH_2-C_6H_5)_3$	+ 23		$P(OC_6H_5)_5$	− 85
	$P[C(CH_3)_3]_3$	+ 62			
	$P(C_6H_5)F_2$	+ 208	$-P=$	$(C_6H_5)_3P=CH_2$	+ 20
	$P(C_2H_5)Cl_2$	+ 196		$(C_6H_5)_3P=C(C_2H_5)_2$	− 11
	$P(C_6H_5)Cl_2$	+ 161			
	$P(C_2H_5)_2Cl$	+ 81		$(CH_3)_3P=O$	+ 36
	$C_6H_5P(OCH_3)_2$	+ 159		$(C_2H_5)_3P=O$	+ 48
	$(C_6H_5)_2P(OCH_3)$	+ 116		$(C_6H_5)_3P=O$	+ 27
	$P(OCH_3)Cl_2$	+ 181		$(C_6H_5)_2PO(OCH_3)$	+ 32
	$P(OCH_3)_3$	+ 141		$C_6H_5PO(OCH_3)_2$	+ 20
	$P(OC_6H_5)_3$	+ 127		$(C_2H_5)_2POCl$	+ 76
	$P[N(CH_3)_2]_3$	+ 123		$C_6H_5POCl_2$	+ 24
				$PO(OC_2H_5)Cl_2$	+ 3
	$(C_6H_5)_2P-CH=CH_2$	− 12		$PO(OC_2H_5)_3$	− 1
	$Cl_2P-CH=CH_2$	+ 159		$PO(OC_6H_5)_3$	− 18
	$P(C\equiv C-C_6H_5)_3$	− 91		$PO[N(CH_3)_2]_3$	+ 27
				$PS(C_6H_5)Cl_2$	+ 80

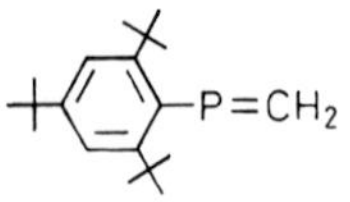

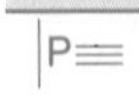

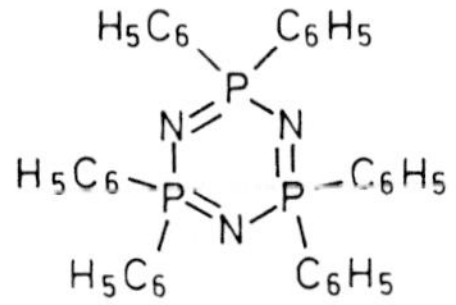

Group	Compound	δ
$\overline{\diagup}P=$		+ 290
	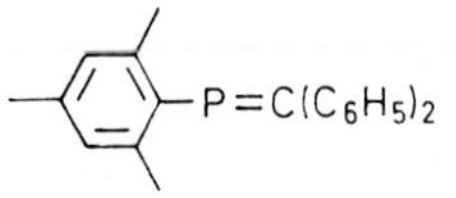	+ 233
	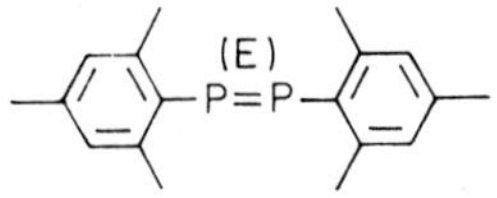	+ 492
		+ 211
	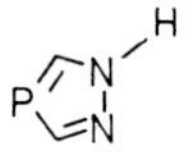	+ 83

+ 9

$-\overset{\mid}{\underset{\diagup}{P}}{}^{\ominus}$

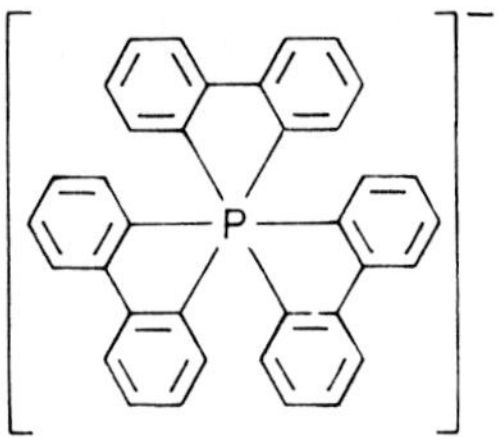

− 181

^{a} Because of the strong solvent dependency only whole-number values are given

Tab. 3.45 $^{31}P,^{31}P$ coupling constants of selected organophosphorus compounds (in Hz)

Compound	$|^1J(P,P)|$	$|^2J(P,P)|$
$[(CH_3)_3C]_2 P—P[C(CH_3)_3]_2$	451	–
$(H_3C)_2 P—P(CH_3)_2$	180	–
$(C_2H_5O)_2 P—P(OC_2H_5)_2$ $\quad\quad\quad \parallel_S \quad \parallel_O$	583	–
$(H_3C)_2 P—P(CH_3)_2$ $\quad\quad \parallel_S \quad \parallel_S$	19	–
$(C_2H_5O)_2 P—CH_2—P(OC_2H_5)_2$ $\quad\quad\quad \parallel_O \quad\quad \parallel_O$	–	< 1
$(C_6H_5)_3 P=CH—P(C_6H_5)_2$	–	150

Of the 15 theoretical transitions one (a combination line) has intensity 0. A maximum of 14 lines is therefore expected. In the ^{31}P spectrum the AB part has 8 lines, in the ^1H spectrum the X part has 4 lines (Fig. 3.86). Two further lines in the X part are so weak that they are lost in the noise. The spin of X (ignoring the spins of A and B) is equally distributed between the two possible orientations. The AB part can therefore be divided into two equally intense sub-spectra and, using the rules for AB spectra, the effective Larmor frequencies v_a^* and v_b^* determined (see p. 79 ff.). In the X part the two intense lines 9 and 12 are separated by $J_A X + J_{BX}$. The separation of the centres of the two sub-spectra is half this amount. The parameters of the ABX system can be obtained from the following formulae:

$$2v_A = v_{1a}^* + v_{2a}^* \quad \text{bzw.} \quad 2v_A = v_{1a}^* + v_{2b}^*$$
$$2v_B = v_{1b}^* + v_{2b}^* \quad\quad\quad 2v_B = v_{1b}^* + v_{2a}^*$$
$$\pm J_{AX} = 2(v_a^* - v_A)$$
$$\pm J_{BX} = 2(v_b^* - v_B)$$

One of the two solutions must be eliminated either from a consideration of the line intensities of from criteria of plausibility of the size of the coupling constants.

The analysis of the spectrum of **213** gives the parameters in Fig. 3.86.

As a final comprehensive example of NMR phosphorin (phosphabenzene, **214**) will be discussed:

214

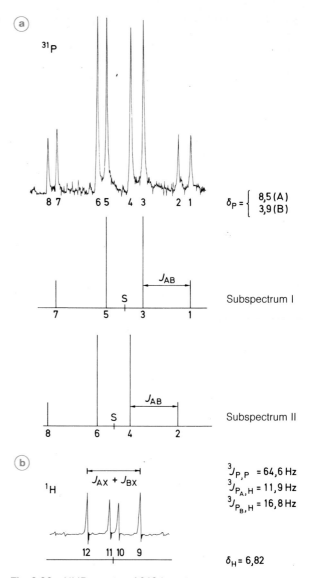

(a) ^{31}P

$\delta_P = \begin{cases} 8{,}5\,(A) \\ 3{,}9\,(B) \end{cases}$

J_{AB} Subspectrum I

J_{AB} Subspectrum II

(b) ^1H

$J_{AX} + J_{BX}$

$^3J_{P,P} = 64{,}6$ Hz
$^3J_{P_A,H} = 11{,}9$ Hz
$^3J_{P_B,H} = 16{,}8$ Hz

$\delta_H = 6{,}82$

Fig. 3.86 NMR spectra of **213** in water
a ^{31}P spectrum
b ^1H spectrum
(Maier, L. (1973), Phosphorus **2**, 229)

Symmetry considerations show that the total NMR of the compound (^1H, ^{13}C, and ^{31}P) depends on 7 chemical shifts and 31 coupling constants. In the **^1H NMR spectrum** the AA'BB'C part of an AA'BB'CX system is observed. The protons show six different couplings between themselves; in addition there are three couplings to phosphorus. Couplings to ^{13}C have no observable effect because of the low natural abundance of ^{13}C.

Tab. 3.46 Chemical shifts for phosphorin (phosphabenzene, **214**)

reffered to TMS		δ-values referred to TMS		referred to H₃PO₄	
H–2	8.61	C–2	154.1	P–1	+211
H–3	7.72	C–3	133.6		
H–4	7.38	C–4	128.8		
H–5	7.72	C–5	133.6		
H–6	8.61	C–6	154.1		

Tab. 3.47 Spin-spin couplings in phosphorin (phosphabenzene, **214**) Values of coupling constants in Hz

J	H–2	H–3	H–4	H–5	H–6	C–2	C–3	C–4	C–5	C–6	P
H–2	—	10.1	1.2	1.2	1.9	157					38
H–3		—	9.1	1.8	1.2		156				8
H–4			—	9.1	1.2			161			3.5
H–5				—	10.1				156		8
H–6					—					157	38
C–2						—					53
C–3							—				14
C–4								—			22
C–5									—		14
C–6										—	53
P											—

In the ³¹P spectrum the X part of the AA'BB'CX system is correspondingly observed. In the coupled ¹³C spectrum the ¹³C,¹H and the ¹³C,³¹P couplings appear. Of the 13 different ¹³C,¹H couplings the direct ^1J(C,H) couplings are easily distinguished.

The determination of the three ¹³C,³¹P couplings is facilitated by proton broadband decoupling.

The chemical shifts of **214** are summarised in Tab. 3.46 and the coupling constants in Tab. 3.47. The ³¹P spectrum is reproduced in Fig. 3.87.

6.3 ¹⁵N NMR Spectroscopy

The isotope ¹⁴N, with a natural abundance of 99.6% has a nuclear spin $I = 1$ and gives broad signals which are of little use for structural determinations. The ¹⁵N nucleus, with $I = 1/2$ is therefore preferred. However, the low natural abundance of about 0.4% and the extremely low relative sensitivity (cf. Tab. 3.1) make measurements so difficult that ¹⁵N NMR spectroscopy was slow to become an accepted analytical tool. A further peculiarity is the negative magnetogyric ratio; in proton decoupled spectra the nuclear Overhauser effect can strongly reduce the signal intensity. DEPT and INEPT pulse techniques (cf. p. 174) are therefore particularly important for ¹⁵N NMR spectroscopy.

The range of ¹⁵N chemical shifts is about 600 ppm wide. If extreme values for metal complexes are included, it extends to over 1400 ppm. Nitromethane is the recommended reference substance, which can be added in a sealed-off capillary. Values are also frequently quoted with respect to a saturated aqueous solution of ammonium chloride or ammonium nitrate. The following ¹⁵N shift values can be used to convert the δ-values:

$$CH_3NO_2 \quad \delta = 0.0$$
$$NH_4Cl \quad \delta = -352.9$$
$$NH_4NO_3 \quad \delta(NH_4^+) = -359.5$$
$$\delta(NO_3^-) = -3.9$$

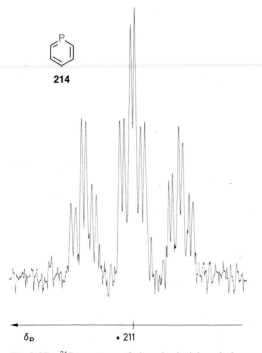

214

Fig. 3.87 ³¹P spectrum of phosphorin (phosphabenzene, **214**) (Ashe, A. J., Sharp, R. R., Tolan, J. W. (1976), J. Am. Chem. Soc. **98**, 5451)

Tab. 3.48　^{15}N chemical shifts of examples of the more important classes of compounds (δ-values referred to CH_3NO_2; for chemically non-equivalent ^{15}N nuclei the δ-values are given in the order of the numbers indicating the atoms)

Class of compound	Example		Solvent	δ
Amines —C—N	Ethylamine	C_2H_5—NH_2	Methanol	− 355.4
	Isopropylamine	$(H_3C)_2CH$—NH_2	Methanol	− 338.1
	tert-Butylamine	$(CH_3)_3C$—NH_2	Methanol	− 324.3
	Diethylamine	$(C_2H_5)_2NH$	Methanol	− 333.7
	Di-tert-butylamine	$[(CH_3)_3C]_2NH$	neat liquid	− 292.8
	Triethylamine	$(C_2H_5)_3N$	Methanol	− 332.0
	Dimethyl-1-propenylamine	H_3C—$CH=CH$—$N(CH_3)_2$	neat liquid	− 349.3
	3-Dimethylaminoacrolein	CHO—$CH=CH$—$N(CH_3)_2$	neat liquid	− 287.5
	Aniline	⟨⟩—NH_2	DMSO	− 320.3
	N-Methylaniline	⟨⟩—NH—CH_3	neat liquid	− 324.0
	N,N-Dimethylaniline	⟨⟩—$N(CH_3)_2$	neat liquid	− 332.2
	Diphenylamine	$(C_6H_5)_2NH$	DMSO	− 288.8
Amides —C—N ‖ O	Formamide	HCO—NH_2	neat liquid	− 267.6
	N-Methylformamide	HCO—NH—CH_3	neat liquid	− 270.1
	N,N-Dimethylformamide	HCO—$N(CH_3)_2$	neat liquid	− 275.2
	Benzamide	C_6H_5—$CONH_2$	DMF	− 279.3
		CH_3O—CO—$N(CH_3)_2$	Chloroform	− 314.2
	Urea	H_2N—CO—NH_2	Water	− 305.0
	Thiourea	H_2N—CS—NH_2	Water	− 273.3
Imines, Oximes, Hydrazones ＼C=N／	N-Methylbenzaldimine	C_6H_5—$CH=N$—CH_3	Chloroform	− 62.1
	N-Phenylbenzaldimine	C_6H_5—$CH=N$—C_6H_5	Chloroform	− 54.1
	Acetone oxime	$(H_3C)_2C=NOH$	Chloroform	− 45.9
	Benzaldehyde oxime	C_6H_5—$CH=NOH$	Choroform	− 26.3
	Benzaldehyde N-phenylhydrazone	C_6H_5-$CH=N$—NHC_6H_5 ₂　　　₁	DMSO	− 237.0 − 54.0
Nitriles —C≡N	Acetonitrile	H_3C—CN	neat liquid	− 137.1
	Benzonitrile	C_6H_5—CN	DMSO	− 121.5
Isonitriles —N=C	Methyl isonitrile	H_3C—$N=C$	neat liquid	− 218.0
	Phenyl isonitrile	C_6H_5—$N=C$	neat liquid	− 200.0
Cyanates, Thiocyanates —X—C≡N	Phenyl cyanate	C_6H_5—O—CN	neat liquid	− 212.0
	Phenyl thiocyanate	C_6H_5—S—CN	neat liquid	− 97.0
Isocyanates, Isothiocyanates —N=C=X	Phenyl isocyanate	C_6H_5—$N=C=O$	neat liquid	− 333.7
	Phenyl isothiocyanate	C_6H_5—$N=C=S$	neat liquid	− 273.1
Azo-compounds, Azoxy-compounds N=N ／ ↓ O	(E)-Azobenzene	H_5C_6 ＼ $N=N$ ＼ 　　C_6H_5	Chloroform	+ 129.0

Tab. 3.48 continued

Class of compound	Example		Solvent	δ
	(Z)-Azobenzene	H_5C_6—N=N—C_6H_5	Chloroform	+ 146.5
	(Z)-Azoxybenzene	H_5C_6—N=N—C_6H_5, O	Chloroform	− 57.1 − 46.7
	(E)-Azoxybenzene	H_5C_6—N=N—C_6H_5, O	Chloroform	− 36.0 − 19.8
Diazo-compounds $\backslash$C=N$_2$	Diazomethane	H_2C—N≡N	Ether	− 96.0 + 7.8
Diazonium salts —C—N$_2^{\oplus}$	Benzenediazonium tetrafluoroborate	C_6H_5—N≡N $BF_4^{\ominus}$	Acetonitrile	− 149.8 − 66.3
Azides —C—N$_3$ —SO$_2$—N$_3$	Methyl azide	H_3C—N—N≡N	Benzene	− 321.2 − 129.7 − 171.0
	Phenyl azide	C_6H_5—N—N≡N	Acetone	− 287.9 − 136.2 − 146.9
	Tosyl azide	H_3C—⟨⟩—SO$_2$—N—N≡N	DMSO	− 240.4 − 146.0 − 138.3
Nitroso-compounds —C—N=O	2-Methyl-2-nitrosopropane Nitrosobenzene	$(H_3C)_3C$—NO C_6H_5—NO	neat liquid neat liquid	+ 578 + 532
Nitrosamines $\backslash$N—NO	Dimethylnitrosamine	$(H_3C)_2N$—NO	neat liquid	− 146.7 + 156.3
Nitrites —C—O—N=O	Butyl nitrite	C_4H_9—O—NO		+ 190.0
Nitro-compounds —C—N$_{O}^{O}$	Nitroethane Nitrobenzene	C_2H_5—NO$_2$ C_6H_5—NO$_2$	DMSO	+ 10.1 − 9.8

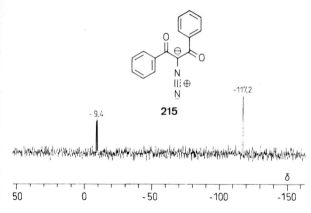

Fig. 3.88   40.5 MHz ^{15}N NMR spectrum of 2-diazo-1,3-diphenyl-1,3-propandione (**215**) in C$_6$D$_6$

In principle ^{15}N NMR spectroscopy has great importance for structural analysis, since N-containing functional groups and N atoms in molecular skeletons are frequently encountered. Tab. 3.84 gives a summary of the shift ranges of the more important classes of compounds. When quoting specific δ-values it should be remembered that the ^{15}N NMR signals often depend strongly on the concentration and temperature, and particularly on the solvent. Intermolecular hydrogen bonds often play an important role.

Fig. 3.88 shows as a specific example the ^{15}N NMR spectrum of 2-diazo-1,3-diphenyl-1,3-propandione (**215**). The inner nitrogen atom gives a signal at $\delta=-117.2$, the outer at $\delta=-9.4$ (referred to CH$_3$NO$_2$).

^{15}N shifts can often show remarkable variations. Thus $\Delta\delta$ in the azene **216** is almost 600 ppm. The charge distribution would, as in the diazo compound **215**, suggest the reversed

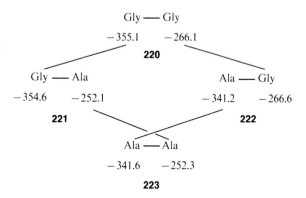

signal assignment; however, the large paramagnetic term, which exists for low-energy electronic transitions (nπ* transitions), is decisive for the chemical shift. The nitrene nitrogen of the aminonitrene therefore appears at very low field.

On formation of the hydrochloride from diethylamine, a low field shift $\Delta\delta$ of less than 4 ppm is observed.

A comparison with the ^{1}H and ^{13}C NMR of ammonium salts shows that the positive charge is essentially located not on the central nitrogen atom, but on the ligands.

Amino acids show the expected dependency of the δ(N)-values on pH:

If an amino acid is built into a peptide, then the δ-value at the N-terminus is largely unaffected, whereas in the peptide bond it is strongly shifted to low field.

The ^{15}N chemical shifts of selected N-heterocycles are given in Tab. 3.49.

The rapid tautomerism due to the intermolecular exchange of protons between two N atoms in azoles can be slowed down by using DMSO as solvent to such an extent that even at room temperature different ^{15}N signals appear. In Tab. 3.49 such an effect is apparent for pyrazole and 1,2,3-triazole, while for 1,2,4-triazole and tetrazole averaged signals are obtained.

Apart from tautomerism between identical structures ^{15}N NMR can also be used to investigate tautomeric equilibria

Tab 3.49 ^{15}N chemical shifts of selected heterocycles (δ-values referred to CH_3NO_2, DMSO as solvent except where otherwise specified)

−393.3 Aziridine[a]	−228.7 cis-2-Methyl-3-phenyloxaziridine[b]	−223.2 trans-2-Methyl-3-phenyloxaziridine[b]	−342.1 Pyrrolidine[c]
−343.2 Piperidine[c]	−351.0 Piperidine hydrochloride[c]	−349.8 Morpholine[c]	−362.2 1-Azabicyclo-[2.2.2]octane[c]
−231.4 Pyrrole[a]	−79.8 −173.1 Pyrazole	−169.0 −169.0 Imidazole	−123.7 Oxazole
+2.2 Isoxazole	−58.0 Thiazole	−81.8 Isothiazole	−134.7 −127.4 −127.4 1,2,4-Triazole
−69.0 −76.0 −69.0 1,2,3-Triazole	−29.2 −16.2 −143.0 1-Methyl-1,2,3-triazole	−50.2 −131.4 50.2 2-Methyl-1,2,3-triazole	−98,3 −5.8 −5.8 −98.3 Tetrazole
−63.2 Pyridine	−164.8 Pyridine hydrochloride	−86.2 Pyridine oxide	+20.3 +20.3 Pyridazine
−84.5 −84.5 Pyrimidine	−46.1 −46.1 Pyrazine	−98.5 −98.5 −98.5 1,3,5-Triazine	−62.0 +2.0 +40.0 1,2,4-Triazine
−247.3 Indole	−267.5 Carbazole[d]	−63.5 Quinoline	−69.3 Isoquinoline

[a] nat liquid, [b] in acetonitrile, [c] in methanol, [d] in diethylether

Tab. 3.49 continued

Phenanthroline — 5-Butyl-2-azidopyrimidine

Barbituric acid — Purine — Adenine

between states of different energy. Thus it can be shown that barbituric acid exists as a urea derivative and is only formally a pyrimidine derivative (Tab. 3.49). Although that can also be established from the ^{1}H and ^{13}C data, the distinction is more difficult in the case of 5-butyl-2-azidopyrimidine. From the ^{15}N spectrum the proportion of the bicyclic form can be determined as 90% (cf. Tab. 3.49). Finally, ^{15}N NMR can be used for investigations on molecular dynamics; inversion at nitrogen is a particularly suitable case.

While isotope effects on ^{15}N and ^{14}N chemical shifts are negligible, the spin–spin coupling constants of ^{15}N and ^{14}N differ both in sign and magnitude.

The following relation applies:

$$J(^{15}N,X) = -1.4027\, J(^{14}N,X) \quad X = {}^1H, {}^{13}C, ...$$

Typical $^1J(^{15}N,^1H)$ coupling constants lie in the range of (-80 ± 15) Hz. Some examples, including exceptions, are tabulated in Tab. 3.50. $^2J(^{15}N,^1H)$ coupling constants are generally less than 2 Hz in magnitude;

only where there are sp^2-hybridised C and/or N atoms do the magnitudes reach 3–12 Hz. In compounds with C=N double bonds the coupling constant can be as large as –16 Hz (Tab. 3.50). The sign of the $^2J(^{15}N,^1H)$ coupling can be positive or negative. The same applies to $^3J(^{15}N,^1H)$ couplings and long range couplings $^nJ(^{15}N,^1H)$. The latter only have significant values when there are multiple coupling pathways.

Couplings between ^{15}N and ^{13}C are difficult to measure without isotopic enrichment. Values of $^1J(^{15}N,^{13}C)$ are generally less than 20 Hz. The sign can be positive or negative. If the value of ^1J is close to zero, it can be exceeded by ^2J or ^3J cou-

plings. Some examples of known $^nJ(^{15}N,^{13}C)$ couplings are given in Tab. 3.51.

^{15}N, ^{15}N couplings will not be treated here. They can only be measured for singly or doubly labelled compounds and are almost totally worthless for structural purposes.

Tab. 3.50 $^nJ(^{15}N,^1H)$ coupling constants (n = 1, 2, 3, 4) of selected compounds

Compound	Solvent	^1J	^2J	^3J	^4J
H$_2$C—NH (H, H)	neat liquid	− 64.5	1.0		
H$_2$C—N—CH$_3$ (H, H)	neat liquid	− 67.0	0.9		
HOOC—C—NH (H, H)	Water	− 74.7	0.5		
O=C—N (H, H)	neat liquid	− 88.3		14.6	
		− 90.7			
O=C—N, H$_2$C (H)	Water	− 88.4		1.3	
		− 90.9			

Tab. 3.50 continued

Compound	Solvent	^1J	^2J	^3J	^4J
$H-\overset{\ominus}{N}-\overset{\oplus}{N}\equiv N$	Ether	-70.2	2.3	2.2	
$H_3C,\ OH \quad C=N \quad H$	Water		-15.9		
$H,\ OH \quad C=N \quad H_3C$	Water		$+2.9$		
$H_2C,\ OH \quad C=N \quad H_5C_6$	Chloroform		-2.0		
$H_5C_6,\ OH \quad C=N \quad H_2C,\ H$	Chloroform		-4.2		
$H_5C_6,\ H \quad C=N \quad H_5C_6$	Pentane	-51.2			
(aniline ring)$-NH$	Chloroform Benzene	-78.0	-1.9	-0.5	
(ring)$-NO_2$	Acetone		-1.9	-0.8	
(pyrrole)	Benzene neat liquid	-96.5	-4.5	-5.4	
(pyridine)	neat liquid		-10.8	-1.5	0.2
(pyridine N-oxide)	Chloroform		0.5	-5.3	1.1

Tab. 3.51 $^nJ(^{15}N,^{13}C)$ coupling constants ($n=1,2,3$) of selected compounds in Hz. (If the sign is not known, the magnitude is given)

$$\underset{1.5\quad 1.2\quad -3.9}{H_3C-CH_2-CH_2-NH_2}$$

$$\underset{6.2}{H_2N-CH_2-COOH}$$

$$\underset{-8.5\ -14.4}{H_3C-\underset{\underset{O}{\|}}{C}-NH_2}$$

$$\underset{-17.5}{H_3C-C\equiv N}$$

$$\underset{1.8}{H_3C}\underset{2.3}{\diagdown}\overset{OH}{\underset{C=N}{\diagup}}{\underset{H}{}}$$

$$\underset{-9.0}{H_3C}\underset{4.0}{\diagdown}{C=N}\underset{H\quad OH}{}$$

(phenyl-NH_2) $-1.9\ -2.7$, -11.5, <0.5

(phenyl-NO_2) $-2.3\ -1.7$, -14.6

(phenyl-$N\equiv N$ $BF_4^{\ominus}$) $3.0\ 2.1$, $10.5\ \oplus$, 5.6

(pyrrole-N) -3.9, -13.0

(pyridine) -3.9, 2.5, 0.7

(pyridine N-oxide) -5.3, 1.4, -15.2

(2-pyridone) 10.5, 5.2, 2.5, <0.5, 11.2

(quinoline) ~0, 2.1, 3.5, 0.9, 3.9, 0.6, 9.3, 2.7, 1.2

6.4 Other Nuclei

Other nuclei with a spin quantum number 1/2 are ^{29}Si, ^{77}Se, ^{117}Sn, ^{119}Sn, ^{195}Pt, ^{199}Hg, ^{203}Tl, ^{205}Tl, and ^{207}Pb.

A combination of the natural abundance and the magnitude of the magnetic moment leads to the following order for the relative **receptivities**:

$$^{205}Tl > {}^{203}Tl > {}^{119}Sn > {}^{195}Pt > {}^{207}Pb > {}^{117}Sn > {}^{199}Hg > {}^{77}Se > {}^{29}Si > {}^{13}C$$

Among nuclei with spin quantum numbers $I > 1/2$ the most important are ^{2}H, ^{7}Li, ^{11}B, ^{14}N, ^{17}O, and ^{33}S. The nuclear quadrupole moments often cause extreme line broadening.

Literature

General Texts

Abragam, A., Goldman, M. (1981) Nuclear Magnetism, Oxford University Press, Oxford.

Abraham, R.J., Fisher, J. (1988), NMR Spectroscopy, J. Wiley, New York.

Abraham, R.J., Loftus, P. (1978), Proton and Carbon-13 NMR Spectroscopy, An Integrated Approach, Heyden, London.

Akitt, J.W. (1983), N.M.R. and Chemistry, An Introduction to Nuclear Magnetic Resonance Spectroscopy, Chapman and Hall, London.

Ault, A., Dudek, G.O. (1978), Protonen-Kernresonanz-Spektroskopie, UTB 842, Steinkopff Verlag, Darmstadt.

Becker, E.D. (1980), High Resolution NMR, Academic Press, New York, London.

Bovey, F.A. (1969), Nuclear Magnetic Resonance Spectroscopy, Academic Press, New York, London.

Breitmaier, E., Voelter, W. (1987) ^{13}C-NMR-Spectroscopy, Methods and Applications, Verlag Chemie, Weinheim.

Canet, D. (1994), NMR-Konzepte und Methoden, Springer, Heidelberg.

Carrington, A., McLachlan, A.D. (1979), Introduction to Magnetic Resonance, Chapman and Hall, London.

Chamberlain, N.F. (1974), The Practice of NMR Spectroscopy (Spectra-Structure Correlations for Hydrogen-1), Plenum Press, New York, London.

Chandrakumar, N., Subramanian, S. (1987), Modern Techniques in High-Resolution FT-NMR, Springer-Verlag, New York.

Clerc, T., Pretsch, E., Sternhell, S. (1973), ^{13}C-Kernresonanzspektroskopie, Akademische Verlagsgesellschaft, Frankfurt/M.

Corio, P.L. (1961), Structure of High-Resolution NMR Spectra, Academic Press, New York, London.

Derome, A.E. (1987), Modern NMR Techniques for Chemistry Research, Pergamon Press, Oxford.

Emsley, J.W., Feeney, J., Sutcliffe, L.H. (1955), High Resolution Nuclear Magnetic Resonance Spectroscopy I, II, Pergamon Press, Oxford, New York.

Ernst, L. (1980), ^{13}C-NMR-Spektroskopie, UTB 1061, Steinkopff Verlag, Darmstadt.

Ernst, R., Bodenhausen, G., Wokaun, A. (1990), Principles of Nuclear Magnetic Resonance in One and Two Dimensions, Clarendon Press, London.

Freeman, R. (1987), A. Handbook of Nuclear Magnetic Resonance, Longman, Harlow.

Friebolin, H. (1988), Ein- und zweidimensionale NMR-Spektroskopie, VCH, Weinheim.

Goldman, M. (1988), Quantum Description of High-Resolution NMR in Liquids, Clarendon Press, London.

Günther, H. (1991), NMR-Spektroskopie, Georg Thieme Verlag, Stuttgart.

Hallap. P., Schütz, H. (1973), Anwendung der ^{1}H-NMR-Spektroskopie, ein Lehrprogramm für Hochschulen, Taschentext 31, Verlag Chemie, Weinheim.

Harris, R.K. (1983), Nuclear Magnetic Resonance Spectroscopy – A Physicochemical View, Pitman, London.

Homans, S.W. (1989), A Dictionary of Concepts on NMR, Clarendon Press, London.

Jackman, L.M., Sternhell, S. (1969), Applications of NMR Spectroscopy in Organic Chemistry, Pergamon Press, Oxford, New York.

Kalinowski, H.-O., Berger, S., Braun, S. (1984), ^{13}C-NMR-Spektroskopie, Georg Thieme Verlag, Stuttgart.

Kleinpeter, E., Borsdorf, R. (1981), ^{13}C-NMR-Spektroskopie in der organischen Chemie, Akademie-Verlag, Berlin.

Levy, G.C., Lichter, R.C., Nelson, G.L. (1980), Carbon-13 Nuclear Magnetic Resonance for Organic Chemists, Wiley Interscience, New York.

Memory, J.D. (1968), Quantum Theory of Magnetic Resonance Spectra, McGraw-Hill, New York.

Michel, D. (1981), Grundlagen und Methoden der kernmagnetischen Resonanz, Akademie-Verlag, Berlin.

Pople, J.A., Schneider, W.G., Bernstein, H.J. (1959) High Resolution Nuclear Magnetic Resonance, McGraw-Hill Book Comp., New York.

Roberts, J.D. (1959), Nuclear Magnetic Resonance, McGraw-Hill Book Comp, New York.

Roberts, J.D. (1961), An Introduction to Spin-Spin Splitting in High Resolution NMR Spectra, Benjamin, New York.

Sillescu, H. (1966), Kernmagnetische Resonanz, Springer-Verlag, Berlin.

Slichter, C.P. (1990), Principles of Magnetic Resonance, Springer-Verlag, Berlin.

Sternhall, S., Field, L.D. (1989), Analytical NMR, J. Wiley, New York.

Stothers, J.B. (1972), Carbon-13 NMR Spectroscopy, Academic Press, New York, London.

Strehlow, H. (1968), Magnetische Kernresonanz und chemische Struktur, Steinkopff Verlag, Darmstadt.

Suhr, H. (1965), Anwendungen der kernmagnetischen Resonanz in der organischen Chemie, Springer-Verlag, Berlin.

Wehrli, F.W., Marchand, A.P. (1988), Interpretation of Carbon-13 NMR Spectra, J. Wiley, New York.

Wehrli, F.W., Wirthlin, T. (1976), Interpretation of Carbon-13 Nuclear Magnetic Resonance Spectra, Heyden, London.

Wiberg, K.B., Nist, B.J. (1969), The Interpretation of NMR Spectra, Benjamin, New York.

Williams, D. (1986), Nuclear Magnetic Resonance Spectroscopy, J. Wiley, New York.

Zschunke, A. (1971), Kernmagnetische Resonanzspektroskopie in der organischen Chemie, Akademie-Verlag, Berlin.

Special Methods and Effects

Bax, A. (1982), Two-Dimensional Nuclear Magnetic Resonance in Liquids, Kluwer Academic Publishers Group, Dordrecht.

Blümich, B., Kuhn, W. (1992), Magnetic Resonance Microscopy, VCH, Weinheim.

Brey, W.S. (1988), Pulse Methods in 1D and 2D Liquid-Phase NMR, Academic Press, New York.

Callaghan, P.T. (1991), Principles of Nuclear Magnetic Resonance Microscopy, Clarendon Press, London.

Croasmun, W., Carlson, R. (1987), Two-Dimensional NMR Spectroscopy, Applications for Chemists and Biochemists, VCH, Weinheim.

Ebert, I., Seifert, G. (1966), Kernresonanz im Festkörper, Geest und Portig, Leipzig.

Emsley, J.W., Lindon, J.C. (1975), NMR Spectroscopy Using Liquid Crystal Solvents, Pergamon Press, Oxford, New York.

Field, L.D., Sternhell, S. (1989), Analytical NMR, Wiley, Chichester.

Freeman, R., Hill, H.D.W. (1975), Dynamic Nuclear Magnetic Resonance Spectroscopy, Academic Press, New York, London

Fukushima, E., Roeder, S.B.W. (1981), Experimental Pulse NMR: a Nuts and Bolts Approach, Addison-Wesley, London.

Fyfe, C.A. (1983), Solid State NMR for Chemists, C.F.C. Press, Ontario.

Hägele, G., Engelhardt, M., Boenigk, W. (1987), Simulation und automatisierte Analyse von Kernresonanzspektren, VCH, Weinheim.

Hofmann, R.A., Forsen, S. (1966), High-Resolution Nuclear Magnetic Double and Multiple Resonance, Pergamon Press, Oxford, New York.

Jackman, L.M., Cotton, F.A. (1975), Dynamic Nuclear Magnetic Resonance Spectroscopy, Academic Press, New York, London

Kaplan, J.I., Fraenkel, G. (1980), NMR of Chemically Exchanging Systems, Academic Press, New York, London.

Kasler, F. (1973), Quantitative Analysis by NMR Spectroscopy, Academic Press, New York, London.

Lepley, A.R., Closs, G.L. (1973), Chemically Induced Magnetic Polarization, Wiley, New York.

Leyden, D.E., Cox, R.H. (1977), Analytical Applications of NMR, Wiley, New York.

Marshall, J.L. (1983), Carbon-Carbon and Carbon-Proton NMR Couplings, Verlag Chemie, Weinheim, Deerfield Beach, Florida, Basel.

Martin, M.L., Delpuech, J.-J., Martin, G.J. (1980), Practical NMR-Spectroscopy, Heyden, London.

Martin, G.E., Zektzer, A.S. (1988), Two-Dimensional NMR Methods for Establishing Molecular Connectivity, VCH, Weinheim.

Mehring, M. (1983), Principles of Resolution NMR in Solids, Springer-Verlag, Berlin.

Müllen, K., Pregosin, P.S. (1976), Fourier Transform NMR Techniques: A Practical Approach, Academic Press, New York, London.

Morrill, T.C. (1987), Lanthanide Shift Reagents in Stereochemical Analysis, VCH, Weinheim.

Munowitz, M. (1988), Coherence and NMR, J. Wiley, New York.

Neuhaus, D., Williamson, M.P. (1989), The Nuclear Overhauser Effect in Structural and Conformational Analysis, VCH, Weinheim.

Noggle, J.H., Schirmer, R.E. (1971), The Nuclear Overhauser Effect, Chemical Applications, Academic Press, New York, London.

Oki, M. (1985), Applications of Dynamic NMR Spectroscopy to Organic Chemistry, Verlag Chemie, Weinheim, Deerfield Beach, Florida, Basel.

Poole, D.P., Farach, H. (1971), Relaxation in Magnetic Resonance, Academic Press, New York, London.

Sandström, J. (1982), Dynamic NMR Spectroscopy, Academic Press, New York.

Schraml, J., Bellama, J.M. (1988), Two Dimensional NMR Spectroscopy, J. Wiley & Sons, New York.

Shaw, D. (1976), Fourier Transform NMR Spectroscopy, Elsevier, Amsterdam.

Sievers, R.E. (1973), Nuclear Magnetic Shift Reagents, Academic Press, New York, London.

Takeuchi, Y., Marchand, A.P. (1986), Applications of NMR Spectroscopy to Problems in Stereochemistry and Conformational Analysis, Verlag Chemie, Weinheim, Deerfield Beach, Florida, Basel.

Ziessow, D. (1973), On-line Rechner in der Chemie, Grundlagen und Anwendung in der Fourierspektroskopie, de Gruyter Verlag, Berlin.

Special Classes of Compounds, Applications

Batterham, T.J. (1973), NMR Spectra of Simple Heterocycles, Wiley, New York.

Berliner, L.J., Reuben, J. (1978), Biological Magnetic Resonance, Bd. I, Plenum Press, New York, London.

Bertini, I., Molinari, H., Niccolai, N. (1991), NMR and Biomolecular Structure, VCH, Weinheim.

Bovey, F.A. (1971), High-Resolution Nuclear Magnetic Resonance of Macromolecules, Academic Press, New York, London.

Bradbury, E.M., Nicolini, C. (1985), NMR in the Life Sciences, NATO Asi Series, A, Vol. 107, Plenum Press, New York

Casy, A.F. (1971), NMR Spectroscopy in Medicinal and Biological Chemistry, Academic Press, New York, London.

Chamberlain, N.F., Reed, J.J.R. (1971), Nuclear-Magnetic-Resonance Data of Sulfur Compounds, Wiley-Interscience, New York.

Damadian, R. (1981), NMR in Medicine, Springer-Verlag, Berlin.

de Certaines, J.D., Bovée, W.M.M.J., Podo, F. (1992), Magnetic Resonance Spectroscopy in Biology and Medicine, Pergamon Press, Oxford.

Dwek, R.A., Campbell, I.D., Richard, R.E., Williams, R.J.P. (1977), NMR in Biology, Academic Press, New York, London.

Dwek, R.A. (1977), Nuclear Magnetic Resonance in Biochemistry, Clarendon Press, London.

Emsley, J.W., Lindon, J.C. (1975), NMR Spectroscopy Using Liquid Crystal Solvents, Pergamon Press, Oxford

Fluck, E. (1963), Die kernmagnetische Resonanz und ihre Anwendung in der anorganischen Chemie, Springer-Verlag, Berlin.

Foster, M.A., Hutchinson, J.M.S. (1987), Practical NMR Imaging, IRL Press, London.

Gadian, D.G. (1981), Nuclear Magnetic Resonance and its Applications to Living Systems, Oxford University Press, Oxford.

Hausser, K.H., Kalbitzer, H.R. (1989), NMR für Mediziner und Biologen, Springer-Verlag, Berlin.

James, T.L. (1975), NMR in Biochemistry, Academic Press, New York, London.

Jardetzky, O., Roberts, G.C.K. (1981), NMR in Molecular Biology, Academic Press, New York, London.

Knowles, P.F., Marsh, D., Rattle, H.W.E. (1976), Magnetic Resonance of Biomolecules, Wiley, London.

Komoroski, R.A. (1986), High Resolution NMR Spectroscopy of Synthetic Polymers in Bulk, Verlag Chemie, Weinheim, Deerfield Beach, Florida, Basel.

LaMar, G.N., Horrocks, W.D., Holm, R.H. (1973), NMR of Paramagnetic Molecules, Academic Press, New York, London

Mann, B.E., Taylor, B.F. (1981), ^{13}C-NMR Data of Organometallic Compounds, Academic Press, New York, London.

Marchand, A.P. (1983), Stereochemical Applications of NMR Studies in Rigid Bicyclic Systems, Verlag Chemie, Weinheim.

Morris, P.G. (1986), Nuclear Magnetic Resonance Imaging in Medicine and Biology, Clarendon Press, London.

Nakanishi, K. (1990), One-dimensional and Two-dimensional NMR Spectra by Modern Pulse Techniques, University Science Books, London.

Pasika, W.M. (1979), Carbon-13 NMR in Polymer Science, Am. Chem. Soc. Symposium Series 103, Washington.

Rabidean, P. (1989), The Conformational Analysis of Cyclohexenes, Cyclohexadienes and Related Hydroaromatic Compounds, VCH, Weinheim.

Roberts, G.C.K. (1993), NMR of Macromolecules, IRL Press, Oxford.

Shulman, R.G. (1979), Biological Applications of Magnetic Resonance, Academic Press, New York.

Slonim, I., Ya, Lyubimov, A.N. (1970), The NMR of Polymers, Plenum Press, New York, London.

Tonelli, A.E. (1989), NMR Spectroscopy and Polymer Microstructure, VCH, Weinheim.

Wehrli, F.W., Shaw, D., Kneeland, J.B., Biomedical Magnetic Resonance Imaging, VCH, Weinheim.

Whitesell, J.K., Minton, M.A. (1987), Stereochemical Analysis of Alicyclic Compounds by C-13 Nuclear Magnetic Resonance Spectroscopy, Chapman and Holl, London.

Wüthrich, K. (1986), NMR of Proteins and Nucleic Acids, J. Wiley & Sons, New York.

Wüthrich, K. (1976), NMR in Biological Research: Peptides and Proteins, North-Holland Publ. Co., Amsterdam.

Nuclei other than ¹H and ¹³C

Axenrod, T., Webb, G.A. (1974), Nuclear Magnetic Resonance Spectroscopy of Nuclei Other than Protons, Wiley, New York.

Berger, S., Braun, S., Kalinowski, H.-O. (ab 1992), NMR-Spektroskopie von Nichtmetallen: Bd. 1 (¹⁷O, ³³S, ¹²⁹Xe), Bd. 2 (¹⁵N), Bd. 3 (³¹P), Bd. 4 (¹⁹F), Georg Thieme Verlag, Stuttgart.

Brevard, C., Granger, P. (1981), Handbook of High Resolution Multinuclear NMR, J. Wiley, New York.

Crutchfield, M.M., Dungan, C.H., Lechter, J.H., Mark, V., van Wazer, J.R. (1967), ³¹P Nuclear Magnetic Resonance, Interscience, New York.

Dungan, C.H., van Wazer, J.R. (1970), Compilation of Reported ¹⁹F NMR Chemical Shifts, Wiley-Interscience, New York.

Evans, E.A., Warrell, D.C., Elridge, J.A., Jones, J.R. (1985), Handbook of Tritium NMR Spectroscopy and Applications, J. Wiley & Sons, New York.

Granger, P., Harris, R.K. (1990), Multinuclear Magnetic Resonance in Liquids and Solids – Chemical Applications, Kluwer, Dordrecht.

Grayson, M., Griffith, E.J. (1967), ³¹P Nuclear Magnetic Resonance, Interscience Publishers, New York.

Harris, R.K., Mann, B.E. (1978), NMR and the Periodic Table, Academic Press, New York, London.

Kintzinger, J.-P., Marsmann, H. (1981), Oxygen-17 and Silicon-29, Springer-Verlag, Berlin.

Lambert, J.B., Riddell, F.G. (1983), The Multinuclear Approach to NMR Spectroscopy, D. Reidel Publishing, Dordrecht.

Lazlo, P. (1983), NMR of Newly Accessible Nuclei, 2 Bd., Academic Press, New York

Levy, G.C., Lichter, R.L. (1979), Nitrogen-15 NMR-Spectroscopy, Wiley, New York.

Martin, G.J., Martin, M.L., Gouesnard, J.P. (1981), ¹⁵N-NMR Spectroscopy, Springer-Verlag, Berlin.

Mason, J. (1987), Multinuclear NMR, Plenum Press, New York.

Mooney, E.F. (1979), An Introduction to ¹⁹F NMR Spectroscopy, Heyden-Sadtler, London.

Nöth, H., Wrackmeyer, B. (1978), Nuclear Magnetic Resonance Spectroscopy of Boron Compounds, Springer-Verlag, Berlin.

Tebby, J.C. (1980), Phosphorus-31 Nuclear Magnetic Resonance Data, CRC Press, Boca Raton.

Verkade, J.G., Quin, L.D. (1987), Phosphorus-31 NMR Spectroscopy in Stereochemical Analysis: Organic Compounds and Metal Complexes, VCH, Weinheim.

Wehrli, F.W. (1974), Nuclear Magnetic Resonance Spectroscopy of Nuclei Other than Protons, Wiley, New York.

Witanowski, M., Webb, G.A. (1973), Nitrogen NMR, Plenum Press, New York, London.

Problems

Bates, R.B., Beavers, W.A. (1981), Carbon-13 NMR Spectral Problems, Humana Press, Clifton, USA.

Breitmaier, E., Bauer, G. (1977), ¹³C-NMR-Spektroskopie, Eine Arbeitsanleitung mit Übungen, Georg Thieme Verlag, Stuttgart

Breitmaier, E. (1990), Vom NMR-Spektrum zur Strukturformel Organischer Verbindungen, B.G. Teubner, Stuttgart.

Duddeck, H., Dietrich, W. (1988), Strukturaufklärung mit moderner NMR-Spektroskopie, Steinkopff Verlag, Darmstadt.

Fuchs, P.L., Bunell, C.A. (1979), Carbon-13 Based Organic Spectral Problems, J. Wiley, New York.

Sanders, J.K.M., Constable, E.C., Hunter, B.K., Pearce, C.M. (1989), Modern NMR Spectroscopy – A Workbook of Chemical Problems, Oxford University Press, Oxford.

Catalogues

Nakanishi, K. (1980), One-dimensional and Two-dimensional NMR Spectra by Modern Pulse Techniques, W.H. Freeman & Co, Oxford.

Bremser, W., Franke, B., Wagner, H. (1982), Chemical Shift Ranges in Carbon-13 NMR Spectroscopy, Verlag Chemie, Weinheim.

Bremser, W., Ernst, L., Franke, B., Gerhards, R., Hardt, A. (1981), Carbon-13 NMR Spectral Data, Verlag Chemie, Weinheim.

Sasaki, S., Handbook of Proton-NMR Spectra and Data, Vol 1–10 and Index, Academic Press, London.

Ault, A., Ault, M.R. (1980), A Handy and Systematic Catalog of NMR Spectra, University Science Books, Mill Valley.

Brügel, W. (1979), Handbook of NMR Spectral Parameters, Heyden, London.

Breitmaier, E., Haas, G., Voelter, W. (1975, 1979), Atlas of Carbon-13 NMR Data, 2 Bde., Heyden, London.

Pouchert, C.J., Campbell, J.R. (1974), The Aldrich Library of NMR Spectra, 11 Bde., Aldrich Chemical Comp., Milwaukee.

Johnson, L.F., Jankowski, W.C. (1972), Carbon-13 NMR Spectra, A Collection of Assigned Coded and Indexed Spectra, Wiley, New York.

Bovey, F.A. (1967), NMR Data Tables for Organic Compounds, Wiley-Interscience, New York.

Simons, W.W. (1967), The Sadtler Handbook of Proton NMR Spectra, Sadtler Research Laboratories, Philadelphia.

Hershenson, H.M. (1965), NMR and ESR Spectra Index, Academic Press, New York.

Howell, M.G., Kende, A.S., Webb, J.S. (1965), Formula Index to NMR Literature Data I, II, Plenum Press, New York.

Bhacca, N.S., Johnson, L.F., Shoolery, J.N. (1962/63). High Resolution NMR Spectra Catalogue, I, II, Varian Associates, Palo Alto.

Series, Periodicals

Diehl, P., Fluck, E., Kosfeld, R. (1969), NMR, Basic Principles Progress, Grundlagen und Fortschritte, Springer-Verlag, Berlin.

Waugh, J.S. (ab 1965), Advances in Magnetic Resonance, Academic Press, New York.

Emsley, J.W., Feeney, J., Sutcliffe, L.H. (ab 1966), Progress in NMR Spectroscopy, Pergamon Press, Oxford.

(ab 1968), Nuclear Magnetic Resonance Abstracts and Index, Preston Techn. Abstr. Comp., Evanston.

Mooney, E.F. (ab 1968), Annual Review of NMR-Spectroscopy, Academic Press, London.

Poole, C.P., Magnetic Resonance Review, Gordon & Breach, London.

Analytical Chemistry Annual Reviews: NMR Spectroscopy.

Levy, G.C. (ab 1974), Topics in Carbon-13 NMR Spectroscopy, Wiley-Interscience, New York.

Harris, R.K. (ab 1972), Nuclear Magnetic Resonance. Specialist Periodical Report. The Chemical Society, London.

Journal of Magnetic Resonance, Academic Press, New York, (ab 1971).

Organic Magnetic Resonance/Magnetic Resonance in Chemistry, Heyden, London (ab 1969).

Bulletin of Magnetic Resonance, Franklin Institute Press, Philadelphia (ab 1979).

Chemical Abstracts Selects: Carbon and Heteroatom NMR, Columbus, Ohio (ab 1979).

4 Mass Spectra

1. Introduction

Although the method of mass spectrometry is relatively old (in 1910, J.J. Thompson was able to separate the ^{20}Ne and ^{22}Ne isotopes), it did not achieve recognition as an important analytical method in organic chemistry until 1960. Two features have helped to bring it to prominence. Firstly, it is possible to determine the relative molecular mass and even the elemental composition of a compound using only the smallest amount of substance. Furthermore, the fragmentation pattern (i.e. the decomposition of the material being analysed under the influence of electron bombardment or other techniques for ion formation) depicted in the mass spectrum allows one to make important deductions about the structure of the compound. In recent years, both of these aspects have been crucial to the development of the application of mass spectrometry in organic chemistry[1].

There are limits to the mass spectrometric determination of the relative molecular mass of a sample. The polarity of a substance is inversely proportional to its volatility. Also, the larger the relative molecular mass, the greater, in general, the number of functional groups and therewith the danger of thermal decomposition upon vaporisation. Therefore, various procedures were developed (e.g. chemical ionisation, field ionisation, field desorption, secondary ion mass spectrometry, fast atom bombardment, electrospray methods), which, compared with electron impact ionisation, allow the relative molecular masses of involatile compounds to be determined in many more cases. In more recent times, efforts have been made to further improve known procedures or to investigate new promising possibilities (see Sec. 8.9, ionisation methods, p. 263). In routine operation, relative molecular masses up to ca. 1200 can be determined using mass spectrometry (see, however, Sec. 8.8, p. 261).

Another aspect that has also received considerable attention is the utilisation of the generally enormous amount of information contained in mass spectra. These efforts have lead to significant advances in instrumentation. Today, fast and dependable spectrometers are available which permit the determination of the empirical formulae of fragment ions and additional equipment has been developed to measure metastable ions or record collision activation spectra. The results from all of these methods of measurement advance our knowledge about mass spectra and, furthermore, ease the deduction of the structures of the compounds being examined. Also, the results from measurements made using isotopically labelled derivatives have contributed strongly to the successful interpretation of spectra. The result is that today we know considerably more about the behaviour of substances in mass spectrometers. Nevertheless, the number of generally applicable rules in proportion to the number of exceptions and special cases has unfortunately become rather small. One can only hope that this proportion will change in the future.

In the following sections, an introduction will be given to the most important aspects of mass spectrometry. Except in those cases where it is specifically stated, the discussion will refer to the method most commonly employed today: electron impact mass spectrometry (abbr. EI).

2. Instrumentation and the Recording of Spectra

First of all, the principle of mass spectrometric separation will be discussed briefly. When accelerated, positively charged particles are in the gas phase, a homogeneous magnetic field will separate them by an amount proportional to their mass. The experimental means of achieving this is complicated and requires further explanation, which will only be given to the extent that is necessary for an adequate understanding of the method by an organic chemist.

2.1 The Principle of the Mass Spectrometer

As indicated in the schematic diagram in Fig. 4.1, the functions of a mass spectrometer may be divided into four sections: sample injection, ion generation, mass separation and ion detection. The ion generation and the processes in the magnetic analyser (mass separation and ion detection) occur under high vacuum in order to minimise undesired collisions between ions and molecules or atoms. In ordinary mass spectrometers, the following pressures are obtained: in the ion generator: 10^{-3} to 10^{-4} Pa; in the magnetic analyser: 10^{-6} to 10^{-7} Pa. A great deal of instrumentation is necessary for the

generation and control of the high vacuum, but this will not be discussed in any detail here.

Sample Injection

It follows from the above explanation that there is a problem in how one gets a sample of a substance at normal pressure into the high vacuum without breaking the vacuum. Principally, there are two types of injection systems. These are the gas inlet and the direct inlet systems.

Gas inlet. Application: for liquid or gaseous samples. A liquid can either be injected with a microsyringe through a septum directly into a previously evacuated reservoir or frozen in a glass vessel (e.g. with liquid nitrogen). The air above the frozen sample can be pumped away and the sample can then be vaporised into the reservoir. In order to minimise the inclusion of gases in the frozen material, it is advisable to thaw and re-freeze the sample at least once while it is under vacuum. The reservoir is fitted with various valves (e.g. to the vacuum pumps, inlet port and ion source), the internal surface is as inert as possible (e.g. glass or enamel) and it can be heated

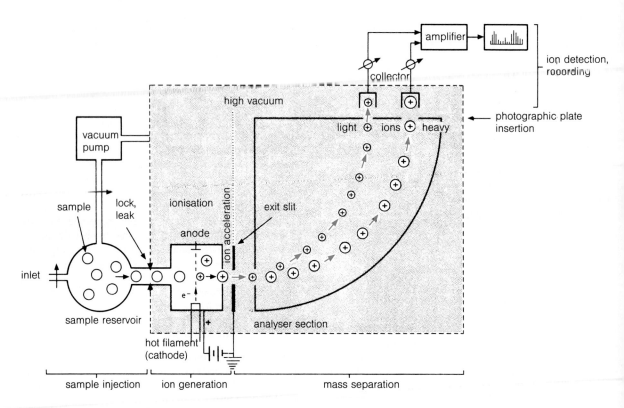

Fig. 4.1 Schematic representation of a mass spectrometer

(max. temperature during continuous service is usually 150°C). The reservoir is connected to the ion source *via* a leak (a hole of a specific size, e.g. a perforated gold foil melted into a glass tube [gold leak]). Gaseous samples can be introduced into the reservoir through a container fitted with a break-seal.

Volatile substances can also be injected directly into the mass spectrometer *via* a gas chromatograph (GC) or a liquid chromatograph (LC, HPLC) (see Sec. 8.11, p. 266).

Direct inlet. Application: crystalline, lacquer-like or viscous liquid samples. The sample is placed in a metal crucible (e.g. gold or aluminium; internal diameter 1 mm), which is fixed on the tip of a heatable probe and the tip of the probe is inserted into a lock-chamber. After evacuating the chamber, the cooled tip of the probe is brought into the ion source and slowly heated until the sample vaporises. In addition, the end of the probe that is in the ion source, which is at a high voltage, must be electrically isolated from the handle. The tip of the probe can also be cooled so as to enable the measurement of easily vaporised samples, to hinder the vaporisation of the sample by the (hot) ion source, or to quickly cool an overheated sample.

Sample requirements. *Via* gas inlet: 0.1 to 1 mg; *via* GC: in the region of 10^{-9} to 10^{-15} g; *via* direct inlet: 0.001 to 0.1 mg for normal measurements. The full amount of the given sample quantities must immediately be available to the instrument and should not, for example, be smeared out as a film on the surface of a large flask!

Ion Generation

From one of the inlet systems (gas or direct inlet), a fine and as constant as possible beam of molecules streams into the ion source where it intersects prependicularly with an electron beam [between a hot filament (cathode) and an anode]. The potential difference between the cathode and the anode can be varied between 0 and, in general, 300 V, which means that the electrons can carry between 0 and 300 eV. For low-voltage spectra, 12 to 15 eV is used and for normal spectra, 60 to 100 eV can be used, although 70 eV is normally employed. The interaction between the electrons and the neutral molecules generates positively charged molecular ions (see pp. 223, 226) according to:

$$M + e^- \rightarrow M^{+\bullet} + 2e^-$$
or less frequently
$$M + e^- \rightarrow M^{2+} + 3e^-$$

Other ionisation procedures are discussed in Sec. 8.9 (see p. 263).

The non-ionised particles are removed from the ion source chamber by the high vacuum pumps, whereas the molecular ions that have been generated are now accelerated and focused. The acceleration of the ions is achieved by applying a potential to the source (the acceleration potential varies with

the type of instrument from 2 to 10 kV) and the final speed is reached at the exit slit (0 V). The focusing of the ions is achieved by an additional electric field. The exit slit permits only the narrow, central, and therefore homogeneous region of the ion beam to pass into the magnetic analyser. The speed of the ions can be calculated as follows:

$$z \cdot U = \frac{m \cdot v^2}{2} \tag{1}$$

$$v = \sqrt{\frac{2 \cdot z \cdot U}{m}} \tag{2}$$

z ionic charge ($= n \cdot e$)
m ion mass
v ion speed
U acceleration potential

Mass Separation

The separation of the ions takes place in the magnetic analyser and is proportional to their mass. The separation occurs in a magnetic field (of the order of magnitude of 1 T), in which, with ions of the same charge, the paths of the lighter ions will be more strongly deflected than those of the heavier particles. In other words, the various ions move along circular trajectories with mass dependent radii. (With double-focusing mass spectrometers, an additional electrostatic analyser, which serves to focus the energy of the ions, is placed between the ion source and the magnetic analyser.) The radius of curvature can be expressed as:

$$r_m = \frac{m \cdot v}{z \cdot B} \tag{3}$$

B magnetic field strength

Equations (1) and (3) can be combined to give the fundamental equation of mass spectrometry (4):

$$\frac{m}{z} = \frac{r_m^2 \cdot B^2}{2 \cdot U} \tag{4}$$

The mass/charge ratio is therefore dependent upon the magnetic field strength, the radius of curvature and the acceleration potential. This equation has allowed the development of specific instrumentation for the direct detection of ions.

Ion Detection

If the acceleration potential and the magnetic field strength are kept constant, Eqn. (4) reduces to Eqn.(5):

$$\frac{m}{z} = k \cdot r_m^2. \quad (k = \text{const.}) \qquad (5)$$

This means that the m/z ratio is directly proportional to the square of the radius of curvature. As a result of this behaviour, it is possible to use a large number of individual collectors for the ion detection, or a photographic plate on which the number of impinging particles will be indicated by a corresponding variation in the degree of darkening of the plate. The distance between the individual dark streaks is then related to the masses of the recorded particles.

If the acceleration potential and the radius of curvature in Eqn. (4) are kept constant, Eqn. (6) is obtained:

$$\frac{m}{z} = k \cdot B^2 \quad (k = \text{const.}) \qquad (6)$$

Thus, for the determination of m/z (earlier m/e) ratios using a fixed radius of curvature it is only necessary to vary (scan) the magnetic field strength. In this case, only one ion detector is required at the exit to the magnetic analyser and an electron multiplier (EM) is used to amplify the very weak ion current.

A mirror galvanometer is used for the actual ion detection. This device directs a beam of UV light onto UV sensitive paper and as the paper moves forward, a spectrum is produced. Usually three traces are recorded simultaneously, which depict the same spectrum at different sensitivity ratios (usually 1:10:100). Additional traces are often also recorded, such as the ticked line of the mass marker. (Mass markers generally function very accurately, however it is necessary to calibrate them against a known mass from time to time. Normally a tick-mark is made at every fifth mass unit.) Another trace which can be drawn records the total ion current. This can be used as a record of the sample pressure developed during the measurement.

Today, the electrical signals are generally recorded during the measurement by a computer, which is directly coupled to the instrument. The data can subsequently be evaluated and printed as desired. Frequently the data are printed as a list of masses which contains both the mass number and the relative abundance of each mass. Furthermore, the computer can also display the data graphically as spectra, similar to those shown throughout this chapter. In contrast to the recording of spectra directly onto photosensitive paper, weak signals ($<1\%$) are not registered under normal conditions. If such signals are to be made visible – e.g. for the recognition of molecular ions with a low relative abundance – additional manipulation of the spectrum must be performed by the operator (e.g. a signal of low abundance, rather than the most abundant one, can be chosen as the base peak).

(see Fig. 4.2, p. 223). For this, the abundances of all signals above a specific mass (c.g. $m/z-20$) can be added together and the sum (e.g. 335) set to 100%. If, for reasons of low abundance, an important signal (e.g. $M^{+\bullet}$) still does not appear in the displayed spectrum when either of these scales are used, the affected region can be drawn using an expanded scale. This section would then be indicated with the label $\times 0.1$ or $\times 0.01$ (equivalent to $\times 10$ or $\times 100$), see Fig. 4.2. Another possibility for the display of signals of low abundance is to use a logarithmic rendering of the abundance of the total spectrum, instead of rel. %. For various reasons (e.g. the over-weighting of weak signals), this latter method is used only rarely.

3. Fragmentation of Organic Compounds

The following discussion contains **generalised remarks** about the behaviour of organic compounds under the influence of electron bombardment (70 eV). For the behaviour of inorganic or organometallic compounds, see the bibliography. With regard to other ionisation methods, which sometimes substantially reduce the fragmentation, see Sec. 8.9 (p. 263).

For reproduction in the literature, the recorded spectrum is drawn so that the most abundant peak of the spectrum (base peak) is set to 100% (relative %) and all other signals are scaled accordingly. If the most abundant signal appears at $m/z = 28$, or a similar mass number, it is advantageous to examine the sample of the compound for the presence of foreign substances (air, solvent). A convenient scale for displaying spectra has been found to be 1 rel. % = 1 mass number = 1 mm. Occasionally, one also finds the percentage fraction of the total ion current (% Σ) marked on the right hand side of the spectrum

The determination of the mass numbers in a spectrum, i.e. the association of each signal with a mass, is achieved either from the output of an automatic mass marker (as in a spectrum recorded and stored by mass on a computer), or by the count

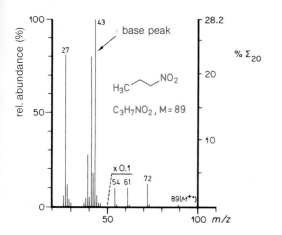

Fig. 4.2 The schematic representation of a spectrum, explained using the spectrum of 1-nitropropane as an example

ing out of easily identified and constantly appearing masses in a computed spectrum (stretched spectrum) [e.g. $12(C^{+\bullet})$, $18(H_2O^{+\bullet})$, $28(N_2^{+\bullet})$, $32(O_2^{+\bullet})$, $40(Ar^{+\bullet})$].

Molecular ion. Aside from a few exceptions (see Sec. 5, p. 246), the signal with the highest mass corresponds to the molecular ion peak.

Exception: so-called $[M + 1]^+$ or $[M + H]^+$ signals, which result from the tendency of H^+ ions to accumulate on molecules (especially prevalent with amines and alcohols). Furthermore, the $M^{+\bullet}$ signal is sometimes not observed, but that from the $[M - R]^+$ ion is seen instead. This occurs when the compound decomposes very easily.

Organic compounds generally consist of carbon, hydrogen, oxygen and nitrogen atoms and sometimes they also contain sulfur, phosphorus or halogen atoms. As shown in Tab. 4.13 (see p. 304), most of these elements are not monoisotopic, but are composed of a mixture of naturally occurring isotopes. Because most organic compounds have a natural origin, the ratio of this mixture of isotopes is also reflected in their mass spectra. Three categories can be defined for the most important elements, i.e. those most commonly occurring in organic compounds:

- **Monoisotopic elements:** ^{19}F, ^{31}P, ^{127}I;

- **Elements with one very abundant isotope:**
  ($> 98\%$): $H(^1H)$, $C(^{12}C)$, $N(^{14}N)$, $O(^{16}O)$;

- **Elements with two abundant isotopes:**
  $S(^{32}S, ^{34}S)$, $Cl(^{35}Cl, ^{37}Cl)$, $Br(^{79}Br, ^{81}Br)$.

Depending upon which of these elements are present in a substance, the molecular ion peak is accompanied by one or more

isotope peaks, which are always found at higher masses. Thus, the molecular ion of C_7H_6ClNO ($M = 155$) is composed as follows (Fig. 4.3):

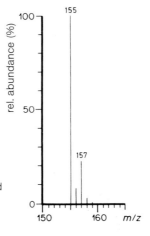

Fig. 4.3 Molecular region of the mass spectrum of C_7H_6ClNO

$m/z = 155$:

$$^{12}C_7 \ ^1H_6 \ ^{35}Cl_1 \ ^{14}N_1 \ ^{16}O_1 \qquad (1)$$

$m/z = 156$:

$$^{12}C_6 \ ^{13}C_1 \ ^1H_6 \ ^{35}Cl_1 \ ^{14}N_1 \ ^{16}O_1 \qquad (2)$$
$$+ \ ^{12}C_7 \ ^1H_5 \ ^2H_1 \ ^{35}Cl_1 \ ^{14}N_1 \ ^{16}O_1 \qquad (3)$$
$$+ \ ^{12}C_7 \ ^1H_6 \ ^{35}Cl_1 \ ^{15}N_1 \ ^{16}O_1 \qquad (4)$$
$$+ \ ^{12}C_7 \ ^1H_6 \ ^{35}Cl_1 \ ^{14}N_1 \ ^{17}O_1 \qquad (5)$$

$m/z = 157$:

$$^{12}C_5 \ ^{13}C_2 \ ^1H_6 \ ^{35}Cl_1 \ ^{14}N_1 \ ^{16}O_1 \qquad (6)$$
$$+ \ ^{12}C_7 \ ^1H_4 \ ^2H_2 \ ^{35}Cl_1 \ ^{14}N_1 \ ^{16}O_1 \qquad (7)$$
$$+ \ ^{12}C_7 \ ^1H_6 \ ^{37}Cl_1 \ ^{14}N_1 \ ^{16}O_1 \qquad (8)$$
$$+ \ ^{12}C_7 \ ^1H_6 \ ^{35}Cl_1 \ ^{14}N_1 \ ^{18}O_1 \qquad (9)$$
$$+ \ ^{12}C_6 \ ^{13}C_1 \ ^1H_5 \ ^2H_1 \ ^{35}Cl_1 \ ^{14}N_1 \ ^{16}O_1 \qquad (10)$$
$$\vdots$$

$m/z = 158$:

$$^{12}C_6 \ ^{13}C_1 \ ^1H_6 \ ^{37}Cl_1 \ ^{14}N_1 \ ^{16}O_1 \qquad (11)$$
$$\vdots$$

The isotope with the highest possible mass can be expected at $m/z = 173$ ($^{13}C_7 \ ^2H_6 \ ^{37}Cl \ ^{15}N \ ^{18}O$). As can be seen by estimation or calculation from the natural abundances of the individual isotopes (see Tab. 4.13, p. 304), the contribution given to the total abundance of an isotope peak by each of the various combinations can be very different. Whereas (1), (2), (8) and (11) represent the main contribution to the corresponding peaks, certain other combinations can be ignored because they have very low abundance; this is especially so in the case of $m/z = 173$.

It is a characteristic of compounds that possess elements with two common isotopes (e.g. Br and Cl), that the type and number of atoms of these elements can be deduced from the intensity ratios of the isotope peaks (see Tab. 4.13, p. 304 and 4.10, p. 294).

The mass number of the molecular ion in compounds of the type $C_u H_v N_w O_x (halogen)_y S_z$ also permits certain information to be derived about the number of N-atoms that are present. When the mass of the molecular ion is an even number, the presence of an even number of N-atoms ($N_0, N_2, N_4, ...$) is indicated. Conversely, an odd number for the mass of the molecular ion points to $N_1, N_3, N_5, ...$ (nitrogen rule).

Furthermore, the molecular ion represents that ion in a spectrum which possesses the smallest appearance potential (AP). In order to remove an electron from a neutral atom or molecule, a minimum amount of energy, the ionisation potential (IP), is required. For organic molecules, this energy lies between 7 and 14 eV (1 eV = 23.04 kcal·mol^{-1} = 96.3 kJ·mol^{-1}). Some examples:

n-hexane	10.17 eV	ethanol	10.48 eV
cyclohexane	9.88 eV	acetaldehyde	10.21 eV
cyclohexene	8.95 eV	acetic acid	10.35 eV
benzene	9.25 eV	methylamine	8.97 eV
anthracene	7.23 eV	aniline	7.70 eV
		trifluoromethane	13.84 eV

Therefore, when just the ionisation energy is available, only the molecular ion can appear as a signal in the mass spectrum. For the creation of fragment ions, an additional dissociation energy must be provided, so that the appearance potential of the fragment ions lies above that of the molecular ion[2].

Clear information about the elemental composition of a molecular ion can be obtained by the determination of its exact mass. This can be achieved by the use of high resolution mass spectrometry. The resolving power, A, of a mass spectrometer is defined by

$$A = \frac{m}{\Delta m}. \qquad (7)$$

According to the 10%-valley definition, two neighbouring signals can be considered to be resolved when they do not overlap each other by more than 10%. (The alteration of the positions of both maxima caused by the 10% overlap will still be within a tolerable limit.) As an example, two signals of equal intensity are shown in Fig. 4.4. In order to separate, for example, $m/z = 950$ from 951, a resolving power of 950 is required: $A = 950/1 = 950$. Low resolution mass spectrometers have a resolving power of between 1000 and 2000. In contrast, the determination of the exact masses of ions requires a much larger resolving power, as can easily be seen in the following example.

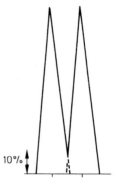

Fig. 4.4 Schematic representation of two neighbouring peaks of equal intensity with a 10% overlap (10%-valley definition)

The elemental compositions given in formulae (2)–(5) (p. 223) correspond, as which can easily be calculated with the help of Tab. 4.13, with the masses

156.017147	(2)
156.020069	(3)
156.010827	(4)
156.018008	(5).

For the separation of these masses, the following resolving powers are required:

$$A_{(2)/(3)} = \frac{156}{0.002922} = 53\,388$$

$$A_{(2)/(4)} = \frac{156}{0.006320} = 24\,684$$

$$A_{(2)/(5)} = \frac{156}{0.000863} = 180\,756$$

$$\vdots$$

From this it can be deduced that in order to record all four signals, a resolving power of ca. 181 000 is needed. The resolving power of a mass spectrometer fitted with a magnetic analyser is particularly limited when the ions are generated by electron impact, because the translational energy of the ions (caused, for example, by the effect of charge) is too inhomogeneous. The insertion of an electrostatic analyser before the magnetic analyser converts a simple mass spectrometer into a double-focusing instrument (cf. Fig. 4.5). The electrostatic analyser focuses the speed and energy of the ions. High resolution mass spectra can only be produced with such equipment.

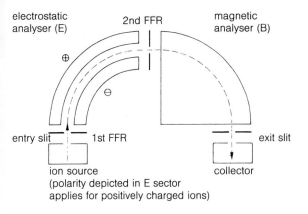

electrostatic analyser (E)

2nd FFR

magnetic analyser (B)

entry slit   1st FFR

exit slit

ion source
(polarity depicted in E sector
applies for positively charged ions)

collector

Fig. 4.5 Schematic representation of a double-focusing mass spectrometer with the EB-configuration (Nier-Johnson geometry).

FFR = field free region

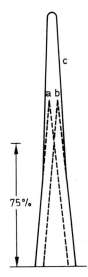

75%

Fig. 4.6 The overlap of two peaks, a and b, of equal intensity when the resolving power is insufficient. The resultant peak, c, will be recorded

Today, commercially available double-focusing instruments guarantee a resolving power of up to 150 000. However, because instruments in routine service can only quickly reach half of this value (due to light contamination of the ion source by the samples), only three signals in our example will be recorded: (3), (4) and (2) + (5). The peaks for (2) and (5) overlap and, depending on their intensity, each influences the recorded mass of the other. If (2) and (5) have equal intensities, the mean of both masses will be recorded (Fig. 4.6). On the other hand, if the intensity of (2) $\gg$ (5), the mass of (2) appears correctly, because (5) can be ignored. Such cases are to be reckoned with constantly and they can lead to the misinterpretation of spectra. Frequently, signals with different intensities overlap each other and the resolving power of the instru-

ment is just sufficient, so that the form of the peak displayed on the oscilloscope can be recognised visually as being either a singlet (ideal peak shape) or the result of two or more superimposed signals. In the latter case, the peaks can usually be separated by increasing the resolving power.

The elemental composition can be calculated from the exactly determined mass number either manually or, more realistically, with the time-saving help of a computer. If, for such calculations, no restrictions were made and it was assumed that any of the elements in the entire periodic table could be present in the unknown compound, then the number of possible combinations is very large. However, based on a knowledge of the origin of the substance and the chemical reactions used to synthesise it, it is nearly always possible to reduce the expected number of elements present in a compound to just a few, so that, in the ideal case, only the composition involving one set of elements must be determined. The same considerations are valid for the determination of the masses of fragment ions, whereby new selection rules, which can deliver additional details about the molecular ion, also come into play (e.g. fragment ions cannot contain any elements other than those in the molecular ion; the number of individual atoms in the fragment ion cannot exceed the number in the molecular ion; for typical fragment ions, it must be possible to find the corresponding cleavage sites, e.g. for $[M - 15]^+$, the fragment $[M - CH_3]^+$ must be used in the calculation and so a methyl group must be present in the molecular ion). In general, however, it is a valid concept that the number of combinations increases as the mass number rises.

Three different procedures are available for the production of high resolution mass spectra. For all of the following procedures, it is absolutely essential that the electric and magnetic fields of the instrument, as well as those from the surroundings, remain constant. Magnetic fields generated by nearby electric railways or tramways can have particularly disturbing effects.

Exposure of photographic plates. In a mass spectrometer tuned for high resolution (narrowed exit slit, see Sec. 2, p. 220), the sample being examined and a reference sample, usually perfluorokerosene (PFK), are vaporised simultaneously while the magnetic field and acceleration potential are held constant. The spectrum is recorded by making multiple exposures of a photographic plate. The high resolution mass numbers can be obtained by measuring the distances and intensities of the individual signals (the intensities are determined by comparing the darkness of the various streaks) with the aid of the reference spectrum. The advantage of the method is that all of the peaks are recorded simultaneously, which is important for thermal reactions, or when only a small amount of substance is available (e.g. for the investigation of metabolites and biological materials). The disadvantage is that the method requires additional instrumentation to conduct the measurement of the intensities and the distances.

Electrical recording (magnetic field scan). In this case the mass spectrum is not recorded on photographic paper (see Sec. 2), but three data channels (ion current, total ion current and variation in the magnetic field strength) are processed as a function of time by an interface and stored and displayed on a computer. The spectra of the sample and the reference substance (PFK) are recorded simultaneously and a "superimposed spectrum" is produced. The spectrum of PFK is recognised by the computer, because it is stored therein, and before it is eliminated, it is used as an aid for the subsequent calculations. A special computer program associates each signal with an exact mass (within a certain margin of error or uncertainty). (This calculation is based on the ratio of the distance between neighbouring PFK signals to the distance between a PFK signal and that of the substance whose mass is to be determined – both values are measured – as well as on the exact mass of the PFK signal, whose value is known.) In this way, the elemental composition of the corresponding ions can be determined and printed out as a list. It is advisable to record several spectra, one after the other, and compare the results. In this way it is possible to eliminate false spectra (e.g. due to the absence of any substance, the presence of only PFK, impurities, spikes, electronic noise or peak deformation).

The advantage of the method lies in its ability to rapidly record and evaluate high resolution spectra. The time required for one scan (ca. 20 s for measurements up to $m/z = 450$) can be a disadvantage when there is only the smallest amount of sample available or when the substance is thermally labile. This type of measurement also has a relatively low resolving power (ca. 10 000; multiplets). The computer printout of a high resolution spectrum is listed and explained on p. 325 (Chapter 5, **16**, spectrum 19). One possibility for overcoming these disadvantages is to record Fourier-transform spectra.

Peak matching method. First of all, by varying the magnetic field, a signal from a reference substance with a known exact mass (m_1) is displayed on a cathode ray oscilloscope in such a way that the half-height width of the peak fills about one third of the oscilloscope screen and the peak is scanned repeatedly at fixed time intervals. By the application of a supplementary potential (by the alteration of the acceleration potential), a signal of known nominal mass, but unknown exact mass (m_2),

can also be projected onto the oscilloscope screen in a similar way. Both of the signals are displayed alternately. The supplementary potential is adjusted so that both signals appear at exactly the same point on the screen. This supplementary potential can be determined exactly and the exact mass can then be calculated from ($B = $ const.):

$$m_1 : m_2 = U_2 : U_1 \qquad (8)$$

$$m_2 = \frac{m_1 \cdot U_1}{U_2} \qquad (9)$$

Because $U_1 = 1$ and m_1 are known, the unknown mass can be determined by division (uncertainty ± 3 ppm). There is a restriction that, depending on the type of instrument, the differences between the masses m_1 and m_2 must not exceed 10 to 20% of the mass of m_1.

The advantage of the method is that the ion signals of the sample in question are visible to the operator, which means that multiplets can be recognised visually, even when the constituent peaks have significantly different intensities. The limits of uncertainty can also be checked and ions of predicted elemental composition can be verified by the suitable application of Eqn. (9). The mass numbers obtained in this way are very exact. Two disadvantages are evident: a longer time, and thus a greater amount of substance, is required for a measurement, which consequently leads to the ion source becoming contaminated more quickly.

General reference for high resolution mass spectrometry: [3].

The molecular ion that has been excited by electron bombardment can now undergo fragmentation reactions (i.e. decomposition reactions). In Sec. 4 it will be assumed that the charge is localised, at least at the moment of the onset of the fragmentation reaction. Preferred sites for localisation are primarily heteroatoms with lone pairs of electrons, although the sites of π-bonds and π-bonded systems are also favoured. The least preferred sites are σ-bonds. As will easily be seen from the examples that are given, this concept is well suited to the interpretation of the spectra of organic compounds. There are also other theories applicable to fragmentation reactions, however, these will not be discussed here.

4. The Main Fragmentation Reactions of Organic Compounds

In this section, the most important, i.e. the most frequently observed, fragmentation reactions of organic compounds will be presented and discussed with the aid of examples.

4.1 α-Cleavage

Analogous reactions from other areas of chemistry (photochemistry): Norrish Type I reaction (α-cleavage).

α-Bonds adjacent to heteroatoms (such as N, O, S) are cleaved preferentially, because the ensuing charge is stabilised by the heteroatom.

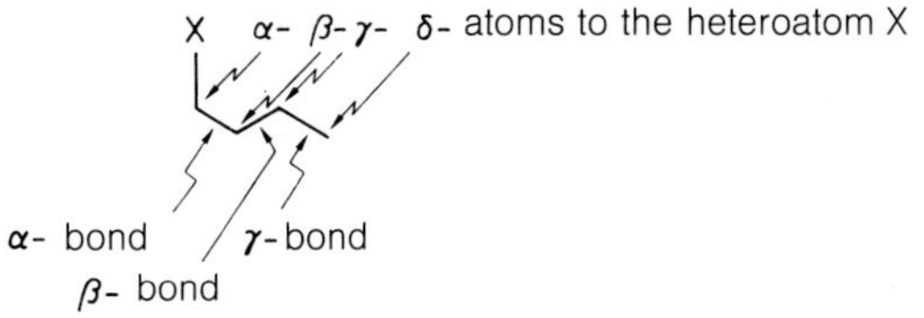

X   α- β- γ- δ- atoms to the heteroatom X

α- bond   γ-bond
β- bond

Aside from a very few exceptions, α-cleavage can occur only once in a decomposition chain (sequential fragmentation reactions). This is because the homolytic cleavage in a cation that has been created by the α-cleavage of a radical cation requires too much energy[a].

The mass spectrum of 2-butanone (**1**, M = 72) is displayed in Fig. 4.7. Two characteristic fragment ions are present: $m/z = 43$ and 57. Their masses differ from that of the molecular ion by 29 and 15 amu[b], respectively, which means that the corresponding fragment ions have been formed by the loss of the radicals $C_2H_5^{\bullet}$ and $CH_3^{\bullet}$, respectively, from the molecular ion. (*A priori* it is conceivable that the loss of $C_2H_5^{\bullet}$ has occurred firstly by the loss of $CH_3^{\bullet}$ ($m/z = 57$) followed by CH_2 (14 amu). The loss of CH_2 from molecular or fragment ions occurs extremely rarely, if at all. Therefore, we can exclude the two step process from consideration in this case. In contrast, the loss of CH_2 during collision activation reactions has been established using special compounds.) The creation of fragment ions

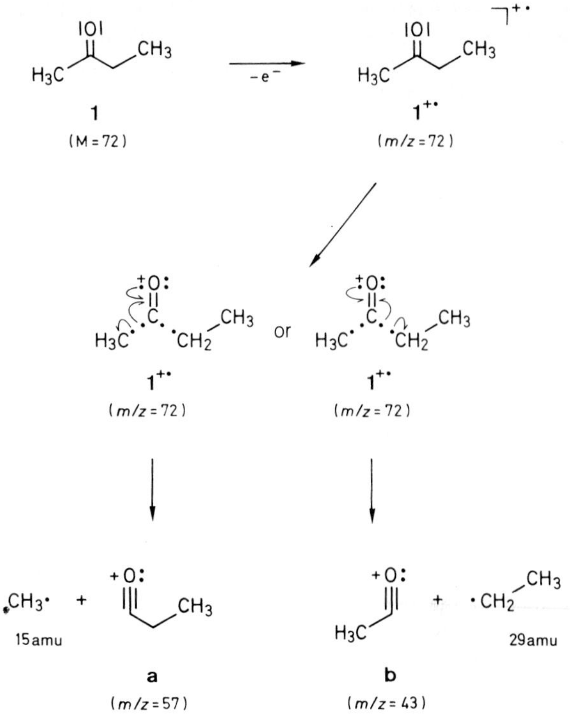

1
(M = 72)

1$^{+\bullet}$
(m/z = 72)

1$^{+\bullet}$
(m/z = 72)   or   **1**$^{+\bullet}$
(m/z = 72)

$_{\bullet}CH_3^{\bullet}$ +   **a**
15 amu   (m/z = 57)

b
29 amu   (m/z = 43)

Scheme 4.1   Detailed style for writing the main fragmentation of 2-butanone (**1**), see Fig. 4.7[d]

1

1$^{+\bullet}$
(m/z = 72)

- CH$_3^{\bullet}$ - C$_2$H$_5$ $\bullet$

a
(m/z = 57)   **b**
(m/z = 43)

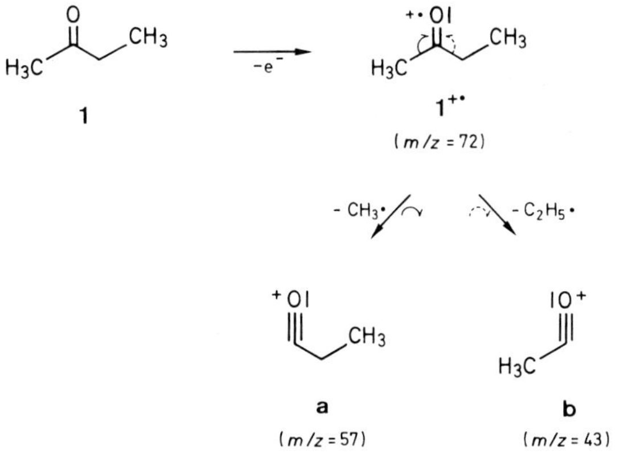

Scheme 4.2   Shortened style for Scheme 4.1

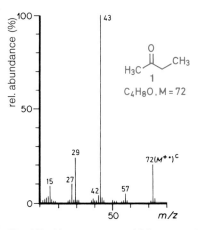
43

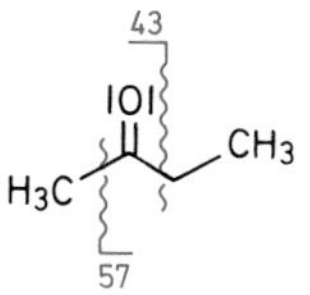

57

Scheme 4.3   Shorthand way of writing the main decomposition of 2-butanone (**1**) with an indication of the fragment ion masses (originating from α-cleavage).

rel. abundance (%)

43

72(M$^{+\bullet}$)^c

29

15   27   42   57

m/z

50

Fig. 4.7   Mass spectrum of 2-butanone (**1**)

---

[a] There are only a few cases described in the literature in which two consecutive α-cleavages have been observed. One example is the aromatic di-(*tert*-butyl) ether.

[b] 1 amu (atomic mass unit) is the constant of atomic mass which is defined as 1/12 of the mass of a ^{12}C atom. – The designation Dalton refers to a mass unit which is defined as the mass of a hypothetical atom of atomic weight 1 on the atomic weight scale.

[c] This is the usual style for representing the radical cation $M^{+\bullet}$

[d] The structures of the fragment ions are written such that the geometry and normal style of representing the original molecular ion is retained. As a result, the geometry of the fragment ions may sometimes be incorrectly represented.

from 2-butanone is formulated in Schemes 4.1, 4.2 and 4.3. In order to explain the style usually employed today for writing mass spectrometric decomposition reactions, the possibilities in this example will be discussed in detail. Under electron bombardment, one electron is ejected from the neutral molecule **1** and the singly charged positive molecular ion **1**$^{+\bullet}$ is created, which is registered on the m/z scale (mass per charge) at 72. By writing [formula]$^{+\bullet}$ it is meant that one is not making any assumption about the location of the charge within the molecular ion (Scheme 4.1). Because both of the fragment ions **a** and **b** result from the localisation of the charge at the O-atom, one chooses the two writing styles formulated in Scheme 4.1 where, in each case, an electron has been removed from one of the two electron pairs on the O-atom to give a single positive charge on this atom. Two bonds are present which are α-bonds to the O-atom (not to the C=O group). It is therefore possible to stabilise the lone electron on the O-atom by pairing it with one electron of the single bond at the carbonyl C-atom. The second electron of the σ-bond remains with the alkyl fragment, in this case the CH_3 (creating $CH_3^{\bullet}$) or $C_2H_5^{\bullet}$ fragments, respectively[e]. These radicals will not be recorded by the mass spectrometer because they are uncharged. The resultant fragment ions are denoted with lower case letters (**a–z**, **aa–zz**, **ba**...) when they are referred to in the text. It proves to be extremely useful if the mass is given in parentheses under the symbol of the fragment ion. Sometimes it is also useful to indicate the heavier neutral fragments with their weight; this is then done as depicted in Scheme 4.1: e.g. $CH_3^{\bullet}$ (15 amu). The ions $m/z = 29$ and 15 originate most often from $m/z = 57$ and 43, respectively; see in this regard Sec. 4.7 (p. 244).

If one only wishes to indicate on a structural formula how the main fragment ions are generated by α-cleavage, the formulation given in Scheme 4.3 should be chosen. Suitably modified schemes can also be used to indicate other cleavage reactions.

A generally important rule for α-cleavage is that with compounds of type **2** the heavier substituent will cleave preferentially in the case where R^1 and R^2 are homologues (see Tab. 4.1). An analogous behaviour is found for compounds of the general type **3** (see Table 4.2). For the α-cleavage of carboxylic acids and their derivatives, see Sec. 4.5 (p. 240). Although the α-cleavage of aliphatic compounds leads directly to the formation of fragment ions, the corresponding alicyclic compounds yield only isomeric molecular ions. Cyclohexanone (**4**; M = 98), for example, is just such a case. The base peak of the spectrum (Fig. 4.8) is $m/z = 55$. It can be shown by labelling experiments that the mechanism given in Scheme 4.4 for the formation of the ion of this mass is correct.

---

[e] The movement of a single electron is indicated by a fish-hook arrow ($\rightarrow$); that of an electron pair is shown by a normal arrow ($\rightarrow$). In principle, the movement of every single electron must be indicated by a fish-hook arrow, as shown in Scheme 4.1. However, because the shorter style used in Scheme 4.2 is just as clear, it is used in preference.

Tab. 4.1 The relative abundances of fragment ions generated by α-cleavage of

X	R¹ R²	M − R¹		M − R²		Compound	M
	all straight-chain	m/z rel. abund. (%)		m/z rel. abund. (%)			
ketones							
O=C	CH_3 C_2H_5	57	6	43	100	2-butanone	72
	CH_3 C_4H_9	85	4	43	100	2-hexanone	100
	C_2H_5 C_3H_7	71	61	57	100	3-hexanone	100
	C_3H_7 C_4H_9	85	75	71	100	4-octanone	128
	C_3H_7 C_6H_{13}	113	66	71	100	4-decanone	156
secondary alcohols							
OH−C−H	CH_3 C_2H_5	59	19	45	100	2-butanol	74
	CH_3 C_3H_7	73	6	45	100	2-pentanol	88
	CH_3 C_4H_9	87	5	45	100	2-hexanol	102
	C_2H_5 C_3H_7	73	41	59	100	3-hexanol	102
secondary thiols							
SH−C−H	CH_3 C_2H_5	75	5	61	100	2-butanthiol	90
	CH_3 C_3H_7	89	2	61	100	2-pentanthiol	104
amines							
NH₂−C−H	CH_3 C_2H_5	58	11	44	100	2-aminobutane	73

Tab. 4.2 The relative abundances of fragment ions generated by α-cleavage of $R^1-CH_2-X-CH_2-R^2$

X	R¹ R²	M − R¹		M − R²		Compound	M
	all straight-chain	m/z rel. abund. (%)		m/z rel. abund. (%)			
ethers							
O	CH_3 C_3H_7	87	2	59	100	butyl ethyl ether	102
	C_2H_5 C_3H_7	87	54	73	100	butyl propyl ether	116
amines							
NH	CH_3 C_2H_5	72	10	58	100	N-ethyl propyl amine	87
	C_2H_5 C_3H_7	86	43	72	100	butyl ethyl amine	115

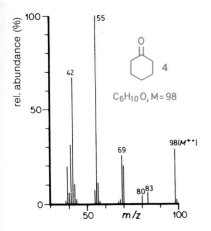

Fig. 4.8   Mass spectrum of cyclohexanone (**4**)

but the ion **c** also appears at the same mass number. In the mass spectrum of 2-methyl- and 3-methylcyclohexanone, $m/z = 69 (= 55 + 14)$ will be recorded in addition to $m/z = 55$. For dimethylcyclohexanone, appropriate ions will be recorded at $m/z = 55$ (no methyl substituents on the 2 and 3 or the 5 and 6 positions, respectively), 69 (one methyl group at one of these positions) and 83 (two methyl groups).

α-Cleavages of other alicyclic compounds occur in a similar manner and form ions that have a structure comparable with ion **c** from cyclohexanone. Thus, the corresponding ion in the mass spectrum of cyclohexanol (**5**; M = 100, Fig. 4.9) is shifted by +2 amu to $m/z = 57$ (**d**) and that from N-ethylcyclohexyl-amine (**6**; M = 127, Fig. 4.10) is found at $m/z = 84$ (**e**). The ethylene acetal of cyclohexanone shows $m/z = 99$ (**f**) as the most abundant fragment ion signal.

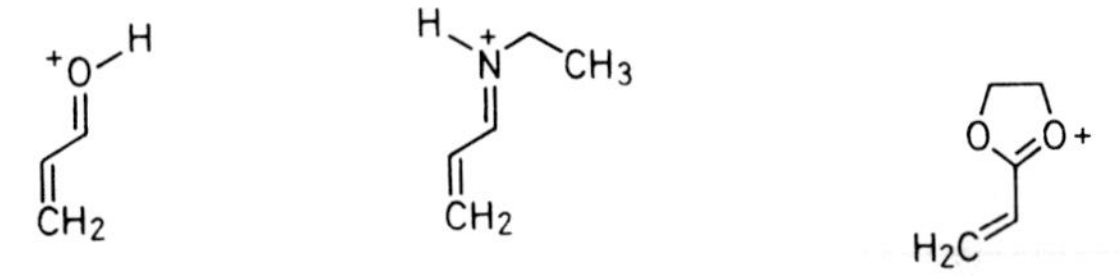

d
(m/z = 57)

e
(m/z = 84)

f
(m/z = 99)

If a carbonyl group is incorporated in larger alicyclic assemblages, fragment ions that originate from α-cleavage adjacent to the carbonyl group will still be detected, although their abundances will be weaker, because other cleavage reactions

A primary radical is present in the isomeric molecular ion. This radical is stabilised by the transfer of an *H*-atom, *via* a six-membered transition state, over from the C-2 position, which has been activated by the C=O group. As a result, a resonance-stabilised radical is created, which is energetically more stable than the initial step. A radical cleavage reaction results in the formation of a propyl radical as well as the ion **c** ($m/z = 55$), in which the multiple bonds are conjugated.

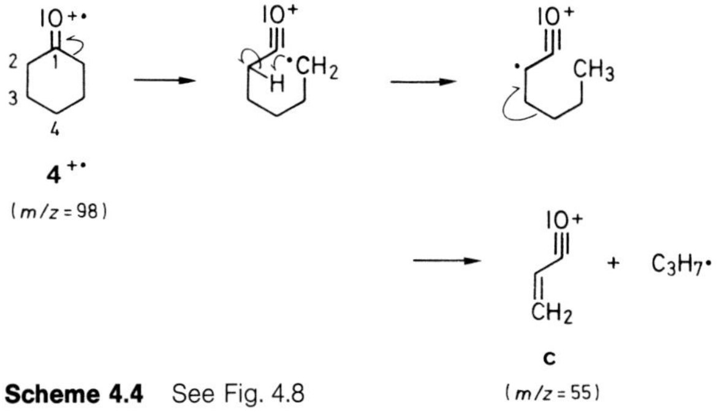

Scheme 4.4   See Fig. 4.8

Depending upon the type of substituent and the site of substitution, alkyl derivatives of cyclohexanone show the ion **c** or a homologue thereof. If, for example, a methyl group is at the 4-position, not only is the molecular ion recorded at $m/z = 112$,

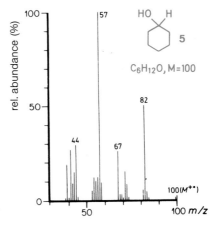

Fig. 4.9   Mass spectrum of cyclohexanol (**5**)

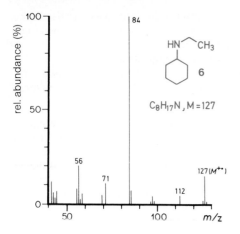

rel. abundance (%)

100 — 84

HN—CH₃

6

$C_8H_{17}N$, M = 127

56

50 —

71

112

127 ($M^{+\cdot}$)

50   100 m/z

Fig. 4.10   Mass spectrum of N-ethylcyclohexylamine (**6**)

$7^{+\cdot}$

($m/z = 318$)

3 ‖ 4

2 ‖ 3

g

i

h

j

f

($m/z = 99$)

k

m

($m/z = 125$)

l

Scheme 4.5   See Fig. 4.11

may occur with a similar probability. On the other hand, the incorporation of an ethylene acetal function instead of a carbonyl group causes a marked preference for α-cleavage at the new group. To illustrate this, the spectrum of 5α-androstan-3-one ethylene acetal (**7**; M=318) is reproduced in Fig. 4.11. The primary main cleavage reaction is the α-cleavage that is controlled by the ethylene acetal residue. However, in contrast to the "model substance" behaviour of the ethylene acetal of cyclohexanone, the two α-bonds adjacent to the functional group (C-2–C-3 and C-3–C-4) are not equivalent, because the "cyclohexane ring" is substituted at C-5 and C-10. As a result, it is apparent that **7** can undergo two α-cleavages. Both of these possibilities have been confirmed by D-labelling experiments and are shown in Scheme 4.5. The cleavage of the α-positioned C-3–C-4 bond produces the isomeric molecular ion **g**, which is comparable with the primary cleavage ion from cyclohexanone (Scheme 4.4). (In Scheme 4.5, the α-cleavage of the bonds C-3–C-4 and C-2–C-3 is depicted by 3 ‖ 4 and 2 ‖ 3, respectively. This is an alternative to the use of various arrows to indicate cleavage possibilities, as was done in Scheme 4.2.) The resonance-stabilised and also isomeric molecular ion **h** is generated by the transfer of an H-atom from the 2-position. This is then converted by the breaking of the C-1–C-10 bond into the ion **f** (m/z=99), which possesses conjugated double bonds. In a similar way, the second α-positioned bond, C-2–C-3, cleaves to give **i** and then **j**. The cleavage of the C-5–C-10 bond in **j** does not, however, result in the loss of a radical, but once again an isomeric molecular ion (**k**) is formed, in which the tertiary radical once more accepts one of the C(6)-H-atoms via a 6-membered transition state. In the thereby obtained ion **l**, the possibility exists for an ideal radical cleavage: the cleavage of the C-7–C-8 bond yields the ion **m** (m/z = 125) with three conjugated double bonds. The spec-

trum in Fig. 4.11 clearly shows that the fragment ions **f** and **m** play a dominating role in the decomposition of **7**.

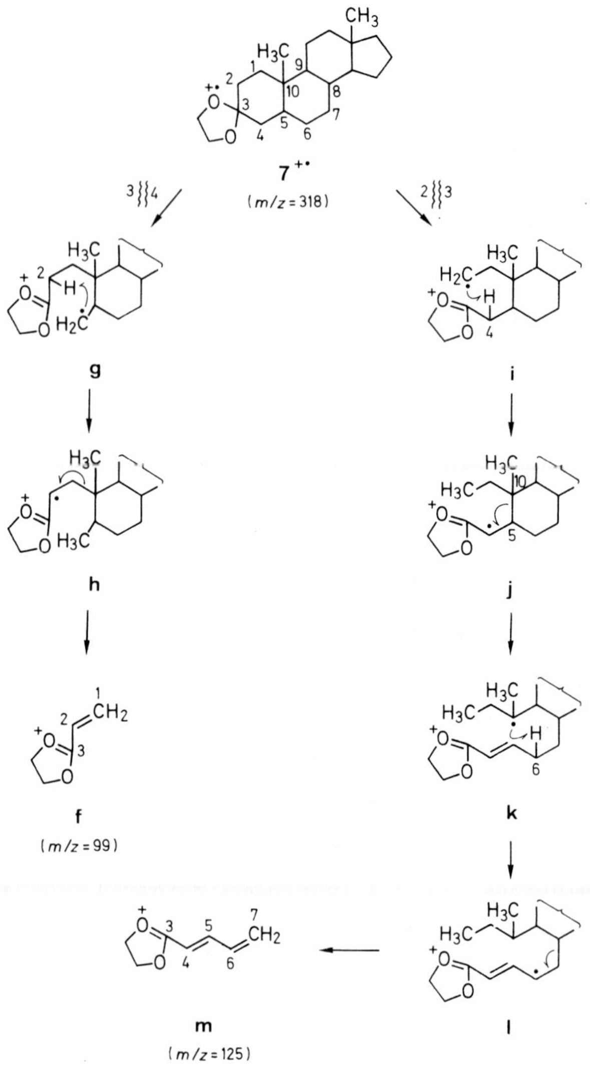

Similarly, the N,N-dimethylamino group, which is frequently found in steroid alkaloids, possesses the same strong α-cleavage properties as the ethylene acetal group. Very abundant fragment ion signals, which correspond with ions of similar structure, are also observed for these compounds. The appearance of various very abundant signals in the spectra of these types of compounds permits the determination of the points of substitution of the α-cleavage directing groups and, as a result, important parts of their structures can be deduced.

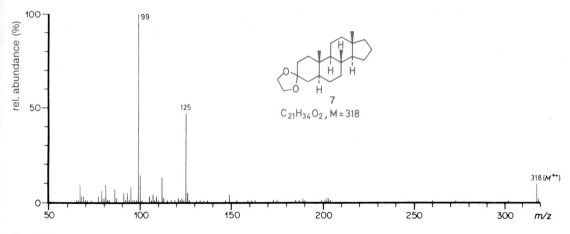

Fig. 4.11   Mass spectrum of 5α-androstan-3-one ethylene acetal (**7**)

As already mentioned, the ethylene acetal group is able to direct decomposition reactions that start with α-cleavage to a much greater extent than the carbonyl group. By comparing the spectra of Figs. 4.8-4.10, it is evident that ketones, secondary alcohols and amines direct the α-cleavage to different degrees. With compounds of the type X–CH$_2$–CH$_2$–Y, where X and Y stand for different functional groups, it is possible to observe directly the effect of X and Y on the α-cleavage. The spectrum of 2-aminoethanol (**8**; M=61, Fig. 4.12) can be used as an example. Cleavage of the C–C bond yields the ions with m/z=30 and 31. As can be seen in Fig. 4.12, the intensity of the signal due to the ion m/z= 30, which is the nitrogen-containing fragment ion, exceeds by far that of the ion m/z=31. Hence it follows that the NH$_2$-group is significantly more charge stabilising and therefore directs the fragmentation more strongly than the aliphatic hydroxy group.

Tab. 4.3   A ranking of the relative degree to which substituents influence charge stabilisation (α-cleavage)

Functional group	Ion abundance	Functional group	Ion abundance
—COOH	1	—I	109
—Cl	8	—SCH$_3$	114
—OH	8	—NHCOCH$_3$	128
—Br	13	—NH$_2$	990
—COOCH$_3$	20	⬡ (dioxolane)	1600
⟩=O	43	—N(CH$_3$)CH$_3$	2100
—OCH$_3$	100		

Table 4.3 lists some measured values, from which the individual functional groups can be compared with each other with respect to the relative abundances of the ions that arise from direct α-cleavage. When these values are only considered in relative terms, they actually give an order of effectiveness of substituents for inducing α-cleavage in equivalent structural entities.

The α-cleavage is the most important primary fragmentation reaction in mass spectrometry. Additional examples are included in the discussion of the other fragmentation reactions.

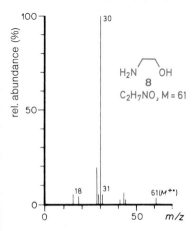

Fig. 4.12   Mass spectrum of 2-aminoethanol (**8**)

4.2 Benzyl and Allyl Cleavage

Aromatic centres, delocalised double bonded systems and even isolated double bonds have an activating influence on suitable benzylic or allylic bonds, in a similar way that a heteroatom influences a bond and leads to α-cleavage.

Benzyl cleavage. The mass spectrum of butylbenzene (**9**; M = 134) is shown in Fig. 4.13. The cleavage of the benzylic C–C bond results in the loss of a propyl radical to yield the main fragment ion, which is seen as the base peak of the spectrum $m/z = 91$ (**n, o, p**) (see Scheme 4.6).

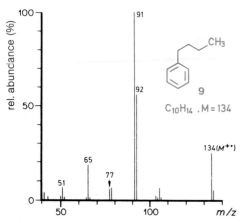

Fig. 4.13   Mass spectrum of butylbenzene (**9**)

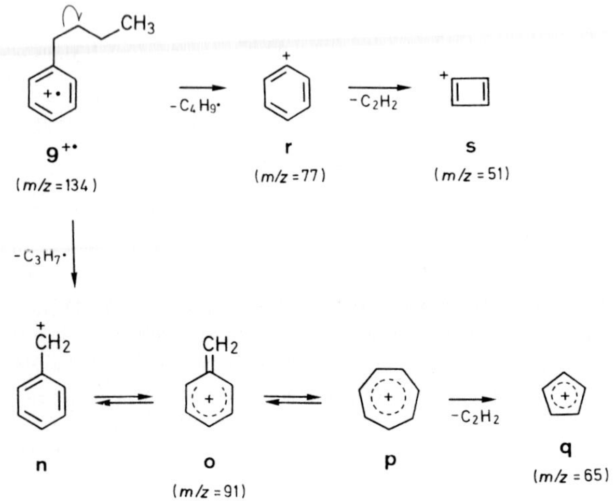

Scheme 4.6   See Fig. 4.13

The high abundance of this signal indicates that the corresponding ion is very stable. The stability is not just a result of the formation of either of the tautomers **n** or **o**, but also originates from the formation of the tropylium ion (**p**, $C_7H_7^+$). The proof that **p** really is the deciding species comes from, among others, the subsequent reaction, which is the loss of C_2H_2 to yield **q** ($m/z = 65$). In the symmetrical tropylium ion (**p**), all of the C-atoms are equivalent to each other, as are all of the H-atoms. In contrast, the ions **n** and **o** possess at least three types of C-atom (CH_2, CH, C) and at least two types of H-atom. If a ^{13}C- or D-labelled compound is now used, it must be possible in the case of the loss of C_2H_2 from **n** and **o** to determine a dependence of the labelled atoms on the original labelling positions, whereas this would not be possible for **p**. Analyses of alkylbenzenes have confirmed the equivalence of the C-atoms in the ion $m/z = 91$, which favours the acceptance of **p**. From the spectrum of butylbenzene, the reverse conclusion can also be drawn, which is not unimportant for structural analyses. That is, that an abundant signal at $m/z = 91$ indicates the presence of a benzyl residue in a compound of unknown structure. Weak signals, however, are less characteristic because the highly stable tropylium ion can also be formed by complicated rearrangements.

At the same time, the spectrum of **9** shows that the phenyl-cleavage (formation of **r**, $m/z = 77$) is significantly less favoured than the benzyl cleavage. **r** also loses acetylene (ion **s**, $m/z = 51$). The pairs of ions, $m/z = 91/65$ and $m/z = 77/51$, are typical for monosubstituted alkyl aromatics. The ion $m/z = 92$ will be discussed in Sec. 4.5 (p. 240).

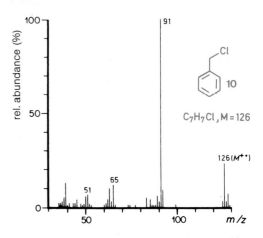

Fig. 4.14   Mass spectrum of benzyl chloride (**10**)

Other monosubstituted *n*-alkylbenzenes also show $m/z=91$ as the base peak (e.g. toluene, ethylbenzene, propylbenzene, pentylbenzene), as do *o*-, *m*- and *p*-xylene.

It is interesting to compare the spectra of benzyl chloride (**10**; M=126, Fig. 4.14) and *o*-chlorotoluene (**11**; M=126, Fig. 4.15). Aside from small differences in the relative abundances, the two spectra are the same.

In line with the above discussion, the spectrum of benzyl chloride appears as expected because a *Cl*-atom is more easily removed from the benzyl position than an *H*-atom. The spectrum of *o*-chlorotoluene, however, is surprising. The *Cl*-atom is directly attached to the phenyl ring and, as a result, the tendency for it to be ejected is small. Thus one would assume that a chlorosubstituted tropylium ion would be created by the loss of H · (from the CH₃ group) and that therefore an analogous behaviour to that of *p*-(chloro)ethylbenzene (**12**; Fig. 4.16)

would be expected. (The peak $m/z = 125$ in the spectrum of *p*-(chloro)ethylbenzene is caused by the chlorotropylium or chlorobenzylium ion.)

However, because the spectra of **10** and **11** are the same, one can assume that **11** rearranges to **10** or, more likely, that they both isomerise to another common species, e.g. **13**, before they fragment (cf. Scheme 4.7). One would expect **13** to have the same mass spectrometric behaviour as **10**, but so far this investigation has not been carried out.

To what extent other systems, which show abundant $m/z=91$ signals (or corresponding derivatives thereof) in the mass spectrometer, isomerise after ionisation, but before their decomposition to cycloheptatriene derivatives, must be clarified in each particular case.

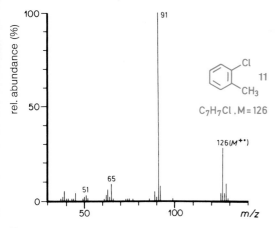

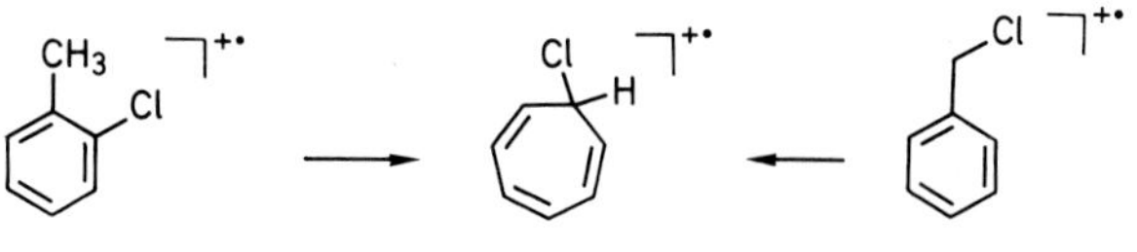

Fig. 4.15   Mass spectrum of *o*-chlorotoluene (**11**)

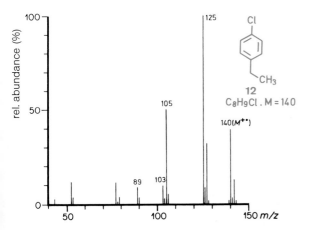

Fig. 4.16   Mass spectrum of *p*-(chloro)ethylbenzene (**12**)

Scheme 4.7   See Figs. 4.14 and 4.15

Furthermore, the appearance of ions of mass 91, or their derivatives, does not automatically allow one to conclude that they are tropylium or benzylium ions. Usually this question must also be investigated in each case.

Allyl cleavages are less pronounced than benzyl cleavages, because the energy stabilisation through the formation of the resulting allyl cation is smaller. Fig. 4.17 shows the mass spectrum of 1-heptene (**14**; M = 98) and Fig. 4.18 depicts that of 4-methyl-1-hexene (**15**; M=98). In both spectra, the base peak is at $m/z=41$ (**t**). Because the allyl-positioned bond in straight chain isomers is the weakest bond and, at the same time, the allyl cation is the most stable ion, $m/z=41$ will be recorded as the most abundant ion. The other cleavage fragment, $m/z= 57$, which is also formed by an allyl cleavage, is, with 27% relative abundance, much weaker. Similarly, the allyl-positioned C-3–C-4 bond in the other isomer is also the most labile bond and so the spectrum of this isomer also shows the allyl cation as the most prominent ion. However, the peak $m/z = 57$ is

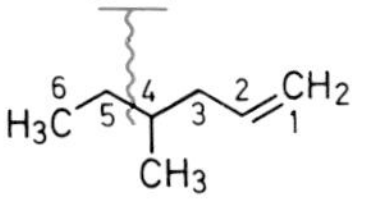

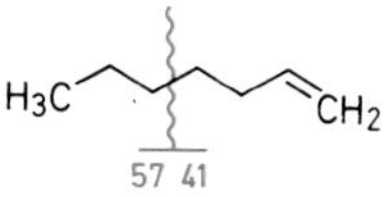

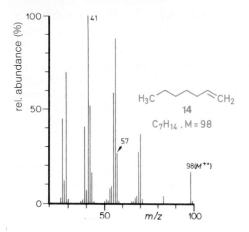

Fig. 4.17   Mass spectrum of 1-heptene (**14**)

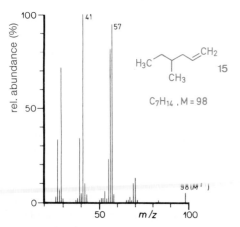

Fig. 4.18   Mass spectrum of 4-methyl-1-hexene (**15**)

recorded as the second most abundant peak with 95% relative abundance, because of the greater stability of the secondary carbocation. (The ions $m/z = 42$ and 56 result from a McLafferty rearrangement, see Sec. 4.5, p. 240.) From these examples, it follows that cleavage at allyl positions is favoured over that at C—C bonds. However, the charge stabilisation of the resulting allyl cation is not reliable and, in addition, competing reactions can occur, which interfere with or even prevent the recognition of the C=C bonds. Particularly unpleasant, moreover, is the shifting of C=C bonds from their original sites. It is therefore better and safer to fix the positions of the C=C bonds by derivatisation and to analyse the derivatives mass spectrometrically. The acetonyl compounds of the corresponding diols have proven themselves to be suitable derivatives. Similarly, for C≡C bonds, the analysis of derivatives (carbonyl compounds produced by the addition of water) is preferable.

4.3 The Cleavage of "Non-activated" Bonds

This section will summarise cleavage reactions in which the bond to be cleaved is not activated by heteroatoms (α-cleavage), phenyl groups (benzyl cleavage) or C=C bonds (allyl cleavage). Fig. 4.19 shows the mass spectrum of hexadecane (**16**; M = 226). This spectrum is typical for unbranched, straight-chain hydrocarbons. The most abundant signals lie in the region corresponding to fragments with three and four C-atoms, i.e. between $m/z = 40$ and 60. With an increasing number of C-atoms, the abundance of the homologous ions decreases almost asymptotically. An $[M - 15]^+$ ion will not be

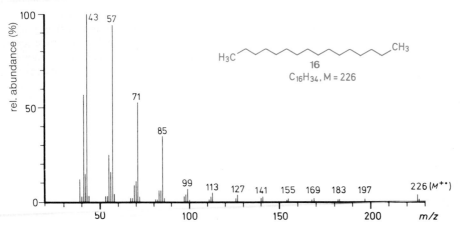

Fig. 4.19   Mass spectrum of hexadecane (**16**)

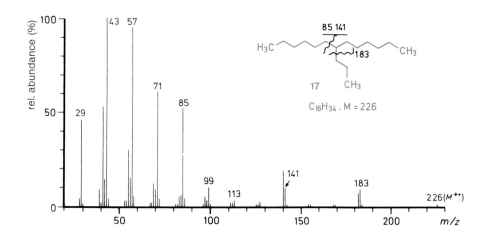

Fig. 4.20 Mass spectrum of 7-propyltridecane (**17**)

recorded, but, as a result of this, the signal due to the molecular ion is very easily recognised. The general shape of the profile of the spectrum (the connection lines between the highest peaks within each of the groups of signals) is typical. It is useful to memorise this profile, because hydrocarbons often appear as impurities in samples (see Sec. 6, p. 251). Signals that stand out by producing such a uniform picture must have a meaning for the structural analysis. The uniformity is indeed essentially the result of the fact that the cleavage of each C–C bond results in the formation of a primary carbocation and a primary radical. An exception exists only with the two terminal C–C bonds, where either $CH_3^{\bullet}$ or CH_3^+ can be formed. If, however, the hydrocarbon is branched by an alkyl chain, the bonds to be cleaved are no longer equivalent. Additional secondary carbocations can now be formed and they are also more readily produced, so that the signals due to secondary carbocations stand clearly above the general profile of the spectrum. As an example, the spectrum of 7-propyltridecane (**17**; M = 226) is shown in Fig. 4.20.

With multiply branched hydrocarbons the spectra become unclear and the analysis turns out to be significantly more difficult.

The mass spectrometric analysis is of great importance, because, with the exception of ^{13}C NMR spectroscopy, other analytical techniques are not suitable for the determination of the structures of higher hydrocarbons. The larger the hydrocarbon part of monofunctional compounds, the more similar their mass spectra are to those of the pure hydrocarbons themselves. By altering the type of functional group, the number of CH_2 groups necessary before such a compound displays this spectral behaviour can be made higher or lower.

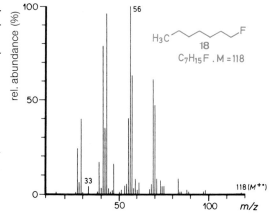

Fig. 4.21 Mass spectrum of 1-fluoroheptane (**18**)

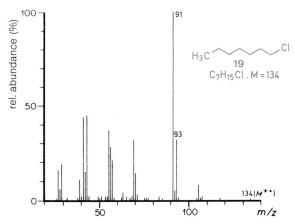

Fig. 4.22 Mass spectrum of 1-chloroheptane (**19**)

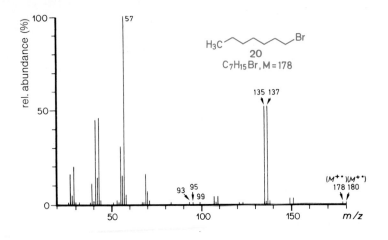

Fig. 4.23 Mass spectrum of 1-bromoheptane (**20**)

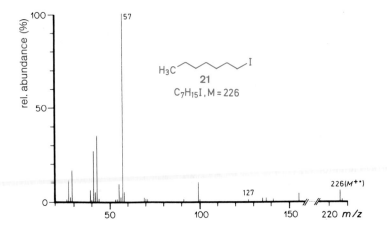

Fig. 4.24 Mass spectrum of 1-iodoheptane (**21**)

The decomposition of aliphatic halohydrocarbons is determined only to a very small extent by α-cleavage adjacent to the halogen atom. Fluorohydrocarbons show ions that have arisen through α-cleavage as single abundant peaks. Conversely, with iodoalkanes, the breaking of the C—X bond with charge localisation on the halogen atom occurs much more frequently and can easily be recognised in low resolution spectra by the large gap in the masses, which is due to the iodine. (If the charge is localised on the alkyl chain, the presence of iodine cannot be established.) Particularly characteristic is the behaviour of 1-chloro- and 1-bromohydrocarbons with at least five linearly ordered methylene groups. They usually form, as the most abundant signal of the spectrum, pentacyclic chloronium and bromonium ions with their characteristic isotopic abundance ratios (see Tab. 4.10, p. 294). To illustrate these

points, the mass spectra of four 1-haloheptanes are given in Figs. 4.21 to 4.24.

$m/z = 91 + 93$ $m/z = 135 + 137$

4.4 The Retro Diels-Alder Reaction (RDA Reaction)

Review article: [4].

Six-membered cyclic systems that contain a double bond can, *via* a concerted decyclisation reaction, dissociate into two fragments, the ene and diene components. The diene component is favoured as the charge carrier, however the ene part is also frequently observed in mass spectra. The cyclohexene ring need not contain any heteroatoms, but it can possess one or more of them. It may also form part of a larger ring system. The RDA reaction can occur in molecular ions as well as in fragment ions in which the double bond in the ring has been formed initially as a result of another fragmentation reaction (e.g. through an α-cleavage). The RDA reaction is a so-called neutral process in which a radical cation is formed either when the starting ion is a radical cation, or when a cation losses a non-radical fragment to form another cation.

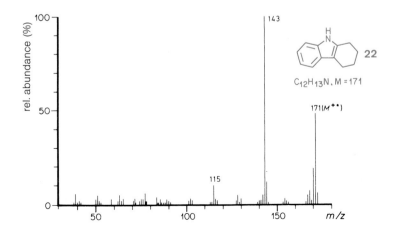

This reaction type will be clarified further by using the mass spectrum of 1,2,3,4-tetrahydrocarbazole (**22**; M = 171) as an example. At the same time, the term "shift technique" should be introduced and the procedure for the clarification of a reaction mechanism demonstrated. Both are important working techniques in mass spectrometry.

The most abundant ion in the mass spectrum of **22** (Fig. 4.25) is $m/z = 143$. Firstly, it was determined by high resolution mass spectrometry that the main fragment ion $m/z = 143$ ($C_{10}H_9N$) differs from the molecular ion by the deficiency of C_2H_4 (28 amu). Furthermore, a metastable signal (m*; cf. Sec. 8.25, p. 277) could be found at $m/z = 119.6$, which confirms the transition $171 \rightarrow 143$, i.e. the ion of mass 143 is formed directly from the molecular ion. (Based on the empirical formula of the molecular ion, another formulation of the fragment ion

(namely $C_{11}H_{11}$) would also be possible; this could, for example, have been formed in the following way:

$$M^{+\bullet} \rightarrow [M-H]^+ \rightarrow [M-H-HCN]^+.)$$

By using measurements from low-voltage and field ionisation spectra, it could be confirmed that the compound did not contain any impurities. Based on the 70 eV spectrum alone, one would have the right to surmise that **22** is contaminated by dehydrogenation products, as evidenced by the signals at $m/z = 169$ (M - 2 H) and 167 (M - 4 H). However, because only $m/z = 171$ (and the corresponding isotope peaks) is found in the field ionisation spectrum, the dehydrogenation products must result from mass spectrometric processes. (If the dehydrogenation products had been present in the sample of the compound, their presence would also be apparent in the UV spectrum. However, the UV spectrum is in agreement with the liter-ature spectrum of a pure sample, which provides an additional confirmation of the results from the field ionisation spectrum.) The methods of analysis described above can be carried out on the unlabelled compound and can therefore be completed without additional synthetic effort. They should always be done before the investigation of the reaction mechanism.

Useful information can also be delivered by the mass spectra of derivatives which exhibit mass spectrometric behaviour analogous to that of the molecule of interest. In the present case, the mass spectra of N-methyl-1,2,3,4-tetrahydrocarbazole (**23**; M = 185 corresponding to 171 + 14) and 1,2,3,4-tetrahydrocarbazol-3-ol (**24**; M = 187 corresponding to 171 + 16) were drawn into the investigation. Both spectra exhibit great similarity with that from **22**, i.e. one intense fragment ion signal is present (for **23**: $m/z = 157$, for **24**: $m/z = 143$), which in both cases is the base peak. The molecular ion peaks are about

Fig. 4.25 Mass spectrum
of 1,2,3,4-tetrahydrocarbazole (**22**)

half as abundant as these fragment ion signals (see Tab. 4.4). In the spectrum of the methyl derivative **23**, not only is $M^{+\bullet}$ shifted by the mass of the substituent, but the fragment ion signal is also shifted by a similar amount; i.e. the fragment ion contains the methyl group and therefore also the N-atom. The situation is different with the hydroxy compound. Indeed the molecular ion is also shifted by +16 amu, however, the fragment ion signal is found at the same mass number as in the

spectrum of 1,2,3,4-tetrahydrocarbazole (**22**) itself. This allows the conclusion to be drawn that during the transition from $M^{+\bullet} \rightarrow m/z = 143$ the atom C-3 is also lost. This method of analysis is known as the Shift Technique (or Biemann Shift). It can be employed when compounds possess the same skeleton, but have different substituents and, aside from small differences in the abundances of the signals, exhibit similar mass spectra. Based on the shifts in the signals (or the absence thereof), conclusions can be drawn about the substituents[f].

For the unequivocal clarification of a reaction mechanism, it is, however, necessary to investigate labelled derivatives. Therefore the following deuterated compounds were synthesised: 4,4-dideutero-1,2,3,4-tetrahydrocarbazole (**22a**; M = 173; by reduction of 1,2,3,4-tetrahydrocarbazol-4-one with LiAlD$_4$ and working up in the presence of H$_2$O); 1,1,3,3-tetradeutero-1,2,3,4-tetrahydrocarbazole (**22b**; M = 175; by boiling the vinylogous amide, 1,2,3,4-tetrahydrocarbazol-4-one, with CD$_3$OD/CH$_3$ONa, neutralisation of the reaction solution with DCl/D$_2$O and reduction of the thus formed 1,1,3,3,9-pentadeutero-1,2,3,4-tetrahydrocarbazol-4-one with LiAlH$_4$) and 1,1,2,3,4,4-hexadeutero-1,2,3,4-tetrahydrocarbazole [**22c**; M = 177; by boiling 1,2,3,4-tetrahydrocarbazol-1-one with DCl/D$_2$O and reduction of the product with zinc amalgam/DCl/D$_2$O (→ 1,1,2,2,4,4,5,6,7,8,9-undecadeutero-1,2,3,4-tetrahydrocarbazole) followed by boiling with HCl/H$_2$O]. The results of the mass spectra of the three compounds are summarised in Table 4.4. (The determination of the D-content of the molecular ions was, because of the strong [M – 1]$^+$ signal, carried out by comparing the field ionisation spectra. The D-content of the fragment ions was determined using the 70 eV spectra (see Sec. 8.6, p. 261). The correct positioning of the isotope insertion was confirmed with ^{1}H NMR spectra.) If one assumes a priori that for the loss of ethylene from (**22**) the six combinations

(A) $H_2\overset{1}{C}=\overset{2}{C}H_2$,   (B) $H_2\overset{1}{C}=\overset{3}{C}H_2$,   (C) $H_2\overset{1}{C}=\overset{4}{C}H_2$

(D) $H_2\overset{2}{C}=\overset{3}{C}H_2$,   (E) $H_2\overset{2}{C}=\overset{4}{C}H_2$,   (F) $H_2\overset{3}{C}=\overset{4}{C}H_2$

are possible, then the spectrum of **22a** permits the elimination of possibilities (C), (E) and (F), the spectrum of **22b** additionally eliminates (B) and that of **22c** eliminates (A), whereby the possibility (D), i.e. the elimination of H$_2$C(2)=C(3)H$_2$, can be taken as proven. Based on these results, two mechanisms can be brought into consideration (Scheme 4.8).

Tab. 4.4 Compounds used in the analysis of the RDA reaction of 1,2,3,4-tetrahydrocarbazole (**22**)

Compound	$M^{+\bullet}$				Fragment ion		
	m/z	difference from $M^{+\bullet}$ of **22**	rel. abundance (%)	m/z	difference from m/z = 143	rel. abundance (%)	
22	171	0	63	143	0	100	
23	185	+ 14	50	157	+ 14	100	
24	187	+ 16	43	143	0	100	
22a	173	+ 2	62	145	+ 2	100	
22b	175	+ 4	62	145	+ 2	100	
22c	177	+ 6	80	147	+ 4	100	

[f] In the case of the two compounds **23** and **24**, when the sites of both substituents are unknown, but the mechanism of the fragmentation reaction is, however, known, the following conclusions can be drawn: the methyl group could be attached at the positions 1,4,5,6,7,8 or 9 and the hydroxy group at positions 2 or 3, without expecting the spectra to be significantly different.

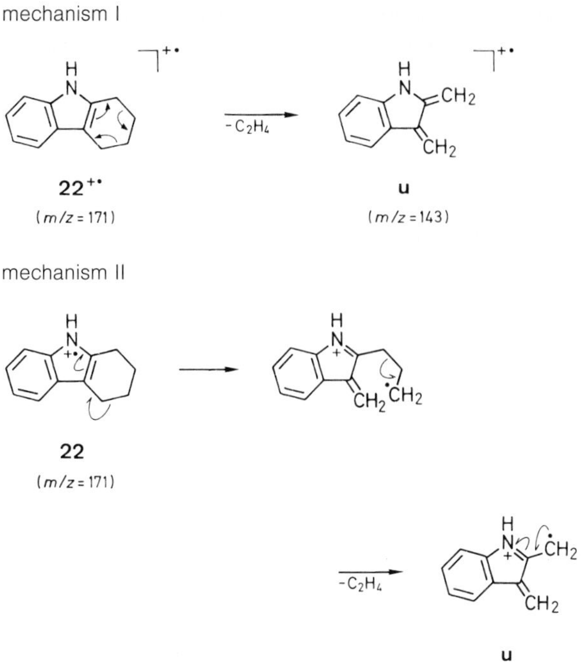

mechanism I

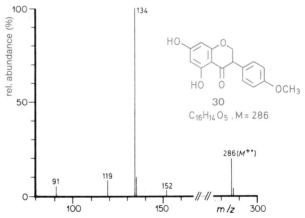

$$22^{+\bullet} \xrightarrow{-C_2H_4} \mathbf{u}$$

22$^{+\bullet}$
($m/z = 171$)

u
($m/z = 143$)

mechanism II

22
($m/z = 171$)

$$\xrightarrow{-C_2H_4}$$

u
($m/z = 143$)

Scheme 4.8   See Fig 4.25

Mechanism I is a concerted RDA reaction, whereas mechanism II is a stepwise process that begins with a vinylogous α-cleavage and leads, *via* a free radical elimination reaction, to the same structural ion **u**. On the basis of the above results, it is not possible to decide between these mechanisms.

The RDA reaction has been demonstrated for a series of additional systems, e.g. 1,2,3,4-tetrahydro-β-carboline (**25**) and 1,2,3,4-tetrahydroisoquinoline (**26**). In all of these cases the ions $m/z = 143$ (from **25**) and 104 (from **26, 27** and **28**), respectively, form the base peaks of the spectra. Conversely, only about one third of the intensity of the main fragment ion signal $m/z = 104$ of the unsubstituted 1,2,3,4-tetrahydronaphthalene (tetralin, **29**) results from an RDA reaction, the remaining two thirds result from a rearrangement of the CH$_2$ groups before the loss of ethylene.

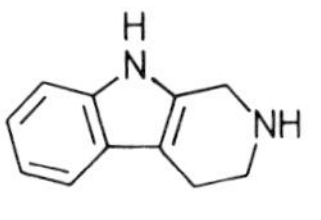

25

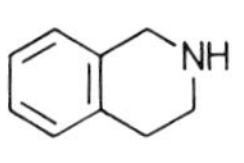

26

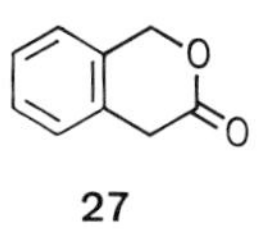

27

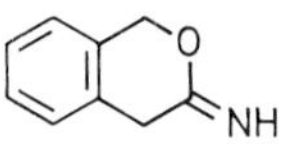

28

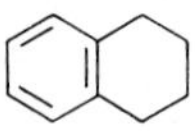

29

Many organic natural substances contain these types of ring systems and the elucidation of their structures is achieved frequently by using the mass spectrometric RDA reaction. Particular examples of this are indole alkaloids (with **22** and **25** as part of the structure) and tetrahydroisoquinoline alkaloids (with **26** as part of the structure). In addition, many natural substances that belong to the flavonoid family (e.g. flavones, isoflavones, rotenoids) also have a central ring which can undergo the RDA reaction. The mass spectrum of 5,7-dihydroxy-4'-methoxyisoflavanone (**30**; M=286) is reproduced in Fig. 4.26. The molecular ion is cleaved *via* an RDA reaction in ring C into two parts, each of which can carry the charge (Scheme 4.9). This reaction produces **v** ($m/z = 152$) as the diene component and, as the ene part, **w** ($m/z = 134$), which forms the base peak of the spectrum. **w** can also lose CH$_3^{\bullet}$ to give $m/z = 119$. From the masses of both fragment ions, one can deduce that ring A (mass of the unsubstituted ion: 120) carries two hydroxy groups and ring B (mass of the unsubstituted ion: 104) has either a methoxy group or, and this cannot be differentiated mass spectrometrically, one hydroxy and one methyl group. This method of determining the distribution of substituents between rings A and B proves to be very useful for the elucidation of structurally unknown compounds. Additional conclusions about the points of substitution are, however, not possible. These must be determined by additional spectroscopic or chemical analyses.

30
C$_{16}$H$_{14}$O$_5$. M = 286

Fig. 4.26   Mass spectrum of 5,7-dihydroxy-4'-methoxyisoflavanone (**30**)

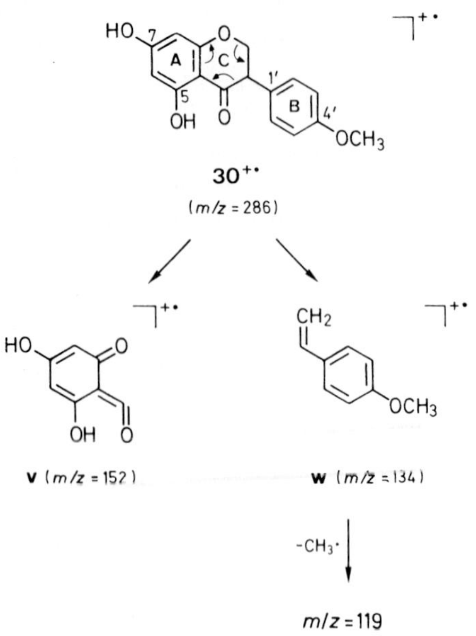

30 $^{+\bullet}$

$(m/z = 286)$

v $(m/z = 152)$ **w** $(m/z = 134)$

$-CH_3\bullet$

$m/z = 119$

Scheme 4.9   See Fig. 4.26[9]

4.5 The McLafferty Rearrangement

Analogous reactions: photochemistry: Norrish Type II reaction; ester pyrolysis, Tschugaev reaction, ene reaction.

This reaction type is also designated as a β-cleavage with an H-atom migration. In this reaction, an H-atom is transferred *via* a 6-membered transition state from the γ-position to another atom which must be at least double-bonded. Simultaneously, a migration of the double bond occurs and a neutral fragment containing the β- and γ-positioned atoms is ejected. The process can occur either in a concerted (I) or stepwise (II) fashion. The acceptor double bond that is necessary for reaction can be either already present in the starting molecule or formed in another fragmentation reaction (e.g. α-cleavage). Among the groups that can undergo a McLafferty rearrangement, a few are: C=O (e.g. carboxylic acids, esters, aldehydes, ketones, amides, lactams, lactones), C=N (e.g. azomethines or Schiff bases, hydrazones, oximes, semicarbazones), S=O (e.g.

g At this point one should allude to the occasional errors in formulation that appear in the literature: if both the ene and diene components appear as ions from an RDA reaction, as in the case of **30**, then the formulation

$30^{+\bullet} \rightarrow$ **v** $(m/z = 152) +$ **w** $(m/z = 134)$

is wrong, because it implies the nonsensical physical condition of $+^{\bullet} = +^{\bullet} + +^{\bullet}$. However, it is acceptable to write

$30^{+\bullet} \rightarrow$ **v** $(m/z = 152)$ or **w** $(m/z = 134)$.

sulfonic acid esters) and C=C (e.g. alkylarenes, alkylheterocycles, benzyl ethers, olefins).

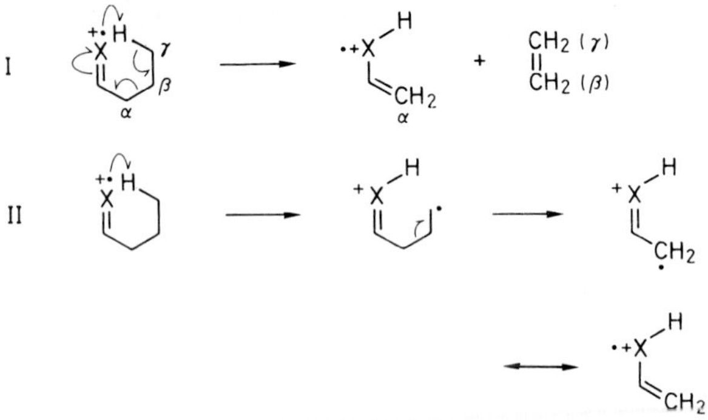

I

II

The McLafferty rearrangement can be illustrated with a couple of examples. Methyl butanoate (**31**; M = 102, Fig. 4.27) forms the ion $m/z = 74$ (**x**) through a McLafferty rearrangement involving the loss of ethylene. Other fragment ions emanating from **31** ($m/z = 31, 59, 71$) can be explained by α-cleavages (Scheme 4.10). The ejection of CO from the ion $m/z=71$ results in $m/z = 43$.

The signals at $m/z = 74$ (McLafferty rearrangement) and 59 (α-cleavage) are characteristic for methyl esters; for ethyl esters, the signals are found at 88 and 73, respectively, etc. The ion from the McLafferty rearrangement of aliphatic carboxylic acids is $m/z = 60$.

$m/z = 60$

y

$(m/z = 87)$

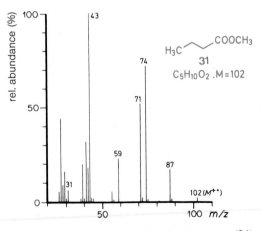

Fig. 4.27   Mass spectrum of methyl butanoate (**31**)

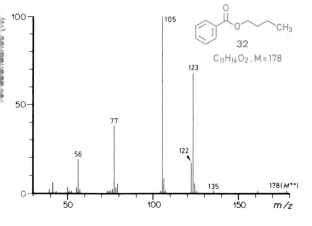

Scheme 4.10   See Fig. 4.27

For higher fatty acid esters, there is an increase in the abundance of the fragment ion **y**, which appears +13 amu higher than the fragment ion from the McLafferty rearrangement [for methyl esters: $m/z = 74$ (**x**) + 87 (**y**)]. Esters of higher alcohols and of aromatic carboxylic acids, e.g. butyl benzoate (**32**; $M = 178$, Fig. 4.28), also show, in addition to the ions from

Scheme 4.11   See Fig. 4.28

Fig. 4.28   Mass spectrum of butyl benzoate (**32**)

the McLafferty rearrangement, the ions of the protonated carboxylic acid. For **32**, the former are **z** ($m/z = 122$) and **aa** ($m/z = 56$) and the protonated benzoic acid ($C_6H_5COOH_2^+$) has the mass 123 (see Scheme 4.11). The ion $m/z = 105$, which is the

most abundant signal of the spectrum, originates from the α-cleavage adjacent to the carbonyl group and is typical for derivatives with a phenylcarbonyl group, such as benzylic acid esters, phenyl ketones, etc. From $m/z = 105$, the ion $m/z = 77$ is formed by the loss of CO and, following that, the ejection of acetylene results in $m/z = 51$ (see Sec. 4.1, p. 226).

McLafferty rearrangement reactions that occur under the influence of C=C bonds are represented, for example, by $m/z = 92$ in the spectrum of butylbenzene (**9**; Fig. 4.13) and by $m/z = 42$ and 56 in those of 1-heptene (**14**; Fig. 4.17) and 4-methyl-1-hexene (**15**; Fig. 4.18).

McLafferty rearrangements in which an alkyl or another residue is rearranged, instead of a γ-H-atom, are rare processes.

4.6 The Onium Reaction

Behind this name lies a reaction type that is observed mainly with fragment cations in which the heteroatom is the charge carrier. Fragment ions of this type are the ox**onium**, ammon**ium**, phosph**onium** and sulf**onium** ions.

Under electron bombardment, *N*-isopropyl-*N*-methylbutyla-mine (**33**; M = 129, Fig. 4.29) loses $CH_3^\bullet$ or $C_3H_7^\bullet$ through α-cleavage, whereby the ammonium ions **ab** ($m/z = 114$) and **ac** ($m/z = 86$) are formed.

The double bond in ion **ab** has a γ-positioned *H*-atom which can rearrange according to McLafferty. The fragment ion thereby formed is **ad** ($m/z = 72$). The other ion from the α-cleavage (**ac**) is unable to undergo a similar decomposition reaction (no γ-*H*-atom). Conversely, both ammonium ions can undergo the onium reaction in which an *H*-atom is transferred from the alkyl residue to the *N*-atom and the alkyl residue is eliminated. Thus C_4H_8 (56 amu) is eliminated from **ab** to give the ion **ae** ($m/z = 58$), while the elimination of C_3H_6 (42 amu) from **ac** produces **af** ($m/z = 44$) (see Scheme 4.12). Because the precise origin of the *H*-atom that is transferred to the hetero-atom is usually unknown (no regiospecificity; deuteration experiments have shown that the various *H*-atoms along an alkyl chain will be transferred to varying degrees), one choos-es, for example for the further decomposition of **ab**, the follow-ing manner of representation:

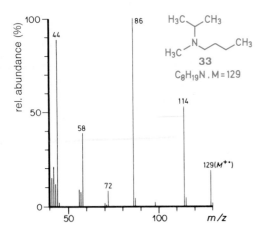

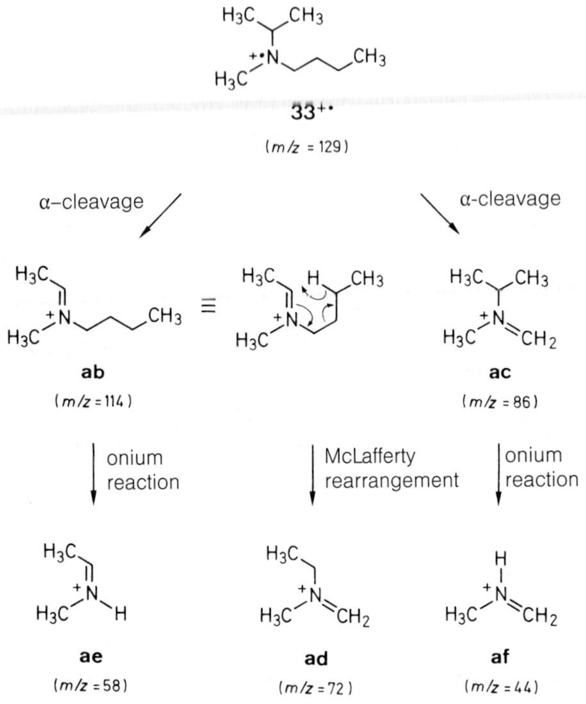

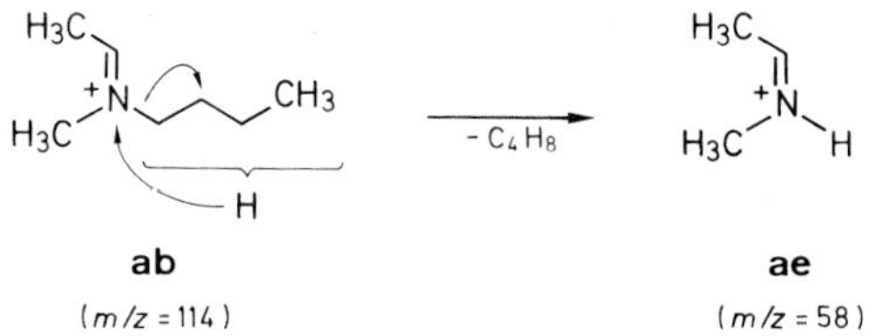

Fig. 4.29 Mass spectrum of *N*-isopropyl-*N*-methylbutylamine (**33**)

Similar decomposition sequences can be established for ethers [see Fig. 4.30, Scheme 4.13, butyl ethyl ether (**34**; M = 102)] and thioethers.

In addition to alkyl substituents (other than CH_3), acyl resi-dues can also undergo the onium reaction. In the mass spec-trum of *N* butylacetamide (*N*-acetylbutylamine, **35**; M = 115, Fig. 4.31), $m/z = 30$ (**ag**) is the base peak. The same base peak is found for the *n*-alkylamine itself. However, the primary amino group in **35** is acetylated and therefore is not free. According to

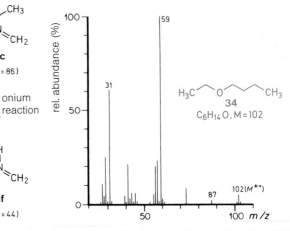

Fig. 4.30 Mass spectrum of butyl ethyl ether (**34**)

Scheme 4.12 See Fig. 4.29

Scheme 4.13 See Fig. 4.30

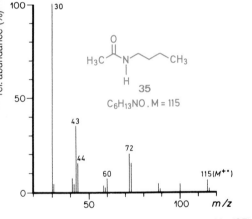

Fig. 4.31 Mass spectrum of N-butylacetamide (**35**)

$C_6H_{13}NO$. M = 115

Scheme 4.14 See Fig. 4.31

Scheme 4.14, **ah** ($m/z = 72$) is formed first of all from the molecular ion through an α-cleavage adjacent to the N-atom. Subsequently, **ag** is produced by an onium reaction in which ketene is lost. At this point it is worth mentioning that N- and O-acetyl compounds can be recognised by the signal at $m/z = 43$. Furthermore, the spectra of N-substituted acetamides are characterised by a signal at $m/z = 60$ (**ai**), which results from a McLafferty rearrangement with an additional H-atom transfer.

Other N-acetyl compounds behave similarly to **35**. It is worth remarking that acyl residues, which do not contain any aliphatic bonded H-atoms, can also be ejected through an onium reaction (with an H-migration!); such residues are, e.g., benzoyl-, benzenesulfonyl-, p-toluenesulfonyl- (= tosyl-).

The most abundant ion $m/z = 149$ (**am**) in the spectra of phthalic acid dialkyl esters, e.g. diethyl phthalate (**36**; M = 222, Fig. 4.32), also has an onium reaction to thank for its existence: the ion that is produced by the ejection of an alkoxy residue, $m/z = 177$ (**ak**), cyclises. The cyclisation results from

Scheme 4.15 See Fig. 4.32

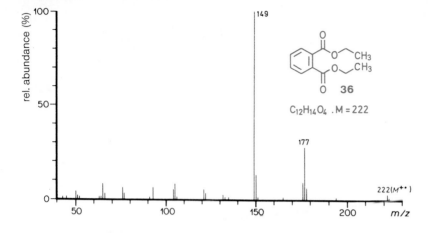

Fig. 4.32 Mass spectrum of diethyl phthalate (**36**)

the influence of the adjacent (*o*-positioned) second ethoxycarbonyl group (**al**) and enables the charge from the charge-carrying carbonyl group to be transferred to the ether *O*-atom. The ion **am** (*m/z* = 149, Scheme 4.15) is then produced *via* an *H*-transfer from the alkyl residue to the *O*-atom (see pp. 253, 302). Frequently, *o*-disubstituted benzene derivatives, when compared with the *m*- and *p*- isomers, show a special mass spectrometric behaviour, which is known as the *ortho* effect, see Sec. 8.15 (p. 269).

In a few cases, the ejection of acyl residues that are bound to *O*- or *N*-atoms occurs directly from the molecular ion. In particular, acyloxybenzenes and *N,N*-diacetylalkylamines belong to such compounds. It must be remembered, however, that these substances are very easily hydrolysed and that, because of this, it is very easy for mixtures of substances (e.g. formation of phenols, *N*-acylalkylamines) to find their way into the analysis.

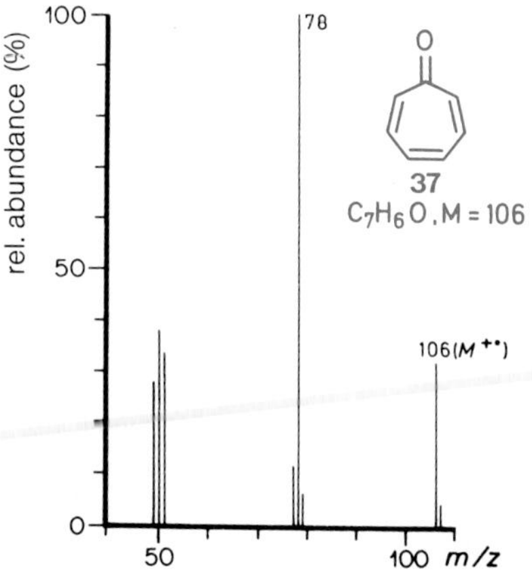

4.7 Loss of CO

Cyclic, highly unsaturated compounds and even ions that have been formed by α-cleavage adjacent to a carbonyl group (see p. 228), have the tendency to eject CO (28 amu). If several CO groups are present in a molecule, they can be eliminated one after another. Usually such a fragmentation reaction is indicated by an (intense) metastable signal (cf. Sec. 8.25, p. 277).

The mass spectrum of tropone (**37**; M = 106) is given in Fig. 4.33. It clearly documents the favoured fragmentation reaction in such systems. The remaining part of the spectrum of **37** resembles that of benzene, which means that the ion resulting from the loss of CO is cyclic.

Fig. 4.33 Mass spectrum of tropone (**37**)

Compounds, whose enol form in solution is much more abundant than the keto form, also lose CO in the mass spectrometer. Phenols are typical of this substance class.

Phenol (**38**; M = 94) itself gives the spectrum shown in Fig. 4.34. The most abundant fragment ion of the spectrum is *m/z* = 66 (**an**), which is formed by the loss of CO. Because this ion is still a pseudo-molecular ion, it can transform to the cyclopentadienyl cation (*m/z* = 65) through the loss of an H• (Scheme 4.16). As an example of a compound with two CO groups, dispiro[4.1.4.1]dodecane-6,12-dione (**39**; M = 192, Fig. 4.35) will be discussed. The base peak of the spectrum is at *m/z* = 96, which is half of the mass of the molecular ion. This peak is not due to a doubly charged molecular ion (isotope peaks at half mass numbers are not present), but represents the fragment ion **ao**. This fragment then loses CO and presumably produces the cyclopentene ion **ap** (*m/z* = 68), which can convert to the cyclopentadienyl cation **aq** (*m/z*=65) by the ejection of a total

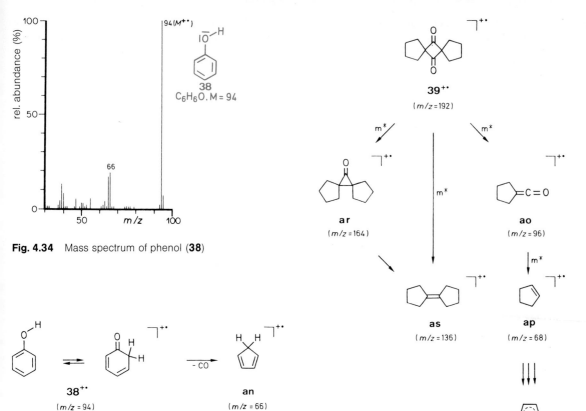

Fig. 4.34   Mass spectrum of phenol (**38**)

Scheme 4.16   See Fig. 4.34

Scheme 4.17   See Fig. 4.35

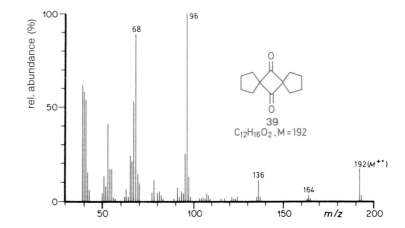

Fig. 4.35   Mass spectrum of
dispiro[4.1.4.1]dodecane-6,12-dione (**39**)

of three H-atoms. In a second decomposition pathway, one CO can initially be ejected from $M^{+\bullet}$ to give **ar** ($m/z = 164$), followed by a second CO to give **as** ($m/z = 136$). Metastable signals show that the formation of **as** can also occur directly from the molecular ion (Scheme 4.17). In this step C_2O_2 can be eliminated either as one fragment or in two rapid consecutive reactions. The loss of both CO groups also occurs to some extent with other diketones (e.g. anthraquinone).

The α-cleavage with its variants of the benzyl and allyl cleavages and the cleavage of non-activated C—C bonds, the RDA reaction, the McLafferty rearrangement, the onium reaction and also, because of the frequent occurrence of carbonyl groups, the loss of CO, very often play a deciding role in the

decomposition of organic molecules. However, the range of decomposition reactions is not completely exhausted. Special functional groups or a particular arrangement of atoms can sometimes bring about special fragmentation reactions. Some examples of such reactions are: loss of water, $S_N i$ reactions and reactions with neighbouring group participation (see Sec. 8.15, p. 269).

5. Thermal Reactions in the Mass Spectrometer

In order to record an electron impact mass spectrum, it is necessary, as described earlier, that the sample of the substance be in the gas phase. Because most organic compounds are liquids or solids at room temperature, they must first be brought into the vapour phase. Furthermore, for particular measurements (e.g. spectra of gases and liquids), it can be important to maintain the samples in the gas phase while they are held in sample reservoirs for relatively long periods of time. Not only does the vaporisation process require an elevated temperature, but the gaseous molecules can also undergo collisions with the walls of the reservoir and with parts of the ion source, which must be maintained at higher temperatures in order to prevent condensation. Collisions of this type lead to an increase in the energy of the molecules and can give rise to catalysed thermal decomposition reactions.

The inlet sections of the mass spectrometer have been improved markedly in recent years. Materials that promoted thermal reactions have been replace by more inert components. The vaporisation device has also been improved significantly. The result of this is a reduction in the number of observed thermal decomposition processes in the mass spectrometer (for this reason, among others, older and newer spectra frequently show characteristic differences). However, many reactions of this type are still observed, so that a knowledge of them is absolutely necessary for the interpretation of mass spectra. Thermal decomposition reactions occur especially when high temperatures are necessary for the vaporisation of the sample, which is the case for compounds with large relative molecular masses (> 400) and/or multiple polar functional groups (e.g. −COOH, −OH, −NH$_2$, −SH). Furthermore, impurities in the sample (e.g. silica gel, aluminium oxide or even activated charcoal) can catalyse thermal reactions.

These thermal reactions have nothing to do with the actual mass spectrometric fragmentation reactions. They occur

before the ionisation and lead, in comparison with the substance being analysed, to heavier, lighter or equally heavy (i.e. isomeric) particles. If two or more particles result from such processes, they will be ionised independently of each other and give superimposed mass spectra.

5.1 The Most Important Types of Thermal Reactions[5]

It can happen that organic substances decompose in an unspecific way into many small and large pieces. This usually occurs with high molecular weight compounds that are introduced into the mass spectrometer *via* the direct inlet system and ionised under electron impact conditions. If one records several mass spectra from such samples, one after the other, these spectra generally bear no great resemblance to one another. One gets the impression that a mixture of several substances must be present whose ion abundances decrease with increasing mass, without there being an obvious end to the spectrum. Aside from this type of general, more or less unspecific decomposition reaction, a few frequently observed general cases are described below with the aid of examples.

Thermal Loss of Smaller Fragments

CO$_2$ (Decarboxylation). In particular, β-oxocarboxylic acids, but also aromatic and other compounds with multiple carboxylate groups, are readily inclined to lose CO$_2$. This is especially so when the β-oxocarboxylic acid is part of a larger molecular assemblage.

CO (Decarbonylation). α-Oxocarboxylic acids and their alkyl esters lose CO to some extent at their distillation temperature.

Such reactions can also occur under the conditions used to record mass spectra. Thus the spectra of ethyl (2-oxocyclohexyl)glyoxylate (**40**; M = 198) and its decarbonylation product, ethyl 2-oxocyclohexanoate (**41**; M = 170), are the same, aside from one signal with only a low abundance at $m/z = 28$ (gas inlet, 200°C).

40   **41**

CH₃COOH. The thermal loss of acetic acid from acyloxy derivatives can have its origin in an ester pyrolysis. This occurs particularly when the newly formed double bond can conjugate with an existing one and causes a reduction in the decomposition temperature. As an example, the mass spectrum of the indole alkaloid, *O*-acetylhervine (**42**; M = 426, ion source temperature 250°C, direct inlet), is described. At the beginning of the measurement, the ratio of $M^{+\bullet}/[M - 60]^{+\bullet h}$ was found to be 0.72. After just 3 min $M^{+\bullet}$ had disappeared and only $m/z = 366$ (**43**) was still observed. Hence it follows that the cause of the behaviour is a thermal ester pyrolysis and not its mass spectrometric equivalent, the McLafferty rearrangement.

42   **43**

Other functional groups can also display a similar behaviour under certain circumstances.

HX (H₂O, HCl, etc.). There are so many examples of the loss of water, hydrochloric acid, etc., that one should not need to go into this here. Not only are hydroxy groups responsible for the loss of water, but also certain *N*-oxides (after an ensuing rearrangement) and amides (lactams), which, for example, can transform into amidines when a transannularly disposed amine *N*-atom is present.

When compounds that contain water of crystallisation, other solvents of crystallisation, or inclusions are heated or vaporised, these included components are released and ionised independently of the compound itself. If a common mo-

lecular ion is recorded, a (thermal) reaction between both molecules must have taken place.

Retro Reactions

Retro Aldol reaction. Special arrangements of atoms must be present in a molecule in order for a retro Aldol reaction to occur. Compounds that possess the general structural element **44** can decompose thermally with the loss of CH₂O (or its equivalent). Examples of this kind have been observed for basic natural substances (alkaloids).

44

Retro Diels–Alder reaction. One of the most frequently observed reactions of alicyclic and heterocyclic six-membered ring systems containing a double bond is the retro Diels–Alder reaction. Sometimes it is not easy to distinguish a thermal reaction, which takes place in the mass spectrometer (inlet), from a mass spectrometric reaction, or a mixture of both reactions. An example of this is the quinone **45** (M=296) which, when introduced into the mass spectrometer through the direct inlet system (170°C), produces a signal corresponding to the molecular ion. In contrast, when the gas inlet system (200°C) is used, a thermal retro Diels–Alder reaction takes place and the mass spectra of both fragments **46** (M=148) and **47** (M = 148) are superimposed upon one another. Preparatively, **46** and **47** can be obtained from **45** by distillation. The addition mass spectrum from **46** and **47** is very similar to the spectrum of **45** obtained by introducing the sample through the gas inlet system and not dissimilar (larger differences in the abundances of the main signals) to that of **45** which was obtained when the sample was introduced into the apparatus through the direct inlet system. Presumably this is an example of the parallel occurrence of a thermal and a mass spectrometric retro Diels–Alder reaction. A large number of similar cases have been observed.

45   **46**   **47**

Isomerisation Reactions

It is clearly apparent that the partial mass spectrum of the *E/Z* isomeric mixture (**48**; M=160, gas inlet, 200°C) shown in Fig. 4.36 is in stark contradiction to its structure: the loss of •CH₃ from the molecular ion can readily be explained, but the very

---

h   60 amu = CH³COOH; the relative abundances of $m/z = 366$ and 426 are 95 and 68%, respectively.

strong loss of $\bullet C_2H_5$ ($m/z = 131$) is not foreseeable (it would mean the cleavage of a C=C bond with an *H*-migration). However, if one assumes that, preparatively, **48** can be isomerised thermally to **49** (this can actually be done by heating for 20 hours in octane), which is possible *via* an aromatic [1,7]-sigmatropic hydrogen shift followed by a cyclisation, then the mass spectrum can be explained very well. Indeed the mass spectrum of **49**, prepared synthetically, is in exceptional agreement with that of **48**.

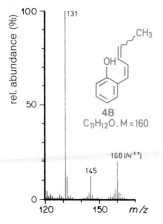

Other isomerisation reactions are also known (e.g. double bond shifts)[5].

Fig. 4.36 Part of the mass spectrum of 2-(1,3-pentadienyl)-phenol (**48**)

Disproportionation, Dehydrogenation and Hydrogenation Reactions

During the measurement of certain classes of compounds (e.g. dihydroquinoxalines, dihydroquinolines), it can happen that the expected molecular ion is not recorded, but, instead, those of the compound's hydrogenation **and** dehydrogenation products. Thus, for 2-(*tert*-butyl)-1,2-dihydroquinoxaline (**50**; M = 188), only the molecular ions of **51** (M = 186) and **52** (M = 190) are found. The reason for this is a thermal disproportionation reaction, whereby one molecule functions as a donor and the other as an acceptor of two *H*-atoms.

The appearance of relatively abundant [M + 2]$^{+\bullet}$ signals is characteristic of the mass spectra of quinones. Conversely,

[M − 2]$^{+\bullet}$ signals are observed in the spectra of hydroquinones. Such peaks have not only been identified in the spectra of *o*-quinones, but also in those of *p*-quinones and quinone monoimines. It can be shown (by administering D_2O) that H_2O, which is present in the mass spectrometer, acts as an *H*-donor.

Pyrolysis of Quaternary Nitrogen Compounds

Protonated salts of organic bases easily decompose thermally into the bases and acids from which they were constituted. These compounds are then ionised independently under electron bombardment in the mass spectrometer. However, the analogous deprotonation of the thermally more stable quaternary nitrogen compounds cannot occur, because, in these compounds, the substituent bound to the nitrogen atom is an alkyl residue.

Under the conditions necessary for the recording of EI mass spectra, a salt cannot be brought into the vapour phase without first transforming the salt into neutral particles. This thermal transformation into neutral molecules occurs according to specific rules, which permit conclusions to be drawn about the relative molecular mass of the salt in question. Quaternary nitrogen compounds of the general formula **53** (X = halogen) can be transformed into neutral molecules in three ways.

Dealkylation. The anion attacks at the alkyl group of the quaternary *N*-atom with the formation of a tertiary amine (*norbase*) and an alkyl halide:

This type of degradation is particularly favoured by iodides, but occurs only rarely with fluorides.

---

[i] CM = cation mass

Thermal Hofmann elimination. The anion attacks at an *H*-atom which is *β*-positioned with respect to the quaternary *N*-atom **54**, whereby a tertiary amine (Hofmann base, **55**) and a hydrogen halide is formed.

Although with the thermal dealkylation of quaternary nitrogen compounds of the general type **53** only one possibility exists for demethylation, four Hofmann eliminations can, in principle, take place under favourable conditions (at least four *H*-atoms *β*-positioned with respect to the *N*⁺-atom and bound to different *C*-atoms). This number can be increased by vinylogous or ethylogous decomposition reactions. The thermal Hofmann elimination occurs primarily with fluorides and much less frequently with bromides and iodides.

Examples of acceptors:

$$R-COOCH_3, \quad N-CH_3, \quad N-H, \quad C_6H_5-OH.$$

The reaction can proceed inter- or intramolecularly. In the case of intermolecular methylation, the $[M+14]^{+\bullet}$, $[M'-14]^{+\bullet}$ and, less frequently, the $[M+28]^{+\bullet}$ and $[M-28]^{+\bullet}$ signals are found, which correspond to a single and double reaction, respectively. The transfer of a CH_3 (15 amu) is then followed by the reverse transfer of an *H*-atom (or another alkyl group), as illustrated in the following example of the reaction of 2-methyl-1,2,3,4-tetrahydroisoquinoline (**57**; M = 147) with methyl cyclohexanoate (**58**; M = 142). Structural elements similar to **57** and **58** can be part of larger molecules and can therefore cause isomerisation reactions within the same molecule or between two molecules which have the same structure.

A side reaction, which is frequently observed with thermal Hofmann eliminations, is the dimerisation of the Hofmann base, which sometimes gives rise to abundant $[2M]^{+\bullet}$ signals.

Substitution reactions. The anion attacks another *C*-substituent of the quaternary *N*-atom and forms the tertiary base **56**, which contains the anion. This rather rare reaction is related to an isomerisation, because the formula weight of the salt is exactly the same as the relative molecular weight of the pyrolysis product. [Under soft ionisation conditions, as, for example, with the field desorption or electrospray ionisation methods, the cations from onium compounds (ammonium-, sulfonium-, phosphonium salts) can be measured directly[6].]

Transalkylation and Alkylation Reactions

When certain structural prerequisites are satisfied, the thermal transfer of an alkyl group (methyl, ethyl) from one functional group to another can occur.

Examples of alkyl donors:

$$R-COOCH_3, \quad R-COOC_2H_5, \quad N^+-CH_3, \quad C_6H_5-OCH_3.$$

Once again it should be pointed out that thermal reactions are less often observed with small molecules; they occur more frequently with heavier molecules, especially when additional polar groups in the molecule hinder its vaporisation in the mass spectrometer, or when the decomposition is accelerated by accompanying impurities.

Furthermore, it should be mentioned that organometallic, organoboron and organosilicon compounds sometimes decompose easily, however their decomposition reactions are of a different nature to those described above.

5.2 Recognition of Thermal Reactions

As infrequent as it is that thermal reactions exhibit uniform behaviour, one can still give a generally valid procedure for their recognition. The fundamental aspects will be discussed below.

Preparative high vacuum distillation or sublimation. If the sample of the substance to be analysed is purified by a bulb-to-bulb distillation (or sublimation) in a glass vessel under high vacuum (at least 0.1 Pa) and the distillate (or sublimate) shows itself in thin layer chromatography to be identical with the undistilled material, then there is no reason to assume that the sample will decompose thermally upon introduction into the mass spectrometer. However, if this treatment of the sample results in the formation of one or more products that are different from the undistilled material, then this reaction can also occur in the mass spectrometer, although not necessarily so, because the vaporisation conditions in the mass spectrometer are considerably more favourable than in the distillation vessel (lower pressure, no condensation necessary, shorter flight path and therefore fewer wall reactions). If one also records the mass spectra of the distillation products and all of the signals due to the distillation products can be detected with the same relative abundance in the spectrum of the undistilled material, then it is probable (but not certain) that the same thermal reaction has also occurred in the mass spectrometer. If the reaction is, for example, a retro Diels-Alder reaction, which is a reaction that can occur thermally as well as mass spectrometrically, then the mass spectrum of the undistilled material can arise from different causes: pure thermal decomposition, pure mass spectrometric decomposition, or a mixture of both processes.

Recording of other mass spectra. If, instead of recording a spectrum at the usual 70 eV, it is recorded at 12 to 15 eV (a low-voltage spectrum), which is just above the ionisation potential of organic compounds, then, for energetic reasons, fragmentation reactions will be reduced and partially suppressed, whereas thermal reactions will remain unaffected (because the conditions necessary for such reactions have remain unchanged). If, for example, two new molecules arise from the thermal reaction, then these will be recorded as molecular ions. (However, this behaviour is not necessarily observed in all cases, because particularly energetically favourable fragmentation reactions can also occur at this low ionisation potential.) When soft ionisation methods are employed (see Sec. 8.9, p. 263), the ionisation conditions are significantly gentler on substances than those used for electron impact ionisation spectra. As a result, considerably fewer thermal reactions are observed under these conditions. However, it should be emphasised that even under the soft conditions required to record, for example, field ionisation and field desorption spectra, thermal reactions have still been found to occur.

Measurement of metastable transition signals. If the thermal reaction leads to the formation of two or more products, then it is not possible to find a transition signal (cf. Sec. 8.25, p. 277) between the "molecular ions" of the pyrolysis products, except if the process also occurs mass spectrometrically.

Recording of multiple spectra. If several mass spectra are recorded one after another under the same measurement conditions and the sample has decomposed in the direct inlet part of the mass spectrometer into at least two different transient compounds, then, when compared with each other, these spectra can exhibit significant differences in the abundances of the signals. The behaviour is similar to that of a mixture of substances in which the components possess different vaporisation temperatures. Newer instruments are equipped with an automatic temperature programmed mode, which allows a partial separation of components that have different volatilities.

Derivatisation. If the functional groups that are suspected of being in the substance are derivatised, then the molecular ion peak must shift by a definite mass difference. Examples: $-COOH \rightarrow -COOCH_3$ (+ 14 amu), $-OH \rightarrow O-COCH_3$ (+ 42 amu), $-CH=CH \rightarrow -CH_2-CH_2-$ (+ 2 amu), see Tab. 4.8 (p. 279). If the relative molecular mass of the starting material is determined to be, for example, $M^{+\bullet}$ and one finds $[M + 58]^{+\bullet}$ for the methyl ester, then the starting material has lost CO_2 (44 amu) during a thermal reaction (decarboxylation). The hydrogenation of a double bond can, for example, prevent a (thermal) retro Diels-Alder reaction.

Analysis of the fragmentation pattern. Sometimes a mass spectrum is obtained which does not appear to be in agreement with the structure of the compound under investigation (see, e.g., **48**, p. 248), or there are mass differences between abundant signals and the heaviest ion (possibly the molecular ion) which cannot be explained by fragmentation reactions (e.g. M-14, M-20), or which can only be explained with great difficulty. In such cases a mixture is probably present, which was either already present from the outset or has arisen through a thermal reaction.

5.3 Prevention of Thermal Reactions in the Mass Spectrometer

For the prevention of thermal reactions, it is important to recognise their cause. In a few cases, the purification of a sample (recrystallisation, filtration) is sufficient to stabilise it. Frequently, however, functional groups are responsible, either directly (e.g. $-COOH$, $-CH=CH-$) or indirectly (by increasing the vaporisation temperature), for the occurrence of a thermal reaction. If this is the case, a derivatisation (which thereby also alters the molecular weight) is unavoidable. The methods for the modification of functional groups are well known (e.g. esterification, reduction, hydrogenation, ether formation). In addition, the procedures to aid the volatilisation of organic substances will be assumed to be known, because they are also required for gas chromatographic analysis. Suitable derivatives are, among others, for **hydroxy groups**: methyl ethers, trimethylsilyl ethers, acetonides, acetates; for **carboxyl groups**: methyl esters; for **amino groups**: acetamides, trifluoroacetamides, N,N-dimethylamides.

As an example of the prevention of thermal decomposition reactions by increasing the volatility, the triaminocarboxylic acid **59** (M = 397) is described.

In the spectrum of **59**, one can only observe a general decomposition of the compound. After esterification (CH_3OH/HCl) and acetylation [$(CH_3CO)_2O$/pyridine], one obtains the derivative **60** (M = 495), which produces a mass spectrum that can be analysed readily.

The following methods, among others, should also be referred to: the CI (see Sec. 8.1, p. 258), FI (see Sec. 8.6, p. 261), FD

59 $R^1 = R^2 = H$
60 $R^1 = COCH_3$, $R^2 = CH_3$

(see Sec. 8.5, p. 261) and FAB (see Sec. 8.4, p. 260) techniques and the method of cation addition spectroscopy (see Sec. 8.10, p. 265). These soft ionisation methods allow compounds to be analysed with a lower probability of thermal decomposition.

6. Mass Spectra of Contaminated Samples and Mixtures

Because there is no fundamental difference, but rather a quantitative one, between a contaminated sample and a mixture, these concepts will not be discussed separately. In order to be able to obtain clear and correct evidence from the mass spectrum of a compound, the sample of the substance must be homogeneous, i.e. pure. This statement is valid without exception, although the majority of samples analysed in a mass spectrometry laboratory unfortunately do not meet these requirements. If contaminated samples are submitted, it is important to know what consequences this can have for the mass spectrometric analysis.

If, for reasons of its volatility, a sample is introduced into the instrument through the gas inlet system, then all volatile components in the sample will be transferred into the reservoir of the mass spectrometer. This has the result that each of the individual components of the sample are ionised independently from one another, but simultaneously, in the ion source and produce superimposed mass spectra. Certain conclusions can indeed be drawn from the spectra, however a quantitative analysis of the composition of the mixture is not possible as long as the structures of the components are unknown and a calibration has not been carried out. With a two component mixture, for example, no conclusions can be drawn about the ratio of the quantities of the individual components, because of the different ionisation probabilities and partial pressures and, as a result, the very different relative abundances of the molecular ion signals. To illustrate this, Fig. 4.37 shows the mass spectrum of N,N-diethyl-1,3-propanediamine (**61**) with "traces" of 1,3-diethylperhydropyrimidine (**62**) and 1,3-diethyl-2-methylperhydropyrimidine (**63**). This mixture of substances resulted when purest N,N-diethyl-1,3-propanediamine (**61**) was dissolved in denatured ethanol (methylated spirits) and

finally evaporated to dryness. The residue produced the illustrated spectrum in which 0.3% of compound **62** gives a molecular ion signal with the same abundance as that from 99.4% of **61**. See also Sec. 6.2, p. 252.

The GC/MS combination is a great help for the analysis of volatile mixtures (see Sec. 8.11, p. 266). For various reasons, however, it is often unavoidable that analyses have to be carried out on sample mixtures that are either difficult to vaporise or cannot be vaporised. For this purpose, in addition to the LC/MS technique, tandem mass spectrometry (see Sec. 8.22, p. 275) coupled with a suitable ionisation procedure can be useful.

If the direct inlet system is used for mixtures of substances, one obtains spectra that are different from those acquired when the gas inlet system is employed. Because the sample is heated in a crucible, the more volatile components will be vaporised first, followed by the less volatile and finally the least volatile components. If two components have the same or very similar vaporisation characteristics, they will be ionised at the same time. Mass spectra are recorded at regular time intervals until all of the components have been vaporised. Depending on the vaporisation characteristics of the components, these spectra overlap to a greater or lesser degree. In the ideal case one can obtain spectra of the pure components, however mixed spectra are normally observed. Because a certain amount of time is necessary to record a mass spectrum and this time can coincide with a sharp decline in the presence of a component, these mixed spectra frequently do not correspond to a clean addition of the spectra of the individual components. Neither quantitative statements nor calibrations are possible in this case. In connection with this it is perhaps

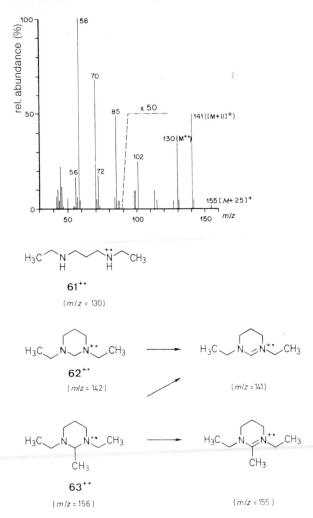

Fig 4.37 The [M + 11]⁺ peak. Mass spectrum of a mixture composed of ca. 99.4% *N*,*N*'-diethyl-1,3-propanediamine (**61**; M = 130), ca. 0.3% 1,3-diethylperhydropyrimidine (**62**; M = 142) and ca. 0.3% 1,3-diethyl-2-methylperhydropyrimidine (**63**; M = 156), gas inlet, 70 eV. The relative amount of each compound in the mixture was determined by GC

lowing sections, a few frequently encountered contaminants will be discussed[7].

6.1 Solvent

Organic substances are frequently contaminated with solvent residues. This can occur in crystals, lacs and undistilled oils. Usually they vaporise before the actual substance. The spectra of a few solvents are illustrated in Tab. 4.11 (p. 297).

6.2 Foreign Substances in Solvents

For various reasons commercial solvents frequently contain impurities or additives, which can subsequently be found again in samples of substances. To be noted in particular are chloroform (stabilised with ca. 2% ethanol), denatured ethanol (most common additives: benzene or methanol), petroleum ether (contains heavier and therefore less volatile hydrocarbons), diverse ethers (stabilised with 2,6-di(*tert*-butyl)-4-methylphenol, see Tab. 4.12, p. 301) and carbon tetrachloride (upon long standing in the presence of light, various products are formed from dichlorocarbene CCl_2). Because solvents are usually used in a great excess, compared with the amount of a substance, any impurities that are present under these circumstances will become quite significant.

The following example convincingly shows how incorrect conclusions can be made during the evaluation of spectra from contaminated samples.

Nearly all commercially available grades of methanol contain varying amounts of formaldehyde, although usually only in fractions of parts per million. Now there are substances which react quantitatively with even the smallest quantities of formaldehyde, e.g. 1,3-propanediamine. *N*,*N*'-diethyl-1,3-propanediamine (**61**) reacts with formaldehyde to form 1,3-diethylperhydropyrimidine (**62**). Although the molecular ion of **61** ($m/z = 130$) produces a very weak signal, the [M − 1]⁺ signal of **62** at $m/z = 141$ is extremely strong. (The $M^{+\bullet}$ ion of **62** also produces a very weak signal.) If both compounds are present as a mixture, their mass spectra will be superimposed. Such a case is shown in Fig. 4.37. In this spectrum, the signal abundances for $m/z = 130$ ($M^{+\bullet}$ from **61**) and $m/z = 141$ ([M − 1]⁺ from **62**) are, in terms of their order of magnitude, the same. However, a gas chromatographic measurement determined the ratio of [**61**]:[**62**] to be 99.4:0.3, from which it can be concluded that **62** is present as an impurity with an extremely small concentration. The main fraction of this "mixture" is therefore not in agreement with the result of a superficial mass spectrometric analysis which shows the presence of the straight-chain diamine **61** and the formaldehyde condensation product **62** as a secondary component in an approximately 1:1 ratio, and a third substance (**63**).

important to point out that the operator stops recording a spectrum when a mass spectrum is obtained which appears to be "sensible". However, if this is only the spectrum of an impurity about which the operator has not been informed, then only time is wasted and the desired information is not forthcoming. For important preliminary information, mass spectra of mixtures (e.g. chromatographic fractions of organic natural materials) can deliver valuable data. The handling of contaminated samples can be facilitated by using a mass spectrometer equipped with a temperature programmed mode. In the fol-

Compound **62** can be recognised by an $[M - 1]^+$ signal at $m/z = 141$. If this signal originated from **61**, it would indicate an $[M + 11]^+$ peak. Such signals are almost always observed in the spectra of 1,3-propane- and 1,2-ethanediamine derivatives when methanol is used as the solvent, because methanol is easily oxidised by the oxygen in the air. Formaldehyde reacts equally well with 1,3-propane- and 1,2-ethanediamines, however it does not react in the same way with 1,4-butanediamines.

The contamination of samples can be prevented most easily by employing specially purified solvents (even when this means more work for the chemist!).

6.3 Foreign Substances in Reagents

A few reagents also contain stabilising agents that must be removed carefully from the sample after a successful reaction. These are, among others, kerosene in $LiAlH_4$ and oil in syntheses involving KH and NaH (Tab. 4.12, p. 303). Substances recovered after NMR experiments can still contain tetramethylsilane (TMS).

6.4 Materials from Laboratory Apparatus

Many pieces of laboratory apparatus are made completely or partially from synthetic polymers. These contain various softeners which can be leached out, particularly by solvents (chloroform, among others, is an ideal leachant). The quantity of softeners leached in this way can sometimes be considerable. Of the plastic components, those that should be mentioned in particular are stoppers of flasks, glass bottles, etc., all types of tubing, particularly that plastic tubing which is coloured like rubber tubing, stopcocks, seals, plastic bottles and containers, and incompletely polymerised substances from ion exchangers. (It should be mentioned that even natural rubber is not inert to organic solvents, however the impurities that are leached from it only contribute to a "general background" in the mass spectrometer and do not produce a spectrum of a characteristic and specific compound.) Other sources of contaminants are, among others, isolating liquids in apparatus (e.g. hydrogenation apparatus), stopcock grease and lubricants from fans (e.g. from in-room air conditioners). In addition, filter papers that have not been specially cleaned, adsorbents and chromatographic materials of all kinds should not be neglected as sources of contamination. From the chemical point of view, the principle component in softeners is phthalic acid diester (see Tab. 4.12, p. 302 and Fig. 4.32, p. 244), which has $m/z = 149$ as the base peak of its mass spectrum (if such a peak is found in a spectrum, it should be assumed that the sample contains a phthalic acid diester; only when evidence has been produced to show that the sample is pure, can one assume that $m/z = 149$ actually represents a fragment ion signal from the sample being analysed!).

6.5 Contaminants from Thin-layer Chromatography Plates

High activity TLC plates that are produced or kept in the laboratory are also excellent absorbers of substances that are present in the laboratory air. TLC plates can absorb, among other things, oil vapours emanating from oil-filled vacuum pumps which are present in most laboratories. During the subsequent extraction of chromatographed substances from the adsorbent, these materials will also be eluted and will once again contaminate the freshly purified compound.

In a few cases it is difficult to differentiate between mixtures that are present from the beginning or those which have just been thermally produced in the mass spectrometer (see also Sec. 5, p. 246).

If one suspects or knows the possible sources of contaminants, then the chemist is easily able to develop procedures for the prevention or removal of impurities. The problem of impurities in samples has been discussed here because of the high detection sensitivity of the mass spectrometer, however it is not solely a mass spectrometric problem. Contaminants can also be the source of erroneous information from other methods of analysis (IR, NMR, UV, ORD, etc.).

7. Isotopic Labelling Reactions

The specific labelling of functional groups or their environment is an experimental technique that is employed frequently for spectroscopic, kinetic, bioorganic or mechanistic investigations. The employment of ^2H (D), ^{13}C, ^{15}N and ^{18}O labelling is particularly favoured. Today, the range of commercially available reagents and compounds which contain these isotopes is very large, so that a multitude of labelling experiments can be conducted. It is the nature of things that, aside from a few examples, the synthesis of ^{13}C- and ^{15}N-labelled compounds usually necessitates a great deal of effort, whereas D-labelling is possible with much less labour.

Labelled compounds are used in order to identify special functional groups or their position within a molecular assemblage, to investigate chemical or biochemical reaction mechanisms, or to clarify the mechanisms of mass spectrometric fragmentation reactions.

Important, constantly recurring reactions are summarised in the following sections. There are a few trivial, but important, comments that should be made at the outset. These are that, under the same conditions, H/D exchange reactions are also D/H exchange reactions, that the air contains a considerable amount of H_2O and that glass and other containers at room temperature are coated with a film of H_2O.

7.1 H/D Exchange Reactions

Acidic Protons

Protons that are bonded to heteroatoms, as in the groups $-NH_2, =NH, -CO-NH_2, -COOH, -OH$ and $-SH$, exchange very easily with deuterons. For this purpose, samples are evaporated several times with D_2O, CH_3OD, etc. under high vacuum (not the vacuum from a water siphon!) and finally introduced into the mass spectrometer, whereby D_2O or CH_3OD is introduced simultaneously into the gas inlet system. The number of exchanged deuterons can be determined from the shift in mass of the molecular ion. Under these conditions the exchange is rarely quantitative. Therefore, the chances of obtaining clear proof that more than three protons have been exchanged are normally not very likely.

Aromatic Protons

Aromatic protons can be exchanged by an electrophilic aromatic substitution with DCl/D_2O, D_3PO_4 or D_2SO_4.

By employing a final washing with H_2O or CH_3OH, the functional groups with acidic protons (e.g. $-OH$), should any be present in the compound, will be converted back to proton-containing residues.

Example: Evaporate the sample [60 mg (4-phenyl)butylamine] three times (high vacuum or dry N_2) in the reaction vessel, each time with 1 ml CH_3OD (for the removal of H_2O), then add 5 ml 38% DCl/D_2O. Heat at 150 °C for 30 h in a sealed tube, then dilute with 10 ml D_2O. Neutralise with anhydrous sodium carbonate (additionally heated); extract with ether; dry the ether extract with sodium carbonate; evaporate to dryness and distil the residue. After repeating the entire procedure once, a nearly quantitative insertion of 5 D-atoms into the aromatic ring is achieved.

Carbonyl Groups With α-Positioned Protons

Due to the formation of enols or enolates, the following transformations can be carried out with acids (DCl, D_2SO_4, D_3PO_4, etc.) or bases ($NaOD, CH_3ONa/CH_3OD, Na_2CO_3/D_2O$, etc.):

$$-CH_2-COOCH_3 \longrightarrow -CD_2-COOCH_3$$

$$-CH_2-COOH \longrightarrow -CD_2-COOD$$

$$\xrightarrow{H_2O} -CD_2-COOH$$

$$-CH_2-CO-N\overset{/}{\underset{H}{}} \longrightarrow -CD_2-CO-N\overset{/}{\underset{D}{}}$$

$$\xrightarrow{H_2O} -CD_2-CO-N\overset{/}{\underset{H}{}}$$

With all of these reactions, one must be careful to ensure that the neutralisation of the reaction solution is done in the absence of proton donors (H_2O, etc.). In order to remove solvent residues and water, the sample should be evaporated with D_2O or CH_3OD before beginning the exchange reaction.

Example: 100 mg of a ketone was dissolved in a solution of 100 mg sodium in 10 ml CH_3OD; boiled under reflux for 2 h, then neutralised with 20% DCl/D_2O; the resulting precipitate was filtered off, washed with D_2O and the product, after repeating the entire procedure twice, purified by a bulb-to-bulb sublimation (160 °C / 1 Pa).

The exchange of protons that are α-positioned with respect to a nitrile group for deuterons can also be accomplished by using KCN/D_2O or KCN/CH_3OD as the base.

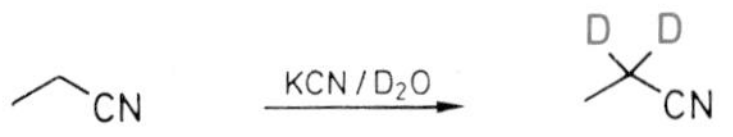

Example: 75 mg 4-phenylbutyronitrile, 81 mg KCN, 4 ml dioxane (all anhydrous) and 3.6 ml D_2O were heated at 165°C under N_2 in a sealed tube for 24 h, then filtered under an inert gas in a bulb; after removal of the solvent, the residue was purified by a bulb-to-bulb distillation (side-product: carbonic acid).

It should also be mentioned that DCl/D_2O has an advantage over many other acidic reagents in that it is readily vaporised and can therefore be separated easily from the compound. The advantage of the exchange reactions just described is that the number of acidic or aromatic protons can be ascertained, which can, from the point of view of a structural analysis, yield important information.

Further Exchange Reactions

An exchange reaction of a completely different type can be used with tertiary *N*-methyl derivatives.

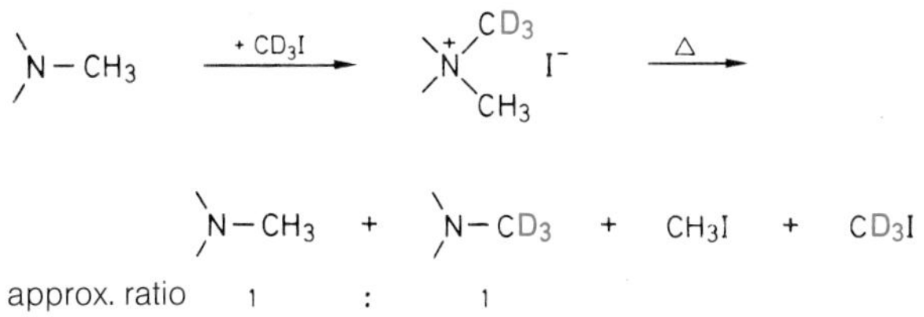

Methylation with CD_3I initially yields the quaternary methyl iodide, which, *via* pyrolysis involving the loss of CH_3I and CD_3I, forms the starting material again together with the tri-deuteromethyl derivative. The ratio of the amounts of each product can depend on steric factors and for achiral products this ratio is 1:1. All of the ions which contain the *N*-methyl group now appear in the mass spectrum as "doublets" (X and X + 3).

The following is an example of an ^{18}O exchange reaction:

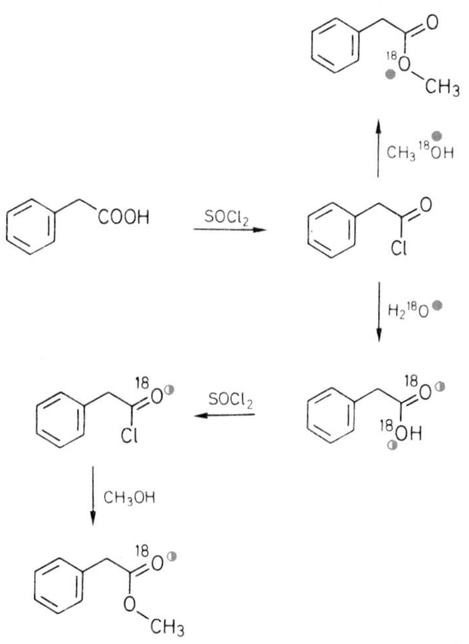

In both of the isotopomeric methyl phenylacetates, the product of the reaction between the acid chloride and $CH_3^{18}OH$ has the same ^{18}O content as the reagent. Conversely, the other isotopomer has only half of the isotope content of $H_2^{18}O$. Each reaction labels a different *O*-atom.

7.2 Transformations of Functional Groups Under Deuterating Conditions

It is frequently necessary to introduce *D*-atoms at specific positions within a molecule, or to obtain evidence for the existence of a particular functional group in a molecule. A few typical reactions are described below.

Reduction Reactions

The degree of deuteration of unsaturated compounds with D_2/catalyst is very dependent upon the type and quality, and thus the activity, of the catalyst. In a few cases the correct uptake of 2 *D*-atoms per reduced C=C bond is observed, whereas in other cases a multiple of the theoretically expected number (i.e. without increasing the degree of hydrogenation of the compound) of *D*-atoms is taken up (presumably due to successive dehydrogenation and rehydrogenation reactions). The conversion of functional groups containing *C*-atoms in high oxidation states into functional groups in which these atoms have lower oxidation states can frequently be accomplished with, for example, $LiAlD_4$ and $NaBD_4$.

The reduction of alkoxycarbonyl groups with $LiAlD_4$ and the subsequent work-up in proton-rich solvents frequently results in the inclusion of a somewhat greater number of D-atoms than is predicted theoretically. Presumably, before the reduction, additional exchange reactions occur to a small extent at the α-position with respect to the carbonyl group.

The reduction of primary and secondary hydroxy groups can be carried out expediently by the reduction of the tosyloxy derivative with $LiAlD_4$.

Tos = $p-CH_3C_6H_4SO_2$

The reduction of ketones with Zn/DCl according to Clemmensen does indeed lead to the formation of the expected dideuteromethylene derivative, but the extent of "over deuteration" (caused by an acid-catalyzed enolisation reaction) is too great. Therefore, this reaction is not recommended. An exception is when the carbonyl group has no α-positioned methyl groups or there has already been a complete CH_2/CD_2 exchange.

Benzyl positioned C=O, C–OR and C–N residues can be transformed with D_2/Pd and, in the case of the first two compound types, with $LiAlD_4/AlCl_3$ into di- or monosubstituted derivatives (CD_2, CDH) with good yields and correct D-insertion.

The diphenyl ether cleavage with Na/ND_3 (synthesised from $Mg_3N_2 + D_2O$) is a suitable method for the specific labelling of the sites of aromatic ether linkages.

The decarboxylation of maleic acid derivatives under deuterating conditions produces monodeuterated derivatives[8]:

7.3 Determination of the Degree of Labelling

The aim of this method is the determination of the heavy isotope content (^2H, ^{13}C, ^{15}N, ^{18}O) of the labelled compound, i.e. the establishment of the extent to which the inclusion of these isotopes in the desired compound has been successful. However, before this mass spectrometric determination is carried out, the following analyses should be undertaken:

a) The labelled compound must be chemically homogeneous (melting point, DC, GC) and behave in regard to these properties exactly like the unlabelled compound whose characteristics are to be investigated.

b) Independently from the synthesis, it should be checked spectroscopically (e.g. by IR, ^{1}H-NMR, ^{13}C-NMR) that the labelled atoms are located at the desired sites in the molecule. Furthermore, a quantitative determination of the D-content can sometimes be carried out by using ^{1}H-NMR spectra. This can be done when the exact integration of a definite region of resonance is possible. In addition, a combustion analysis where still available, together with an IR spectroscopic determination of the D_2O or HDO concentration, delivers very good results for the total D-content. As a complement to the mass spectrometric determination of the D-content, these procedures give important additional information from which erroneous conclusions are less likely to be drawn.

These analyses are essential in order to obtain the greatest possible value from the information that is available in the mass spectra of labelled compounds.

For the mass spectrometric determination of the degree of labelling of a compound, the molecular ion peak is brought into play. In the ideal case, this peak should have a nicely visible intensity and not be accompanied by the $[M - H]^{+\bullet}$, $[M - 2 H]^{+\bullet}$ or the $[M + H]^+$ signals. When more abundant accompanying signals are present, overlapping signals will occur for the labelled compound and these signals will make a quantitative evaluation of the spectrum impossible. On the other hand, the margin of error will be too great when evaluating very weak signals that only just appear above the background or general noise of the spectrum. However, if behaviour of this kind is present, the following alternative possibilities are available: the evaluation of low-voltage spectra (in general, in low-voltage spectra measured at 12 to 15 eV, the $[M-H]^{+\bullet}$ and $[M-2H]^{+\bullet}$ signals are weaker and the molecular ion signals are more intense with respect to the rest of the spectrum); the evaluation of the spectra of derivatives, which, under certain circumstances, have a more favourable presentation of the molecular region; or the evaluation of high resolution partial spectra, which have been written down on paper and analysed in conjunction with the elemental composition. In the latter case, a clear separation of the molecular ion signals from the other signals can be obtained. This procedure is limited, however, by the resolving power of the instrument and the abundances of the ions. With this method of analysis, the spectrum of the labelled compound is compared with that of the unlabelled compound. It is therefore important that both spectra are measured under the same recording conditions. (Due to the memory effect, it is essential to take care that the first sample to be measured has passed completely through the mass spectrometer before starting on the second sample.) At least three partial spectra of the molecular ion region are recorded for each substance. Finally, the peak abundances are determined (for practical reasons in mm or %) and separately averaged for each of the isotopomers, so that two sets of spectra are obtained that can be evaluated as described below. [The explanation is given using the example of N-(2-phenylethyl)formamide ($C_9H_{11}NO$, M = 149) and its 1-^{13}C-labelled isotopomer. For the synthesis of the latter, a ca. 90% enriched ^{13}C reagent was employed.]

Measurement Results

A) Unlabelled Compound

In the molecular ion region only the signals at $m/z = 149$ and 150 have an abundance greater than 1 rel.%; the average values (from five individual measurements) were $m/z = 149$ (100.00%) and 150 (11.29%).

B) Labelled Compound

The corresponding averaged signal abundances in the molecular ion region are: $m/z = 149$ (10.94%), 150 (100.00%) and 151 (11.11%).

Because there is no signal at $m/z = 148$ for the unlabelled compound, one can presume that $m/z = 149$ in the spectrum of the labelled compound is due to the molecular ion of the unlabelled fraction. The signal at $m/z = 150$ is partially due to the first isotope peak of the unlabelled compound, however the

main contribution to this signal comes from the synthetic, singly labelled isotopomer. The ion that produces the first isotope peak of this particle at $m/z = 151$ naturally contains two ^{13}C-atoms and the ratio between the signals at $m/z = 150$ and 151 is proportional to that of the corresponding signals for the unlabelled compound. In this way the averaged spectrum of the unlabelled compound can be subtracted proportionally from that of the labelled compound:

m/z	149	150	151
labelled	10.94	100.00	11.11
unlabelled	10.94 (100)	1.24 (11.29)	
	0	98.76	11.11
		98.76	11.15
		0	−0.04

If the fraction of the unlabelled compound (10.94) and that of the singly labelled isotopomer (98.76) are normalised to 100%, the degree of labelling is obtained as ^{13}C$_0$: 10% (9.97), ^{13}C$_1$: 90% (90.03). The natural ^{13}C content has not been taken into account in these values. This has been subtracted with the spectrum of the unlabelled compound.

The small "negative" value under $m/z = 151$ can be neglected. Frequently, however, much larger positive or negative values are observed, which make the calculation of the isotope content difficult or even impossible. The causes of such absurdities are frequently impurities or different recording conditions for the two samples (different $[M+H]^+$ and $[M-H]^+$ fractions). It is often possible to reduce or circumvent these difficulties by the evaluation of low-voltage spectra, as described above.

For additional examples and other methods, refer to the literature[9].

In order to establish the degree of labelling of fragment ions, a procedure analogous to that for the evaluation of molecular ions is employed. However, overlapping signals occur more frequently. This problem can sometimes be overcome by the evaluation of high resolution spectra. In low-voltage spectra and the spectra of derivatives, it is possible to find altered ratios of isotope inclusion, which do not correspond with the results from 70 eV spectra, because different fragmentation reactions may occur.

8. Additional Methods and Concepts

Other important concepts and methods of mass spectrometry, which were not discussed in the previous sections, will be explained briefly here. With regard to additional keywords, see the bibliography.

8.1 Chemical Ionisation (CI)

A mass spectrometric ionisation method involving ion/molecule reactions. In principle, a reactant gas (e.g. hydrocarbons, H_2, H_2O, NH_3, alcohols, noble gases) is ionised by electron impact (gas pressure ca. 1 kPa). In the case of methane, the ion $[CH_4]^{+\bullet}$ is formed, which reacts with other methane molecules, e.g.:

$$[CH_4]^{+\bullet} \; + \; CH_4 \; \longrightarrow \; [CH_5]^+ \; + \; CH_3^\bullet$$

(A whole series of additional ions, such as $C_2H_5^+$, $C_3H_5^+$, CH_3^+, etc., are also formed.) At the same time, the molecule to be analysed, M, is also present, although at a much lower concentration. This then reacts with the protonated methane (Brønsted acid), whereby a proton transfer occurs in the gas phase:

$$[CH_5]^+ \; + \; M \; \longrightarrow \; [M+H]^+ \; + \; CH_4$$

$[M + H]^+$ undergoes decomposition reactions and produces a CI spectrum.

If other types of reactant gases are chosen (e.g. noble gases, CO_2, N_2), then a charge exchange occurs (abbr.: CE), instead of the protonation of M:

$$[He]^{+\bullet} \; + \; M \; \longrightarrow \; [M]^{+\bullet} \; + \; He$$

Additional reaction types are electrophilic addition ($M + X^+ \rightarrow MX^+$) and anion extraction ($AB + X^+ \rightarrow B^+ + AX$).

Depending upon the choice of reactant gas, it is possible to control the formation of fragment ions, as shown in Figs. 4.38 and 4.39, which depict spectra obtained from lysine methyl ester (**64**). From this it follows that CI is a soft ionisation method and can therefore be used as an alternative to electron impact ionisation.

Literature review: [10].

An EI and a CI mass spectrum from the same compound can be compared by referring to Figs. 4.56 and 4.57, pp. 271, 272.

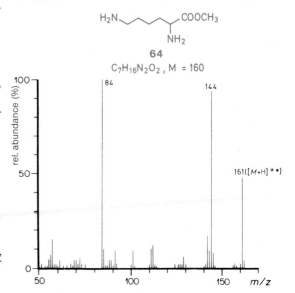

Fig. 4.39 CI mass spectrum of lysine methyl ester (**64**), reactant gas: methane

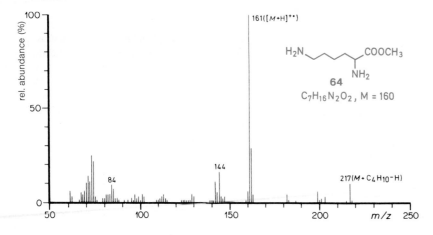

Fig. 4.38 CI mass spectrum of lysine methyl ester (**64**), reactant gas: 2-methylpropane (isobutane)

8.2 Direct Chemical Ionisation (DCI)

[synonymous terms: in-beam electron ionisation, direct exposure chemical ionisation, flash volatilisation and plasma desorption (not to be confused with the $^{252}_{98}Cf$ plasma desorption technique)]

On the tip of the probe (see p. 221) there is a wire loop (e.g. Pt, Re, W) into which a drop of a dissolved substance is placed, just as is done for field desorption (FD) spectra. The tip is inserted into the mass spectrometer and, after the solvent has been evaporated under vacuum, the thin film of substance on the loop is measured under CI conditions. The spectra resemble both those that are measured under FD conditions (abundant pseudo-molecular ion) and CI spectra (protonation of M, electrophilic addition). The DCI spectrum of glucose (M = 180), with NH_3 as the reactant gas, is shown in Fig. 4.40. To explain the peaks: $198 = [M + NH_4]^+$, $215 = [M + NH_4 + NH_3]^+$. The character of the spectrum is strongly dependent on the temperature of the ion source.

Reference: [11].

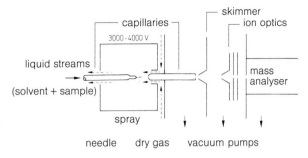

Fig. 4.41 Schematic representation of the electrospray ion source. (Finnigan MAT instrument TSQ-700). The ions, generated by electrospray at atmospheric pressure, are admitted to the mass analyser through a glass capillary, skimmer and ion optics. (Reproduced with the kind permission of Finnigan MAT, Bremen)

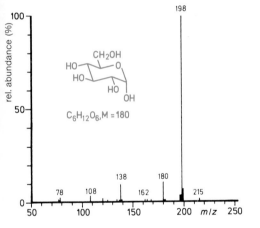

Fig. 4.40   DCI spectrum of glucose (M = 180), reactant gas: ammonia (reproduced with the kind permission of Finnigan MAT, Bremen)

8.3 Electrospray Ionisation (ESI)

In this ionisation method, a solution of a substance is sprayed (flow rate 1 to 20 µl/min) through a capillary into a chamber, cf. Fig. 4.41. A dry gas flows in the opposite direction to this spray mist. A potential of a few kV is applied between the capillary and the chamber wall (cylindrical electrode). Charged droplets are produced which become smaller as the solvent (e.g. CH_3OH/H_2O, CH_3CN/H_2O) vaporises. Driven by the electric field, the charged droplets move through a glass capillary (ca. 0.5 mm inside diameter) into the pre-analyser region. By focusing the droplets with an electrostatic lens system, they are directed into the analyser of the mass spectrometer (e.g. quadrupole mass spectrometer). The other spray ionisation methods, namely atmospheric pressure ionisation (API)[12] and ion spray[13], are similar.

This procedure results in singly and multiply charged ions, $[M + nH]^{n+}$ and $[M - nH]^{n-}$, where n can be of the order of 100 with suitable molecules. At the same time other molecular ions will also be recorded, which differ successively from each other by one less unit of charge (e). Because m/z ($z = n \cdot e$) is recorded, the instrument's mass scale for displaying the spectra, which is limited to a mass number of ca. 4000, can be used to identify masses higher than 100 000. Neighbouring ion signals, which differ by one charge unit, are used for the calculation of the molecular ions. To illustrate this, the ESI mass spectrum of interleukin 6 (M = 20903) is shown in Figs. 4.42 and 4.43.

The original spectrum contains several differently charged molecular ions (see indicated charges), which, of course, represent only **one** singly charged molecular ion of mass 20903 (Fig. 4.43). The process (relationship between Figs. 4.42 and 4.43) is comparable with the multiple images obtained from **one** person who is standing between two parallel mirrors.

ESI mass spectrometry is a powerful analytical method, because it allows one to analyse the molecular ions of polar and higher molecular compounds in aqueous solution.

Reference: [14].

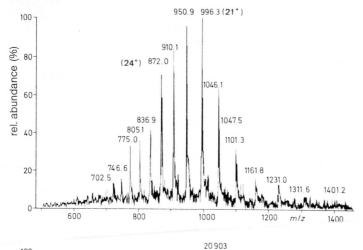

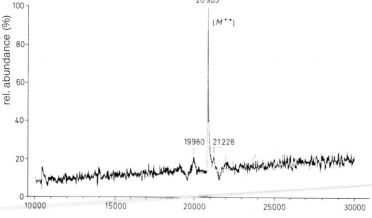

8.4 Fast Atom Bombardment (FAB)

(also known as liquid secondary ion mass spectrometry, LSIMS)

This is an ionisation method for organic molecules that are either difficult to vaporise or cannot be vaporised. The principle of the method is that fast neutral atoms are shot onto a thin film of the sample in the ion source of a mass spectrometer. The sample ions thereby formed are accelerated, focused and finally analysed with the usual instrumental optics.

The fast neutral atoms (usually argon and less frequently xenon) are generated by a so-called atom gun. Within the gun, $Ar^{+\bullet}$ ions are initially generated by charge separation and then accelerated (5 to 10 keV). They then collide with neutral Ar-atoms in a collision chamber, whereby a charge exchange occurs without the loss of very much kinetic energy. This results in a beam of fast Ar^0 atoms. (The fast particles are printed in bold.)

$$Ar^0 \longrightarrow Ar^{+\bullet} + e^-$$
$$Ar^{+\bullet} + Ar^0 \longrightarrow Ar^0 + Ar^{+\bullet}$$

The atom beam is then directed onto the sample film. The sample itself is embedded in a matrix (glycerine is used frequently, although other substances are also suitable, e.g. 3-nitrobenzyl alcohol, thioglycerine), which is placed on a flattened copper tip. The preparation of the sample requires experimental skill and experience.

Upon the arrival of the fast *Ar*-atoms at the surface of the sample, (pseudo)-molecular and fragment ions are formed, which originate from both the substance being analysed and the matrix. (Sometimes pyrolytic processes also occur, which lead to additional ions.) Because the spectrum of the matrix is largely known, this does not disturb the interpretation too severely. However, mutual matrix/sample interactions are also known, which are very dependent upon the nature of the sample. Therefore, it is not possible to correct for these interactions without additional effort. When positive ions are being measured, $[M + H]^+$ and $[M + Na]^+$ ions are usually formed (when measuring negative ions, $[M + H]^-$ ions occur). At the same time, however, $[M + glycerine]^+$ ions also appear. For the determination of the relative molecular weight of an unknown substance, it is advisable, by the addition of sodium chloride or potassium chloride, etc., to generate ions whose masses allow one to home in more easily on the molecular ion of the unknown sample. FAB has been used successfully for the analysis of organic acids ($-COOH$, $-SO_3H$, $-OPO_3H$) and salts, polypeptides (e.g. the α-amino acid containing peptide, melittin with M = 2844.8), oligosaccharides (e.g. γ-cyclodextrin = cyclooctylamylose with M = 1296.4), nucleotides, etc.

The physical explanation of this ionisation process is not completely clear.

Reference: [15].

8.5 Field Desorption (FD)

Under the influence of a strong electric field, positive ions are desorbed from an (activated) wire, onto which a nearly involatile sample has been placed. These ions are then analysed mass spectrometrically. This method is frequently suitable for the determination of the relative molecular weight of polar compounds.

For an example spectrum, see cation addition mass spectrometry (Sec. 8.10, p. 265).

Reference: [16].

8.6 Field Ionisation (FI)

An ionisation method for molecules that employs an extremely high electric field (10^9 to 10^{10} V · m^{-1}). The ionisation occurs at the anode, which is a pointed tip, sharp blade or a very thin wire. Usually the anode is activated before the mea-

surement by surrounding it with a blanket of the finest needles. The method produces, in comparison with electron impact ionisation, more abundant molecular ions and less fragment ions[16].

8.7 Field Ionisation Kinetic (FIK)

A method for the analysis of the kinetic behaviour of ions, which are generated by the field ionisation method (FI). This method allows ions with a lifetime of 10^{-8} to 10^{-11} s to be temporally resolved and studied. It is suitable for the analysis of, among others, the competitive decomposition reactions of radical cations[17].

8.8 Measurement of High Masses

Biologically interesting molecules with high molecular weights, such as peptides, carbohydrates, glycopeptides, glycolipids, nucleic acids, etc., are being isolated in chemical laboratories in increasing numbers. The elucidation of the structures of such substances using spectroscopic methods generally meets with fundamental difficulties. The compounds, which are usually only available in small quantities, frequently contain many structural units that are the same or very similar. These then produce overlapping signals in UV, IR and NMR spectra, which can therefore be difficult to analyse. These structural elements often also exist in several conformations, which further increases the extent to which signals overlap. In these situations the method of mass spectrometric structural analysis comes into its own, in particular for the determination of the molecular weight. In recent times, several methods have become available which permit the molecular ions of these so-called biomolecules to be determined. Electron impact ionisation, for which the vaporisation of the sample is a prerequisite, facilitates the determination of molecular weights up to ca. 1500. This limit is set by the thermal lability of the (bio)molecule. For the measurement of such molecules, it is only possible to employ mild ionisation procedures. The highest molecular weights that have been determined to date are about 25 000 Daltons (FAB), 45 000 Daltons (PD) and 300 000 Daltons (ESI and LDI). Aside from the fact that it is difficult to imagine such enormous gaseous ions, the advance of mass spectrometry into this field has brought with it a series of new problems to be solved. For the mass marking in the high mass region, alkali metal halides, especially CsI, can be used, which form cluster ions of high mass. During such measurements the resolving ability of the instrument should be as high as possible. However, a high resolving power diminishes the ion gain at the detector. Therefore the resolving power must be set as high as is necessary (for peak separation), but as low as possible (to maintain ion abundance).

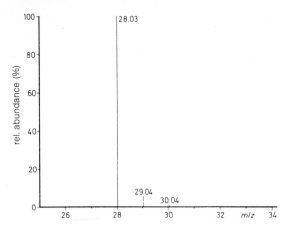

Fig. 4.44 Computer calculation of the isotope pattern of the molecular ion of C_2H_4

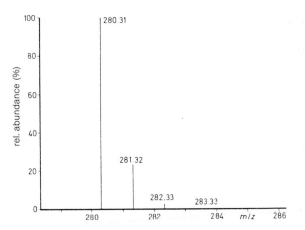

Fig. 4.45 Computer calculation of the isotope pattern of the molecular ion of $C_{20}H_{40}$

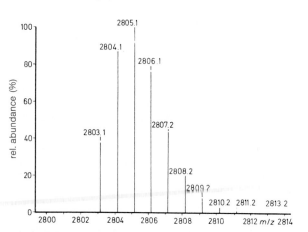

Fig. 4.46 Computer calculation of the isotope pattern of the molecular ion of $C_{200}H_{400}$

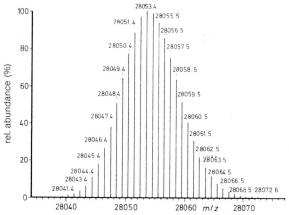

Fig. 4.47 Computer calculation of the isotope pattern of the molecular ion of $C_{2000}H_{4000}$

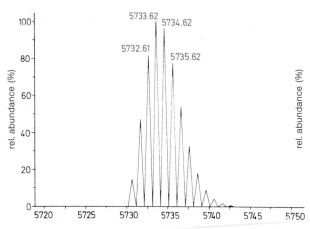

Fig. 4.48 Molecular region of the mass spectrum of bovine insulin ($M^{+\bullet} = 5729.6009$); computer calculation of the $[M+H]^+$ signal with a resolving power of 6000

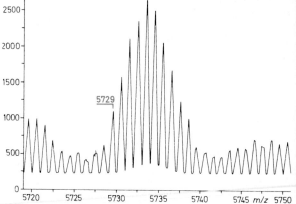

Fig. 4.49 Molecular region of the FAB mass spectrum of bovine insulin; averaged spectrum from 15 individual spectra, computer smoothed (both spectra recorded on MAT 90; reproduced with the kind permission of Finnigan MAT, Bremen)

Tab. 4.5 Comparison of molecular ions from C_2H_4, $C_{20}H_{40}$, $C_{200}H_{400}$ and $C_{2000}H_{4000}$ (from computer calculations)[a]

Composition	Molecular weight[b]	Exact mass of the molecular ion[c]	Absolute abundance[4]	Most abundant signal in the molecular ion region	Fig.
C_2H_4	28.05376	28.031300	97.7337	28	4.44
$C_{20}H_{40}$	280.5376	280.31300	79.5138	280	4.45
$C_{200}H_{400}$	2805.376	2803.1300	10.1023	2805	4.46
$C_{2000}H_{4000}$	28053.76	28031.300	1×10^{-8}	28053	4.47

[a] Calculations from Dr. R. Schubert, Finnigan, MAT, Bremen.
[b] The calculation of the molecular weights is based on the relative atomic masses, cf. Tab. 4.13 (p. 304 ff.).
[c] The exact mass of the molecular ion was calculated from the mass of the principle isotope, see Tab. 4.13 (p. 304 ff.).
[d] The value of the absolute abundance was normalised to 100. It refers to the ion for which the exact mass has been given, relative to all molecular ions.

The masses of highly charged ions (e.g. resulting from electrospray ionisation) fall in a lower mass region (up to ca. 2000). Therefore, these masses can be determined very accurately (at 20 000 Daltons to better than 0.02%).

A special problem is the recognition of the exact molecular mass in the spectrum. Using Tab. 4.5 and the schematic spectra in Figs. 4.44-4.47, the nature of this problem is described in terms of the hydrocarbon C_nH_{2n}. Naturally occurring carbon and hydrogen both consist of two isotopes which form the so-called isotope peaks in a spectrum (in both cases at 1 amu heavier than the principle isotope). This was explained in more detail in Sec. 3. The abundances of these isotope peaks from "normal" low molecular weight molecules, which consist of C, H, N, O, are always smaller than that of the molecular ion peak. However, as soon as a certain number of atoms is exceeded (this number is governed by the isotopic abundance; for carbon this is 90 atoms and for chlorine it is 3, cf. Tab. 4.10), the first isotope peak becomes more abundant than the molecular ion signal. If the molecule contains additional atoms that are composed of several isotopes, then the appearance of the relevant molecular ion region in the spectrum becomes very complex and can only be analysed quickly with the aid of a computer. One should also not disregard the fact that, among others, $[M+H]^+$ signals can still occur, which complicates the interpretation further.

As a demonstration example, the molecular region (resolving power 6000) of the FAB spectrum of bovine insulin is reproduced in Figs. 4.48 and 4.49 [$C_{254}H_{377}N_{65}O_{75}S_6$, molecular weight = 5733.5739; $M^{+\bullet}$ (nominal mass): 5727; $M^{+\bullet}$ (exact mass of principle isotope): 5729.6009. The molecular region contains $[M+H]^+$ ions and isotope peaks]. The left-hand spectrum was recorded directly onto photo-sensitive paper, while the right-hand one was smoothed by a computer. The mass spectrometric molecular weight determination and the derivation of the structure from the mass spectrum of an unknown compound with a high relative molecular mass is enormously more difficult than the analogous determination for low molecular weight compounds.

Reference: [18].

8.9 Ionization Methods

Even when mass spectrometry was first used on organic molecules, the thermal lability of many compounds proved to be a hindrance to the determination of the relative molecular mass. Various improvements to the sample injection process for EI mass spectrometers have contributed substantially to the reduction of this disadvantage. However, for fundamental reasons, this problem cannot be eliminated entirely. Whenever the above-mentioned conditions are used for the measurement, the organic sample must be vaporised before it is ionised and thereby rendered accessible for the mass spectrometric analysis. Therefore, there has been no lack of experimental effort to develop ionisation procedures which do not

Tab. 4.6 Alternative ionisation methods

Sample	Ionisation method (Abbreviation)	Explanation, see p.
vaporisable[a]	electron impact (EI)	220
	chemical ionisation (CI)	258
difficult to vaporise or cannot be vaporised	field ionisation (FI)	261
	atmospheric pressure ionisation (API)	259
	electrospray ionisation (ESI)	259
	field desorption (FD)	261
	laser desorption/ionisation (LDI)	268
	secondary ion mass spectrometry (SIMS)	273
	fast atom bombardment (FAB)	260
	direct chemical ionisation(DCI)	259
	thermal desorption (TD)	277
	thermospray ionisation (TSI)	277

[a] including GC analyses

require that organic samples be in the vapour phase before their ionisation. Recently, these endeavours have become particularly topical because the desire of organic and bio-organic chemists to investigate mass spectrometrically the structures of higher molecular and biologically relevant materials is becoming increasingly stronger. Compounds of this kind (such as polypeptides, oligosaccharides, glycosides, nucleotides, etc.) nearly always contain several polar functional groups, which make it impossible to vaporise the compound without pyrolysis occurring.

Another justification of alternative ionisation procedures to electron impact ionisation is that there are various classes of compounds which, under EI conditions, either do not yield a molecular ion or only one whose abundance is too low.

In addition to electron impact ionisation, chemical ionisation is also suitable for vaporisable samples. Because the latter can be operated with many collision gases, which lead to various types of spectra, this method has a very large scope. Furthermore, there is now a very large arsenal available for use with

Tab. 4.7 Frequently employed mass spectrometric ionisation methods

ionisation method (abbreviation)	ionising particle	Types of ions	Possible additional signals	Normal mass region max. up to ca.	Possibility of thermal decomposition
Electron impact ionisation **(EI)**	e^-	$M^{+\bullet}$ and fragment ions	–	3500	yes
Chemical ionisation **(CI)**	charged reactant gas, e.g. CH_5^+, NH_4^+ $Ar^{+\bullet}$	e.g. with NH_4^+: $M^{+\bullet}$, $[M + H]^+$, $[M + NH_4]^+$ and clusters	reactant gas and reactant gas clusters	3500	yes
Fast atom bombardment **(FAB)**	e.g. Ar^0 high kinetic energy	e.g. $[M + H]^+$, $[M + Na]^+$ $([M + K]^+)$ and clusters e.g. $[2 M + H]^+$	signals from matrix clusters, e.g. $[2\ glycerine + H]^+$	3500	very rare
Electrospray ionisation **(ESI)**	none (electrostatic)	$[M + H]^+$, $[M + Na]^+$, $([M + K]^+)$ and clusters	–	100 000	no
Thermospray ionisation **(TSI)**	frequently $CH_3CO_2NH_4$	$[M + H]^+$, $[M + NH_4]^+$	sometimes solvent clusters	3500	no

samples that are either difficult to vaporise or cannot be vapo-rised. Tab. 4.6 summarises the ionisation methods normally available in organic chemistry departments. The methods given in the table can be supplemented by other procedures, which are not yet as widely used at present as they probably should be. One of these is $^{252}_{98}$Cf plasma desorption (PD; ioni-sation by bombardment of a sample situated on a supporting plate with nuclear particles)[19].

It should also be said that each ionisation method has its own special characteristics and requirements with regard to selec-tivity, speed of the analysis, necessary quantity of a sample and its preparation, etc. However, because of its widespread use, the extensive spectral material available and, finally, because of the problem free recording of the spectra of "simple" or-ganic compounds, electron impact ionisation is still the most important method; cf. summary in Tab. 4.7.

Frequently, the mass spectrometer is arranged so that several ionisation methods can be employed. However, the change-over from one ionisation method to another is not always as

simple as just throwing a switch. More often, longer term rebuilds, followed by adjustments, etc., are necessary. As a user, one should understand that it is not possible to perform measurements with every method at all times.

8.10 Cation Addition Mass Spectroscopy

By adding alkali-metal salts to polar organic compounds (M) it is possible, under FD conditions, to obtain so-called cluster ions of the general formula [M + alkali-metal]$^+$. All alkali-metal cations can be used. The tetraphenylborate anion, $[B(C_6H_5)_4]^-$, is particularly suitable as the counter-ion of the alkali-metal cation. The advantage of the method is that it faci-litates the determination of the relative molecular masses of polar or thermally labile compounds. The [M + alkali-metal]$^+$ signal appears as the most abundant signal in the spectrum. To illustrate this, the spectrum of loroglossin (**65**), a natural pro-duct of plant origin, is shown in Fig. 4.50. Although the com-

Possibility for on-line combination with	Advantages	Disadvantages
GC	– fragment ion signals = structural information. – largely correct abundances of the isotope signals.	– $M^{+\bullet}$ is sometimes absent. – (very) polar substances cannot be measured.
GC	– suppression of fragmentation which results in more abundant ions in the M region.	– very polar substances cannot be measured. – in cases of uncertainty it is possible to differentiate between [M + H]$^+$ and e.g. [M + NH$_4$]$^+$ by changing the reactant gas. – incorrect abundances of the isotope signals.
–	– measurement of polar substances.	– reduced solubility of substances in the matrix (frequently used: glycerine). – fragmentation is rare.
LC or HPLC	– multiply charged ions often produced (structurally dependent). – measurement of high molecular weight substances in solution.	– reduced choices for types of solvents. – big differences in the ionisation of particular classes of substances. – fragmentation very rare.
LC or HPLC	– measurement of polar substances in aqueous solutions. – fragment ions sometimes occur.	– reduced choices for types of solvents. – the presence of a vaporisable electrolyte is necessary.

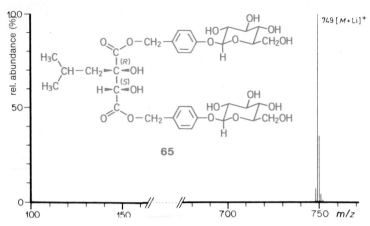

65

Fig. 4.50　Li⁺ ion addition spectrum of loroglossin (65)

pound decomposes under EI conditions, an $[M - 2\,H_2O]^{+\bullet}$ signal can be observed under FD conditions. In the Li⁺ ion addition spectrum, the only peak to appear is that of the $[M + Li]^+$ ion at $m/z = 749$.

Even without the addition of alkali-metal cations, it is also found that under soft ionisation conditions $[M + Na]^+$ and, less frequently, $[M + K]^+$ ions are recorded. These alkali-metal cations appear to be ubiquitous.

Reference: [20].

8.11 The Coupling of Other Instruments to Mass Spectrometers

Because of the very sensitive detection limit for small amounts of substances, the mass spectrometer is used as a detector for gas chromatographs (GC) and liquid chromatographs (LC). Of course, eluate samples which have been separated by GC and LC can be collected and subsequently analysed individually by mass spectrometry. This latter method can occasionally have certain advantages for the solution of special problems (e.g. identification of the components of fragrances, or the recording of other types of spectra).

Today, the **GC/MS combination** is a standard method of organic chemistry. Both types of column, i.e. packed and capillary columns, can be coupled to mass spectrometers. When a packed column is combined with a mass spectrometer, it is necessary to have a separator at the junction of the instruments in order to reduce the amount of carrier gas.

A discussion of the chromatographic process is not possible within the scope of this book. We shall start with the assumption that a mixture is well separated into its components by a chromatograph (GC or LC). The column of the gas chromatograph is located in a thermostatted chamber (adjustable up to 200°C). The end of the column leads into the ion source of the mass spectrometer, where the succession of components emerging from the column are directly ionised on-line and recorded. The connection, or interface, between the gas chromatograph and the mass spectrometer has a special importance. The substances that have been separated in the chromatograph should not become mixed again while traversing the interface. For this reason, the component connecting the instruments should be as short as possible and should be heated (to ca. 20°C above the GC temperature) in order to avoid partial condensation.

The time required while a single component of a substance emerges from a packed column can be up to ca. 1 min; with a capillary column this time can be of the order of 1 s. The time factor is therefore very important for the recording of the spectra. In order to collect all of the components mass spectrometrically, as many scans as are necessary will be carried out automatically and stored in a computer [magnetic field instrument: scan time 1 s, reset time 2 s, total ca. 3 s for one mass decade (ca. $m/z = 300$ to 30)].

After completion of the measurement, the chromatogram is computed (abscissa: sequence numbers of the mass spectra or the time in minutes; ordinate: concentration of the individual components, measured with the total ion current detector (TIC) instead of the flame ionisation detector (FID) customarily used in GC analyses).

The evaluation of the spectra, i.e. the determination of the structures of the components, is then conducted, preferably by using a computer to compare the spectra with those in a spectral library.

The measurement of a spectrum with the GC/MS technique only requires exceptionally small quantities of the individual components (nano- to femtogram region), cf. Sec. 2.1. Naturally, all of the limitations that apply to gas chromatography also apply to the GC/MS combination: only thermally stable compounds can be measured.

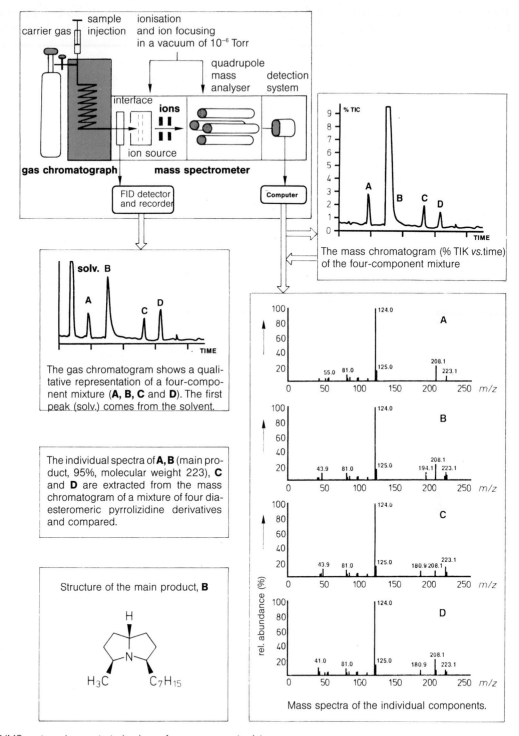

Fig. 4.51 The GC/MS system demonstrated using a four-component mixture

For full details about this method, as well as additional advantages and disadvantages, refer to the specialised literature[21].

The principle of the GC/MS combination is illustrated by the following example (see Fig. 4.51).

Consider a mixture of four compounds, one of which is definitely the synthesised pyrrolizidine alkaloid. The gas chromatogram – detected with a flame ionisation detector (FID) – shows five signals, one of which is the solvent. A mass spectrometric analysis yields the TIC chromatogram (total ion current) which corresponds with the FID chromatogram. The mass spectra were recorded automatically at 1 s intervals, independent of the appearance of any compounds. The best spectrum of each component is illustrated (A, B, C and D). The results show that all four compounds have the same molecular weight ($M = 223$), however there are small, almost insignificant differences in the abundances of the fragment ions. This evidence, together with the support of other analytical results, identifies the substances as four diastereomers.

A further GC/MS example is given in Chapter 5, exercise 7.

There are special practical difficulties with the **directly coupled LC/MS**. The column chromatography is carried out with pure solvents or mixtures of solvents. However, for the mass spectrometric analysis, these solvents (usually strongly polar, sometimes buffered) must be removed, because, in comparison with the solvent, only a tiny amount of material is of interest. Because the LC method is a very important analytical tool in chemical laboratories, a great deal of effort has recently been put into the development of a usable directly coupled LC/MS. Three methods are available commercially.

The moving belt method. The LC eluate is continuously placed onto an endless rotating ribbon (or wire). After the evaporation of the solvent (IR heater, vacuum), a thin film of the relatively involatile sample remains behind on the ribbon. The ribbon then enters the high vacuum region where the sample is vaporised (IR heater) and introduced into the mass spectrometer. Before the ribbon is reused, it runs through cleaning processes. These are not always completely successful and residues can build up with each cycle of the ribbon. This makes the previously obtained LC separation apparently ineffective.

Polar compounds are readily separated with the aid of LC or high-performance liquid chromatography (HPLC) and they are usually thermally labile. Suitable ionisation methods for these compounds are those which were developed for substances that are either difficult to vaporise or cannot be vaporised, i.e. FAB, FD, LD and SIMS[22].

The thermospray ionisation method (TSI). This second method of analysing products from the directly coupled LC/MS will be explained in more detail in Sec. 8.24, p. 277.

The third procedure is the ionisation of particles by **electrospray** (ESI). This method was presented in Sec. 8.3, p. 259.

Reference: [22].

8.12 Laser Desorption / Ionisation Mass Spectrometry (LDI)

The interaction of a UV laser beam with matter forms positively and negatively charged ions, which desorb from a surface and can then be analysed mass spectrometrically.

The method was developed for the analysis of, among other things, tissue samples and allows the smallest regions to be specifically analysed with the microscopically adjustable laser beam (LAMMA® = laser microprobe mass analyser). (Pseudo)-molecular ions are formed (e.g. $M^{+\bullet}$, $[M + H]^+$ and $[M - H]^-$, as well as M^{2+}, M^{3+} or cluster ions such as $[2M]^+$, $[3M]^{2+}$, $[2M]^{3+}$, etc.). For the mass spectrometric analysis, a time-of-flight (TOF) mass spectrometer is employed with, for example, a 3 keV acceleration potential. The laser is, for example, a Nd-YAG laser with a wavelength of 266 nm and a pulse frequency of 10 ns. The sample being analysed absorbs energy in one of three ways: directly (a chromophore is available) or, as occurs particularly with aliphatic compounds, the metal support underneath the sample absorbs the energy initially and then transfers it to the molecule (e.g. the formation of $[valine \cdot Ag]^+$), or the matrix in which the sample to be examined has been embedded initially absorbs the UV radiation (MALDI = matrix assisted laser desorption ionisation). Nicotinic acid or aminobenzoic acid, for example, serve as suitable matrices. LDI is a soft ionisation method. It is possible to determine molecular weights up to ca. 300 000 Daltons and the actual amount of substance required is ca. 10^{-17} mol, i.e. 50 fmol. The remaining unused material can be recovered.

Reference: [23].

8.13 Multiply Charged Ions

As mentioned in Sec. 2 (p. 220), singly charged molecular ions are formed by the electron bombardment of molecules. However, at the same time, even if considerably more rarely, processes do take place in which two and, still less frequently, three electrons are removed from the molecule. This results in the formation of doubly and triply charged particles, respectively. Because the ions are recorded according to the m/z or $m/n \cdot e$ scales (in most cases $n = 1$, so that m/e is valid), doubly and triply charged ions will be registered at $m/2 \cdot e$ and $m/3 \cdot e$, respectively, i.e. they appear at one half and one third of their actual masses, respectively. If the relative molecular masses are even numbers, then, for doubly charged ions, the signals will be recorded at integer m/e values and they cannot be distinguished from those of singly charged fragment ions without additional effort. However, because the first isotopomer of an ion with an even numbered relative molecular mass has an odd numbered mass, the first isotope peak from a doubly charged ion will appear at a half mass number. Hence it is possible to find the position of the doubly charged molecular ions quite easily, because these ions have the same isotope ratios as singly charged molecular ions. The appearance of the doubly charged molecular ion can be used as a good criterion for the checking of correctly counted out mass spectra. The appearance potentials of doubly charged ions lie well above those of singly charged ions (ca. 20 to 30 eV higher). Upon lowering the ionisation potential (low-voltage spectra), the peaks due to multiply charged ions disappear. This is a property which can also be used to identify such ions. One frequently finds doubly charged molecular ions with aromatic compounds and polyolefins.

As well as doubly charged molecular ions, doubly charged fragment ions can also appear, for which the same rules are valid. In a few cases [e.g. bis(benzyltetrahydroisoquinoline) alkaloids from oxyacanthines[24]] the base peak actually results from a doubly charged fragment ion. During the investigation of the mechanism of a fragmentation, it should be remembered that the mass differences must be doubled as well. The calculations for metastable peaks ($m_M^{2+} \rightarrow m_T^{2+}$, $m_M^{2+} \rightarrow m_{T1}^{+}$ $+ m_{T2}^{+}$, etc.) can be carried out with the formula given under "metastable peaks" in section 8.25 (p. 277). Multiply charged ions appear during electrospray ionisation and are the principle reason that high masses can be determined by this method (cf. Sec. 8.8, p. 259). Occasionally doubly charged ions are also observed in FD spectra.

8.14 The Memory Effect

If residues of a substance still remain behind in the ion source region of a mass spectrometer after a measurement (e.g. by condensation on cooler parts) and if signals due to these residues reappear during the next measurement, then one speaks of a memory effect (ME). By recording a background spectrum before every measurement, proof of the presence of undesirable sample residues can be obtained. The problem can be eliminated by heating the ion source or by mechanical cleaning.

8.15 Neighbouring Group Participation Reactions

Bifunctional Alkanes

The mass spectrometric decomposition of di- and polyfunctional alkanes is essentially characterised by two fundamentally different reactions: on the one hand the structures of fragment ions are explained by the independent fragmentation of every single functional group; on the other hand, however, quite a large number of fragment ions are formed by the mutual interaction of two or more functional groups. Such reactions have been found with a relatively large number of α,ω-disubstituted alkanes, e.g. ω-hydroxycarboxylic acid esters, ω-methoxycarboxylic acid esters, ω-oxocarboxylic acid esters, ω-hydroxyethylene acetals, α,ω-diaminoalkanes, N,N'-diacyl-α,ω-diaminoalkanes, ω-aminocarboxylic acid esters and ω-aminophenylalkanes. The number of methylene groups between the two functional groups almost always plays a deciding role in the extent to which the fragmentation reaction occurs by neighbouring group participation (cyclic transition states, ring formation reactions).

As an example of this special behaviour, the mass spectrum of 1,4-bis(acetylamino)butane (66; M=172) is reproduced in Fig. 4.52 and discussed with the aid of Scheme 4.18.

The fragment ions at $m/z = 129$ (M−COCH₃) are typical, as are the two cyclic ions with $m/z = 112$ and 70. Monofunctional systems show a different behaviour (see, e.g., N-butylacetamide, **35**, Fig. 4.31).

66⁺·
$m/z = 172$

− ĊOCH₃

$m/z = 129$

$m/z = 112$

− NH₃ | m*

m*
− CH₂CO

$m/z = 70$

Scheme 4.18 See Fig. 4.52

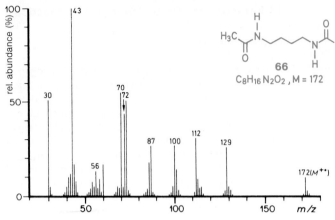

Fig. 4.52 Mass spectrum of 1,4-bis(acetylamino)butane (= N,N'-diacetylputrescine, **66**)

Mass spectrometric S_Ni reactions also belong to this class of fragmentation reactions. Literature review: [25].

The *ortho* Effect

This is a special case of mass spectrometric neighbouring group participation reactions which is observed with *ortho*-disubstituted benzene derivatives (or with *peri*-disubstituted naphthalene derivatives). Frequently, a mass spectrometric differentiation between *o*-, *m*- and *p*-isomeric benzene derivatives is impossible. The spectra of the *m*- and *p*-isomers are generally identical, but they can be distinctly different from that of the *o*-isomer. However, a prerequisite for this behaviour is that both of the adjacent substituents undergo, by mutual interaction, reactions that neither of the substituents would be involved in when alone. This indicates that "atypical" fragmentation reactions occur for the special *ortho* arrangement. These reactions are recognisable by the positions of the signals which cause spectra of *o*-isomers to have a different appearance.

A typical example of the similarities (*m*- and *p*-isomers) and differences (*m*- and *p*- versus the *o*-isomer) in the spectra is shown by nitrotoluene: *o*-nitrotoluene (**67**; M = 137, Fig. 4.53), *m*-nitrotoluene (**68**; M = 137, Fig. 4.54) and *p*-nitrotoluene (**69**; M = 137, Fig. 4.55).

The *m*- and *p*-isomers give typical fragment ions for the nitro compounds: $[M - 16]^+$: $m/z = 121$, $[M - 30]^+$: after rearrangement to the nitrite ester $m/z = 107$ and $[M - 46]^+$: $m/z = 91$, Scheme 4.19. For the *o*-compound **67**, these signals are indeed present, but now $m/z = 120$ (**a**) also appears and is the most abundant peak of the spectrum. This corresponds to the loss of OH• from the molecular ion (Scheme 4.20). Other ions that have a higher abundance in the spectrum of **67** are $m/z = 92$

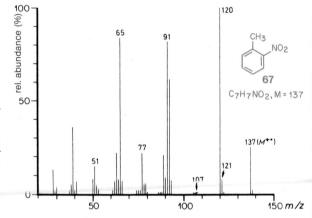

Fig. 4.53 Mass spectrum of *o*-nitrotoluene (**67**)

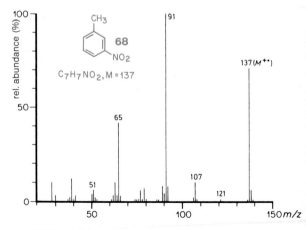

Fig. 4.54 Mass spectrum of *m*-nitrotoluene (**68**)

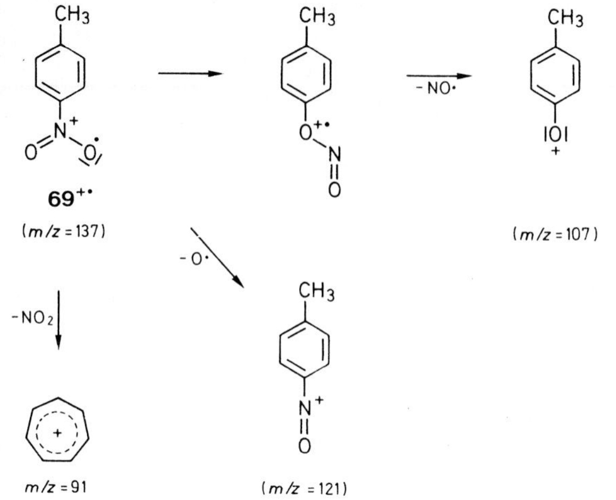

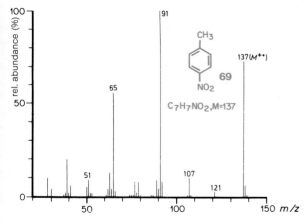

69⁺•

$(m/z = 137)$

$(m/z = 107)$

$m/z = 91$

$(m/z = 121)$

Scheme 4.19   Cf. Figs. 4.54 and 4.55

67⁺•

$(m/z = 137)$

a

$(m/z = 120)$

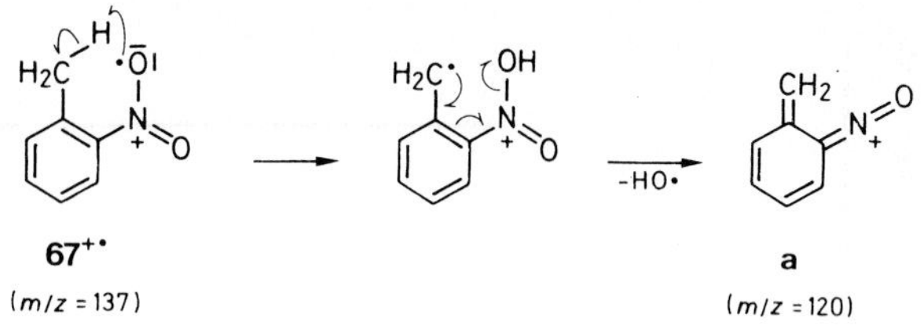

Scheme 4.20   See Fig. 4.53

and 77. These result from the loss of CO from **a** or the loss of HCN from $m/z = 92$.

The mass spectra of aliphatic nitro compounds rarely contain the molecular ion. Usually the most abundant signals to be recorded are the ions $[M-30]^+$ and/or $[M-46]^+$. However, if the spectra are recorded using the chemical ionisation (CI, see Sec. 8.1, p. 258) or fast atom bombardment (FAB, see Sec. 8.4, p. 260) techniques, the (pseudo)-molecular ion will appear. This can be seen by comparing the spectra of 3-(1-nitro-2-oxo-cyclododecyl)propanal (**70**; M = 283) recorded under EI (Fig. 4.56) and CI conditions (Fig. 4.57).

A benzene derivative, for which the spectra of the o-, m- and p-isomers are the same within the accuracy and reproducibility of the measurements, is xylene. The spectrum of m-xylene (**71**; M = 106), which also represents those of the other isomers, is reproduced in Fig. 4.58. The most important fragment ion is $m/z = 91$, which is formed from the molecular ion by the loss of a methyl group (and by a rearrangement).

In connection with this discussion of aromatic compounds, it should also be mentioned that in a few cases the mass spectra

Fig. 4.55   Mass spectrum of p-nitrotoluene (**69**)

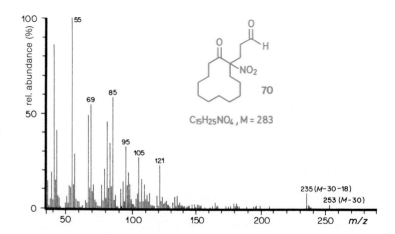

Fig. 4.56   EI mass spectrum of 3-(1-nitro-2-oxocyclododecyl)propanal (**70**)

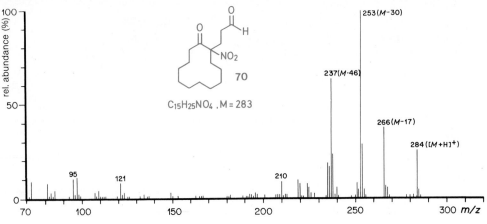

Fig. 4.57 CI mass spectrum of 3-(1-nitro-2-oxocyclododecyl)propanal (**70**); reactant gas: 2-methylpropane (isobutane)

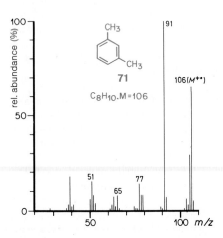

Fig. 4.58 Mass spectrum of *m*-xylene (**71**)

8.16 Photoionisation (PI)

The ionisation of molecules occurs by irradiation with energetic photons:

$$M + h\nu \rightarrow M^{+\bullet} + e^-$$

The method is particularly suitable for the exact determination of the ionisation potential [also cf. laser desorption / ionisation mass spectrometry (Sec. 8.12, p. 268)].

8.17 Quadrupole Mass Analysers

Quadrupole mass analysers serve to separate ions. The separation is achieved through deflection of the different masses by means of electric fields. Four parallel metal rods, arranged symmetrically about the *z*-axis, form the heart of the device. The rods diametrically opposed to each other are electrically connected. A constant potential U and a modulated radio-frequency potential $(V_0 \cdot \cos \omega t)$ are applied (Fig. 4.59) to the rods. The ions, which are injected along the *z*-axis (field axis), oscillate in the *x*- and *y*-directions.

When a particular potential is applied, a specific ion of a certain mass m will execute a stable oscillation and, after passing through the rod system, reach the detector, while under the same conditions, the other masses will be screened out. In this way a mass separation is achieved. A mass spectrum is obtained by varying U and V_0, while maintaining the U/V_0 ratio. The recorded mass m is proportional to V_0.

of *p*-substituted compounds can also differ from those of the *m*-isomers. This is because only the *p*-isomers are capable of forming resonance stabilised ions upon the loss of functional groups and the corresponding signals are therefore particularly abundant. *p*-Methoxybenzene derivatives, for example, are capable of such reactions.

Comprehensive literature: [26].

The *peri* Effect

The phenomenon of the mass spectral behaviour of 1,8-disubstituted (*peri*-substituted) naphthalene derivatives is mostly summarised under the discussion of the *ortho* effect.

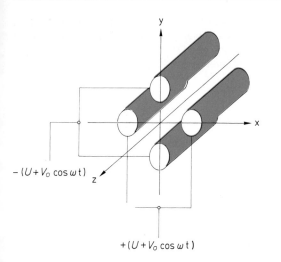

$-(U + V_0 \cos \omega t)$

$+(U + V_0 \cos \omega t)$

Fig. 4.59   Schematic representation of a quadrupole mass analyser

Although only a maximum m/z ratio of 2000 can be achieved with quadrupole analysers and they are only suitable for use as low resolution devices, they are very popular and in widespread use. They are relatively easy to build, relatively inexpensive and can be operated without a lot of experience. It is possible to combine the quadrupole analyser with most of the ionisation procedures and currently it is most frequently used in combination with coupled GC/MS instruments operating under EI or CI conditions.

Reference: [27].

8.18 Secondary Ion Mass Spectrometry (SIMS)

A beam of energetic primary ions (e.g. $Ar^{+\bullet}$ of 2 to 10 keV) is used to generate positively and negatively charged ions from a sample which has been placed on a metal plate (e.g. Ag). The acceleration, focusing, separation and detection of these ions occurs in the mass spectrometer in the usual way. In addition to $M^{+\bullet}$ and $M^{-\bullet}$, the principle ions to be recorded are $[M + H]^+$, $[M - H]^-$, $[M + Na]^+$ (this ion appears even without the special addition of a salt) and $[M + Ag]^+$, as well as fragment ions. The metal ions originate from the metal surface or from impurities. With the help of this method, mass spectra of organic compounds that are either involatile or have only a low volatility (e.g. ammonium salts, peptides, oligosaccharides, glycosides) can be recorded and used for the determination of the relative molecular masses of the compounds, as well as for the elucidation of their structures.

Reference: [28].

8.19 Spectral Libraries

Mass spectra are suitable for digitalisation (mass number *vs.* relative abundance) and storage. In this way spectral libraries can be built up on tapes or disks. It is sensible to store only the mass spectra of structurally known compounds of the highest purity. When spectra of new, structurally unknown samples are recorded, especially those of multicomponent mixtures from GC/MS, a computer can be used to compare these spectra with the spectral library and thereby identify the substances. Spectral libraries with several tens of thousands of spectra can be obtained commercially (see bibliography). Aside from the technical aspects relating to computers, there are a few basic points which should be mentioned. The reproducibility of mass spectra recorded with the same ionisation method is not very great. It is dependent upon the type of instrument, the purity of the sample, crystallinity, etc. Spectral libraries sometimes contain an abundance of fantastic unique structures that are hardly ever measured a second time as an unknown substance. On the other hand, simple compounds are sometimes missing. For the time-saving and effective use of computer comparisons, it is also worth striving to develop one's own spectral library. However, the time-consuming process of building such a library of one's own is only worthwhile if it can also be used, i.e. when this method of substance identification is employed frequently (e.g. in forensic chemistry, the analysis of fragrances, etc.). Of course, special spectral libraries are necessary when different ionisation methods are employed, because even the spectra from the same compound can be different.

8.20 Stereoisomers

Optical Antipodes

The mass spectra of optical isomers are identical and independent of the number of chiral centres (achiral recording conditions of the mass spectra!). The same is true of racemic mixtures.

Geometric Isomers

E,Z-isomers (trans-, cis-isomers). The mass spectra of *E,Z*-isomers can be, but do not have to be, different. This depends primarily on whether or not the fragmentation reaction involves the double bond or the functional group at the double bond or in its immediate vicinity. If the fragmentation occurs outside the sphere of influence of the double bond, the spectra will be the same, otherwise differences can be observed. These differences are usually evident in distinctly different abundances of individual signals; rarely are completely different mass spectra observed. As examples, the mass spectra of maleic acid (**72**; M = 116) and fumaric acid (**73**; M = 116) are repro- duced in Figs. 4.60 and 4.61.

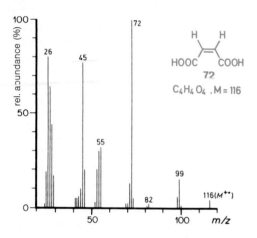

Fig. 4.60 Mass spectrum of maleic acid (**72**)

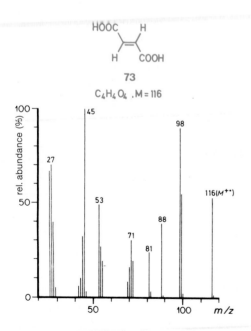

Fig. 4.61 Mass spectrum of fumaric acid (**73**)

Decarboxylation ($m/z = 72$) plays a dominant role in the mass spectrometric decomposition of maleic acid (**72**), whereas in the case of fumaric acid (**73**), the loss of water ($m/z = 98$) and the decarbonylation ($m/z = 88$) of the molecular ion are predominant.

Diastereoisomers

Depending on the distances between two functional groups, diastereoisomers can behave in a mass spectrometer in similar manner to *E,Z*-isomers. According to the type of compound and the fragmentation reaction, the spectra of diastereoisomers can display either almost insignificant or considerable differences. This behaviour is also dependent upon the location of the primary fragmentation. For example, the mass spectra of *cis*- (**74**) and *trans*-1,4-cyclohexanediol (**75**) are very different with regard to the loss of water from the molecular ion. In the spectrum of the *cis*-compound, the loss of water amounts to 1.8% Σ and a subsequent 1,4-elimination (determined by a D-labelling experiment) consumes about half of this (0.9% Σ). In the spectrum of the *trans*-compound, 8.1% Σ is due to the loss of water, but, compared with **74**, an eight times greater proportion (7.3% Σ) is involved in the 1,4-elimination. This find- ing is in good agreement with the geometric arrangement of the hydroxy groups in the boat form. Similar to this, if not quite so pronounced, is the loss of ammonia from *cis*- and *trans*-cyclohexanediamines and their derivatives.

74   **75**

Review article: [29].

8.21 Collision Activation (CA)

(also known as collision induced dissociation: CID)

This is a method for the analysis of the structures of ions. When ions, which possess a high translational energy (a few hundred eV), impinge upon gaseous neutral atoms or molecules, the ions are electronically excited at the expense of the translational energy. They then undergo decomposition reactions and produce a spectrum which is characteristic of the structure and energy content of the ions (CA spectrum). CA spectra of the same ions (same elemental composition) from different sources are identical (including the abundances and half-height widths of the lines).

It is particularly useful to record CA spectra in conjunction with those ionisation methods that produce only the molecular ion of a compound (e.g. DCI, ESI, FAB and FD spectra). These spectra then yield fragment ion signals from which structural information may be obtained. For such an apparatus the following sequence is chosen: ion source – magnetic sector – collision chamber with collision gas – electrostatic sector – electron multiplier. Another sequence is shown in Fig. 4.62 (p. 276).

Comprehensive literature: [30].

8.22 Tandem Mass Spectrometry

This refers to two mass spectrometers that are arranged one behind the other, so that the method is also known as mass spectrometer/mass spectrometer (MS/MS). This combination opens up an additional area of mass spectrometric information. The procedure is as follows. A sample is ionised (all types of ionisation are possible) and gives a mass spectrum in the first mass spectrometer (MS 1). If one is now interested in a particular type of ion (fragment or molecular ion), this can be selected and diverted into a collision chamber (cf. Sec. 8.21). As a result of collisions with the gas that exists therein, the kinetic energy of these ions is partially transformed into vibrational energy, which causes the ions to fragment. These fragments then enter the second mass analyser (MS 2) where they are separated and analysed. In this way it is possible to obtain structural information about the type of ion that was selected. The method is suitable for structural analyses and the analysis of mixtures (selection of individual molecular ions), even when the sought-after substance therein is only present in minute quantities (e.g. analysis of biological materials or metabolites). The resulting profusion of data can be processed readily by a computer and this has allowed MS/MS to become one of the most efficient analytical instruments.

Fig. 4.62 depicts the principle of the method. First of all, a mixture of substances, consisting of three types of molecules, A, B and C, is introduced into the first mass spectrometer (MS 1). This results in the production of a mixed spectrum of all components and their fragments, the signals of which are superimposed on one another. In order to analyse a particular molecular ion more closely, in this case B^+, the instrument is adjusted so that only this type of ion is detected and all of the others are screened out. The selected ions are then directed into a collision chamber where they interact with an inert gas (e.g. xenon). As a result of the collision activation, B^{+*} undergoes a fragmentation into specific particles, $B^+ - X$, $B^+ - Y$, $B^+ - Z$, etc., which are characteristic for a particular structure. B^{+*} means that the ion B^+ possesses additional energy (transla-

tional and kinetic energy). This fragmentation is recorded and the mass spectrum MS 2 is produced, from which the compound can be identified, usually by comparing the spectrum with a spectral library.

As an example of this method, the proof of the presence of trichlorodibenzodioxin (M=286) in a contaminated coal sample will be demonstrated.

The total spectrum of the mixture, as obtained from MS 1, gives absolutely no information about the presence of the indicated impurity. However, if the desired mass is selected and, after fragmentation in the collision chamber, analysed in MS 2, then a spectrum is obtained (Fig. 4.62), which is in complete agreement with that from an authentic sample of the dioxin. For fundamental reasons, the isotope peaks are absent. When selecting one type of ion, only one mass, $m/z = 286$ (i.e. $^{12}C_{12}{}^1H_5{}^{16}O_2{}^{35}Cl_3$), can be taken into account. All other isotope combinations are therefore excluded.

A further illustrative example of the use of the MS/MS combination has been published. The naturally occurring polypeptide, eglin c (M=8092.02), is composed of 70 amino acid units. For reasons of identification, it had to be compared with a preparation that had been synthesised using gene technology. The synthetic product, in spite of almost identical biological, immunological and chromatographic characteristics, had a molecular mass which was 42 amu higher. (All measurements with FAB-MS, matrix: thioglycerine.) What caused the difference (CH_2CO or C_3H_6) and on which atom was the substituent located? In order to answer this question, the enzymatic hydrolase was generated from each of the preparations with trypsin and the mixture (which at any one time contained seven cleaved peptides) was analysed mass spectrometrically without a chromatographic separation having been performed. The cleaved peptide that contained the N-terminal amino acid from eglin 3 was 42 amu heavier in the synthetic compound. The molecular ion from this cleaved peptide was then selected from the mixture with MS 1 and treated in a collision chamber with an inert gas. Its mass spectrum was subsequently observed in MS 2. By analysis of the fragmentation pattern, the N-terminal amino acid was identified as the carrier of the substituent, which proved to be a $CH_3CO-(N)$ residue (determined by CD_3CO labelling). The quantity of sample required for the experiment was 20 µg.

Reference: [31].

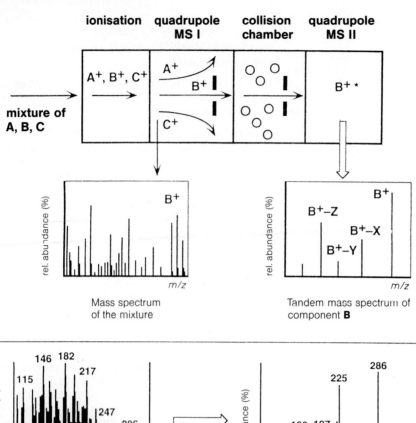

Mass spectrum
of the mixture

Tandem mass spectrum of
component **B**

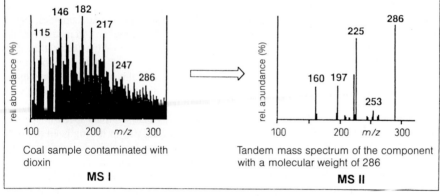

Coal sample contaminated with
dioxin

MS I

Tandem mass spectrum of the component
with a molecular weight of 286

MS II

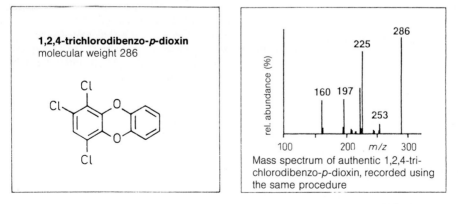

1,2,4-trichlorodibenzo-*p*-dioxin
molecular weight 286

Mass spectrum of authentic 1,2,4-tri-
chlorodibenzo-*p*-dioxin, recorded using
the same procedure

Fig. 4.62   Fundamental outline of a tandem mass spectrometer together with an example

8.23 Thermal Desorption Mass Spectrometry (TD)

(sometimes this process is also known as thermal ionisation)

Organic salts (ammonium, arsonium, oxonium salts) – but also neutral organic molecules when in the presence of Na^+, K^+, etc. – can be brought directly into the gas phase in the ion source of a mass spectrometer if the electron beam is switched off and the temperature is elevated. The acceleration, separation and analysis of the ions that are formed by this method are carried out in the usual manner. The range of applications found for this very new method is not yet reviewable.

Reference: [32].

8.24 The Thermospray Ionisation Procedure[28] (TSI)

A solution (frequently used solvent: CH_3CN/H_2O or CH_3OH/H_2O, where at least 10% should be H_2O), together with an additional vaporisable electrolyte (e.g. 0.1 M CH_3COONH_4), is sprayed under pressure from a hot capillary (inner diameter ca. 0.015 cm, flow rate 0.5-2 cm^3/min, recommended: the end of a liquid chromatography column, LC) into a heated antechamber (1-10 Torr) of the ion source of a mass spectrometer so that a mist of the finest droplets is formed (Fig. 4.63). In contrast to electrospray ionisation, an external electric field is not applied in order to form the ions. The electrolyte causes the formation of an equal number of positively and negatively charged droplets (charge exchange).

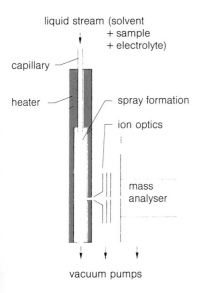

liquid stream (solvent + sample + electrolyte)

capillary

heater

spray formation

ion optics

mass analyser

vacuum pumps

Fig. 4.63 Schematic representation of a thermospray ionisation inlet system

These droplets lose their solvent molecules in the vacuum and the ions thus formed are then analysed by an attached mass spectrometer.

The electrolyte ions can ionise molecules of the substance of interest according to the usual CI processes ($[M+NH_4]^+$, $[M+Na]^+$, also $[M+H]^+$ and sometimes cluster ions with the solvent $[M+CH_3CN+H]^+$).

The advantage of this method lies in its ability to bring polar and thermally labile compounds into the gas phase without having to use a direct vaporisation process.

Reference: [33].

8.25 Metastable Signals

(also known as signals of metastable ions or metastable peaks)

Ions that survive from the ion source to the collector without decomposing must have a lifetime of at least 10^{-5} s (e.g. molecular ions). If they have a significantly shorter lifetime (of the order of 10^{-6} s or shorter), then they decompose while still in the ion source, are correctly accelerated according to their mass and are recorded at the collector as fragment ions. Ions with a lifetime of between 10^{-5} and 10^{-6} s decompose between the ion source and the collector. These ions are known as metastable ions. Of special importance are decompositions which take place in the first field free region (cf. Fig. 4.5, p. 225). These metastable ions have experienced the full acceleration of the mother ion (m_M), but they decompose before they enter the electrostatic analyser. Because the acceleration corresponded to the heavier mass of the mother ion, the speed of the daughter ions (m_D) is slower than that attained by normally accelerated ions of the same mass. Thus they are deflected more severely in the magnetic field and appear in the mass spectrum at masses which are too "small". The signals from metastable ions in the spectrum are readily distinguishable from those of the other ions. They are recorded as broad peaks, which sometimes spread over several mass numbers, and which are usually of a lower abundance. The position of the peak can be calculated as follows [derived from Eqns. (2) and (4), Sec. 2:

$$m^* = \frac{m_D^2}{m_M}$$

As an example, the loss of ethylene (28 amu) from 1,2,3,4-tetrahydrocarbazole (**22**; M=171) produces a metastable peak at $m/z=119.6$ (cf. p. 237), which can be calculated as follows:

$$m^* = \frac{143^2}{171} = 119.6$$

This signal therefore proves that the ion of mass 143 is formed directly from the molecular ion. In general, it can be said that the appearance of metastable peaks does not exclude other processes which lead to the formation of an ion of the same mass. On the other hand, the absence of m^* peaks does not exclude the existence of a particular fragmentation step. Metastable peaks are important aids for the deduction of fragmentation pathways. Sometimes more metastable ion signals are recorded in the spectrum than can be assigned to consecutive decomposition cascades. Unfortunately metastable peaks will not be registered when the mass spectra are recorded with the aid of a computer, because the programs are designed to suppress broad signals of low intensity.

As an alternative possibility, the so-called "linked scan" should be described. In this procedure, the magnetic field B and the deflection potential V of the electrostatic analyser are scanned simultaneously while a specific relationship is maintained between them. In particular, three procedures should be mentioned:

a) If the scan is conducted with the ratio B/V=const., then all daughter ions of a given mother ion appear in the spectrum.

For example, it has been reported that the ions $C7H_7NO+$ (m/z=121), $C_7H_6NO^+$ (m/z=120), $C_7H_7O^+$ (m/z=107) and $C_7H_7^+$ ($m/z = 91$) are formed from the molecular ion of o-nitrotoluene (**67**; $m/z = 137$, cf. Fig. 4.53, p. 270).

b) If the ratio B/V^2=const. is chosen for the scan, then the origin of the daughter ions is recorded. In the present example of o-nitrotoluene it can be shown, for example, that the daughter ion at m/z=120 originates from the ion m/z=137.

c) Finally, if the function $B^2/V^2(1 - V)$=const. is used for the scan, then all ions which lose neutral fragments of the same mass are recorded (in the present case, for example, 16 corresponds to O and CH_4).

By using the techniques that have been described, it is possible to obtain an insight into the molecular structure of an unknown compound. However, one must constantly bear in mind that a particular fragment ion is not necessarily formed in only one way, but that many decomposition pathways can sometimes be involved (... *many roads lead to Rome* ...).

Reference: [34].

9. **Tables for Use in Mass Spectrometry**

9.1 List of Frequently Occurring Ions and Characteristic Mass Differences Resulting From Mass Spectrometric and Chemical Reactions

The mass differences and ions given in this summary do not represent a complete listing. They should provide an indication (not a proof) of possible structures or structural elements, whereby normal mass spectrometric behaviour or a normal chemical or thermal transformation reaction should be considered rather than a special case.

Notes:
- For fragment ion signals, no differentiation has been made between F^+ and $F^{+\bullet}$.
- *Means that one should look for the characteristic isotope pattern (Tab. 4.10, p. 294).
- The given mass differences (M±) refer not only to the molecular ion, but also to the fragment ion.
- If "e.g." is given beside an entry, isomeric structural elements can easily by deduced.
- Numbers in ( ) are important cross references; see under the corresponding mass number.

Tab. 4.8 Frequently occurring ions and characteristic mass differences

Mass	M ± x		m/z mass difference as a result of chemical and thermal transformation reactions
1	+ H – H	with amines, nitriles from aldehydes, primary and secondary alcohols, cyclic amines, ethers, nitriles, sometimes aromatic derivatives	The mass region $m/z = 1$ to ≈ 10 is generally not recorded under normal circumstances + 1
2	– H$_2$ ± H$_2$	with saturated primary alcohols (R–CH$_2$–OH$^{+\bullet}$ → R–CH=O$^{+\bullet}$), N-Oxides quinones, hydroquinones	+ 2 – 2 Δ, disproportionation, dehydrogenation
3	– H$_3$	with saturated primary alcohols (R–CH$_2$–OH$^{+\bullet}$ → R–C≡O$^{+\bullet}$)	
4			+ 4 – 4
7	– 7	FD spectra: Li	
11	+ 11	typical for aliphatic di- and polyamines (see p. 252)	

Tab. 4.8 continued

Mass	M ± x		m/z mass difference as a result of chemical and thermal transformation reactions	
12			C^+	typical for all *C*-derivatives (possible starting point for counting out the spectrum)
			– 12	
14		homologues	CH_2^+, N^+, CO^{2+}	
			+ 14	
			– 14	
				Δ, *N*-, *O*-methylation, transesterification with homologous alcohols
15	– CH₃	unspecific, frequently CH₃ groups are present in some form	CH_3^+	unspecific
16	– NH₂	from primary amides, amines, sulfonamides	CH_4^+, NH_2^+, O^+	
	– O	from diarylsulfoxides, nitro derivatives (46, 30), *N*-oxides, sulfones	+ 16	
			– 16	
17	– NH₃ – OH	from amines, diamines from alcohols, benzylalcohol derivatives, carboxylic acids, *N*-oxides, oximes, sulfoxides; also all *O*-derivatives	NH_3^+, OH^+	

Tab. 4.8 continued

Mass	M ± x		m/z mass difference as a result of chemical and thermal transformation reactions	
18	– H$_2$O	from aldehydes, alcohols, ethers, carboxylic acids, lactones, N-oxides; also all O-derivatives	H$_2$O$^+$	unspecific (from compounds, solvents, air; possible starting point for counting out the spectrum)

Mass	M ± x		m/z	
19	– F – H$_3$O	F-derivatives (H$_2$O + H), as for M – 18	F$^+$ + 19	F-derivatives −CN ⟶ −COOH
20	–HF	F-derivatives	HF$^+$	F-derivatives
22			CO$_2^{2+}$	
23	– 23	FD spectra: Na		
26	– C$_2$H$_2$ – CN	from aromatic hydrocarbons (91, 77, 65, 51) from aromatic nitriles	C$_2$H$_2^+$ CN$^+$	unspecific
27	–C$_2$H$_3$ – HCN	occassionally from end-positioned vinyl derivatives from aromatic amines, aromatic N-heterocycles, aromatic nitriles (92, 65)	C$_2$H$_3^+$ HCN$^+$	
28	– C$_2$H$_4$ – CO	e.g. from ethyl esters (McLafferty rearrangement), cyclohexene derivatives, 1-tetralone derivatives (RDA), O- and N-ethyl derivatives (onium reaction) from aldehydes, quinones, O-heterocycles, lactams, unsaturated lactones, phenols, α-cleavage products from carbonyl derivatives	C$_2$H$_4^+$ CO$^+$ N$_2^+$ + 28 – 28	unspecific from the air −CH$_2$OH ⟶ −COOCH$_3$ −NH$_2$ ⟶ −N(CH$_3$)$_2$ also Δ −NH$_2$ ⟶ −NH(CHO) −COOCH$_3$ also Δ
29	– C$_2$H$_5$ – CHO	e.g. from ethyl derivatives from aromatic aldehydes, aromatic methoxy derivatives	C$_2$H$_5^+$ CHO$^+$	e.g. from ethyl derivatives from aldehydes

Tab. 4.8 continued

Mass	M ± x		m/z mass difference as a result of chemical and thermal transformation reactions	
30	$-CH_2O$	from cyclic ethers, aromatic methoxy derivatives	$C_2H_6^+$ CH_4N^+:	unspecific $CH_2=NH_2^+$ from secondary acylamides, primary amines, amines (onium reaction)
	$-NO$	from nitro derivatives (46, 16)	NO^+	from nitrosoamines
			-30	$-NO_2 \longrightarrow -NH_2$
31	$-CH_3O$	from primary alcohols, methyl ethers, methyl esters	CH_3O^+:	$H_2C=OH^+$ from primary alcohols, ethers (onium reaction) from methyl esters, CH_3OH
32	$-CH_4O$ $-S$	from methyl esters from S-derivatives (sometimes)	O_2^+ S^+ solvent:	from the air from S-derivatives CH_3OH
33	$-CH_5O$ $-HS$	$=(H_2O + CH_3)$ from isothiocyanates, thiols	CH_2F^+ HS^+	from aliphatic F-derivatives from S-derivatives
34	$-H_2S$	from thiols	H_2S^+ -34	from S-derivatives $\rangle-Cl \longrightarrow \rangle-H^*$
35	$-Cl$	from Cl-derivatives *	Cl^+	from Cl-derivatives, quaternary ammonium chlorides, hydrochlorides *
36	$-HCl$	from Cl-derivatives *	HCl^+ 36	from Cl-derivatives, quaternary ammonium chlorides, hydrochlorides * $\overset{H}{\underset{}{\rangle}}\overset{Cl}{\underset{}{\langle}} \longrightarrow /-\backslash^*$
37	$-Cl$	from Cl-derivatives *	Cl^+	from Cl-derivatives, quaternary ammonium chlorides, hydrochlorides *
38	$-HCl$	from Cl-derivatives *	HCl^+	from Cl-derivatives, quaternary ammonium chlorides, hydrochlorides *
39	-39	FD spectra: K	$C_3H_3^+$	from alkynes, sometimes aromatic compounds
40			Ar^+	from the air
41	$-C_3H_5$ $-C_2H_3N$	from alicyclic compounds CH_3CN from aromatic N-methyl heterocycles	$C_3H_5^+$: $C_2H_3N^+$: solvent	$CH_2=CH-CH_2^+$ from allyl derivatives $CH_2=C=NH^+$ from oximes CH_3CN from C-methyl-N-heterocycles CH_3CN
42	$-C_3H_6$ $-C_2H_2O$	e.g. from butylcarbonyl derivatives, aromatic propyl ethers, via McLafferty rearrangement, from O- and N-propyl derivatives (onium reaction) from α, β-unsaturated cyclohexanone derivatives, 2-tetralone derivatives (RDA), acetoacetic acid derivatives, aromatic O- and N-acetyl derivatives, enols and enamine acetates	$C_2H_4N^+$: $C_2H_2O^+$: $C_3H_6^+$ $+42$	$CH_2=\overset{+}{N}=CH_2$ from cyclic amines from acetyl derivatives $\rangle-OH \longrightarrow \rangle-O\overset{O}{\underset{CH_3}{\diagup}}$ $-NH_2 \longrightarrow -N\overset{H}{\underset{}{\diagdown}}\overset{O}{\underset{CH_3}{\diagup}}$

Tab. 4.8 continued

Mass	M ± x		m/z mass difference as a result of chemical and thermal transformation reactions	
43	– C$_3$H$_7$	from propyl-, isopropyl derivatives	C$_3$H$_7^+$	e.g. from propyl derivatives
	– C$_2$H$_3$O	CH$_3$CO from N-acetyl derivatives, aldehydes, methyl ketones, (CH$_3$ + CO) from aromatic methyl ethers	C$_2$H$_5$N$^+$: C$_2$H$_3$O$^+$ CHNO$^+$:	$^\bullet$CH$_2$–$\overset{+}{N}$H=CH$_2$ from cyclic amines from O- and N-acetyl derivatives, methyl ketones NH=C=O$^+$ from O-derivatives
44	– C$_2$H$_4$O	from aldehydes (McLafferty rearrangement	C$_2$H$_6$N$^+$:	e.g. H$_2$C=$\overset{+}{N}$H–CH$_3$ from acylamides, amines
	– CO$_2$	from anhydrides, carboxylic acids, carbonic acid esters	C$_2$H$_4$O$^+$ CH$_2$NO$^+$: CO$_2^+$	from cyclobutanol derivatives, CH$_2$=CHOH$^+$ from aldehydes, vinyl ethers O=C=NH$_2$ from primary carboxylic acid amides e.g. from the air

$+ 44$

$- 44$ also Δ

Mass	M ± x		m/z	
45	– C$_2$H$_7$N	e.g. from N,N-dimethylamino derivatives	C$_2$H$_5$O$^+$:	H$_3$C–CH=$\overset{+}{O}$H from 2-alkanol derivatives, ethyl ethers
	– C$_2$H$_5$O	from O-ethyl derivatives (ethers, esters)		H$_2$C=$\overset{+}{O}$CH$_3$ from methyl ethers C$_2$H$_5$O$^+$ from ethyl esters
	– CHO$_2$	from carboxylic acids, lactones	CHO$_2^+$: CHS$^+$:	COOH$^+$ from carboxylic acids HC≡S$^+$ from disulfides, aromatic and unsaturated S-heterocycles
46	– C$_2$H$_6$O	from ethyl esters (H$_2$O + C$_2$H$_4$), from long chain primary alcohols	CH$_2$S$^+$ solvent:	from thio ethers C$_2$H$_5$OH

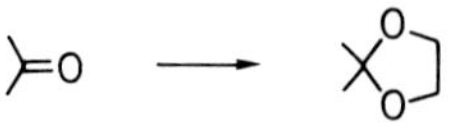

$+ 46$

Mass	M ± x		m/z	
	– NO$_2$	from nitro derivatives (30, 16)		
47			CH$_3$S$^+$:	CH$_2$=$\overset{+}{S}$H from thio ethers; primary thiols
48	– SO	from diarylsulfoxides		
49			CH$_2$Cl$^+$	from Cl-derivatives *
50			C$_4$H$_2^+$ CH$_3$Cl$^+$	from o-disubstituted phenylcarbonyl derivatives (76) from quaternary methylammonium chlorides *
51			C$_4$H$_3^+$ CH$_2$Cl$^+$	from aromatic compounds (77) from Cl-derivatives *
52			C$_4$H$_4^+$ CH$_3$Cl$^+$	from aromatic compounds (78) from quatenary methylammonium chlorides *

Tab. 4.8 continued

Mass	M ± x		m/z mass difference as a result of chemical and thermal transformation reactions
55	– C₄H₇	from (aromatic) butyl esters, alicyclic compounds	$C_3H_3O^+$: $H_2C=\!\!\!\!=\!\!\!\equiv O^+$ from cyclopentanone- and cyclohexanone derivatives $C_4H_7^+$
56	– C₄H₈	e.g. from butyl esters	$C_4H_8^+$ from alkenes, alkenols butyl esters
	– C₂O₂	from diketones, unsaturated lactones	$C_3H_6N^+$: $H_2C=\!\!\!\!=\!\!\!\overset{+}{N}H_2$ from cyclopentyl- and cyclohexyl- amino derivatives
			$C_2H_2NO^+$: $H_2C=\overset{+}{N}=C=O$ from isocyanates
57			$C_4H_9^+$ from alkanes
			$C_3H_7N^+$: e.g. $H_2C=\overset{+}{N}{<}^{CH_3}_{\cdot CH_2}$ from cyclic amines
			$C_3H_5O^+$: $H_2C=\!\!\!\!=\!\!\!\overset{+}{O}H$ from cyclopentanol and cyclohexanol derivatives
			$C_2H_5-C\equiv O^+$ from ethyl ketones, propionic acid derivatives
58			$C_3H_8N^+$: e.g. $H_2C=\overset{+}{N}{<}^{CH_3}_{CH_3}$ from amines
			$C_3H_6O^+$: $H_3C-C{<}^{OH^+}_{CH_2}$ from methyl alkyl ketones
			solvent: acetone
59	– C₂H₃O₂	from methyl esters	$C_3H_7O^+$: $HO=\!\!<^{CH_3}_{CH_3}$ from 2-methyl-2-alkanol derivatives
			$C_2H_5-CH=OH^+$ from 3-alkanol derivatives, propyl ethers
			$C_2H_5NO^+$: $H_2C=CH-N{<}^{H}_{OH}]^{+\cdot}$ from oximes
			$H_2C{=}{<}^{OH}_{NH_2}]^{+\cdot}$ from primary carboxylic acid amides
			$C_2H_3O_2^+$: $COOCH_3^+$ from methyl esters
60	– C₂H₄O₂	CH₃COOH from O-acetyl derivatives	$C_2H_4O_2^+$: $H_2C{=}{<}^{OH}_{OH}]^{+\cdot}$ from aliphatic carboxylic acids
			solvent: acetic acid, propanol
			– 60 from o-acetyl derivatives Δ
61			$CH_2CH_2SH^+$ aliphatic thiols

Tab. 4.8 continued

Mass	M ± x		m/z mass difference as a result of chemical and thermal transformation reactions
63			$CH_3O_3^+$: $HO=\overset{+}{C}\underset{OH}{\overset{OH}{<}}$ from carboxylic acid dialkyl esters
			CH_3SO^+ from alkylsulfoxides
65			$C_5H_5^+$: from aromatic alkyl derivatives (91), N-heterocycles (92), aromatic amines (92)
66			$H_2S_2^+$: HSSH$^+$ from disulfides
69			$C_4H_5O^+$: $H_2C=\overset{CH_3}{\underset{\equiv O^+}{<}}$ from 2- or 3-methylcyclo-pentanone or -hexanone derivatives
			CF_3^+ from trifluoromethyl and trifluoroacetyl derivatives, PFK
70			$C_4H_8N^+$: from 2-substituted pyrrolidine derivatives
			$C_2H_2NS^+$: $H_2C=N=C=S^+$ from isothiocyanates
71			$C_5H_{11}^+$: from alkanes
			$C_4H_7O^+$: $H_3C-CH_2-CH_2-C\equiv O^+$ from butanoic acid esters, propyl ketones
			from 2-substituted tetrahydrofuran derivatives
72	– C_2O_3	from aromatic anhydrides	$C_4H_{10}N^+$: e.g. $H_2C=\overset{+}{N}\overset{C_2H_5}{\underset{CH_3}{<}}$ from amines
			$C_4H_8O^+$: $H_2C=\overset{\lceil OH\rceil^{+\bullet}}{\underset{C_2H_5}{<}}$ from ethyl alkyl ketones
			solvent: 2-butanone
73	– $C_3H_5O_2$	•CH2COOCH$_3$ from methyl esters •COOC2H$_5$ from ethyl esters	C_4H_9O+: e.g. $H_3C-CH=\overset{+}{O}-C_2H_5$ from ethers
			C_3H_9Si+: $(CH_3)_3Si^+$ from trimethylsilyl derivatives, TMS
			$C_3H_5O_2^+$: COOC$_2$H$_5^+$ from ethyl esters CH$_2$–COOCH$_3^+$ from methyl esters
			solvent: dimethylformamide
74			$C_3H_6O_2^+$: $H_2C=\overset{\lceil OH\rceil^{+\bullet}}{\underset{OCH_3}{<}}$ from methyl esters
			solvent: diethyl ether

Tab. 4.8   continued

Mass	M ± x		m/z mass difference as a result of chemical and thermal transformation reactions
75			$C_3H_7O_2^+$: $H_3C-\overset{+}{O}=CH-OCH_3$ from dimethyl acetals
			$C_3H_7S^+$: $C_2H_5-\overset{+}{S}=CH_2$ from ethylalkylsulfides
			$C_2H_7OSi^+$: $HO=\overset{+}{Si}\begin{smallmatrix}CH_3\\\\CH_3\end{smallmatrix}$ from O-trimethylsilyl derivatives
76			$C_6H_4^+$: from O-disubstitited phenyl(carbonyl) derivatives, anthraquinone derivatives (50)
			solvent: CS_2
			+76 $\searrow=O \longrightarrow$ (dithiolane)
77	$-C_6H_5$	from phenyl derivatives	$C_6H_5^+$: from phenyl derivatives (155, 105, 51)
78			$C_6H_6^+$: from derivatives (52)
			solvent: benzene, dimethylsulfoxide
			-78 $\searrow-Br \longrightarrow \searrow-H$
79	$-Br$	from Br-derivatives *	Br^+ from Br-derivatives, quaternary ammonium bromides, hydrobromides * solvent: pyridine
80			$C_5H_6N^+$: from N-alkylpyrrole derivatives
			from pyridine derivatives
			HBr^+ from Br-derivatives, quaternary ammonium bromides, hydrobromides *
81	$-Br$	from Br-derivatives *	$C_5H_5O^+$: $\langle O \rangle-CH_2^+$ from furan derivatives
			Br^+ from Br-derivatives, quaternary ammonium bromides, hydrobromides *
82			$C_5H_8N^+$: from dihydropyrrole derivatives
			HBr^+ from Br-derivatives, quaternary ammonium bromides, hydrobromides *

Tab. 4.8 continued

Mass	M ± x		m/z mass difference as a result of chemical and thermal transformation reactions
83			$CHCl_2^+$ from $CHCl_3$

84

$C_5H_{10}N^+$:   from N-ethylcyclopentyl and cyclohexyl derivatives (56)

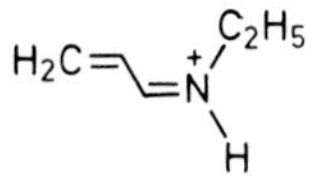

$C_5H_{10}N^+$:   from 2-substituted piperidine derivatives

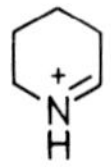

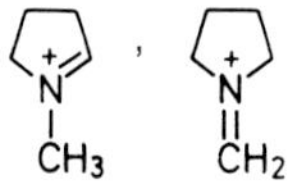

from pyrrolidine derivatives

solvent: CH_2Cl_2

+ 84

85

$C_6H_{13}^+$ from alkanes
$C_5H_9O^+$: $H_3C-(CH_2)_3-C\equiv O^+$ from butyl ketones, valeric acid esters

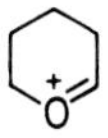

from 2-substituted pyrane derivatives

$C_4H_5O_2^+$:   from 4-substituted γ-lactones

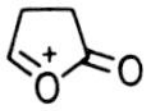

86 solvent: hexane

87

$C_5H_{11}O^+$: e.g. $C_3H_7-\overset{+}{O}=CH-CH_3$ from ethers

$C_4H_7O_2^+$:   from methyl esters

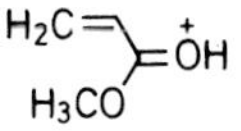

$CH_2COOC_2H_5^+$ from ethyl esters

88 solvent: dioxane, ethyl acetate

90 solvent: glycol dimethyl ether

− 90

91 − C_7H_7 from benzyl derivatives

$C_7H_7^+$:   from benzyl derivatives (155, 65)

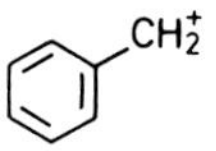

$C_4H_8Cl^+$:   from 1-chloroalkane derivatives *

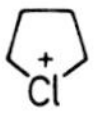

Tab. 4.8 continued

Mass	M ± x	m/z mass difference as a result of chemical and thermal transformation reactions
92		$C_7H_8^+$ from alkylbenzene derivatives
		$C_6H_6N^+$ from alkylpyridine derivatives (65) solvent: toluene
93		$C_4H_8Cl^+$: from 1-chloroalkane derivatives *
		CH_2Br^+ from alkylbromides *
94		$C_6H_6O^+$: $C_6H_5{-}OH^+$ from phenyl ethers, aromatic sulfones
		$C_5H_4NO^+$: $C{\equiv}O^+$ from pyrrolecarbonyl derivatives
95		$C_5H_3O_2^+$: $C{\equiv}O^+$ from furylcarbonyl derivatives
		CH_2Br^+ from alkylbromides *
96		+ 96
97		$C_5H_5S^+$: from alkylthiophene derivatives
98		$C_6H_{12}N^+$: from piperidine derivatives
99		$C_5H_7O_2^+$: from 5-substituted δ-lactones
		from ethylene acetals
101		$C_5H_9O_2^+$: from dimethyl acetals
102		solvent: diisopropyl ether
103		$C_8H_7^+$: $C_6H_5{-}CH{=}CH^+$ from cinnamic acid derivatives (131, 77)

Tab. 4.8   continued

Mass	M ± x	m/z mass difference as a result of chemical and thermal transformation reactions

---

104

$C_8H_8^{+\cdot}$: $C_6H_5-CH=CH_2^+$ from phenylethyl derivatives (131, 77)

$C_8H_8^{+\cdot}$: from 1,2,3,4-tetrahydro-(hetero)-naphthalene derivatives (RDA), o-methyldiphenylmethane derivatives

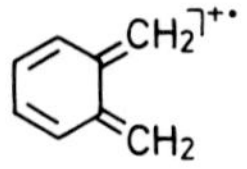

$C_7H_4O^{+\cdot}$: from 2-substituted benzoic acid derivatives (76)

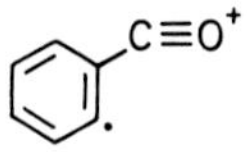

+ 104

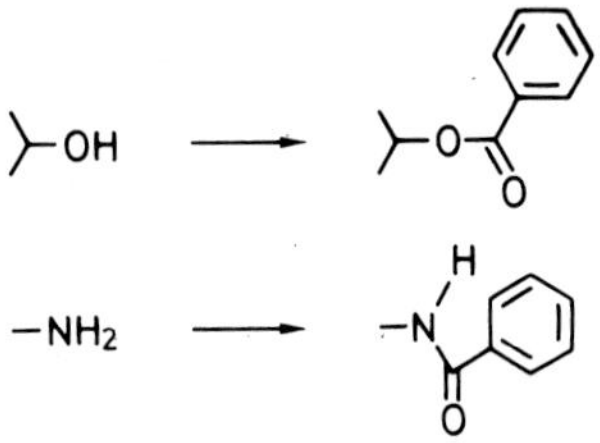

105

$C_8H_9^+$:   e.g. from alkyltoluenes

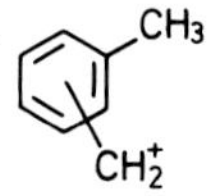

$C_7H_5O^+$: $C_6H_5-C\equiv O^+$ from phenylcarbonyl derivatives (123, 122, 77)

$C_6H_5N_2^+$: $C_6H_5-\overset{+}{N}\equiv N$ from aromatic azo-compounds

---

107 $C_7H_7O^+$ from benzylalcohols hydroxylated in the ring

---

110 $C_7H_{12}N^+$: e.g. from dimethyl-aminosteroids

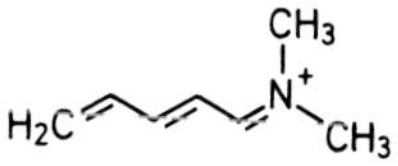

---

111 $C_5H_3OS^+$: from thiophenecarbonyl derivatives

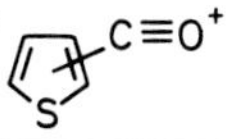

---

115 $C_9H_7^+$: from bicyclicaromatic and hetero-aromatic compounds, aromatic ketones

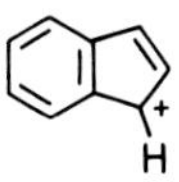

---

117   solvent: CCl_4 (CCl_3^+!) *

---

118   solvent: $CHCl_3$ *

---

119 $C_8H_7O^+$: from (methylphenyl)carbonyl derivatives

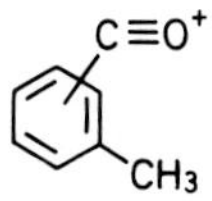

Tab. 4.8   continued

Mass	M ± x		m/z mass difference as a result of chemical and thermal transformation reactions
120			$C_7H_4O_2^{+\cdot}$: from γ-pyrone and γ-pyranone derivatives
121			$C_8H_9O^+$: from (methoxyphenyl)alkyl derivatives
			$C_7H_5O_2^+$: from (hydroxyphenyl)carbonyl derivatives
122			$C_7H_6O_2^+$: $C_6H_5-COOH^{+\bullet}$ from benzoic acid esters (105, 77)
123			$C_7H_7O_2^+$: $C_6H_5-COOH_2^+$ from benzoic acid esters (105, 77)
125			$C_7H_9O_2^+$: from ethylene acetals
127	– I	from I-derivatives	$C_7H_{11}O_2^+$: from dimethyl acetals
			I^+ from I-derivatives
128			HI^+ from I-derivatives
130			$C_9H_8N^+$: from indole and indoline derivatives (typical for indole alkaloids; 144)
			+ 130
131			$C_9H_7O^+$: $C_6H_5-CH=CH-C\equiv O^+$ from cinnamic acid derivatives (103, 77)
			$C_5H_7S_2^+$: from ethylenedithio acetals
135			$C_4H_8Br^+$: from 1-bromoalkane derivatives *
137			$C_4H_8Br^+$: from 1-bromoalkane derivatives *
139			$C_7H_7OS^+$ from tosyl derivatives (155, 91)

Tab. 4.8 continued

Mass	M ± x		m/z mass difference as a result of chemical and thermal transformation reactions
142			CH_3I^+ from quaternary methylammonium iodides
144			$C_{10}H_{10}N^+$: from indole and indoline derivatives (usually together with m/z = 143; typical for indole alkaloids; 130)
149			$C_9H_9O_2^+$: $C_6H_5-\overset{+}{C}H-CH_2-COOH$ from β-substituted dihydrocinnamic acid derivatives $C_8H_5O_3^+$: from phthalic acid esters
154			+ 154
155	– $C_7H_7O_2S$ from tosyl derivatives		$C_7H_7O_2S^+$: from tosyl derivatives (139, 91)
156			$C_2H_5I^+$ from quaternary alkyl ammonium iodides
157			$C_7H_9S_2^+$: from dithio acetals
160			$C_9H_6NO_2^+$: From N-alkylphthalimides
164			solvent: tetrachloroethylene *
179			solvent: hexamethylphosphoramide (HMPA)
205			$C_{14}H_{21}O^+$: from 2,6-di-(tert-butyl)-4-methylphenol
220			$C_{15}H_{24}O^+$ 2,6-di-(tert-butyl)-4-methylphenol
256			S_8^+ elemental sulfur (224, 192, 160, ...32)

9.2 Mass Differences Between the Reactant and Product of Frequently Used Chemical Reactions

Tab. 4.9 Mass differences between reactant and product

Partial structure of reactant → product		Mass difference (amu), type of reaction	Partial structure of reactant → product		Mass difference (amu), type of reaction
Alcohols					
>—OH	>—OCH₃	+ 14 ether formation	>=O	(dithiolane)	+ 76 thioacetalisation
>—OH	>—O—C(CH₃)=O	+ 42 acetylation	>=O	—OH,H	+ 2 reduction
>—OH	(tetrahydropyranyl)	+ 84 acetalisation	>=O	—H,H	− 14 reduction
>—OH	>—O—C(=O)(phenyl)	+ 104 benzoylation	—CHO	—COOH	+ 16 oxidation
>—OH	>—O—SO₂—(C₆H₄)—CH₃	+ 154 tosylation	(O=)C—COOCH₃	—COOCH₃	− 28 decarbonylation
(OH, H)	(alkene)	− 18 elimination	CH₂=C(=O)	(H,H)C—C(=O)	− 12 hydrolysis
>—OH	>—H	− 16 reduction	**Carboxylic acids, esters, lactones**		
>—OH	>=O	− 2 oxidation	—COOH	COOCH₃	+ 14 esterification
—CH₂OH	—COOH	+ 14 oxidation	—COOCH₃	—COOC₂H₅	+ 14 transesterification
—CH₂OH	—COOCH₃	+ 28 esterification	—COOCH₃	—C(CH₃)₂—OH	± 0 methylation
Ketones, aldehydes			—COOCH₃	—C(=O)Cl	+ 4[a] acid chloride formation
>=O	>—OCH₃	+ 14 enol ether formation	(lactone)	—OH, —COOH	+ 18 hydrolysis
>=O	—OH, CH₃	+ 16 methylation	(lactone)	—OH, —CH₂OH	+ 4 reduction
>=O	—OCH₃, OCH₃	+ 46 acetalisation	—COOCH₃	—CH₂OH	− 28 reduction
>=O	(dioxolane)	+ 44 acetalisation	>—COOH	>—H	− 44 decarboxylation

Tab. 4.9 continued

Partial structure of reactant → product	Mass difference (amu), type of reaction
Nitrogen compounds	
$-NH_2$ → $-N(CH_3)CH_3$	+ 28 methylation
$-NH_2$ → $-N(H)CHO$	+ 28 formylation
$-NH_2$ → $-N(H)COCH_3$	+ 42 acetylation
$-NH_2$ → $-N(H)COCF_3$	+ 96 trifluoroacetylation
$-NH_2$ → $-N(H)CO-C_6H_5$	+ 104 benzoylation
$-NH_2$ → phthalimide (phthalimido)	− 130 phthalide formation
$-NH_2$ → $-N(H)SO_2-C_6H_4-CH_3$	− 154 tosylation
$>C=NH$ → $>C=O$	+ 1 hydrolysis
$-N<$ → $-N^+(-O^-)$	+ 16 oxidation
$-CN$ → $-C(=O)NH_2$	+ 18 hydrolysis
$-CN$ → $-COOH$	+ 19 hydrolysis
$-CN$ → $-CH_2-NH_2$	+ 4 reduction
$O=C(NH_2)$ (amide) → $(H)(H)C(NH_2)$	− 14 reduction
$-NO$ → $-NH_2$	− 14 reduction
$-NO_2$ → $-NH_2$	− 30 reduction

Partial structure of reactant → product	Mass difference (amu), type of reaction
Sulfur compounds	
$>S$ → $>S=O$	+ 16 oxidation
$>S=O$ → $>SO_2$	+ 16 oxidation
1,3-dithiolane → $>CH_2$ (CH2)	− 90 reduction
Halogen compounds	
$>C-Cl$ → $>C-H$	− 34[a] reduction
$>C(Cl)(H)$ → $>C=C<$	− 36[a] elimination
$>C-Cl$ → $>C-OH$	− 18[a] hydrolysis
$>C-Cl$ → $>C-OCH_3$	− 4[a] methanolysis
$>C-Br$ → $>C-H$	− 78[a] reduction
Carbon compounds	
$>C=C<$ → $>C(H)-C(H)<$	+ 2 hydrogenation
$>C=C<$ → $>C(H)-C(OH)<$	+ 18 addition
$>C=C<$ → $>C(H)-C(Cl)<$	+ 36[a] addition
$>C=C<$ → epoxide	+ 16 epoxidation
$-C\equiv C-$ → $-CH_2-CH_2-$	+ 4 hydrogenation
$-C\equiv C-$ → $>C(=O)-CH_2-$	+ 18 addition

[a] The values refer to ^{35}Cl or ^{79}Br (cf. Tab. 4.10).

9.3 Isotopic Ratios in Compounds Containing Cl and Br

Tab. 4.10   Isotopic ratios

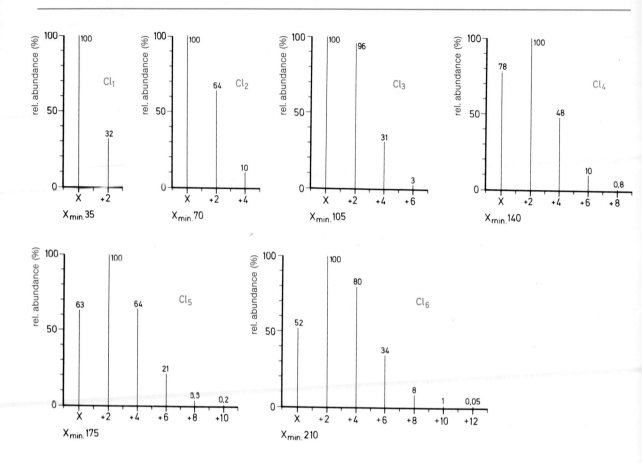

Tab. 4.10 continued

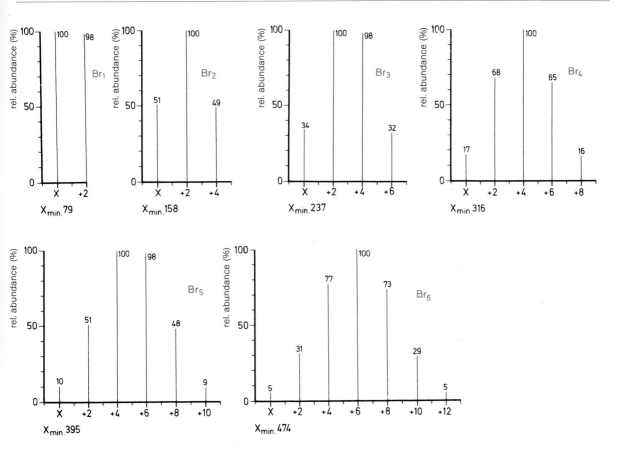

Tab. 4.10 continued

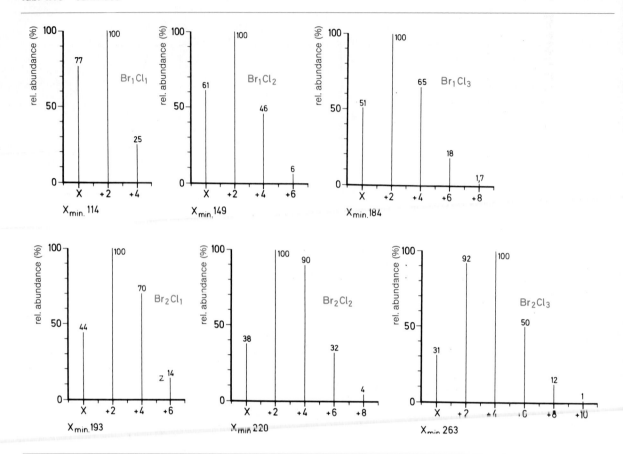

9.4 Mass Spectra of Solvents

Tab. 4.11   *(see ref.[10])

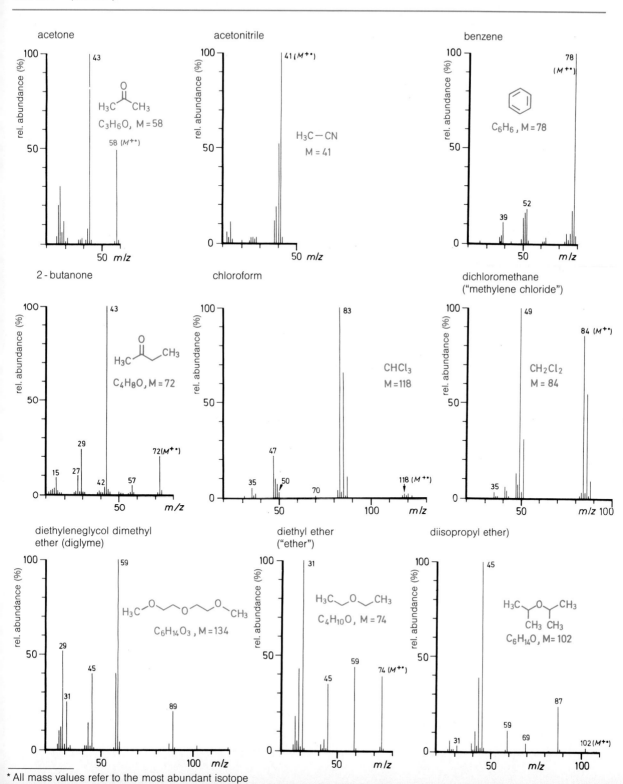

* All mass values refer to the most abundant isotope

Tab. 4.11 continued

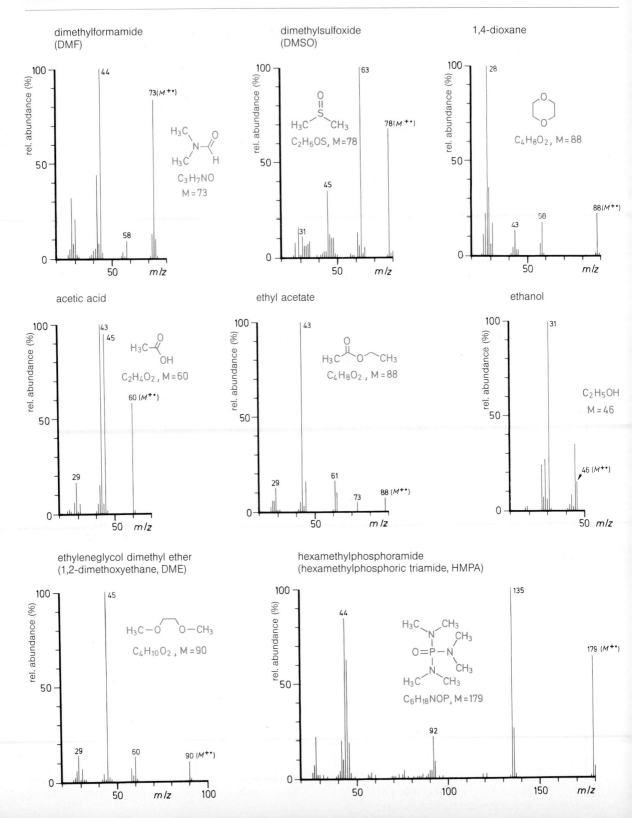

Tab. 4.11 continued

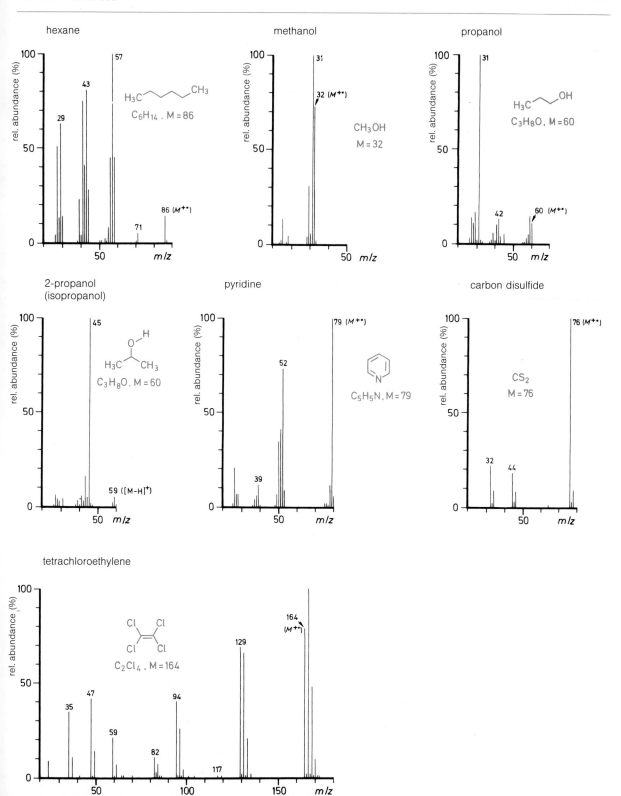

Tab. 4.11 continued

tetrachloromethane
(carbon tetrachloride)

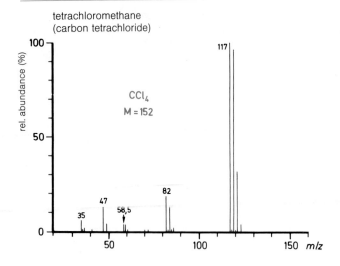

CCl$_4$
M = 152

tetrahydrofuran
(THF)

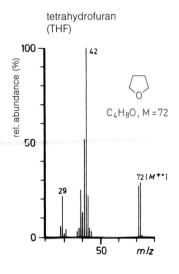

C$_4$H$_8$O, M = 72

tolvene

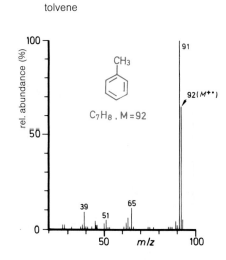

C$_7$H$_8$, M = 92

9.5 Mass Spectra of Volatile Contaminants

Tab. 4.12   *(see ref.[10])

deuterochloroform
(from NMR)

tetramethylsilane
(TMS, from NMR)

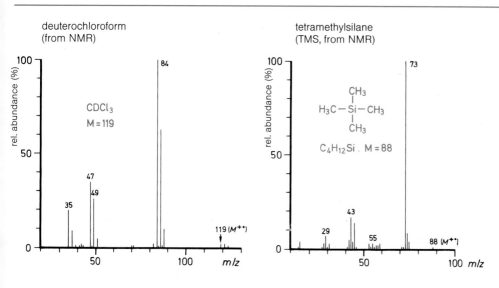

1,2-di-(*tert*-butyl)-4-methylphenol
(stabiliser from ethers)

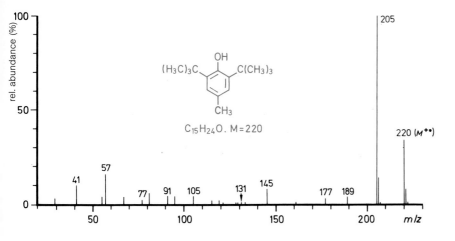

* All mass values refer to the most abundant isotope

Tab. 4.12 Continue

dibutyl phthalate (softener)

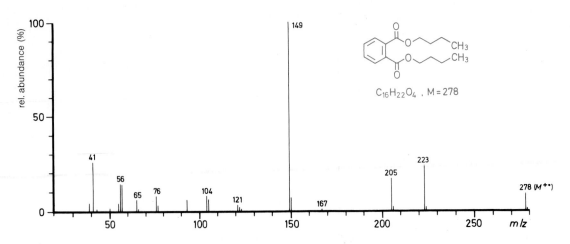

elemental sulfur

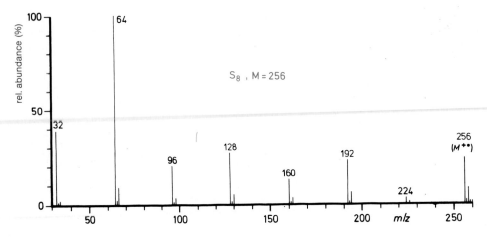

Tab. 4.12 continued

oil from NaH/oil dispersions

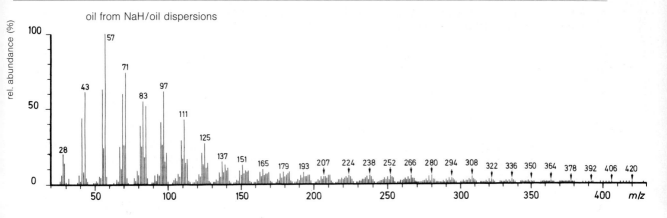

kerosene from LiALH₄

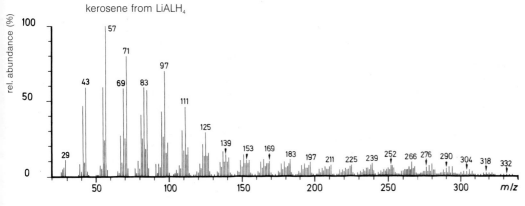

9.6 Mass Numbers and Abundances of the Isotopes of Naturally Occurring Elements

Tab. 4.13 Mass numbers (MN), atomic numbers (AN), relative abundances of the isotopes of naturally occurring elements and atomic masses (ordered alphabetically according to the atomic symbol)[35]

Element	AN	MN	Mass	rel. abundance (%)	rel. atomic mass	Element	AN	MN	Mass	rel. abundance (%)	rel. atomic mass
Ag	47	107	106.905095	51.839	107.8682	Cr	24	50	49.946046	4.345	51.9961
		109	108.904754	48.161				52	51.940510	83.789	
Al	13	27	26.981541	100	26.981539			53	52.940651	9.501	
Ar	18	36	35.967546	0.337	39.948			54	53.938882	2.365	
		38	37.962732	0.063		Cs	55	133	132.905433	100	132.90543
		40	39.962383	99.600		Cu	29	63	62.929599	69.17	63.546
As	33	75	74.921596	100	74.92159			65	64.927792	30.83	
Au	79	197	196.966560	100	196.96654	Dy	66	156	155.924287	0.06	162.50
								158	157.924412	0.10	
B	5	10	10.012938	19.9	10.811			160	159.925203	2.34	
		11	11.009305	80.1				161	160.926939	18.9	
Ba	56	130	129.906277	0.106	137.327			162	161.926805	25.5	
		132	131.905042	0.101				163	162.928737	24.9	
		134	133.904490	2.417				164	163.929183	28.2	
		135	134.905668	6.592							
		136	135.904556	7.854		Er	68	162	161.928787	0.14	167.26
		137	136.905816	11.23				164	163.929211	1.61	
		138	137.905236	71.70				166	165.930305	33.6	
Be	4	9	9.012183	100	9.012182			167	166.932061	22.95	
Bi	83	209	208.980388	100	208.98037			168	167.932383	26.8	
Br	35	79	78.918336	50.69	79.904			170	169.935476	14.9	
		81	80.916290	49.31		Eu	63	151	150.919860	47.8	151.965
								153	152.921243	52.2	
C	6	12	12.000000	98.90	12.011						
		13	10.000055	1.10		F	9	19	18.998403	100	18.9984032
Ca	20	40	39.962591	96.941	40.078	Fe	26	54	53.939612	5.8	55.847
		42	41.958622	0.647				56	55.934939	91.72	
		43	42.958770	0.135				57	56.935396	2.2	
		44	43.955485	2.086				58	57.933278	0.28	
		46	45.953689	0.004							
		48	47.952532	0.187		Ga	31	69	68.925581	60.1	69.723
Cd	48	106	105.906461	1.25	112.411			71	70.924701	39.9	
		108	107.904186	0.89		Ge	32	70	69.924250	20.5	72.61
		110	109.903007	12.49				72	71.922080	27.4	
		111	110.904182	12.80				73	72.923464	7.8	
		112	111.902761	24.13				74	73.921179	36.5	
		113	112.904401	12.22				76	75.921403	7.8	
		114	113.903361	28.73		Gd	64	152	151.919803	0.20	157.25
		116	115.904758	7.49				154	153.920876	2.18	
Ce	58	136	135.90714	0.19	140.115			155	154.922629	14.80	
		138	137.905996	0.25				156	155.922130	20.47	
		140	139.905442	88.48				157	156.923967	15.65	
		142	141.909249	11.08				158	157.924111	24.84	
Cl	17	35	34.968853	75.77	35.4527			160	159.927061	21.86	
		37	36.965903	24.23		H/D	1	1	1.007825	99.985	1.00794
								2	2.014102	0.015	
Co	27	59	58.933198	100	58.93320	He	2	3	3.016029	0.000138	4.002602
								4	4.002603	99.999862	

Tab. 4.13 continued

Ele-ment	AN	MN	Mass	rel. abun-dance (%)	rel. atomic mass	Ele-ment	AN	MN	Mass	rel. abun-dance (%)	rel. atomic mass
Hf	72	174	173.940065	0.162	178.49	Na	11	23	22.989770	100	22.989768
		176	175.941420	5.206		Nb	41	93	92.906378	100	92.90638
		177	176.943233	18.606		Nd	60	142	141.907731	27.13	144.24
		178	177.943710	27.297				143	142.909823	12.18	
		179	178.945827	13.629				144	143.910096	23.80	
		180	179.946561	35.100				145	144.912582	8.30	
Hg	80	196	195.965812	0.14	200.59			146	145.913126	17.19	
		198	197.966760	10.02				148	147.916901	5.76	
		199	198.968269	16.84				150	149.920900	5.64	
		200	199.968316	23.13		Ne	10	20	19.992439	90.51	20.1797
		201	200.970293	13.22				21	20.993845	0.27	
		202	201.970632	29.80				22	21.991384	9.22	
		204	203.973481	6.85		Ni	28	58	57.935347	68.27	58.69
Ho	67	165	164.930332	100	164.93032			60	59.930789	26.10	
								61	60.931059	1.13	
I	53	127	126.904477	100	126.90447			62	61.928346	3.59	
In	49	113	112.904056	4.3	114.82			64	63.927968	0.91	
		115	114.903875	95.7							
Ir	77	191	190.960603	37.3	192.22	O	8	16	15.994915	99.762	15.9994
		193	192.962942	62.7				17	16.999131	0.038	
								18	17.999159	0.200	
K	19	39	38.963708	93.2581	39.0983	Os	76	184	183.952514	0.02	190.2
		40	39.963999	0.0117				186	185.953852	1.58	
		41	40.961825	6.7302				187	186.955762	1.6	
Kr	36	78	77.920397	0.35	83.80			188	187.955850	13.3	
		80	79.916375	2.25				189	188.958156	16.1	
		82	81.913483	11.6				190	189.958455	26.4	
		83	82.914134	11.5				192	191.961487	41.0	
		84	83.911506	57.0							
		86	85.910614	17.3		P	15	31	30.973763	100	30.973762
La	57	138	137.907114	0.09	138.9055	Pb	82	204	203.973037	1.4	207.2
		139	138.906355	99.91				206	205.974455	24.1	
Li	3	6	6.015123	7.5	6.941			207	206.975885	22.1	
		7	7.016005	92.5				208	207.976641	52.4	
Lu	71	175	174.940785	97.441	174.967	Pd	46	102	101.905609	1.020	106.42
		176	175.942694	2.59				104	103.904026	11.14	
								105	104.905075	22.33	
Mg	12	24	23.985045	78.99	24.3050			106	105.903475	27.33	
		25	24.985839	10.00				108	107.903894	26.46	
		26	25.982595	11.01				110	109.905169	11.72	
Mn	25	55	54.938046	100	54.93805	Pr	59	141	140.907657	100	140.90765
Mo	42	92	91.906809	14.84	95.94	Pt	78	190	189.959937	0.01	195.08
		94	93.905086	9.25				192	191.961049	0.79	
		95	94.905838	15.92				194	193.962679	32.9	
		96	95.904676	16.68				195	194.964785	33.8	
		97	96.906018	9.55				196	195.964947	25.3	
		98	97.905405	24.13				198	197.967879	7.2	
		100	99.907473	9.63		Rb	37	85	84.911800	72.165	85.4678
N	7	14	14.003074	99.634	14.00674			87	86.909184	27.835	
		15	15.000109	0.366		Re	75	185	184.952977	37.40	186.207
								187	186.955765	62.60	

Tab. 4.13 continued

Ele-ment	AN	MN	Mass	rel. abun-dance (%)	rel. atomic mass	Ele-ment	AN	MN	Mass	rel. abun-dance (%)	rel. atomic mass
Rh	45	103	102.905503	100	102.90550	Te	52	120	119.904021	0.096	127.60
Ru	44	96	95.907596	5.52	101.07			122	121.903055	2.60	
		98	97.905287	1.88				123	122.904278	0.908	
		99	98.905937	12.7				124	123.902825	4.816	
		100	99.904218	12.6				125	124.904435	7.14	
		101	100.905581	17.0				126	125.903310	18.95	
		102	101.904348	31.6				128	127.904464	31.69	
		104	103.905422	18.7				130	129.906229	33.80	
						Th	90	232	232.038054	100	232.0381
S	16	32	31.972072	95.02	32.066	Ti	22	46	45.952633	8.0	47.88
		33	32.971459	0.75				47	46.951765	7.3	
		34	33.967868	4.21				48	47.947947	73.8	
		36	35.967079	0.02				49	48.947871	5.5	
Sb	51	121	120.903824	57.3	121.75			50	49.944786	5.4	
		123	122.904222	42.7		Tl	81	203	202.972336	29.524	204.3833
Sc	21	45	44.955914	100	44.955910			205	204.974410	70.476	
Se	34	74	73.922477	0.9	78.96	Tm	69	169	168.934225	100	168.93421
		76	75.919207	9.0							
		77	76.919908	7.6		U	92	234	234.040947	0.0055	238.0289
		78	77.917304	23.6				235	235.043925	0.7200	
		80	79.916521	49.7				238	238.050786	99.2745	
		82	81.916709	9.2							
Si	14	28	27.976928	92.23	28.0855	V	23	50	49.947161	0.250	50.9415
		29	28.976496	4.67				51	50.943963	99.750	
		30	29.973772	3.10		W	74	180	179.946727	0.13	183.85
Sm	62	144	143.912009	3.1	150.36			182	181.948225	26.3	
		147	146.914907	15.0				183	182.950245	14.3	
		148	147.914832	11.3				184	183.950953	30.67	
		149	148.917193	13.8				186	185.954377	28.6	
		150	149.917285	7.4							
		152	151.919741	26.7		Xe	54	124	123.90612	0.10	131.29
		154	153.922218	22.7				126	125.904281	0.09	
Sn	50	112	111.904823	0.97	118.710			128	127.903531	1.91	
		114	113.902781	0.65				129	128.904780	26.4	
		115	114.903344	0.36				130	129.903510	4.1	
		116	115.901744	14.53				131	130.905076	21.2	
		117	116.902954	7.68				132	131.904148	26.9	
		118	117.901607	24.22				134	133.905395	10.4	
		119	118.903310	8.58				136	135.907219	8.9	
		120	119.902199	32.59							
		122	121.903440	4.63		Y	39	89	88.905856	100	88.90585
		124	123.905271	5.79		Yb	70	168	167.933908	0.13	173.04
Sr	38	84	83.913428	0.56	87.62			170	169.934774	3.05	
		86	85.909273	9.86				171	170.936338	14.3	
		87	86.908890	7.00				172	171.936393	21.9	
		88	87.905625	82.58				173	172.938222	16.12	
								174	173.938873	31.8	
Ta	73	180	179.947489	0.012	180.9479			176	175.942576	12.7	
		181	180.948014	99.988							
Tb	65	159	158.925350	100	158.92534						

Tab. 4.13 continued

Element	AN	MN	Mass	rel. abundance (%)	rel. atomic mass
Zn	30	64	63.929145	48.6	65.39
		66	65.926035	27.9	
		67	66.927129	4.1	
		68	67.924846	18.8	
		70	69.925325	0.6	
Zr	40	90	89.904708	51.45	91.224
		91	90.905644	11.22	
		92	91.905039	17.15	
		94	93.906319	17.38	
		96	95.908272	2.80	

Literature

Combined Use of Spectroscopic Methods in Organic Chemistry

Baker, A. J., Cairns, T., Eglinton, G., Preston, F. J. (1967, 1975), More Spectroscopic Problems in Organic Chemistry, Heyden, London.

Pretsch, E., Clerc, T., Seibl, J., Simon, W. (1989), Tabellen zur Strukturaufklärung organischer Verbindungen mit spektroskopischen Methoden, 3rd ed., Springer-Verlag, Berlin.

Sternhell, S., Kalman, J. R. (1986), Organic Structures from Spectra, John Wiley & Sons, Chichester.

Negative Ions

Bowie, J. H. (1984), The Formation and Fragmentation of Negative Ions Derived from Organic Molecules, Mass Spectrom. Rev. 3, 161.

Budzikiewicz, H. (1983), Selected Reviews on Mass Spectrometric Topics: Negative Ions, Mass Spectrom. Rev. 2, 515.

Budzikiewicz, H. (1986), Negative Chemical Ionization (NCI) of Organic Compounds, Mass Spectrom. Rev. 5, 345.

Gregor, I. K., Guilhaus, M. (1984), Mass Spectrometry of Metalorganic Negative Ions, Mass Spectrom. Rev. 3, 39.

More Detailed References and Textbooks

Asamoto, B. (1988), Fourier Transform-Mass Spectrometry for Industrial Problem Solving, Spectroscopy 3, 38.

Beynon, J. H. (1960), Mass spectrometry and its Applicaton to Organic Chemistry, Elsevier, Amsterdam.

Biemann, K. (1962), Mass Spectrometry - Organic Chemical Application, McGraw-Hill, New York.

Bremser, W., Neudert, R. (1987), Automation in the Spectroscopic Laboratory - Solutions and Perspectives, Eur. Spectros. News 75, 10.

Chapman, J. R. (1993), Practical Organic Mass Spectrometry, 2nd ed., John Wiley & Sons, Chichester.

Constantin, E., Schnell, A. (1990), Mass Spectrometry, Ellis Horwood, Hemel Hempstead.

Davis, R., Frearson, M. (1987), Mass Spectrometry, Wiley & Sons, Chichester.

Dawson, P. H. (1986), Quadrupole Mass Analyzer: Performance, Design and Some Recent Appications, Mass Spectrom. Rev. 5, 1.

Field, F. H., Franklin, J. L. (1970), Electron Impact Phenomena, Academic Press, New York, London.

Howe, I., Williams, D. H., Bowen, R. D. (1981), Mass Spectrometry; Principles and Applications, 2nd ed., McGraw-Hill, New York.

Kiser, R. W. (1965), Introduction to Mass Spectrometry and its Applications, Prentice-Hall, Englewood Cliffs, New Jersey.

Lehrle, R. S., Parker, J. E. (1972), Time-of-Flight Mass Spectrometry, in Mass Spectrometry, MTP International Review of Science, Physical Chemistry, Series one (Maccoll, A., Ed.), Butterworth, London.

Levsen, K. (1978), Fundamental Aspects of Organic Mass Spectrometry, Vol. 4, Verlag Chemie, Weinheim.

Maccoll, A. (Ed.) (1972), Mass Spectrometry, MTP International Review of Science, Physical Chemistry, Series one, Butterworth, London.

McFadden, W. H. (1973), Techniques of Combined Gas Chromatography/Mass Spectrometry: Applications in Organic Analysis, Wiley-Interscience, New York.

McLafferty, F. W. (Ed.) (1963), Mass Spectrometry of Organic Ions, Academic Press, New York, London.

Merritt, C., McEwen, C. N. (Eds.) (1980), Mass Spectrometry, Parts A & B, Marcel Dekker Inc., New York, Basle.

Millard, B. J. (1978), Quantitative Mass Spectrometry, Heyden & Son, London.

Schlunegger, U. P. (1975), Detection of Fragment Genesis in the Mass Spectrometer: DADI-Mass Spectrometry as an Aid in the Structural Analysis of Organic Compounds, Angew. Chem. Int. Ed. Engl. 14, 679.

Schröder, E. (1991), Massenspektrometrie - Begriffe und Definitionen, Springer-Verlag, Berlin.

Wahrhaftig, A. L. (1972), Theory of Mass Spectra, in Mass Spectrometry, MTP International Review of Science, Physical Chemistry, Series one (Maccoll, A., Ed.), Butterworth, London.

Spectral Catalogues

Cornu, A., Massot, R. (1976), Compilation of Mass Spectral Data, Heyden, London, Vols. I & II. (Contains the spectra of 10 000 compounds).

Eight Peak Index of Mass Spectra (1975), Mass Spectrometry Data Center, Royal Society of Chemistry, Aldermaston (3rd ed. 1986: 67 000 spectra).

McLafferty, F. W., Stauffer, D. B. (1991), Important Peak Index of the Registry of Mass Spectral Data, 3 Vols., Wiley & Sons, Chichester.

Stenhagen, E., Abrahamsson, S., McLafferty, F. W. (1969), Atlas of Mass Spectral Data, 3 Vols., Interscience, New York. (A collection of the mass spectra of several thousand organic compounds.)

Wiley/NBS (1989), Registry of Mass Spectral Data, 7 Vols., Wiley & Sons, New York (a collection of over 150 000 spectra from more than 90 000 compounds).

Special Classes of Compounds and Techniques

Alkaloids

Budzikiewicz, H. (1991), Mass Spectrometry of Amino Steroids and Steroidal Alkaloids, Mass Spectrom. Rev. 10, 79.

Hesse, M. (1974), Indolakaloide, in Progress in Mass Spectrometry, Vol. 1, Part 1: Text, Part 2: Spectra, Verlag Chemie, Weinheim.

Hesse, M., Bernhard, H. O. (1975), Alkaloide außer Indol-, Triterpen- und Steroidalkaloiden, in Progress in Mass Spectrometry, Vol. 3, Verlag Chemie, Weinheim.

Antibiotics

Borders, D. B., Carter, G. T., Hargreaves, R. T., Siegel, M. M. (1985), Recent Applications of Mass Spectrometry to Antibiotic Research, Mass Spectrom. Rev. **4**, 295.

Biochemistry

Jackson, A. H. (1977), Mass Spectrometry of Biochemical Materials, Endeavour New Series **1**, 75.

Waller, G. R. (Ed.) (1972), Biochemical Applications of Mass Spectrometry, Wiley-Interscience, New York; Waller, G. R., Dermer, O. C. (Eds.) (1980), First supplementary volume, Wiley-Interscience, New York.

Carotinoids

Budzikiewicz, H. (1991), Selected Reviews on Mass Spectrometric Topics: Carotinoids, Mass Spectrom. Rev. **10**, 329.

Environmetal Analysis

Budzikiewicz, H. (1993), Selected Reviews on Mass Spectrometric Topics: Environmental Contaminants, Mass Spectrom. Rev. **12**, 205.

Rosen, J. (Ed.) (1987), Applications of New Mass Spectrometry Techniques in Pesticide Chemistry, Wiley & Sons, Chichester.

Explosives

Budzikiewicz, H. (1993), Selected Reviews on Mass Spectrometric Topics: Explosives, Mass Spectrom. Rev. **12**, 207.

Yinon, J. (1982), Mass Spectrometry of Explosives: Nitro Compounds, Nitrate Esters, and Nitramines, Mass Spectrom. Rev. **1**, 257.

Yinon, J., Mass Spectrometry of Explosives, in Yinon, J. (Ed.) (1987), Forensic Mass Spectrometry, CRC Press, Boca Raton.

Flavonoids

Budzikiewicz, H. (1987), Selected Reviews on Mass Spectrometric Topics: Flavones, Chromones, and Related Compunds, Mass Spectrom. Rev. **7**, 114.

Geochemistry

Budzikiewicz, H. (1988), Selected Reviews on Mass Spectrometric Topics: Organic and Petroleum Geochemistry, Mass Spectrom. Rev. **7**, 463.

Teeter, R. M. (1985), High-resolution Mass Spectrometry for Type Analysis of Complex Hydrocarbon Mixtures, Mass Spectrom. Rev. **4**, 123.

Hydrocarbons

Borders, D. B., Carter, G. T., Hargreaves, R. T., Siegel, M. M. (1985), Recent Applications of Mass Spectrometry to Antibiotic Research, Mass Spectrom. Rev. **4**, 295.

Budzikiewicz, H. (1993), Selected Reviews on Mass Spectrometric Topics: Carbohydrates, Mass Spectrom. Rev. **12**, 141.

Komori, T., Kawasaki, T., Schulten, H.-R. (1985), Field Desorption and Fast Atom Bombardment Mass Spectrometry of Biologically Active Natural Oligoglycosides, Mass Spectrom. Rev. **4**, 255.

Reinhold, V. N., Carr, S. A. (1983), New Mass Spectral Approaches to Complex Carbohydrate Structure, Mass Spectrom. Rev. **2**, 153.

Inorganic Chemistry

Adams, F., Gijbels, R., Grieken, R. van (1988), Inorganic Mass Spectrometry, Wiley & Sons, Chichester.

Boumans, P. W. (1987), Inductively Coupled Plasma Emission Spectrometry, Wiley & Sons, Chichester.

Budzikiewicz, H. (1993), Selected Reviews on Mass Spectrometric Topics: Inductively Coupled Plasma Mass Spectrometry (ICP-MS), Mass Spectrom. Rev. **12**, 205.

Ion Cyclotron Resonance

Budzikiewicz, H. (1993), Selected Reviews on Mass Spectrometric Topics: Ion Cyclotron Resonance (ICR) Mass Spectrometry, Mass Spectrom. Rev. **12**, 206.

Lehmann, T. A., Bursey, M. M. (1976), Ion Cyclotron Resonance Spectrometry, Wiley-Interscience, New York.

Iridoids

Popov, S. S., Handjieva, N. V. (1983), Mass Spectrometry of Iridoids, Mass Spectrom. Rev. **2**, 481.

Lipids, Fatty Acids

Lin, Y. Y., Smith, L. L. (1984), Chemical Ionization Mass Spectrometry of Steroids and other Lipids, Mass Spectrom. Rev. **3**, 319.

Budzikiewicz, H. (1991), Selected Reviews on Mass Spectrometric Topics: Lipids, Mass Spectrom. Rev. **10**, 453.

Nucleosides, Nucleotides

Borders, D. B., Carter, G. T., Hargreaves, R. T., Siegel, M. M. (1985), Recent Applications of Mass Spectrometry to Antibiotic Research, Mass Spectrom. Rev. **4**, 295.

Budzikiewicz, H. (1993), Selected Reviews on Mass Spectrometric Topics: Nucleosides and Nucleotides, Mass Spectrom. Rev. **12**, 140.

Organometallic Compounds

Budzikiewicz, H. (1993), Selected Reviews on Mass Spectrometric Topics: Organoelement Compounds, Mass Spectrom. Rev. **12**, 140.

Charalambous, J. (1975), Mass Spectrometry of Metal Compounds, Butterworth, London.

Eller, K., Schwarz, H. (1991), Organometallic Chemistry in the Gas Phase, Chem. Rev. **91**, 1121.

Freiser, B. S. (1994), Selected Topics in Organometallic Ion Chemistry, Acc. Chem. Res. **27**, 353.

Gregor, I. K., Guilhaus, M. (1984), Mass Spectrometry of Metalorganic Negative Ions, Mass Spectrom. Rev. **3**, 39.

Litzow, M. R., Spalding, T. R. (1973), Mass Spectrometry of Inorganic and Organometallic Compounds, Elsevier, Amsterdam.

Müller, J. (1972), Decomposition of Organometallic Complexes in the Mass Spectrometer, Angew. Chem. Int. Ed. Engl. **11**, 653.

Peptides

Biemann, K., Papayannopoulos, I. A. (1994), Amino Acid Sequencing of Proteins, Acc. Chem. Res. **27**, 370.

Borders, D. B., Carter, G. T., Hargreaves, R. T., Siegel, M. M. (1985), Recent Applications of Mass Spectrometry to Antibiotic Research, Mass Spectrom. Rev. **4**, 295.

Budzikiewicz, H. (1993), Selected Reviews on Mass Spectrometric Topics: Amino Acids, Peptides, and Proteins, Mass Spectrom. Rev. **12**, 207.

McNeal, C. J. (Ed.) (1988), The Analysis of Peptides and Proteins by Mass Spectrometry, Wiley & Sons, Chichester.

Polymers

Budzikiewicz, H. (1984), Selected Reviews on Mass Spectrometric Topics: Synthetic Polymers, Mass Spectrom. Rev. **3**, 587.

Lattimer, R. P., Harris, R. E. (1985), Mass Spectrometry for Analysis of Additives in Polymers. Mass Spectrom. Rev. **4**, 369.

Schulten, H.-R., Lattimer, R. P. (1984), Applications of Mass Spectrometry to Polymers, Mass Spectrom. Rev. **3**, 231.

Smith, C. G., Mahle, N. H., Park, W. R. R., Smith, P. B., Matin, St. J. (1985), Analysis of High Polymers - Mass Spectrometry, Anal. Chem. (Reviews), 259 R.

Porphyrins

Gallegos, E. J., Sundararaman, P. (1985), Mass Spectrometry of Geoporphyrins, Mass Spectrom. Rev. **4**, 55.

Steroids

Budzikiewicz, H., Djerassi. C., Williams, D. H. (1964), Structural Elucidation of Natural Products by Mass Spectrometry, Vol II, Steroids, Triterpenes and Related Classes, Holden-Day, San Francisco.

Grote, H., Spiteller, G. (1977), Location of Functional Groups with the Aid of Mass Spectrometry - Mass Spectra of Trimethylsilylated Androstan-3,16,17β-triols, Org. Mass Spectrom. **4**, 216, and references therein.

Lin, Y. Y., Smith, L. L. (1984), Chemical Ionization Mass Spectrometry of Steroids and Other Lipids, Mass Spectrom. Rev. **3**, 319.

Terpenoids (see carotenoids and iridoids)

Budzikiewicz, H. (1991), Selected Reviews on Mass Spectrometric Topics: Terpenoids, Mass Spectrom. Rev. **10**, 332.

Enzell, C. R., Wahlberg, I., Ryhage, R. (1984 and 1986), Mass Spectra of Tobacco Isoprenoids, Mass Spectrom. Rev. **3**, 395; **5**, 39.

References Cited in the Text

[1] IUPAC (1973, 1976), Recommendations for Nomenclature of Mass Spectrometry, Butterworth, London.

[2] Morrison, J. D. (1972), Ionisation and Appearance Potentials, in Mass Spectrometry, MTP International Review of Science, Physical Cemistry, Series one (Maccoll, A., Ed.), Butterworth, London.

[3] Biemann, K. (1970), High Resolution Mass Spectrometry, in Topics in Organic Mass Spectrometry (Burlingame, A. L., Ed.), Wiley-Interscience, New York.

[4] Turecek, F., Hanus, V. (1984), Retro-Diels-Alder-Reaction in Mass Spectrometry, Mass Spectrom. Rev. **3**, 85.

[5] Budzikiewicz, H. (1969), Z. Anal. Chem. **244**, 1.

Veith, H. J. Hesse. M. (1969), Helv. Chim. Acta **52**, 2004.

Anh, N. T. (1972), Die Woodward-Hofmann-Regeln und ihre Anwendung, Verlag Chemie, Weinheim.

[6] Gierlich, H. H. Röllgen, F. W., Borchers, F., Levsen, K. (1977), Org. Mass Spectrom. **12**, 387.

Weber, R., Borchers, F., Levsen, K., Röllgen, F. W. (1978), Z. Naturforsch. Sect. A, **33**, 540.

Veith, H. J. (1983), Mass Spectrometry of Ammonium and Iminium Salts, Mass Spectrom. Rev. **2**, 419.

[7] Spiteller, M., Spiteller, G. (1973), Massenspektrensammlung von Lösungsmitteln, Verunreinigungen, Säulenbelegmaterialien und einfachen aliphatischen Verbindungen, Springer-Verlag, Berlin.

Ende, M., Spiteller, G. (1982), Contaminants in Mass Spectrometry, Mass Spectrom. Rev. **1**, 29.

[8] Thomas, A. F. (1971), Deuterium Labeling in Organic Chemistry, Appleton-Century-Crofts, New York.

[9] Pretsch, E., Clerc, T., Seibl, J., Simon, W. (1986), Tabellen zur Strukturaufklärung organischer Verbindungen mit spektroskopischen Methoden, Springer-Verlag, Berlin, Heidelberg, New York.

Biemann, K. (1962), Mass Spectrometry − Organic Chemical Application, McGraw-Hill, New York.

Budzikiewicz, H. (1992), Massenspektrometrie − eine Einführung, Verlag Chemie, Weinheim.

Birkenfeld, H., Haase, G., Zahn, H. (1969), Massenspektrometrische Isotopenanalyse, VEB Deutscher Verlag der Wissenschaften, Berlin.

Seibl, J. (1970), Massenspektrometrie, Akademische Verlagsgesellschaft, Frankfurt/M.

Krueger, H. W., Reesman, R. H. (1982), Carbon Isotope Analyses in Food Technology, Mass Spectrom. Rev. **1**, 205.

Heumann, K. G. (1986), Selected Reviews on Mass Spectrometric Topics: Isotope Dilution Mass Spectrometry, Mass Spectrom. Rev. **5**, 343.

Fresenius, W., Lüderwald, I. (Eds.) (1988), Element Trace Analysis by Mass Spectrometry, Fresenius Zeitschr. Anal. Chemie, 103-222.

[10] Budzikiewicz, H. (1991), Selected Reviews on Mass Spectrometric Topics: Chemical Ionization Mass Spectrometry, Mass Spectrom. Rev. **10**, 519.

[11] Cotter, R. J., Yergey, A. L. (1982), Spectra **8**, 33.

[12] Asselin, M. J. F., Paré, J. J. R. (1981), Org. Mass Spectrom. **16**, 275.

[13] Bruins, A. P., Corey, T. R., Hennion, J. D. (1987), Ion Spray Interface for Combined Liquid Chromatography/Atmospheric Pressure Ionization Mass Spectrometry, Anal. Chem. **59**, 2642.

[14] Fenn, J. B., Mann, M., Meng, C. K., Wong, S. F., Whitehouse, C. M. (1990), Electrospray Ionization − Principles and Practice, Mass Spectrom. Rev. **9**, 37.

[15] Barber, M., Bordoli, R. S., Sedgwick, R. D., Tyler, A. N. (1981), Nature (London) **293**, 270.

[16] Beckey, H.D., Schulten, H.-R. (1975), Angew. Chem. Int. Ed. Engl. **14**, 403.

Beckey, H. D. (1977), Principles of Field Ionization and Field Desorption Mass Spectrometry, Pergamon-Press, Oxford.

Wood, G. W. (1982), Field Desorption Mass Spectrometry: Applications, Mass Spectrom. Rev. **1**, 63.

Beckey, H. D., Comes, F. J. (1970), Techniques of Molecular Ionization, in Topics in Organic Mass Spectrometry (Burlingame, A. L., Ed.), Wiley-Interscience, New York.

Beckey, H. D. (1971), Field Ionization Mass Spectrometry, Pergamon Press, Oxford.

Budzikiewicz, H. (1991), Selected Reviews on Mass Spectrometric Topics: Field Ionization and Field Desorption, Mass Spectrom. Rev. **10**, 331.

[17] Schwarz, H., Levsen, K. (1978), Nachr. Chem. Tech. Lab. **26**, 136.

Nibbering, N. M. M. (1984), Mechanistic Studies by Field Ionization Kinetics, Mass Spectrom. Rev. **3**, 445.

[18] Dell, A., Taylor, G. W. (1984), High-Field-Magnet Mass Spectrometry of Biological Molecules, Mass Spectrom. Rev. **3**, 357.

Budzikiewicz, H. (1990), Masse Zehntausend - Hunderttausend - eine Million: Wo liegen heute die Grenzen der organischen Massenspektrometrie, in Instrumentalized Analytical Chemistry and Computer Technology (Günter, W., Matthes, J. P., Perkampus, H.-H., Eds.), GIT-Verlag, Darmstadt.

Fenn, J. B., Mann, M., Meng, C. K., Wong, S. F., Whitehouse, C. M. (1989), Electrospray Ionization for Mass Spectrometry of Large Biomolecules, Science **246**, 64.

[19] MacFarlane, R. D. (1980), ^{252}Cf-Plasma Desorption Mass Spectrometry (PDMS), in Biochemical Applications of Mass Spectrometry (Waller, G. R., Dermer, O. C., Eds.), First supplementary volume, Wiley-Interscience, New York. p. 1209.

Sundquvist, B., MacFarlane, R. D. (1985), ^{252}Cf-Plasma Desorption Mass Spectrometry, Mass Spectrom. Rev. **4**, 421.

Budzikiewicz, H. (1993), Selected Reviews on Mass Spectrometric Topics: Plasma Desorption Mass Spectrometry, Mass Spectrom. Rev. **12**, 206.

Junclas, H., Schmidt, L., Fritsch, W.-W., Kohl, P. (1991), ^{252}Cf-Plasma Desorption MS, Internat. Laboratory **2**, 25.

[20] Veith, H. J. (1976), Li$^+$Ion Addition – A Careful Method for Ion Formation in Field Desorption Mass Spectrometry, Angew. Chem. Int. Ed. Engl. **15**, 695.

Veith, H. J. (1978), Org. Mass Spectrom. **13**, 280.

Röllgen, F. W., Schulten, H.-R. (1975), Org. Mass Spectrom. **10**, 660.

[21] Ställberg-Stenhagen, S., Stenhagen, E. (1970), Gas Liquid Chromatography - Mass Spectrometry Combination, in Topics in Organic Mass Spectrometry (Burlingame, A. L., Ed.), Wiley-Interscience, New York.

Gudzinowicz, B. J., Gudzinowicz, M. J., Martin, H. F. (1976), Fundamentals of Integrated GC-MS, Marcel Dekker Inc., New York.

Budzikiewicz, H. (1993), Selected Reviews on Mass Spectrometric Topics: GC/MS, Mass Spectrom. Rev. **12**, 397.

[22] Arpino, P. (1989), Combined Liquid Chromatography Mass Spectrometry. Part I. Coupling by Means of a Moving Belt Interface, Mass Spectrom. Rev. **8**, 35.

Caprioli, R. M. (Ed.) (1990), Continuous-Flow Fast Atom Bombardment Mass Spectrometry, Wiley & Sons, Chichester.

Brown, M. A. (Ed.) (1990), Liquid-Chromatography/Mass Spectrometry, Verlag Chemie, Weinheim.

[23] Vestal, M. L. (1986), Eur. Spectros. News **63**, 22.

Karas, M., Bahr, U., Ingendoh, A., Hillenkamp, F. (1989), Laser-Desorption Mass Spectrometry of 100 000 - 250 000 Dalton Proteins, Angew. Chem. Int. Ed. Engl. **28**, 760.

Karas, M., Bahr, U., Hillenkamp, F. (1989), UV Laser Matrix Desorption/Ionization Mass Spectrometry of Proteins in the 100 000 Dalton Range, Int. J. Mass Spectrom. Ion Processes **92**, 231.

Lubman, D. M. (Ed.) (1990), Lasers and Mass Spectrometry, Oxford University Press, Oxford.

Budzikiewicz, H. (1993), Selected Reviews on Mass Spectrometric Topics: Laser Mass Spectrometry, Mass Spectrom. Rev. **12**, 397.

[24] Hesse, M., Bernhard, H. O. (1975), Alkaloide außer Indol-, Triterpen- und Steroidalkaloiden, in Progress in Mass Spectrometry, Vol. 3, Verlag Chemie, Weinheim.

[25] Bosshardt, H., Hesse, M. (1974), Mass Spectrometric Interactions between the Functional Groups of Multisubstituted Alkanes, Angew. Chem. Int. Ed. Engl. **13**, 252.

[26] Schwarz, H. (1978), Some Newer Aspects of Mass Spectrometric *ortho* Effects, Topics in Current Chemistry **73**, 231.

[27] Finnigan, R. E. (1994), Quadrupole Mass Spectrometers, Anal. Chem. **66**, 969 A.

March, R. E., Hughes, R. J. (1989), Quadrupole Storage Mass Spectrometry, Wiley & Sons, New York.

[28] Day, R. J., Unger, S. E., Cooks, R. G. (1980), Anal. Chem. **52**, 557 A.

Scheifers, S. M., Hollar, R. C., Busch, K. L., Cooks, R. G. (1982), Intern. Lab. **5**, 12.

Vickerman, J. C., Brown, A., Reed, N. M. (Eds.) (1989), Secondary Ion Mass Spectrometry – Principles and Applications, Clarendon-Press, Oxford.

Wilson, R. G. (1989), Secondary Ion Mass Spectrometry, Depth Profiling and Bulk Impurity Analysis, Wiley & Sons, Chichester.

Benninghoven, A., Rudenauer, F. G., Werner, H. W. (1988), Secondary Ion Mass Spectrometry, Basic Concepts, Instrumental Aspects, Applications and Trends, Wiley & Sons, Chichester.

Briggs, D., Brown, A., Vickerman, J. C. (1989), Handbook of Static Secondary Ion Mass Spectrometry, Wiley & Sons, Chichester.

[29] Budzikiewicz, H. (1993), Selected Reviews on Mass Spectrometric Topics: Stereochemistry, Mass Spectrom. Rev. **12**, 140.

Mandelbaum, S. (1983), Stereochemical Effects in Mass Spectrometry, Mass Spectrom. Rev. **2**, 223.

Splitter, J. S., Turecek, F. (Eds.) (1994), Application of Mass Spectrometry to Organic Stereochemistry, Verlag Chemie, Weinheim.

[30] Levsen, K., Schwarz, H. (1976), Collisional Activation Mass Spectrometry – A New Probe for Structure Determination of Ions in the Gaseous Phase, Angew. Chem. Int. Ed. Engl. **15**, 509.

Levsen, K., Schwarz, H. (1983), Gas-phase Chemistry of Collisionally Activated Ions, Mass Spectrom. Rev. **2**, 77.

[31] Yost, R. A., Fetterolf, D. D. (1983), Tandem Mass Spectrometry (MS/MS) Instrumentation, Mass Spectrom. Rev. **2**, 1.

Crow, F. W., Tomer, K. B., Gross, M. L. (1983), Mass Resolution in Mass Spectrometry/Mass Spectrometry, Mass Spectrom. Rev. **2**, 47.

Richter, W. J., Raschdorf, F., Märki, W. (1985), in Mass Spectrometry in the Health and Life Sciences (Burlingame, A. L., Castagnoli, N., Eds.) Elsevier, Amsterdam, p. 193.

Terlouw, J. K., Schwarz, H. (1987), Gasification and Characterisation of Molecules by Neutralization - Reionization Mass Spectrometry (NRMS), Angew. Chem. Int. Ed. Engl. **26**, 805.

McLafferty, F. W. (Ed.) (1983), Tandem Mass Spectrometry, Wiley-Interscience, New York.

Schwarz, H. (1989), Tandem-Massenspektrometrie (MS/MS), Analytiker-Handbuch, p. 199.

Bush, K. L., Glish, G. L., McLuckey, S. A. (1988), Mass Spectrometry/Mass Spectrometry – Techniques and Applications of Tandem Mass Spectrometry, Verlag Chemie, Weinheim.

Budzikiewicz, H. (1991), Selected Reviews on Mass Spectrometric Topics: Tandem Mass Spectrometry, Mass Spectrom. Rev. **10**, 519.

Goldberg, N., Schwarz, H. (1994), Neutralization - Reionization Mass Spectrometry: A Powerful "Laboratory" To Generate and Probe Elusive Neutral Molecules, Acc. Chem. Res. **27**, 347.

Gross, M. L. (1994), Tandem Mass Spectrometric Strategies for Determining Structure of Biologically Interesting Molecules, Acc. Chem. Res. **27**, 361.

McLafferty, F. W. (1994), High-Resolution Tandem FT Mass Spectrometry above 10 kDa, Acc. Chem. Res. **27**, 379.

[32] Cotter, R. J., Yergey, A. L. (1981), Anal. Chem. **53**, 1306.

[33] Vestal, M. L. (1986), Eur. Spectros. News **63**, 22.

[34] Cooks, R. G., Beynon, J. H., Caprioli, R. M., Lester, G. R. (1973), Metastable Ions, Elsevier, Amsterdam.

Fraefel, A., Seibl, J. (1985), Selective Analysis of Metastable Ions, Mass Spectrom. Rev. **4**, 151.

[35] Wapstra, A. H., Bos, K. (1977), The 1977 Atomic Mass Evaluation, Part I. Atomic Mass Table, Atomic Data and Nuclear Data Tables **19**, 177.

Holden, N. E., Martin, R. L., Barnes, I. L. (1984), Isotopic Compositions of the Elements 1983, Pure Appl. Chem. **56**, 675.

Commission on Atomic Weights and Isotopic Abundances (1988), Atomic Weights of the Elements 1987, Pure Appl. Chem. **60**, 841.

Journals and Periodicals

Mass Spectrometry Bulletin (since 1966), Mass Spectrometry Data Centre, Aldermaston, Royal Society of Chemistry; published monthly. A journal with the titles of publications having mass spectrometric content.

Organic Mass Spectrometry (1968-1994), Wiley & Sons, Chichester; international journal, published monthly.

Biomedical and Environmental Mass Spectrometry (since 1973), Wiley & Sons, Chichester; international journal, published monthly.

Advances in Mass Spectrometry (since 1958), The Institute of Petroleum, London; published every three years. Contains the reports, abstracts and discussions from the mass spectrometric conference that takes place every three years.

Specialists Periodical Report (since 1970), Mass Spectrometry, The Chemical Society, London; gives a critical evaluation of publications having mass spectrometric content that have appeared in the previous year.

Mass Spectrometry Reviews (since 1982), Wiley & Sons, New York; international journal, published every two months with review articles on current mass spectrometric topics.

Rapid Communications in Mass Spectrometry (since 1987), Wiley & Sons, Chichester; published monthly.

Journal of the American Society for Mass Spectrometry (since 1990), Elsevier Science Publishers, Amsterdam; published every two months. Original communications with mass spectrometric content.

Journal of Mass Spectrometry, (JMS) (since Vol. 30, 1995), Wiley & Sons, Chichester; international journal.

Example 1

Objective

The structure of the unknown compound **1** is to be determined from its spectroscopic data (spectra 1 to 4).

Conditions for Recording the Spectra
1   UV:

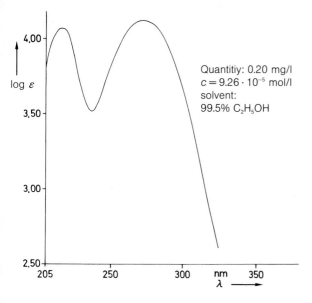

Quantitiy: 0.20 mg/l
$c = 9.26 \cdot 10^{-5}$ mol/l
solvent:
99.5% C_2H_5OH

Answer

The signals with the highest mass numbers in the mass spectrum of the unknown compound are at $m/z = 215$ and 217. These peaks have approximately equal abundances, which suggests the presence of *one Br*-atom in the molecule (see

Tab. 4.10, p. 294). This is supported by the appearance of a similar pair of signals at $m/z = 169$ and 171. No further bromine-containing ions are present. – The appearance of the doublet at the highest mass number in the mass spectrum does not unequivocally prove the existence of bromine (unless high resolution is used); it is also conceivable that $M^{+\bullet}$ is not shown and that two fragment ions, which differ by two mass numbers and have the same abundance, are recorded as the signals with the highest mass numbers, e.g. $[M-17]^+$ and $[M-15]^+$. – Further evidence for bromine is $m/z = 136$, which is also the base peak of the spectrum. This signal is a "singlet" and therefore does not contain bromine; the corresponding ion is formed by the loss of bromine from the molecular ion.

In order to simplify calculations involving *Br*-containing ions and thus ease the interpretation of the mass spectrum, it is advisable to use only the ion which contains the lighter bromine isotope (^{79}Br). Therefore, in the current example we shall assume that $m/z = 215$ is $M^{+\bullet}$.

The mass of the molecular ion indicates the presence of another heteroatom in addition to bromine, namely nitrogen (odd value for the relative molecular mass, see p. 224). Where is the nitrogen situated in the molecule? The mass spectrum also gives information about this: the signals at $[M-16]^{+\bullet}$ ($m/z = 199$), $[M-46]^+$ ($m/z = 169$), as well as $[M-Br-16]^+$ ($m/z = 120$) and $[M-Br-30]^+$ ($m/z = 106$) strongly suggest the presence of a (C)-NO_2 group (see p. 271). This supposition is confirmed by the IR spectrum: two intense bands at $v = 1530$ and 1345 cm^{-1} are characteristic of nitro groups. Aliphatic nitro groups absorb at 1560 cm^{-1}; if they are conjugated their absorption is observed at smaller wavenumbers. This latter case corresponds with the current example. The absorption at 1610 cm^{-1} gives an indication of the type of double bond that is conjugated with the nitro group: aromatic. The other aromatic bands at 1600, 1450 and 1400 cm^{-1} are either absent or

2 IR: KBr

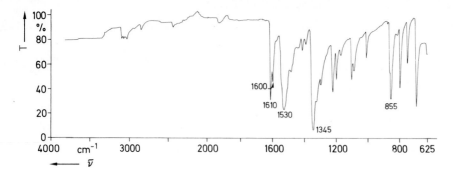

3 ¹H NMR: CDCl₃, 90 MHz, internal TMS; note the expanded region

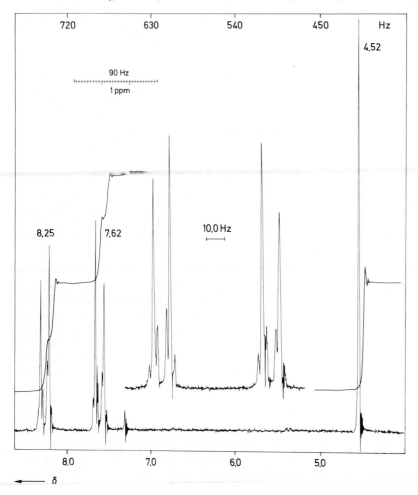

4 MS: 70 eV, direct inlet

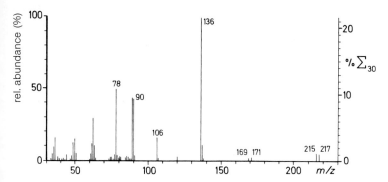

too weak to be considered as significant. However, the aromatic C–H out-of-plane vibration between 3300 and 3100 cm^{-1} is present. This aromatic compound should be *para*-disubstituted because of a strong band at 855 cm^{-1}. (During the determination of the degree of substitution of an aromatic compound, one must take care that this region is not obscured by the absorption of the solvent itself.) The results obtained so far can be summarised by the following partial formula **A**:

We already know that the residue R must contain Br. The portrayed formula with R = ^{79}Br has a mass of 201 and therefore differs by 14 amu (i.e. in this case by only CH$_2$) from the found value (m/z = 215). By the process of elimination, this allows one to write down the somewhat speculative structure **B**. It is speculative because the structural elements of CH$_2$ and a 1,4-disubstituted aromatic ring have not yet been proven. They have only been made probable. Conversely, the presence of Br and NO$_2$ can be taken as certain. How can structure **B** now be confirmed or corrected? The UV and ^{1}H NMR spectra, as well as a complete interpretation of the mass spectrum, are suitable for this purpose.

The illustrated UV spectrum has a very great similarity with that of *p*-nitrotoluene [λ_{max} = 272 nm (log ε = 3.99) in ethanol].

The ^{1}H NMR spectrum is also in agreement with structure **B**. Three groups of signals can be recognised: a singlet at δ = 4.52 ppm, a doublet-like signal at 7.62 and its mirror image at 8.25. The integration shows that these three absorptions have the intensity ratios 1 : 1 : 1. A singlet is to be expected for the pro-

tons on the methylene group and, according to the rule of Shoolery, its chemical shift can be calculated to be 4.45 ppm, which correlates well with the found value (4.52 ppm). The highly symmetrical AA'BB' system with a maximum of 24 lines can be expected for the aromatic protons. The chemical shift of the protons that are *ortho* to the nitro group is 8.25 ppm (calculated 8.21 ppm) and that of the *meta* protons is 7.62 ppm (calculated 7.52 ppm). Accordingly, the correct proton ratios amount to 2 : 2 : 2 H.

For the sake of completeness, the last task for the elucidation of this structure should be to explain the main signal of the mass spectrum, which has not yet been assigned. The high abundance of m/z = 136 ([M – Br]$^+$) can be attributed to the benzyl positioning of the *Br*-atom and the formation of the ion **a**. The ion m/z = 78 (no contamination from benzene!) can also be explained by the loss of CO from **b** (m/z = 106, see p. 244). The structure of the ion m/z = 78 can either be a benzene ring or **c**, however this question will not be discussed here.

This example clearly shows that the posed structural problem can be solved unequivocally with the aid of the spectra that were given; even less spectral information would have been sufficient.

Example 2

Objective

In order to use a case that is as simple as possible for this demonstration, we have chosen "methyl propyl ketone".

A bottle labelled with the name of this substance was taken from the laboratory in order to record the spectra. These are depicted below (spectra 5 to 7). Is it the desired compound?

Conditions for recording the spectra

UV: in heptane; $\lambda_{max} = 280$ nm (log $\varepsilon = 1.22$)

5 **IR:** CCl₄, microcell 0.2 mm

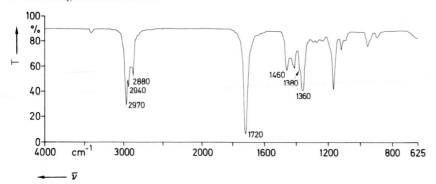

6 **¹H NMR:** CDCl3, 90 MHz, no signals above 4.3 ppm; note the expanded region

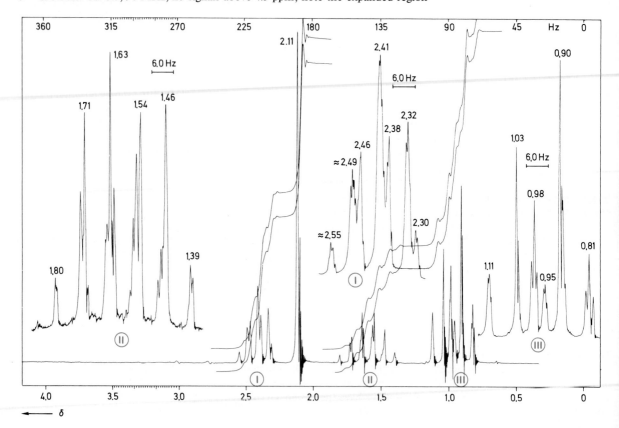

7 MS: 70 eV, gas inlet

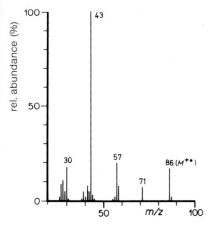

Answer

The band in the IR spectrum at $v = 1720$ cm^{-1} can be recognised as a carbonyl absorption which can be attributed to a saturated ketone. Furthermore, methyl and methylene stretching vibrations are present (1380 and 1360 cm^{-1}). Therefore, the IR spectrum is in agreement with the structure.

Also in accordance with the structure are the appearance of the molecular ion at $m/z = 86$ and the next most abundant ion at $m/z = 43$ (h$_3$C—C≡O$^+$). Conversely, it is difficult to comprehend the relatively high abundance of the ion at $m/z = 57$, which corresponds to $[M - C_2H_5]^+$. However, the ion $m/z = 58$ is once again in complete agreement with the expected fragmentation (loss of ethylene from the molecular ion through a McLafferty rearrangement) of methyl propyl ketone (2-pentanone). On the whole, therefore, the mass spectrum is in order and in agreement with the structure, except for the loss of C$_2$H$_5$•, which can neither be neglected nor explained.

This first doubt is confirmed by the analysis of the ^{1}H NMR spectrum and it becomes clear that there is "something rotten" with this sample. To begin with, the four types of protons that are expected are indeed found:

$$H_3C - \overset{\overset{\displaystyle O}{\|}}{C} : 2{,}11; \quad \overset{\overset{\displaystyle O}{\|}}{C} - CH_2 : \approx 2{,}4 :$$

$$\overset{\overset{\displaystyle O}{\|}}{C} - CH_2 - CH_2 : \approx 1{,}6 : \quad CH_2 - CH_3 : \approx 1{,}0 \text{ ppm}$$

However, some of the multiplicities are peculiar and from the point of view of the structure, the integration ratios are clearly wrong. The mid-points of the signals mentioned above,

2.4/2.11/1.6/1.0 ppm, give 3.04 : 3.00 (reference) : 2.01 : 4.65 H-atoms upon integration, instead of the expected ratios of 2 : 3 : 2 : 3 H. After closer examination, the signal at ≈ 1.0 ppm turns out to be two overlapping triplets. The first triplet is composed of the signals at 1.11, 1.03 and 0.95 ppm, whereas the second consists of 0.98, 0.90 and 0.81 ppm. Both triplets have a coupling constant of ≈ 7 Hz. The signal at 2.4 ppm can only be interpreted as an addition signal, which is composed of a triplet (2.49, 2.41 and 2.32 ppm) and a quartet (2.55, 2.46, 2.38 and 2.30 ppm). On the other hand, the six line signal at ≈ 1.6 ppm (determined by a comparison of the ratios of the individual signals; found ratios: ≈ 1 : 5 : 10 : 10 : 5 : 1) and the singlet at 2.11 ppm appear to be "clean".

Based on the analyses so far, the sample appears to be methyl propyl ketone (**2**) which is contaminated with an isomer. This isomer also possesses a methylene group (absorption at ≈ 2.4 ppm) next to the carbonyl group. However, the methylene group is directly bonded to a methyl group (determined from the multiplicity of the signal at ≈ 2.4 ppm and the chemical shift of the methyl group: 1.03 ppm). Because the relative molecular mass is the same as that of methyl propyl ketone, the impurity must be diethyl ketone (**3**; 3-pentanone). Other isomers that could possibly be present would be pentanal (**4**), methyl isopropyl ketone (**5**) and a hydrocarbon, e.g. heptane (**6**). Compound **4** can be excluded because of the absence of an aldehyde proton in the ^{1}H NMR spectrum.

For **6**, the intensities of the signals at ≈ 1.6 and ≈ 1.0 ppm would be enlarged. However, this is not found. Instead, the intensities of the signals at ≈ 2.4 and ≈ 1.0 ppm have the ratio 2 : 3. This argument also excludes **5**, because the above ratio would be 1 : 6 for this compound. Compound **3** is also consistent with the peak at $m/z = 57$ ($[M - C_2H_5]^+$) in the mass spectrum.

The ^{1}H NMR spectrum also provides information about the relative amounts of each substance. If one subtracts, for example, two H-atoms from the integrated value of 3.04 H (reference: 2.11 ppm ≡ 3 H) for the signal at ≈ 2.4 ppm, then a residue of 1.04 H-atoms remains. This value corresponds to the four α-positioned H-atoms adjacent to the carbonyl group in

3-pentanone (**3**). From this, the relative amounts of the components present in the mixture can be calculated: the proton ratio is 2 (**2**) to 1.04 (**3**), or 1 (**2**) to 0.26 (**3**) when the same number of protons, namely one, is taken into consideration. This gives the fractions (1.26 = 100%) of **2** and **3** in the mixture as 79 : 21%. (78% **2** and 22% **3** is obtained if one repeats the calculation using the signal at ≈ 1.0 ppm.)

In a case such as this it is advisable to confirm the finding by analysing the two component mixture gas chromatographically and determining the relative amounts of the compo-

nents by integration. The structures of the pure components can be confirmed either through mass spectrometric analyses (with the GC/MS combination) or by mixing each component with an authentic sample.

Example 3

Objective

The structure of the unknown compound **7** is to be determined (spectra 8 to 11).

Conditions for Recording the Spectra

UV: in C_2H_5OH; $\lambda_{max} = 281$ nm (log $\varepsilon = 1.2$)

8 **IR:** in CCl_4

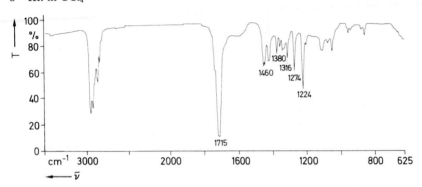

9 **^{1}H NMR:** CDCl$_3$, 90 MHz, internal TMS

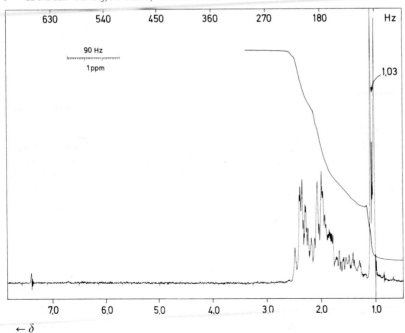

10 ^{13}**C NMR:** ^{1}H broadband decoupled, CDCl$_3$, internal TMS

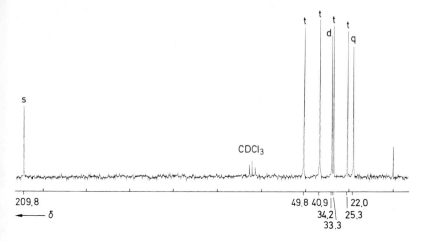

209.8   49.8 40.9 | 22.0
◄——— δ   34.2   25.3
                  33.3

11 MS: 70 eV, gas inlet

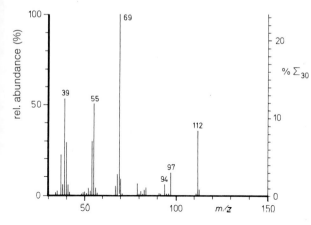

Answer

The substance displays a weak absorption in the UV spectrum at $\lambda_{max} = 281$ nm, which indicates the presence of a carbonyl group.

No signals are present in the absorption region for aromatic and vinyl protons of the ^{1}H NMR spectrum. The only absorptions that are observed are due to aliphatic protons. A doublet at $\delta = 1.03$ ppm is the only signal whose multiplicity could be overlooked. If one assumes that this methyl absorption (CH–**CH**$_3$) arises from three protons, then the multiplet corresponds with 9 (exactly 9.22) H, so that there must be a total of n · 12 *H*-atoms available.

The band at 1715 cm^{-1} in the IR spectrum indicates a saturated ketone. It cannot be due to an ester group because strong

bands in the 1000–1200 cm^{-1} region and the corresponding ^{1}H NMR absorptions are missing. The signal at 1380 cm^{-1} can be interpreted as a band from a CH$_3$ deformation vibration.

The mass spectrometrically determined relative molecular mass is $m/z = 112$. If one subtracts the already known structural elements (C=O, 28 amu; CH–CH$_3$, 28 amu) from this, the remaining structural elements together have a total mass of 56 amu. This value is exactly divisible by 14 to give the number 4, which indicates the presence of 4 CH$_2$ groups, 2 CH$_2$ groups and an additional carbonyl group, or 2 CH$_2$ groups and 2 *N*-atoms or other similar combinations. How can a decision be made between these possibilities without high resolution mass spectra and how can the elemental composition be determined? Regardless of what is proposed, one double bond equivalent (double bond or ring) must be present in the above cases. However, as can be seen from the absence of an absorption in the vinyl region of the ^{1}H NMR spectrum, C–C bonds are not present. An additional C=O group (e.g. **8**) or an N=N bond (e.g. **9** or **10**) would lead one to expect UV spectra in which the enol form of a keto group was clearly visible (conjugation of two double bonds). For **9** and **10**, one would expect mass spectra that feature retro Diels-Alder reactions (for **9**:

8   **9**   **10**

[M − 42]$^{+•}$ and for **10**: [M − 56]$^{+•}$). However, neither of these are found (other groups, such as −CO−CO− and −N=N−CO−, can also be excluded by using similar arguments). Therefore, the most promising variant that remains is the structural element with four CH_2 groups. This is further supported by the integral of the ^1H NMR spectrum (see above), which also excludes the presence of two methyl groups. From this we can conclude that compound **3** can only be one of the three methylcyclohexanones **11, 12** or **13**.

11   **12**   **13**

Unfortunately, the number of *H*-atoms that are α-positioned with respect to the carbonyl group cannot be determined exactly, so that only the mass spectrum can be used to test this hypothesis.

The mass spectrometric behaviour of cyclohexanone has already been discussed (see p. 229). The main fragment ion is $m/z = 55$, for which the structure **d** was deduced.

d
$(m/z = 55)$

e
$(m/z = 69)$

f
$(m/z = 69)$

This ion must also be present in the mass spectra of compounds **11, 12** and **13**. In addition, the ions **e** and **f**, respectively, which have the same mass, must be present in the spectra of **11** and **12**. Because $m/z = 69$ is the base peak of the spectrum of the substance being analysed (**7**), **13** can be excluded from further consideration. The choice is now reduced to just **11** and **12**.

On the basis of the given spectra, including the chemical shifts of the protons in the ^1H NMR spectra, it is not possible to decide between **11** and **12**. Additional information is needed in order to accomplish this. Suitable indications can be obtained by:

- Comparing the spectra of authentic samples. (The IR, ^1H NMR and mass spectra of **11** and **12** are different and can be used for an unequivocal characterisation. However, they are not suitable for the purpose of being unequivocally associated with one of the two compounds.)

- The surest method is the base or acid catalysed D-exchange (see p. 255). In the case of **11**, this must result in the insertion of 3 D-atoms, whereas 4 D-atoms are included for **12** (ascertained by a subsequent mass spectrometric determination).

A ^{13}C NMR spectrum was also recorded (spectrum 10), in which 7 signals (1 s, 1 d, 4 t and 1 q) are visible. This is in agreement with the structure of a methylcyclohexanone. The resonance positions that were found, as well as those that were calculated for **11** and **12** are listed in Table 5.1. The best correlation with the sample being analysed is obtained with 3-methyl-

Table 5.1 Exercise 3: ^{13}C NMR spectrum, values in ppm

found			calculated		calculated			
			11			**12**		
δ_c	multi-plicity	C-atom	calculation	δ_c	C-atom	calculation	δ_c	
209.8	s	1	208.5[a]	= 208.5	1	208.5[a]	= 208.5	
49.8	t	6	40.4[a] + 0.0 (γ-equiv)[c]	= 40.4	2	40.4 + 9.0	= 49.4	
40.9	t	3	26.5[a] + 9.0 (β-equiv)[c]	= 35.5	6	40.4 − 0.2	= 40.2	
34.2	d	2	40.4[a] + 6.0 (α-equiv)[c]	= 46.4	3	26.5 + 6.0	= 32.5	
33.3	t	5	26.5[a] − 0.2 (δ-equiv)[c]	= 26.3	4	23.8 + 9.0	= 32.8	
25.3	t	4	23.8[a] + 0.0 (γ-equiv)[c]	= 23.8	5	26.5 + 0.0	= 26.5	
22.0	q	7	23.1[b] − 9.4 (β-CH₂)[d] + 3.0 (β-CO)[d]	= 16.7	7	23.1 + 9.4 (γ-CH₂) − 3.0 (γ-CO)	= 22.6	

[a] Values for cyclohexanone, p. 158
[b] Value for methylcyclohexanone
[c] Increments for dimenthylcyclohexane
[d] Shift from aliphatic ketones (β-C)

cyclohexanone (**12**). A difference that can be determined very quickly is the doublet, which amounts to 12 ppm in the case of **11**. The other values sometimes also have large differences.

The sample being analysed is indeed **12**. The measured values for **11** are: C–1, 211.9 (s); C–2, 45.2 (d); C–3, 36.3 (t); C–4, 25.3 (t); C–5, 28.1 (t); C–6, 41.8 (t); C–7, 14.8 (q) ppm.

Example 4

Objective

When a solution of indole-3-acetaldehyde (**14**) in chloroform is evaporated to dryness in a rotary evaporator, a mixture of products is obtained, which, aside from **14**, contains an unknown material (**15**) with the spectral properties shown below (spectra 12 to 14). What is the structure of **15**? How can one prevent its formation in order to obtain a greater yield of **14**?

Conditions for Recording the Spectra

UV: in C_2H_5OH; $\lambda_{max}=221$ nm (log $\varepsilon=4.50$), 278 (4.01); shoulders at 289 (3.93), 272 (4.00)

12   IR: CHCl₃

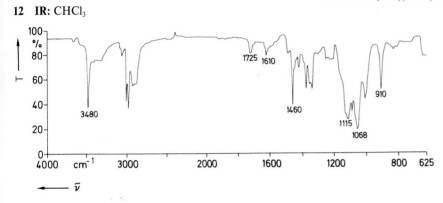

13   ¹H NMR: CDCl₃, 60 MHz, internal TMS

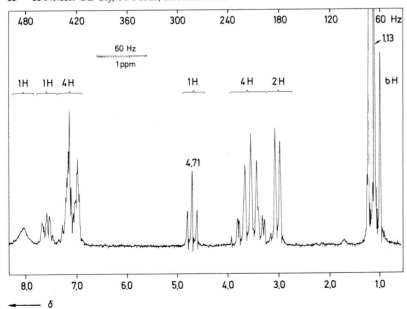

14 **MS:** 70 eV, direct inlet

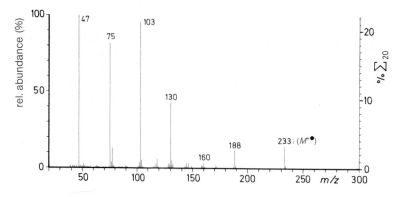

Answer

What is the most rational way to approach this problem? The following individual questions should be clarified in the order given:

1. Has a reaction occurred between **14** and either the solvent (CHCl$_3$) or the hydrochloric acid therein (from the decomposition of CHCl$_3$)?

2. Has an oxidation taken place? A reaction between **14** and the air is possible because the air was not excluded.

3. Which structural elements of the starting material have been retained in the product and which are missing? The structure should be derivable from this information.

The answer to the first question can be ascertained quickly. If CHCl$_3$ or HCl have reacted with **14**, the new compound (**15**) should contain chlorine. However, the mass spectrum does not give the slightest indication of the presence of chlorine (see Tab. 4.10, p. 294).

The second question is concerned with an increase in the relative molecular mass if oxygen has been added (M of **14** = 159; M + 14 [CH$_2$ → C=O], M + 16 [C−H → C−OH]), or with an oxidative dimerisation (159 + 159 − 2 = 316). However, neither consideration produces any suitable results, even when one combines various processes of this type. The relative molecular mass of **15** (M = 233) is not derivable in this way.

Therefore, the problem must be solved by answering the third question.

The UV spectra of **14** and **15** are essentially the same. Also in agreement with this is the mass spectrometric finding that the signal at $m/z = 130$ is present as an abundant peak; the corresponding ion **g** comprises the indole part.

A strong absorption is missing from the carbonyl region of the IR spectrum. This means that the aldehyde group in **14** is not present in **15**. Instead of this, one recognises strong absorptions in the "fingerprint" region between 1200 and 1000 cm^{-1}, which suggest the presence of ether bonds.

Signals for one NH and five aromatic protons are registered in the ^{1}H NMR spectrum between 8.3 and 6.9 ppm. A triplet at 1.13 ppm, which integrates for six protons, is conspicuous. The accompanying methylene protons (4 H) absorb between 3.9 and 3.2 ppm, from which it follows that two −O−CH$_2$−CH$_3$ residues are present in the molecule. A triplet (1 H) appears at low field (4.71 ppm). This originates from an acetal H-atom. This assignment can be verified by analysing the coupling constants and fully interpreting the spectrum. Independently of the ^{1}H NMR spectrum, one can arrive at the same result by using the mass spectrum (see scheme). The acetal group is dominant in the fragmentation (see p. 230).

14

g

(m/z = 130)

Now that the structure of the unknown substance **15** has been clarified, the question of its formation must be answered. This also presents no difficulty. Chloroform contains ethanol for stabilisation. In the presence of HCl, ethanol reacts with an aldehyde to produce diethyl acetal. This reaction can best be prevented by removing the ethanol from the chloroform before its use.

Fragmentation scheme

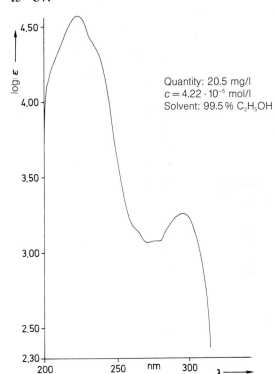

Example 5

Objective

In this case the structure of the unknown compound **16** is to be determined (spectra 15 to 19).

Conditions for Recording the Spectra

15   UV:

Quantity: 20.5 mg/l
$c = 4.22 \cdot 10^{-5}$ mol/l
Solvent: 99.5 % C_2H_5OH

16   IR: $CHCl_3$ and film (inset)

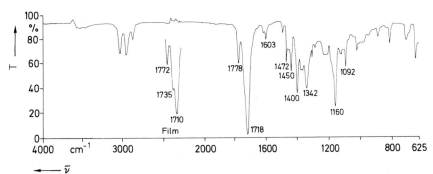

17 **¹H NMR:** CDCl₃, 90 MHz, internal TMS

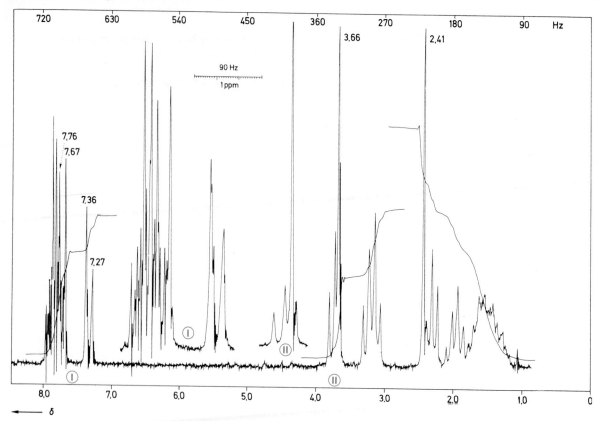

18 **MS:** 70 eV, direct inlet

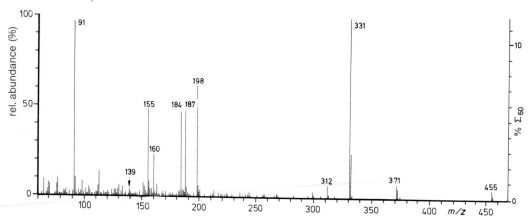

19 High Resolution MS: M/ΔM: 20 000

$M^{+\bullet}$ by FD–MS: 486 ($C_{25}H_{30}N_2O_6S$)

Base peak 10435/mass 331

Peak no.	I/base (%)	Mass	Difference (milli-mass)	C	H	N	O	S
				12	1	14	16	32
28	97.14	91.0524	−2.4	7	7	0	0	0•
			1.6	2	7	2	2	0
29	10.11	92.0580	−0.5	2	8	2	2	0
36	8.23	98.0962	−0.7	6	12	1	0	0
42	5.31	104.0252	−0.9	7	4	0	1	0
50	5.41	111.9921	−2.1	0	4	2	3	1
51	13.74	112.9984	−0.3	3	1	2	3	0
			2.4	0	3	1	6	0
66	6.22	130.0294	0.2	8	4	1	1	0
			2.8	5	6	0	4	0
69	5.12	133.0073	0.1	3	5	2	2	1
87	7.45	150.9986	−0.5	2	3	2	6	0
			2.0	6	3	2	1	1
88	5.33	152.1069	−0.6	9	14	1	1	0
			2.1	6	16	0	4	0
91	48.63	155.0159	−0.8	7	7	0	2	1
92	5.31	155.0247	2.9	6	5	1	4	0
			−0.5	3	9	1	4	1•
94	8.87	156.1016	−0.8	8	14	1	2	0
100	23.81	160.0394	−0.4	9	6	1	2	0•
			2.3	6	8	0	5	0
103	6.81	162.9991	0.0	3	3	2	6	0
			2.5	7	3	2	1	1
120	46.01	184.1337	−0.0	10	10	1	2	0•
128	47.82	188.0703	−0.8	11	10	1	2	0•
			1.8	8	12	0	5	0
129	5.28	189.0752	−1.0	8	13	0	5	0
			1.4	12	13	0	0	1
135	61.43	198.0581	2.6	12	8	1	2	0
			−0.8	9	12	1	2	1•
			1.8	6	14	0	5	1
136	6.58	199.0606	−2.7	12	9	1	2	0
			−0.0	9	11	0	5	0
			2.4	13	11	0	0	1
165	6.92	312.1278	1.6	21	16	2	1	0
			−1.8	18	20	2	1	1
			0.8	15	22	1	4	1•
169	100.00	331.1665	0.7	18	23	2	4	0•
			−2.6	15	27	2	4	1
170	24.44	332.1708	−2.8	18	24	2	4	0
			2.3	19	26	1	2	1
172	7.65	371.1038	0.6	22	15	2	4	0
			−2.7	19	19	2	4	1•
173	5.15	372.1061	0.3	23	18	1	2	1
175	5.28	455.1647	0.7	24	27	2	5	1•

Notes to the list of elements (computer printout)

1. The following elements were entered: ^{12}C, ^{1}H, ^{14}N, ^{16}O, ^{32}S (columns 5–9); ^{13}C entries were foregone.

2. The printed masses must have an abundance of at least 5% of the base peak; weaker signals were not printed out.

3. Error deviations of more than ±3 millimasses between the found masses (column 3) and the masses calculated from the given elemental composition are not listed [the latter masses are not given, only the differences (column 4)].

4. Column 2 contains the relative abundance.

5. Column 1 gives the consecutive numbers for all peaks, even when these are not included in the list because of the conditions mentioned in Note 2.

6. The parameters mentioned in Notes 1–3 can be freely chosen by the operator, i.e. the type and number of the elements, minimum abundance cut-off and the error deviation.

Answer

A quick look at the spectra shows that we are apparently dealing with a complicated compound. It is an oily intermediate product from a multiple step synthesis. The relative molecular mass is not shown in the electron impact mass spectrum (spectrum 18), however $M^{+\bullet} = 486$, which corresponds to $C_{25}H_{30}N_2O_6S$, was ascertained from the FD mass spectrum. How should one now proceed in order to interpret the complex spectra or to at least obtain important information about functional groups and structural elements?

To begin with, the presence of an S-atom in the unknown molecule is conspicuous. The structure of the molecule can possibly be ascertained by using the signals from the fragment ions that contain sulfur. S-containing signals are $m/z = 455, 371, 312, 198$ and 155. The ion $m/z = 155$ is given in the list of frequently appearing fragment ions in Table 4.8 (see p. 279) and originates from toluenesulfonates. The elemental composition of the ion $m/z = 155$ is in agreement with this possibility. If this is true, then $m/z = 91$ (tropylium ion) must also be present as an abundant fragment ion signal, as well as $m/z = 139$. These signals are indeed present in the spectrum. In addition, 155 amu are lost from the molecular ion to give $m/z = 331$, which does not contain sulfur. Is it possible to use the other spectra to confirm this supposition? Yes. In the IR spectrum, absorption bands are recorded at 1342 (asymmetric SO_2 stretching vibration; region: $1370-1330$ cm^{-1}) and 1160 cm^{-1} (region: $1180-1160$ cm^{-1}). Furthermore, an aromatic band and a C–H vibration (in-plane) are observed at 1603 and 1092 cm^{-1}, respectively. These are typical bands for a tosyl residue. In the ^{1}H NMR spectrum, the singlet of the aryl CH$_3$ group is recorded at 2.41 ppm. The aromatic protons appear as an AA'BB' system (see enlarged region) with two doublet-like signals at ≈ 7.31 ppm (*meta*-position from the SO_2 group) and 7.71 ppm (*ortho*-positioned). Therefore, the presence of the structural element *p*-toluenesulfonyl (**C**) has been clarified.

$$-\overset{\overset{\displaystyle O}{\|}}{\underset{\underset{\displaystyle O}{\|}}{S}}-\!\!\left\langle\!\!\!\bigcirc\!\!\!\right\rangle\!\!-CH_3$$

C

How is this group linked to the rest of the molecule, i.e. is a C–S, O–S or N–S bond present? This question is still open after the analysis of the IR spectrum. We must search in the mass spectrum for signals that contain the residue **C** plus additional, but as few as possible, components. The only suitable peak of this kind is $m/z = 198$ (**C** + C$_2$H$_5$N). Therefore, the presence of a *p*-toluenesulfonate (O–S bond) can now be excluded, which leaves only a sulfone (C–S) or a sulfonamide (N–S) to be considered. This problem must be put aside for a while, because an unequivocal decision is not possible at this stage.

Let us now turn to some of the other functional groups. The heaviest fragment ion in the mass spectrum is $m/z = 455$, which differs from the "molecular ion" by the absence of OCH$_3$. A methyl group that is bonded to an O-atom appears in the ^{1}H NMR spectrum at $\delta = 3.66$ (singlet). Tab. 3.12 shows that this O-atom must itself be directly bonded to a C=O or a C=C group. However, methoxy groups at C=C bonds are never or very rarely expelled under electron bombardment. One can conclude from this that the presence of a methyl ester is highly possible. This is supported by the presence of an IR ester carbonyl band at 1735 cm^{-1} (recorded as a film, saturated ester). Strengthened by this information we can closely examine the mass spectrum again. The ion with $m/z = 371$ corresponds to the loss of C$_6$H$_{11}$O$_2$ from the "molecular ion". If one assumes that this is an ester group which is expelled together with other structural elements, then four CH$_2$ residues are attached to the ester (C$_6$H$_{11}$O$_2$ minus COOCH$_3$ = C$_4$H$_8$). Because absorptions due to C–CH$_3$ and CH–CH$_3$ are missing from the ^{1}H NMR spectrum (from chemical considerations, –CH$_2$–CH$_3$ or a ring are not possible), the structural element **D** must be present in the molecule. The triplet-like signal at ≈ 2.3 ppm would appear to correspond to the methylene protons that are α-positioned relative to the methoxycarbonyl group.

$$-CH_2-CH_2-CH_2-CH_2-COOCH_3$$

D

In order for the residue **D** to be ejected from the molecular ion, a particularly favourable activation must exist (α-cleavage relative to a heteroatom, double bond, etc.), because smaller fragments of this chain, such as **D**–CH$_2$, etc., are not expelled.

The structural elements **C** and **D** together comprise C$_{13}$H$_{18}$O$_4$S (270 amu), which when subtracted from the elemental composition of the molecular ion, leave a still "unknown" residue of C$_{12}$H$_{12}$N$_2$O$_2$ (216 amu). It is particularly noticeable that four heteroatoms are present in this residue and that the C/H ratio suggests a heavily unsaturated molecular segment. Undoubtedly the strong infra-red absorptions in the carbonyl region at $v = 1778$ and 1718 cm^{-1} (as a film: 1772, 1710 cm^{-1}) belong to this still unknown structural fragment. The identification of these bands will be possible after the analysis of the as yet unassigned signals in the mass spectrum at $m/z = 160$ (C_9H_6-NO_2) and 188 ($C_{11}H_{10}NO_2$). Based on Table 4.8, the ion of mass 160 can be assigned the structure **h**, which indicates that the structural element **E**, an *N*-alkylphthalimide residue, is present in the unknown compound.

G

h

E

In the absorption region for five-membered imide rings (*sec*-amide), one band can be observed in each of the ranges from 1790–1720 and 1710–1670 cm^{-1}.

The ^{1}H NMR spectrum also confirms the structural element **E**: "doublet-like" aromatic signals are found between 8.0–7.7 ppm (4 H; the total aromatic region integrates for $6+2=8$ protons) and the methylene protons next to the phthalimide *N*-atom exhibit a chemical shift of ≈ 3.75 ppm. Because this latter signal is a triplet (J $\approx$ 7 Hz), an additional neighbouring methylene group must be present. The signal at $m/z = 188$ in the mass spectrum confirms this supposition. The ion of mass 188 contains two CH$_2$ residues more than $m/z = 160$, so that it can be assigned the structure **i** and the corresponding structural element **F**. By addition of the atoms that constitute the structural elements **C**, **D** and **F**, one obtains $C_{24}H_{28}NO_6S$, which means that there is still no information available about the nature of the group CH$_2$N nor about how **C**, **D** and **F** are linked to one another.

i

F

The quartet-like signal at $\approx$ 3.2 ppm, which integrates for four protons, has not yet been assigned. Because the methylene protons next to the COOCH$_3$ and imide groups have already been assigned, the protons causing this signal must be adjacent to the as yet undefined *N*-atom or the SO$_2$ group. In order to arrive at an "acceptable" compound for **16** (i.e. no incomplete valences), in which the signals for the two methylene groups are shifted to a lower field, the only possible way of linking the three components is through the structural element **G**. The quartet-like ^{1}H NMR signal must therefore be two overlapping triplets.

When one summarises the results obtained so far, only two possible structures are obtained for **16**, namely **17** and **18**.

The distinction between **17** and **18** cannot be achieved with the IR and UV spectra. The IR spectrum is consistent with

17

18

both isomers. The UV spectrum corresponds to the addition of both of the substituted benzene chromophores. The information that can be obtained from an analysis of the signals between $\approx$ 2 and 1 ppm in the ^{1}H NMR spectrum is also insufficient to differentiate between the two structures. Therefore, one must try to complete the solution of this exercise by continuing the interpretation of the mass spectrum.

The decomposition of **17** is depicted in the fragmentation scheme. Two pathways have been determined: $m/z = 486$ ($M^{+\bullet}$) → 371 , 198 and $m/z = 486$ ($M^{+\bullet}$) → 312 → 198. In both cases an α-cleavage occurs adjacent to the sulfonamide *N*-atom and the resulting ion decomposes further to $m/z = 198$ via a McLafferty rearrangement. Analogous cleavage reactions with **18** would lead to ions with the masses 385 and 298, neither of which, however, appear in the spectrum of **16**. The subsequent ion $m/z = 198$ could nevertheless be formed from $m/z = 385$ and 298. However, because both of the homologous ions are missing from the spectrum, **17** is the more probable structure for **16**.

Fragmentation scheme

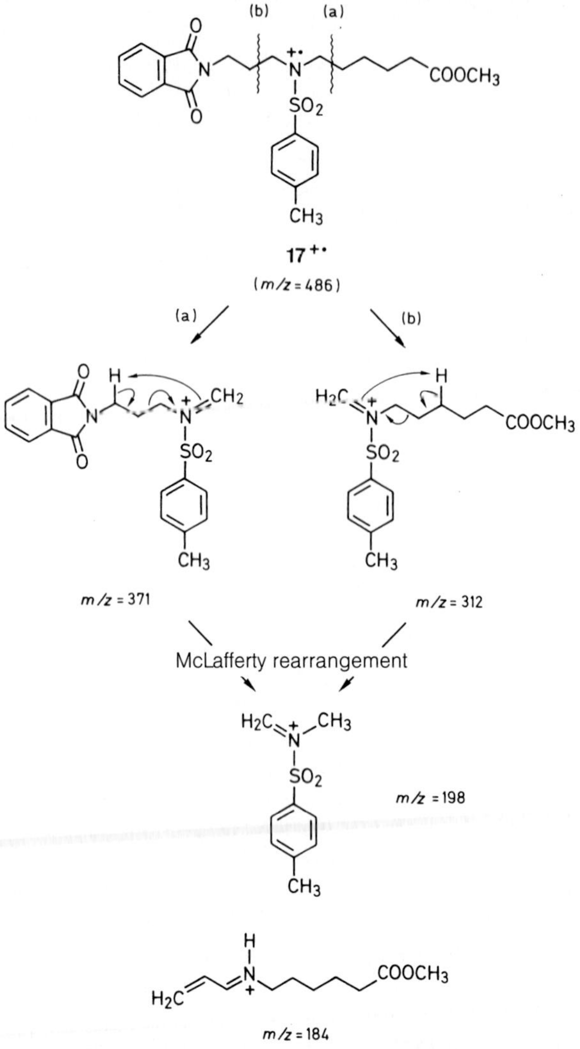

(b) (a)
COOCH3
SO2
CH3
17+•
(m/z = 486)
(a)
(b)
CH2
H2C
COOCH3
m/z = 371
m/z = 312
McLafferty rearrangement
H2C CH3
m/z = 198
COOCH3
H2C
m/z = 184

The ion $m/z = 184$ has not yet been mentioned. It is formed by a fragmentation reaction involving neighbouring group participation, but this will not be discussed in any more detail here. In reality, the unknown compound does indeed possess the structure of **17**.

As was indicated at the start and as is now evident from the structure, **16** is a synthetic product. It does not often occur that the structure of a synthetic product is totally unknown. This example is to be taken more as an exercise than as a real structure elucidation. However, it does show how successful the combination of the spectroscopic methods can be. A structural unit that is tentatively deduced by one spectroscopic method can be confirmed and possibly expanded by the other methods. If conflicting arguments arise, these are to be taken into consideration and their reliability should be evaluated. In any case, a structure that has been elucidated by spectroscopic methods can only be taken as unequivocally determined when all contradictions have been eliminated.

Example 6

Objective

In order to synthesise 1,4-cyclotetradecanedione (**20**), 4-nitrocyclotetradecanone (**19**) was treated for 3 hours with an excess of $TiCl_3$ in a solution of sodium methoxide in methanol. Subsequently, the mixture was acidified with an aqueous acid and extracted with CH_2Cl_2. Additional work-up steps were then carried out (1. K_2CO_3/H_2O; 2. saturated aq. NaCl; 3. dried with Na_2SO_4; 4. chromatographed on silica gel/CH_2Cl_2). Aside from the expected product **20** (ca. 85% yield), another colourless product, **21**, was produced (ca. 7%; mp 150°C), which had an unknown structure. Deduce the structure of **21** by using the known origin of the compound and its spectral data (spectra 20 to 23).

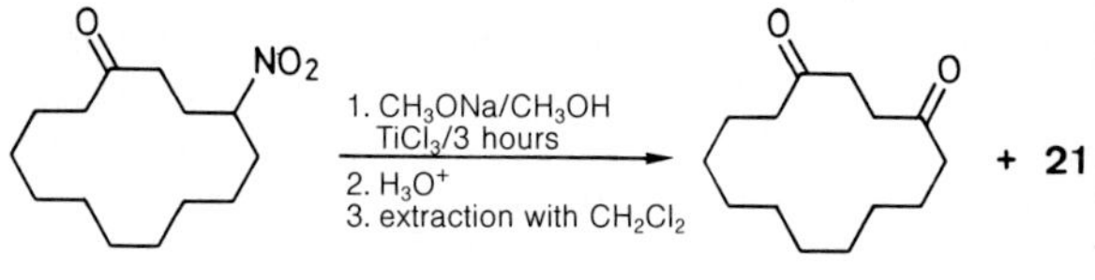

NO2
1. CH3ONa/CH3OH
TiCl3/3 hours
2. H3O+
3. extraction with CH2Cl2
+ 21

19 **20**

Conditions for Recording the Spectra

20   IR: CHCl$_3$

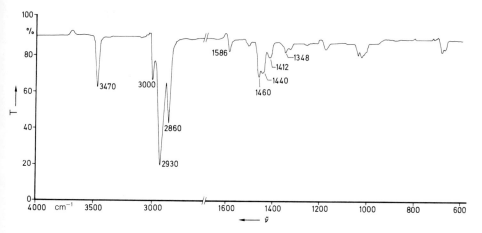

21 ^{1}H NMR: (Bruker AC 300): internal TMS, D$_6$-DMSO

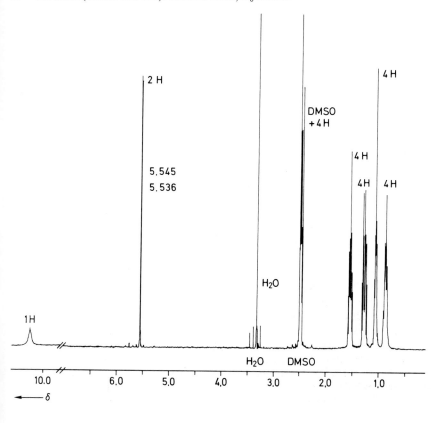

22 ^{13}**C NMR:** (Varian XL-200): internal TMS, D$_6$-DMSO, ^{1}H broadband decoupled

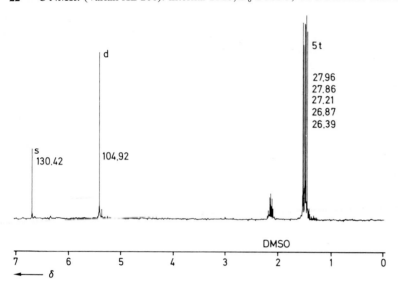

23 EI-MS: (MAT 90): 70 eV

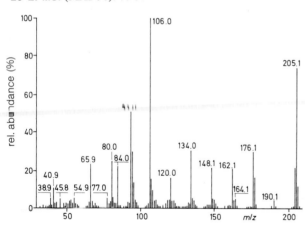

Answer

Both the starting material (**19**; M = 255) and compound **21** contain a nitrogen atom (odd mass for the molecular ion at $m/z = 205$). However, the nitro group is no longer present (in the IR spectrum, the intense bands at ca. 1550 and 1350 cm^{-1} are missing). Two things are conspicuous in the NMR spectra: absorptions for two vinyl protons (5.54 ppm) are present in the ^{1}H NMR spectrum and the ^{13}C NMR spectrum exhibits only seven signals (5 CH$_2$, 1 CH and 1 C), whereas 14 C-atoms were present in the starting material (**19**). After (lengthy) consideration, one comes to the conclusion that the (mild) reaction conditions would not cause seven C-atoms (including a few substituents) to be ejected and at the same time reduce the molecular weight by only 50 amu (transformation of **19** → **21**). A way

out of this dilemma is offered by the assumption that the signals in the ^{13}C NMR spectrum are all doubled, so that the C-atoms are present as magnetically equivalent pairs. This is in agreement with the assumption of a plane of symmetry in **21** and means that the nitrogen atom therein must be symmetrically positioned. The reducing conditions that were used in the reaction do not influence the carbonyl group, however the presence of the carbonyl group would prevent the nitrogen atom from being arranged "symmetrically". Therefore, the unknown compound must be the pyrrole derivative **21** (infrared N–H band at 3470 cm^{-1}). The interpretation of the remaining spectral data, which you should undertake yourself, confirms this assumption.

In order to clarify the reaction mechanism for the formation of **21**, it must be assumed that the starting material **19** is reduced to the imine **19a**. The isomer of the latter (**19b**) cyclises and finally produces the pyrrole derivative **21** after the elimination of water.

Example 7

Objective

The pure substance **22** was refluxed in C_2H_5OH/C_2H_5ONa for 1 hour. Subsequently, the mixture was cooled, neutralised with HCl/C_2H_5OH and distilled, whereby two fractions were collected: fraction 1 up to ca. 100°C, fraction 2 over 100°C. Each fraction was individually analysed with GC/MS (see chapter 4, section 8.11). Deduce the structure of **22** and the alcoholysis products.

Experimental

Fraction 1 was scanned from $m/z = 25-100$ with a cycle time of 0.52 s. A 25 m capillary column (SE 54), which was held at a constant temperature of 35°C, was used for the GC (Varian 3400). A section of the reconstructed ion chromatogram (RIC) is reproduced as a function of time. The three components which have the best mass spectra are marked with the scan number (upper number) and the retention time.

Fraction 2 was scanned from $m/z = 35-180$ with a cycle time of 0.6 s. The capillary column (cf. fraction 1) was heated from 100 to 200°C at a rate of 10°C/min.

The EI mass spectra were recorded with a Finnigan MAT 95 instrument at 70 eV (Dr. R. Schubert, Finnigan MAT, Bremen is thanked for all of the measurements).

Answer

Based on the reaction and the work-up, one can assume that the reagent ethanol ($M = 46$) is still present in the mixture. By comparing spectra, spectrum 25, scan 209 (fraction 1) can be identified as that of ethanol. Furthermore, one can presume that ethanol has reacted with compound **22** and that OC_2H_5 is present in one or more of the other substances.

24 GC of fraction 1. The three components are marked with their scan numbers and retention times

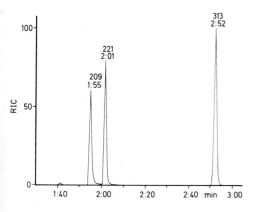

25 Fraction 1, MS scan 209

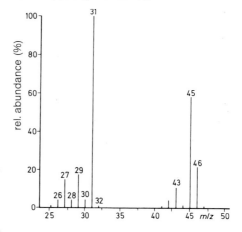

26 Fraction 1, MS scan 221

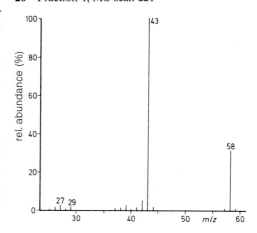

27 Fraction 1, MS scan 313

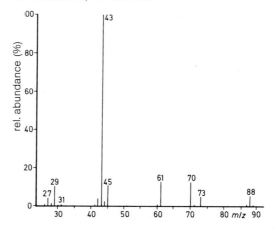

28 GC of fraction 2; see caption of **24**

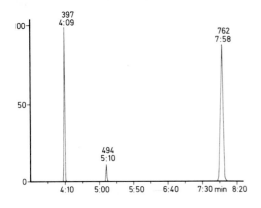

29 Fraction 2, MS scan 397

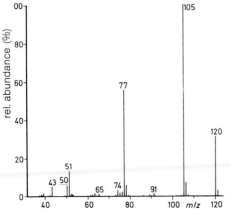
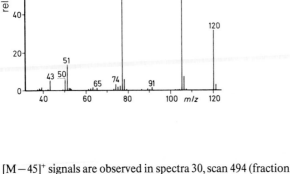

30 Fraction 2, MS scan 494

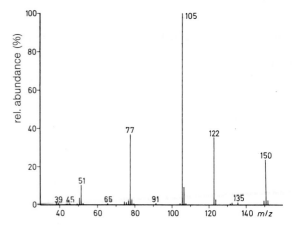

31 Fraction 2, MS scan 762

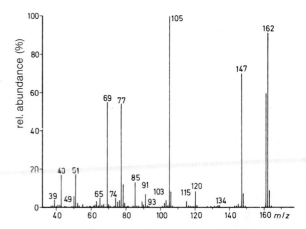

$[M-45]^+$ signals are observed in spectra 30, scan 494 (fraction 2) and 27, scan 313 (fraction 1). With the aid of Table 4.8, the substance in spectrum 30, scan 494 can easily be identified as ethyl benzoate (**23**) ($m/z = 105$ corresponds to $C_6H_5-C\equiv O^+$, which is confirmed by $m/z = 77$ and 51; $m/z = 120$ corresponds to $C_6H_5-COOH^+$, which results from a McLafferty rearrangement of the ester).

Because $m/z = 105$ is also present in spectrum 29, scan 397 (fraction 2), but the molecular ion of the corresponding compound is only 15 amu heavier, this substance could be acetophenone (**25**; M = 120).

As already mentioned, the compound that gives spectrum 27, scan 313 (fraction 1) also contains an ethoxy group. By using a similar approach to that which was employed for the deduction of **23**, this compound is identified as ethyl acetate (**26**; M

= 88, cf. Table 4.11). Spectrum 26, scan 221 (fraction 1) comes from acetone (**24**; M=58). Finally, the last structure with M = 162 [spectrum 31, scan 762 (fraction 2)] can be assumed to be compound **22** because of the intense fragment ion signals at $m/z = 105$ (162-105 = 57, which corresponds to the loss of $^\bullet CH_2-CO-CH_3$ from the molecular ion), 147 ($[M-CH_3]^+$) and 120 ($M^{+\bullet} -$ketene).

Another way of deducing **22** is to analyse the masses in the chemical reaction. Apparently the reaction between **22** and ethanol resulted in the formation of two pairs of products, which by the addition of the molecular weights gives:

$$
\begin{array}{cccc}
162 & + & 46 & = 105 + 58 \\
(\mathbf{22}) & & (C_2H_5OH) & (\mathbf{23}) \quad (\mathbf{24}) \\
& & & = 120 + 88 \\
& & & (\mathbf{25}) \quad (\mathbf{26})
\end{array}
$$

22
($m/z = 162$)

23   **24**
($m/z = 150$)   ($m/z = 58$)

25   **26**
($m/z = 120$)   ($m/z = 88$)

In chemical hindsight, the reaction that has just been discussed is a (known) alcoholysis of an asymmetrically substituted 1,3-diketone which has not run to completion.

Index

Concepts

Types of Compounds and Functional Groups

Entries in **bold** refer to fuller treatments. The double-headed arrows (⇥) refer to compound types and the normal arrows to specific compounds.

Specific Compounds

References to illustrated spectra are given in **bold**.